Organic Synthesis Highlights II

Edited by Herbert Waldmann

VCH

Further Titles of Interest

Drauz, K./Waldmann, H. (eds.)
Enzyme Catalysis in Organic Synthesis. A Comprehensive Handbook
Two Volumes. ISBN 3-527-28479-6

Fuhrhop, J./Penzlin, G.
Organic Synthesis. Concepts, Methods, Starting Materials
Hardcover. ISBN 3-527-29086-9
Softcover. ISBN 3-527-29074-5

Mulzer, J./Altenbach, H.-J./Braun, M./
Krohn, K./Reissig, H.-U.
Organic Synthesis Highlights
ISBN 3-527-27955-5

Nicolaou, K.C./Sorensen, E.
Classics in Total Synthesis
Hardcover. ISBN 3-527-29231-4
Softcover. ISBN 3-527-29284-5

Nogradi, M.
Stereoselective Synthesis. A Practical Approach
Second, Thoroughly Revised and Updated Edition
Hardcover. ISBN 3-527-29242-X
Softcover. ISBN 3-527-29243-8

© VCH Verlagsgesellschaft mbH, D-69451 Weinheim (Bundesrepublik Deutschland), 1995

Distribution:
VCH, P.O. Box 10 11 61, D-69451 Weinheim, Federal Republic of Germany
Switzerland: VCH, P.O. Box, CH-4020 Basel, Switzerland
United Kingdom und Ireland: VCH, 8 Wellington Court, Cambridge CB1 1HZ United Kingdom
USA und Canada: VCH, 220 East 23rd Street, New York, NY 10010-4606 USA
Japan: VCH, Eikow Building, 10-9 Hongo 1-chome, Bunkyo-ku, Tokyo 113 Japan

ISBN 3-527-29200-4 (Hardcover)
ISBN 3-527-29378-7 (Softcover)

Organic Synthesis Highlights II

Edited by Herbert Waldmann

Weinheim · New York · Basel · Cambridge

Professor Dr. Herbert Waldmann
Institut für Organische Chemie
der Universität
Richard-Willstätter-Allee 2
D-76128 Karlsruhe
Germany

This book was carefully produced. Nevertheless, authors, editors and publishers do not warrant the information contained therein to be free of errors. Readers are advised to keep in mind that statements, data, illustrations, procedural details or other items may inadventently be inaccurate.

Published jointly by
VCH Verlagsgesellschaft, Weinheim (Federal Republic of Germany)
VCH Publishers, New York, NY (USA)

Editorial Director: Dr. Ute Anton
Production Manager: Dipl.-Wirt.-Ing. (FH) Bernd Riedel

The cover illustration shows the polycyclic ring system of morphine, surrounded by the primary rings. Strategic bonds determined in a retrosynthetic analysis are highlighted (see p. 358).

Library of Congress Card No. applied for
A catalogue record for this book is available from the British Library

Die Deutsche Bibliothik – CIP-Einheitsaufnahme

Organic synthesis highlights / ed. by Herbert Waldmann. –
Weinheim ; New York ; Basel ; Cambridge ; Tokyo : VCH, 2. (1995)
ISBN 3-527-29200-4 [Geb. Ausg.]
ISBN 3-527-29378-7 [Kart. Ausg.]
NE: Waldmann, Herbert [Hrsg.]

Printed on acid-free and chlorine-free paper

Composition: Mitterweger Werksatz GmbH, D-68723 Plankstadt
Printing: Strauss Offsetdruck GmbH, D-69509 Mörlenbach
Bookbinding: J. Schäffer GmbH & Co. KG., D-67269 Grünstadt
Cover design: Graphik & Text Studio Zettlmeier-Kammerer, D-93164 Laaber-Waldetzenberg
Printed in the Federal Republic of Germany

Preface

Organic synthesis is a powerful art, which has a strong impact on numerous branches of science. Its methodology can be applied to the construction of manifold compounds. Thus, it opens up new avenues of research in medicinal chemistry for developing alternate and better drugs. It fuels research in biology, biochemistry and bioorganic chemistry by making substrates and inhibitors of enzymes and ligands for receptors accessible. Also, without the possibility of preparing tailor-made compounds with a predesigned molecular architecture, the development of new materials and all the challenges which may be loosely described by the phrase "molecular recognition" could not be addressed appropriately. This impact is based on the continuous improvement of the methods of organic synthesis, which must prove their efficiency in the construction of target compounds of an ever-increasing complexity.

Like its congener (*Organic Synthesis Highlights*, J. Mulzer et al. VCH, Weinheim 1990), *Organic Synthesis Highlights II* presents a collection of forty articles which provide an overview of the most recent and important accomplishments in organic synthesis. They are based on contributions made from 1988 to 1993 by a team of young authors from universities and industry to the short review section "Synthese im Blickpunkt" ("Focus on Synthesis") of the *Nachrichten aus Chemie, Technik und Laboratorium*, the members' journal of the Gesellschaft Deutscher Chemiker (German Chemical Society). For their publication in this book, all articles were updated and revised. The selection of the individual topics reflects to a certain extent the points of view of the authors, concerning "important" and "less important" developments, and — of course — their own interests. However, the fact that a total of ten scientists, working in different areas of organic chemistry, have contributed to the book, guarantees the desired diversity.

Organic Synthesis Highlights II is subdivided into two parts. Part I describes the development of new methods and reagents and covers, for instance, new results from the fields of asymmetric synthesis, organometallic chemistry and biocatalysis. Part II details the application of such techniques in the development of new routes to different classes of natural products and to individual target compounds, for instance, calicheamicin γ_1^I and rapamycin.

The short reviews found in this book provide valuable and up to date information for researchers active in organic chemistry. The articles cover the most recent trends in the field, give short introductions to new areas of research and they contain a good collection of references, which suggest alternate solutions to prevailing problems or lead deeper into specific topics. In addition, it is my personal experience with the congener of the book *(vide supra)* which makes me recommend *Organic Synthesis Highlights II* to my colleagues who are involved in the education of advanced students, and, of course, to the students themselves, as a rich source for the preparation of seminars and lectures.

Karlsruhe, March 1995   Herbert Waldmann

Contents

C. Silicon in Organic Synthesis

D. Enzymes in Organic Synthesis

E. Cyclization Reactions

F. General Methods and Reagents for Organic Synthesis

Part II. Applications in Total Synthesis

A. Synthetic Routes to Different Classes of Natural Products and Analogs Thereof

B. Synthesis of Individual Natural Products

List of Contributors

Dr. R. Bohlmann
Institut für Arzneimittelchemie
Pharmazeutische Chemie III
Schering AG
Postfach 65 03 11
D-13342 Berlin
(Germany)

Dr. T. Brumby
Schering AG
Pharma Forschung
Postfach 65 03 11
D-13342 Berlin
(Germany)

Prof. Dr. K. H. Dötz
Institut für Organische Chemie
und Biochemie der Universität
Gerhard-Domagk-Str. 1
D-53121 Bonn
(Germany)

Dr. H. P. Fischer
Ciba-Geigy Ltd.
Research Relations, PP 2.201
R-1047.3.26
CH-4002 Basle
(Switzerland)

Dr. P. Hammann
Hoechst AG
Allgemeine Pharma Forschung
Lead Discovery Gruppe H 780
Postfach 80 03 20
D-65926 Frankfurt/Main
(Germany)

Dr. R. Henning
Bayer AG
Pharma Research
Chemistry Science Labs
PH-FE F CWL
D-42096 Wuppertal
(Germany)

Prof. Dr. M. Maier
Institut für Organische Chemie
Martin-Luther-Universität
Halle-Wittenberg
Weinbergweg 16
D-06120 Halle/Saale
(Germany)

Prof. Dr. D. Schinzer
Institut für Organische Chemie der
Technischen Universität
Postfach 33 29
D-38023 Braunschweig
(Germany)

Dr. G. Sedelmeier
Ciba-Geigy Ltd.
Verfahrensforschung
K-684.233
CH-4002 Basle
(Switzerland)

Prof. Dr. H. Waldmann
Institut für Organische Chemie
Universität Karlsruhe
Richard-Willstätter-Allee 2
D-76128 Karlsruhe
(Germany)

List of Abbreviations

AIBN	azobisisobutyronitrile
APA	aminopenicillanic acid
BINAP	2,2'-bisdiphenylphosphino-1,1'-binaphthyl
Boc	*tert*-butoxycarbonyl
BTMSA	bis(trimethylsilyl)acetylene
COD	1,5-cyclooctadiene
CP	cyclopentadienyl
mCPBA	*m*-chloroperbenzoic acid
CSA	camphorsulfonic acid
CSI	chlorosulfonylisocyanate
DABCO	diazabicyclo[2.2.2]octane
DAST	(diethylamino)sulfurtrifluoride
DBU	1,8-diazabicyclo[5.4.0]undec-7-ene
DCC	dicyclohexylcarbodiimide
DCPE	bis(dicyclohexylphosphino)ethane
DEPC	diethyl phosphorcyanidate
DET	diethyl tartrate
DIBAH	diisobutylaluminium hydride
DMAD	dimethyl acetylenedicarboxylate
EPC	enantiomerically pure compound
FMOC	9-fluorenylmethoxycarbonyl
HMPTA	hexamethyl phosphoric acid triamide
LDA	lithium diisopropylamide
MOM	methoxymethyl
NBS	*N*-bromosuccinimide
NMO	*N*-methylmorpholine-*N*-oxide
TBAF	tetrabutyl ammonium fluoride
TBS, TBDMS	*tert*-butyldimethylsilyl
TBSOTf	*tert*-butyldimethylsilyl triflate
TMS	trimethylsilyl
TOSMIC	tosylmethyl isocyanide

Part I. New Methods and Reagents for Organic Synthesis

A. Asymmetric Synthesis

The Sharpless Epoxidation

Dieter Schinzer

The synthesis of enantiomerically pure compounds is one of the most important goals in organic synthesis and is a major target in industrial syntheses of physiologically active compounds. Even more important are asymmetric transformations under catalytic conditions and high turn over rates because fewer by-products are obtained, which are in many cases a problem for the environment. This chapter will only focus on one particular reaction, which is available under catalytic conditions and is also used industrially: the Sharpless epoxidation.

In 1980 K. B. Sharpless and T. Katsuki published the first paper on this important reaction. [1] The reaction can be operated under simple conditions and all components required for the in situ preparation of the active catalyst are commercially available: titanium tetraisopropoxide, diethyl tartrate (DET), and *t*-butyl hydroperoxide as the oxidizing agent. Two characteristics are important about these reactions:

1) Many substrates give very high asymmetric induction;

2) The chirality obtained depends only on the chirality of the diethyltartrate. The oxygen will be transfered only from one enantiotopic face of the allylic alcohol, independent of the structure of the substrate. Therefore, the absolute configuration of the product can always be predicted (Scheme 1). [2, 3]

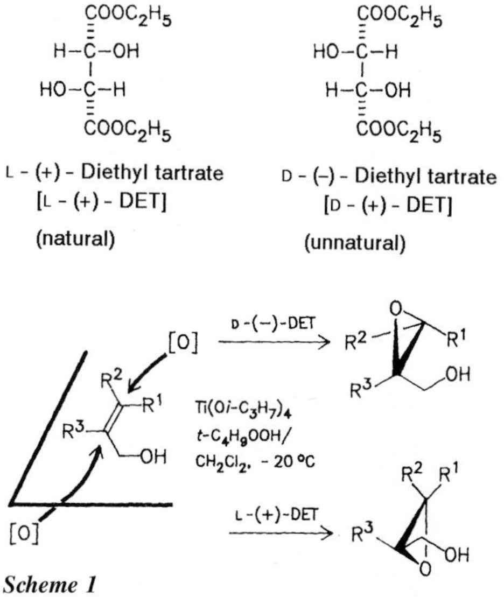

L - (+) - Diethyl tartrate
[L - (+) - DET]
(natural)

D - (-) - Diethyl tartrate
[D - (+) - DET]
(unnatural)

Scheme 1

The catalyst is probably obtained by two exchange reactions of isopropoxide with DET and in solution is dimeric *(1)* (Scheme 2).

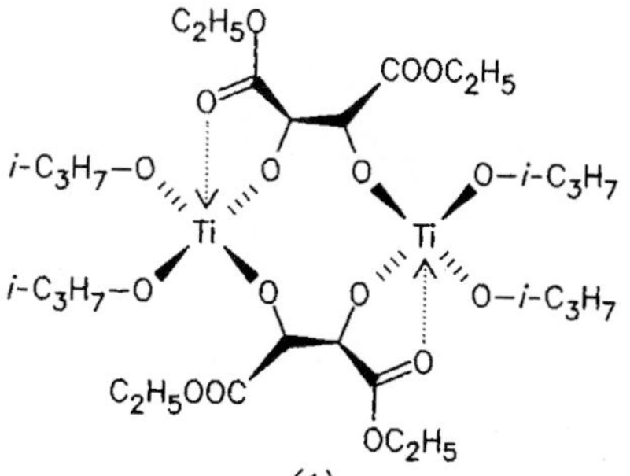

Scheme 2   *(1)*

In a cyclic process the allylic alcohol is fixed, and the oxygen is transfered to the double bond (Scheme 3).

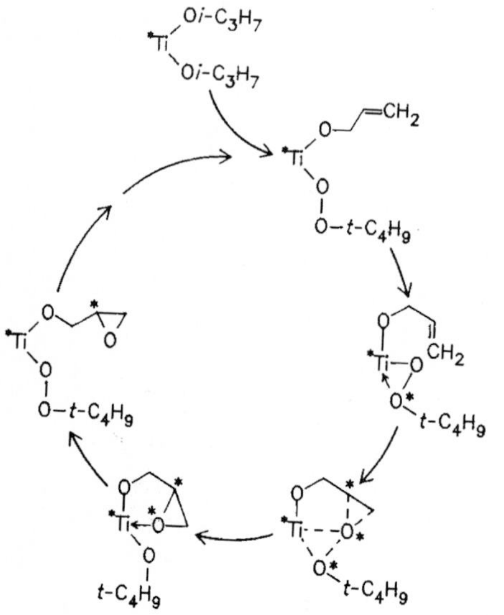

Scheme 3

The reaction is completely chemoselective and only allylic double bonds are epoxidized (Scheme 4).

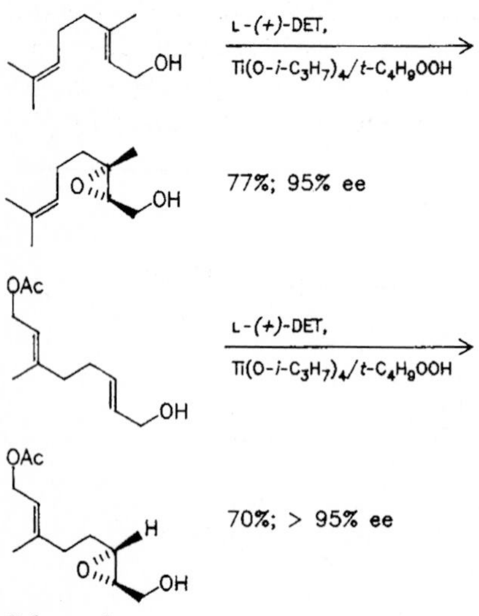

Scheme 4

As early as 1981 Sharpless et al. described an important extension of the epoxidation: the so-called kinetic resolution of secondary allylic alcohols. [4] This type of alcohol is an important building block in many complex syntheses. The enantioselective differentiation of the two antipodes is based on the difference of the relative rate constants. Starting from a racemate, one enantiomer is epoxidized faster, which means that the other enantiomer remains as the allylic alcohol. Even small differences in relative rate constant provide quite high enantiomeric excess.

This effect can be used for many problems in synthesis and yields extremely high enantiomeric excess (99.999999 % *ee*) for secondary allylic alcohols. A very good example was published by Overman et al. in their synthesis of pumiliotoxine C. [5] *(S)-(–)-2-Me-thylpentene-3-ol (2)*, used as starting material, was obtained after a kinetic resolution in a purity >98 % *ee*. Therefore, it is basically available in unlimited amounts! After benzylation of the alcohol and ozonolysis of the double bond, *(3)* is diastereoselectively transformed into the allylic alcohol by addition of vinyl magnesium bromide. In situ addition of propionic acid chloride yields the desired ester *(4)*. After deprotonation at low temperature with LDA, the kinetic enolate is quenched with *t*-butyldimethylsilyl chloride to give the silylenol ether *(5)*. Compound *(5)* is rearranged via an Ireland–Claisen rearrangement [6] to *(6)*. In this sequence a trisubstituted double bond is constructed via a chair-like transition state. The ester group is reduced to the aldehyde and finally transformed into a triple bond using the Corey–Fuchs procedure. [7] Compound *(8)* represents the key intermediate for the coupling with the cyclic fragment, which is based on the amino acid proline (Scheme 5). [5]

Greene et al. published an enantioselective synthesis of the side chain of taxol *(9)*. [8, 9] Starting with allylic alcohol *(10)*, which was epoxidized under standard conditions, the

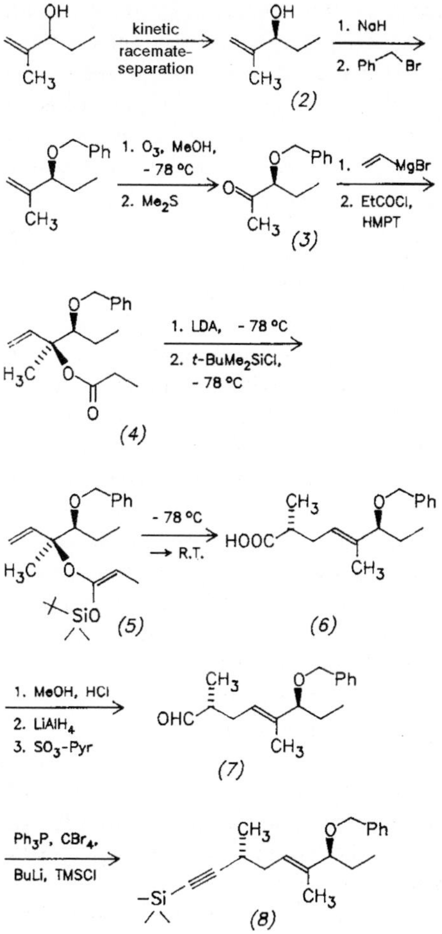

Scheme 5

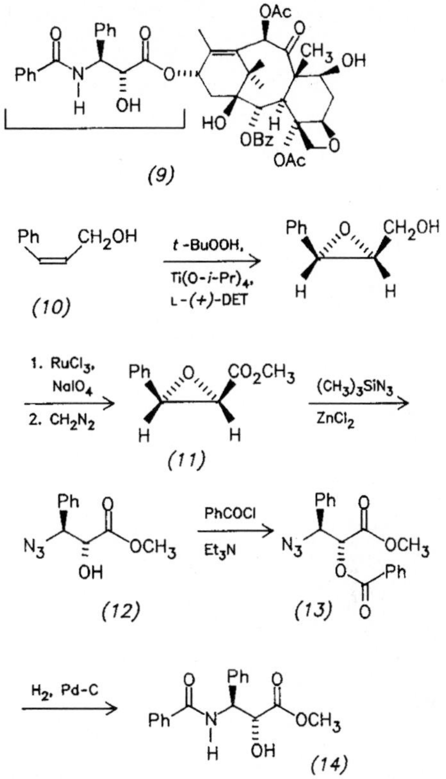

Scheme 6

material and was transformed in five steps to the aldehyde *(17)*. First, a reduction was carried out, followed by a base-catalyzed double bond shift *[(15) → (16)]*, hydrolysis, hydrogenation, and subsequent oxidation to *(17)* (Scheme 7).

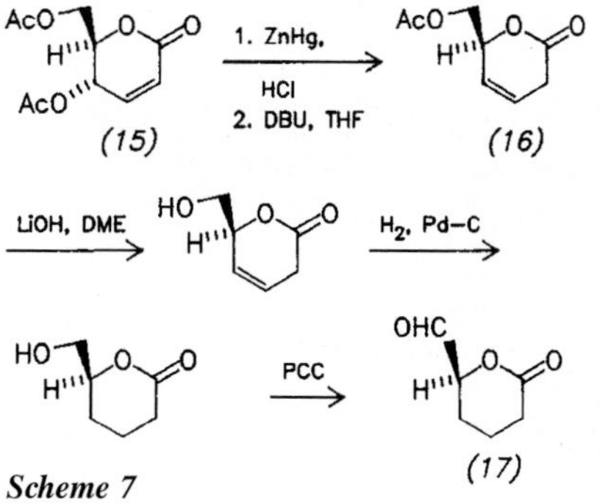

Scheme 7

required alcohol was obtained with the correct configuration. The alcohol was oxidized to the acid and addition of diazomethane gave ester *(11)*. Regio- and stereoselective opening of the epoxide with trimethylsilyl azide in the presence of ZnCl₂ yielded the desired acyclic azide *(12)*. The azido benzoate *(13)* was directly reduced to give the taxol side chain *(14)* (Scheme 6).

The first practicable total synthesis of leucotriene B₅ (by E. J. Corey et al.) also used an symmetric epoxidation as the key step. [10, 11] Lactone *(15)* was used as starting

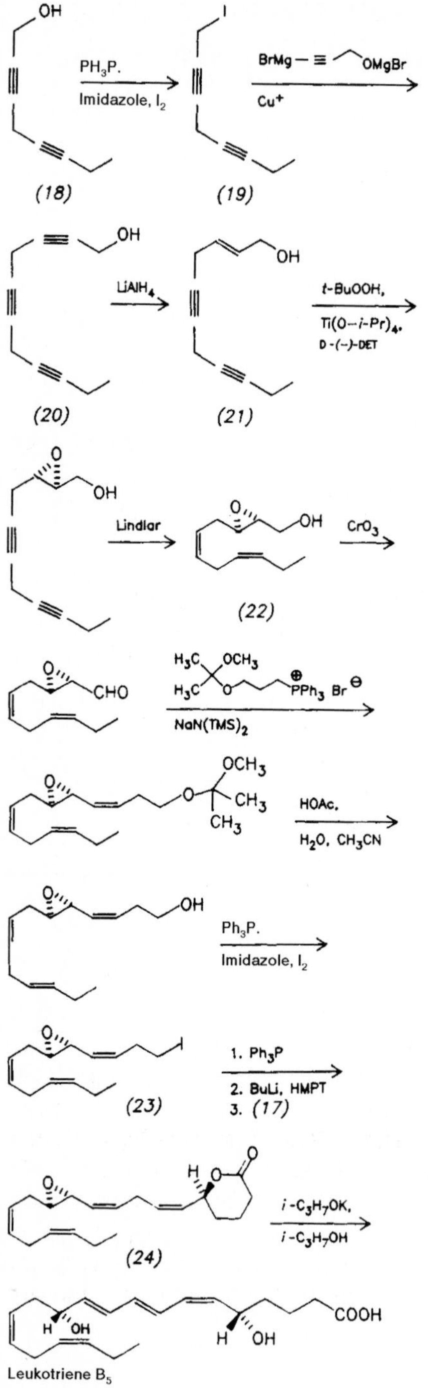

PH$_3$P.
Imidazole, I$_2$

(18)

BrMg —≡ OMgBr
Cu⁺

(19)

LiAlH$_4$

(20)

t-BuOOH,
Ti(O—i-Pr)$_4$,
D-(–)-DET

(21)

Lindlar

CrO$_3$

(22)

CHO

H$_3$C OCH$_3$
H$_3$C O
⊕PPh$_3$ Br⊖
NaN(TMS)$_2$

OCH$_3$
O—C—CH$_3$
CH$_3$
HOAc,
H$_2$O, CH$_3$CN

OH
Ph$_3$P.
Imidazole, I$_2$

1. Ph$_3$P
2. BuLi, HMPT
3. (17)

(23)

(24)

COOH
H OH   H OH
Leukotriene B$_5$

Scheme 8

In a sequence of steps the acyclic part *(23)* was synthesized from octa-2,5-diine-1-ol *(18)*. In the first step the iodide is synthesized from the alcohol. It is coupled in a copper-catalyzed reaction to give *(20)*. Reduction with LAH provided the starting material for the Sharpless epoxidation. Hydrogenation with Lindlar catalyst followed by Collins oxidation provided *(22)*, which was transformed into *(23)*. The final step is the coupling of the subunits *(17)* and *(23)* to give *(24))*. Hydrolysis of *(24)* yielded leucotriene B$_5$ (Scheme 8).

In connection with their studies of cycloadditions Jäger et al. described asymmetric epoxidation of divinyl carbinols *(25)*. [12] Indeed, *(25)* was diastereoselectivley epoxidized with (+)-diethyl tartrate. Opening of *(26)*, tosylation, and acetonide formation yields crystalline *(27)* as a pure diastereomer (Scheme 9).

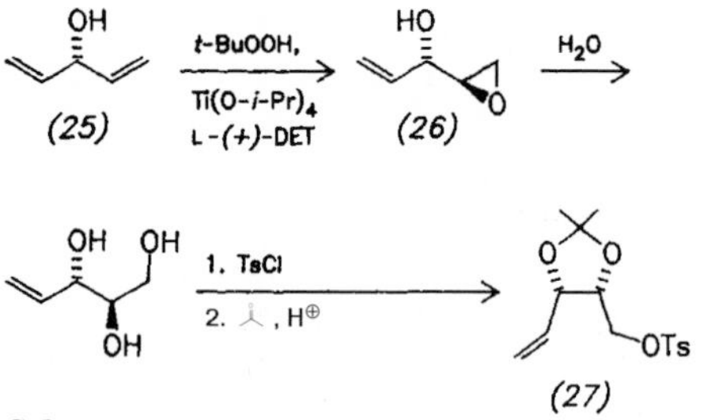

OH
t-BuOOH,
Ti(O-i-Pr)$_4$
L-(+)-DET

(25)

HO

(26)

H$_2$O

OH OH
OH

1. TsCl
2. ⋏, H⊕

(27)

OTs

Scheme 9

A related reaction was described by Schmidt et al. in an efficient synthesis of desoxyhexoses. [13] Starting with racemic or meso-diglycols a kinetic resolution was used to obtain *(28)*. The starting material for this elegant transformation can be obtained by a reductive dimerization of crotonaldehyde. The epoxide *(28)* was treated with sodium hydridobis(2-methoxyethoxy)aluminate (Red-Al) to give triol *(29)* in a regioselective manner. The ozonolysis yielded 4,6-desoxy-L-*xylo*-hexose *(30)* in only five steps (Scheme 10).

All the reactions presented so far are stoichiometric processes, that is, stoichiometric amounts of the titanium-tartrate reagent must

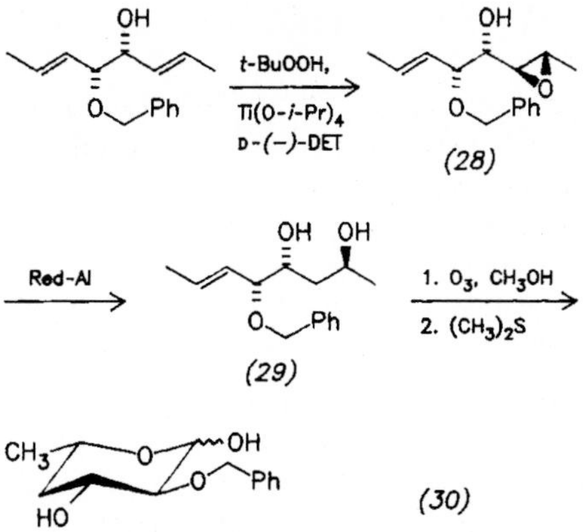

(28)

(29)

(30)

Scheme 10

Scheme 12

be used. In the presence of molecular sieves the titanium reagent can be applied in catalytic amounts, which makes the whole process even more attractive (Scheme 11). [14, 15]

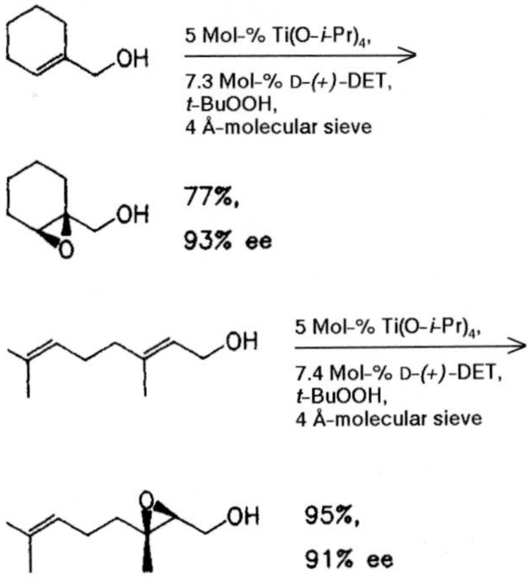

Scheme 11

The best selectivity can be obtained by the use of diisopropyltartrate. The only drawback to this beautiful reaction is the limitation to allylic alcohols as substrates.

Highly enantioselective oxidations of *trans*-olefins have been achieved in a fairly general sense via osmium-catalyzed asymmetric dihydroxylation by Sharpless et al. [16] They have used a procedure based on a cinchona alkaloid catalyst (Scheme 12).

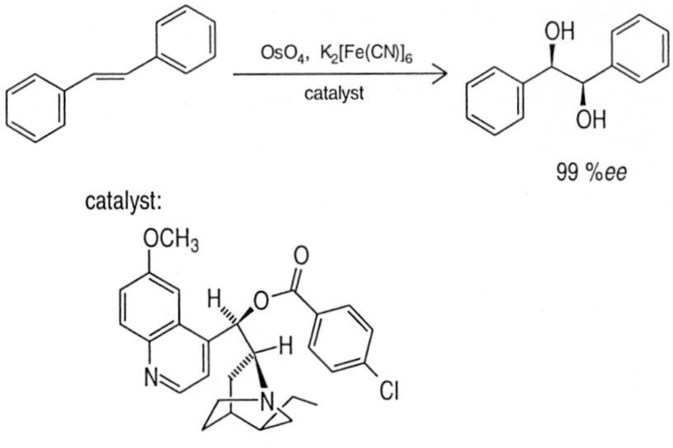

Recently, Jacobsen et. al. published an asymmetric epoxidation of *cis*-olefins with a manganese catalyst (Scheme 13). [17]

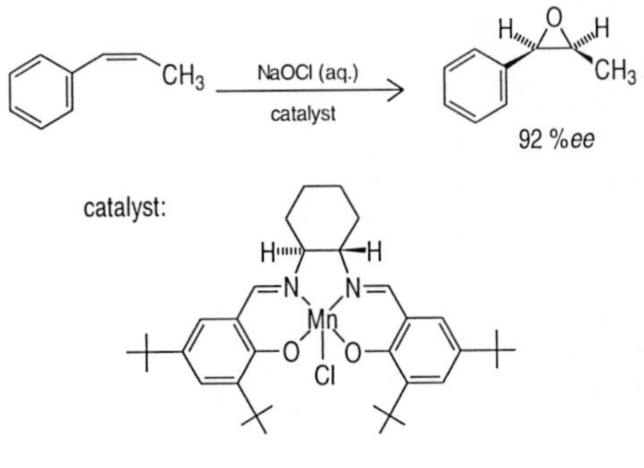

Scheme 13

In this account only a few aspects of this beautiful chemistry could be covered. The number of examples and applications in synthesis could be extended to an enormous amount. The high and even very high selectivities obtained indicate high potential for this reaction in the future. The catalytic variation in particular will be of great interest. No other reaction in organic synthesis with such an impact on chemistry has been found in the last decade.

References

[1] T. Katsuki, K. B. Sharpless, *J. Am. Chem. Soc.* **1980**, *102*, 5974.

[2] B. E. Rossiter in J. D. Morrison (Ed.): *Asymmetric Synthesis*, 5. Academic Press, **1985**.

[3] A. Pfenninger, *Synthesis* **1986**, 89.

[4] V. S. Martin, S. S. Woodard, T. Katsuki, Y. Yamada, M. Ikeda, K. B. Sharpless, *J. Am. Chem. Soc.* **1981**, *103*, 6237.

[5] L. E. Overman, N.-H. Lin, *J. Org. Chem.* **1985**, *50*, 3669.

[6] R. E. Ireland, R. H. Müller, A. K. Willard, *J. Am. Chem. Soc.* **1976** 98, 2868.

[7] E. J. Corey, P. L. Fuchs, *Tetrahedron Lett.* **1972**, 3769.

[8] J.-N. Denis, A. E. Greene, A. A. Serra, M.-J. Luche, *J. Org. Chem.* **1986**, *51*, 46.

[9] D. Schinzer, *Nachr. Chem. Tech. Lab.* **1989**, *37*, 172.

[10] E. J. Corey, S. G. Pyne, W. Su, *Tetrahedron Lett.* **1983**, 4883.

[11] E. J. Corey, P. B. Hopkins, J. E. Munroe, A. Marfat, S. Hashimoto, *J. Am. Chem. Soc.* **1980**, *102*, 7986.

[12] B. Häfele, D. Schröter, V. Jäger, *Angew. Chem.* **1986**, *98*, 89.
Angew. Chem. Int. Ed. Engl. **1986**, *25*, 87.

[13] U. Küfner, R. R. Schmidt, *Angew. Chem.* **1986**, *98*, 90.
Angew. Chem. Int. Ed. Engl. **1986**, *25*, 89.

[14] R. M. Hanson, K. B. Sharpless, *J. Org. Chem.* **1986**, *51*, 1922.

[15] Y. Gao, R. M. Hanson, J. M. Klunder, S. Y. Ko, H. Masamune, K. B. Sharpless, *J. Am. Chem. Soc.* **1987**, *109*, 5765.

[16] H. L. Kwong, C. Sorato, Y. Ogino, H. Chen, K. B. Sharpless, *Tetrahedron Lett.* **1990**, *31*, 2999.

[17] E. N. Jacobsen, W. Zang, A. R. Muci, J. R. Ecker, L. Deng, *J. Am. Chem. Soc.* **1991**, *113*, 7063.

Enantioselective *cis*-Dihydroxylation

Herbert Waldmann

Methods for the stereoselective oxidation of olefins are of great interest to organic synthesis since the epoxides and alcohols generated thereby are valuable intermediates for further syntheses. Allylic alcohols can be converted to the corresponding oxiranes by means of the Sharpless epoxidation [1] and improved methods are also being developed for the analogous transformation of simple unfunctionalized olefins. [2] In addition, effective reagents for the highly enantioselective *cis*-dihydroxylation were recently introduced.

The addition of OsO_4 to double bonds to give vicinal diols belongs to the standard methodology of organic synthesis. Based on the finding [3] that pyridine accelerates this transformation several research groups investigated chiral nitrogen ligands of osmium in

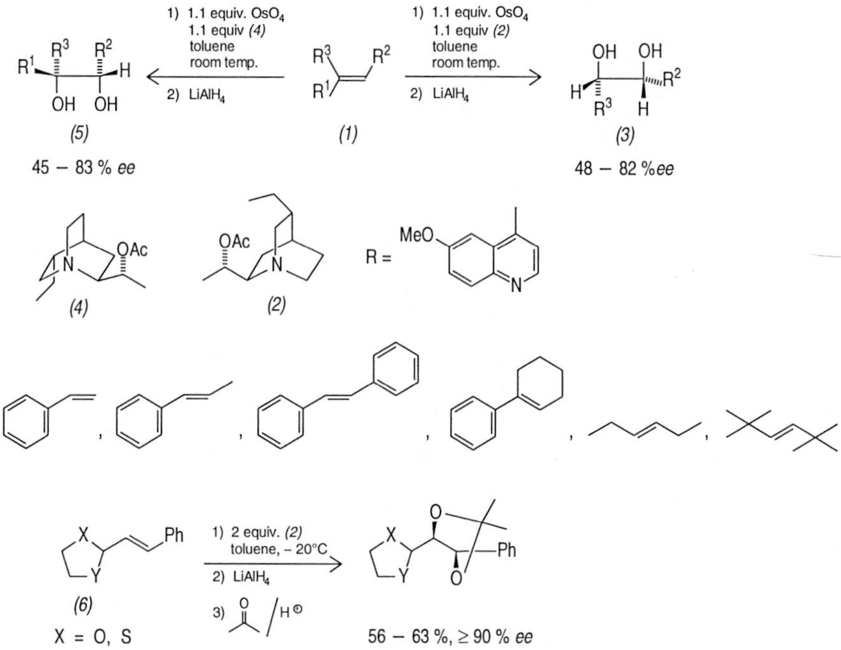

Scheme 1. Enantioselective *cis*-dihydroxylation employing *O*-acetylated dihydrocinchona alkaloids as chiral ligands according to Sharpless et al. [4].

order to develop enantioselective dihydroxy-lations. Thus, Sharpless et al. [4] found that the olefins (1) are converted to the diols (3) and (5) with moderate *ee* values in the presence of the dihydroquinidine acetate (2) and the dihydroquinine acetate (4), respectively (Scheme 1). If the acetals (6) of cinnamic aldehyde are used the enantiomeric excess even exceeds 90%. [5] The cinchona alkaloids

(2) and (4) are diastereomers; however, in these transformations they display a quasi enantiomeric behavior.

Narasaka et al. [6] introduced the chiral diamine (8) derived from tartaric acid as a stereo-differentiating ligand of osmium. In the presence of this mediator *cis*-dihydroxylation of the phenyl substituted olefins (7) proceeds with an enantiomeric excess of 35–90%

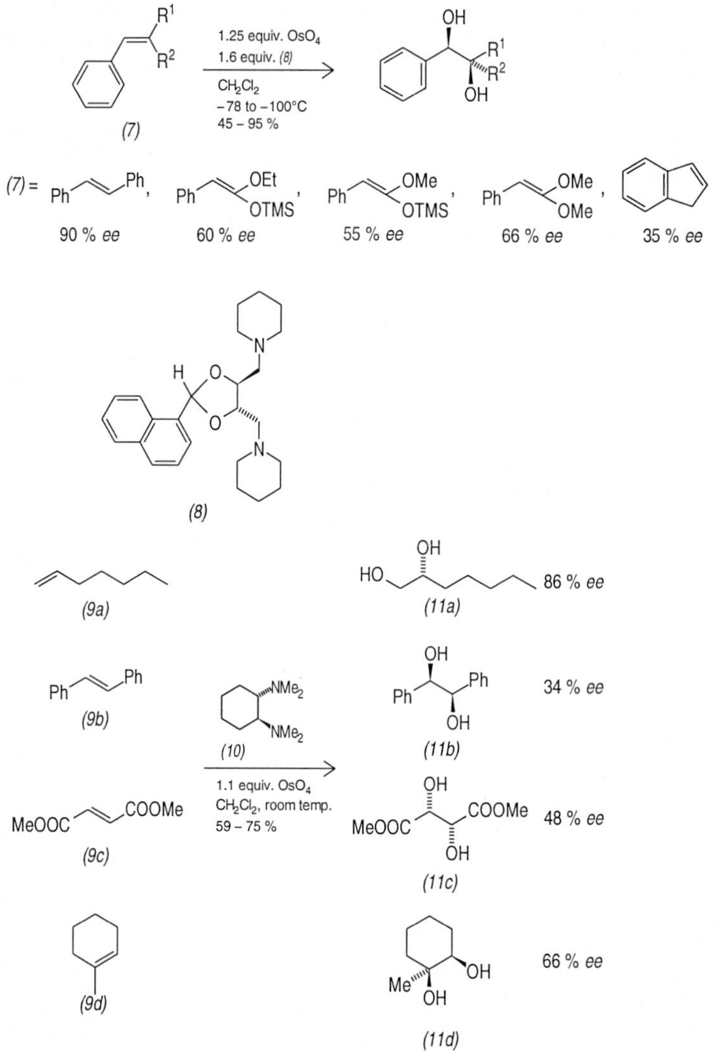

Scheme 2. Enantioselective *cis*-dihydroxylation employing the diamines (8) and (10) as chiral ligands according to Narasaka et al. [6] and Snyder et al. [7].

(Scheme 2). In the absence of an aromatic substituent the enantioselectivity remains low. The C_2 symmetric diamine *(10)* was investigated by Snyder et al. [7] Amine *(10)* makes the alcohols *(11)* available with 34–86 % ee. The influence of the C_2 symmetry [8] incorporated in suitable ligands is more pronounced if the diamines *(12)*, *(18)* and *(19)* are employed. According to Tomioka et al. [9] in the presence of the bis(diphenylpyrrolidine) *(12)* monosubstituted and *E*-configured disubstituted olefins are converted into vicinal diols with uniformly high enantioselectivity. For alkenes carrying three substituents, for example, *(14)*, the *ee* value is lower. The reaction sequence has proved its efficiency, for instance, in the synthesis of anthracyclin antibiotics. [9a, c, f] Thus, the ketone *(16)*, obtained by selective deoxygenation of the diol *(15)*, was transformed into 4-demethoxydaunomycine *(17)* by means of well-established techniques. To explain the stereoselectivity of the hydroxylation step the authors assume that the reaction proceeds as a [2+2]-cycloaddition leading to the metallacycle *(13)* which then rearranges to the respec-

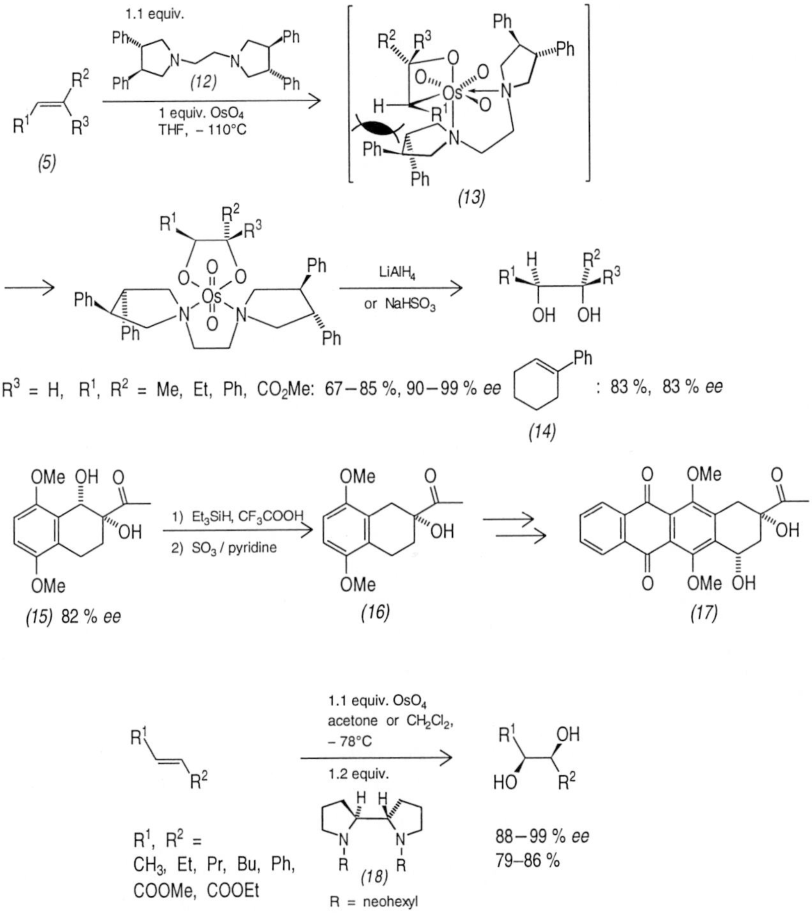

1.1 equiv.
(12)
1 equiv. OsO4
THF, – 110°C
(5)
(13)
LiAlH4
or NaHSO3
R3 = H, R1, R2 = Me, Et, Ph, CO2Me: 67–85 %, 90–99 % ee
: 83 %, 83 % ee
(14)
1) Et3SiH, CF3COOH
2) SO3 / pyridine
(15) 82 % ee
(16)
(17)
1.1 equiv. OsO4
acetone or CH2Cl2,
– 78°C
1.2 equiv.
R1, R2 =
CH3, Et, Pr, Bu, Ph,
COOMe, COOEt
(18)
R = neohexyl
88–99 % ee
79–86 %

Scheme 3. Enantioselective *cis*-dihydroxylation employing the C_2-symmetric diamine ligands *(12)* and *(18)* according to Tomioka et al. [9] and Hirama et al. [10].

tive osmate ester. [9c, e] In *(13)* the substituents R^1 to R^3 occupy the sterically most advantageous positions. This model also accounts for the unexpected result that the sense of the asymmetric induction in some cases is reversed if 3,5-xylyl substituents are present in *(12)* instead of the phenyl groups. [9d] Hirama et al. [10] employed the bis-pyrrolidine *(18)* as mediator of chirality in the *cis*-dihydroxylation of monosubstituted and *E*-configured alkenes (Scheme 3). The respective reactions proceed with high stereoisomeric excess, which is also influenced by the solvent used. Olefins that are conjugated with an aromatic ring give the best results in toluene, in the other cases the use of dichloromethane is recommended.

The addition of OsO_4 to *E*-alkenes also proceeds with high enantioselectivity if the chiral

(19)

(20)

75–90 %, 82–98 % *ee*

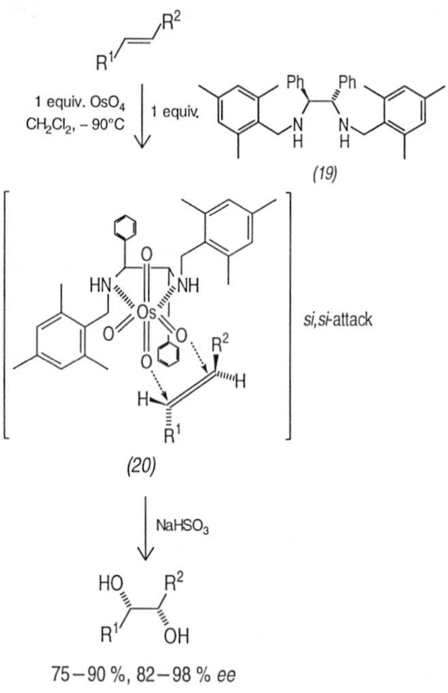

R^1, R^2 = H, CH_3, Et, p-MeO–C_6H_5, C_6H_5, MeOOC,
(CH₂)₂OTBDPS, *t*-BuO–C(O)–NH–CH₂–

Scheme 4. Enantioselective *cis*-dihydroxylation employing the diamine ligand *(19)* according to Corey et al. [11].

diamine *(19)* which was developed by Corey et al. [11] is used as the stereodirecting ligand of the metal atom (Scheme 4). It causes a substantial acceleration of the osmylation and, like the amines *(12)* and *(18)*, it can be recovered in high yield. Corey et al. assume that the *cis*-dihydroxylation proceeds in the sense of a [3+2]-cycloaddition in which the C_2-symmetric complex *(20)* determines the stereoselection. In this complex the amino groups of the diamine ligand occupy equatorial positions, with the phenyl and the neighbouring mesityl groups in an *anti*-orientation. The nitrogen atoms donate electrons to the σ* orbitals of the *trans*-Os–O bonds and, thereby, the equatorial oxygen atoms become more nucleophilic than the corresponding oxygen atoms occupying axial positions. Consequently, one equatorially and one axially oriented oxygen behave as reacting centers in the cycloaddition. The olefin is then coordinatively bound to one of the edges of the octahedron in such a way that the steric interactions between R^1, R^2 and the mesityl substituents are minimized. Due to the C_2-symmetry of *(20)* the edges of the octahedral complex are identical. The notion that a nucleophilic and an electrophilic oxygen are involved in the process also explains the observed acceleration of the dihydroxylation by the amine ligand.

Although the processes described above proceed with excellent stereoselectivity and although the chiral auxiliary reagents can be recovered, they make the use of equimolar amounts of the volatile, toxic, and expensive OsO_4 necessary. For applications on a larger scale, therefore, similarly effective, but catalytic, methods must be developed. Sharpless et al. [12–14] succeeded in reaching this goal for the reagent system originally developed by this group (see above, Scheme 1). First, they found that the *cis*-dihydroxylation in the presence of the nitrogen bases *(2)* and *(4)* can be carried out with catalytic amounts of OsO_4 if *N*-methylmorpholine-*N*-oxide (NMO) is employed for the reoxidation of the consumed osmium. [12a] In this process the chiral amine accelerates the oxidation reaction ("ligand accelerated catalysis"). If the olefin is added slowly, the disturbing influence of an undesired competing catalytic cycle, which proceeds only with low enantioselectivity, is reduced and the *ee* values reach levels that are otherwise only accessible in the presence of equimolar amounts of the cinchona alkaloid. [12c] However, the decisive breakthrough was achieved by the findings that a) the undesired second catalytic cycle is eliminated completely if $K_3Fe(CN)_6$ is used together with K_2CO_3 for the reoxidation of Os(VI) in a two-phase system composed of *tert*-butanol and water; that b) the volatile OsO_4 can be replaced by the solid, non-volatile $K_2Os(VI)O_2(OH)_4$ (potassiumosmate dihydrate); and that c) under these conditions the dihydroxylation proceeds at 0–25 °C and without slow addition of the olefin with higher selectivity than with subsequent addition of the alkene and reoxidation by means of NMO. In conclusion, in the catalytic cycle thereby developed (Scheme 5) Os(*VI*) (*21*) is oxidized in the aqueous phase to Os(*VIII*) (*22*) which then passes into the organic phase as OsO_4 to form a complex with the ligand. The latter then converts the olefin to the monoglycol ester (*23*) which is hydrolyzed rapidly. The diol and the ligand thereby released remain in the organic phase, whereas Os(*VI*) passes into the aqueous phase where it is reoxidized to Os(*VIII*). The most important features of this catalytic process are the rapid hydrolysis of the ester (*23*) (in the case of tetrasubstituted, trisubstituted or 1,2-disubstituted olefins the hydrolysis of the intermediary osmate ester is improved by the addition of methanesulfonamide) and the separation of the hydroxylation step and the reoxidation step into two phases. In a homogeneous

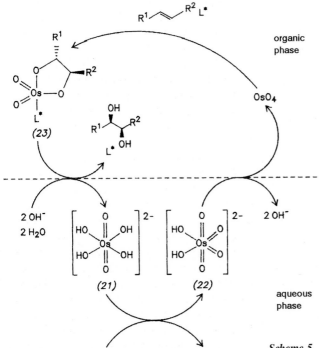

Scheme 5. The catalytic cycle of the enantioselectively catalyzed *cis*-dihydroxylation according to Sharpless et al. [12].

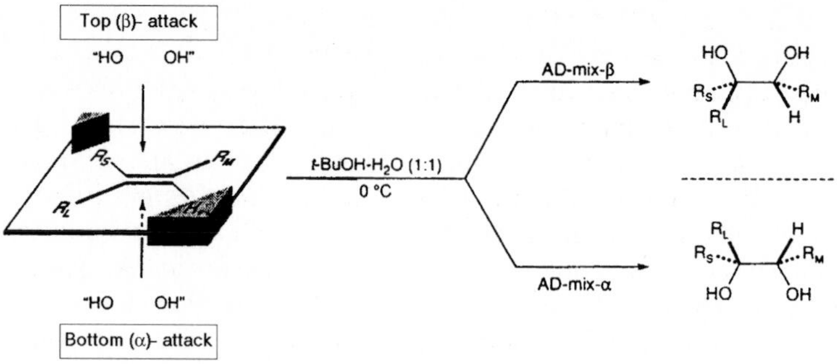

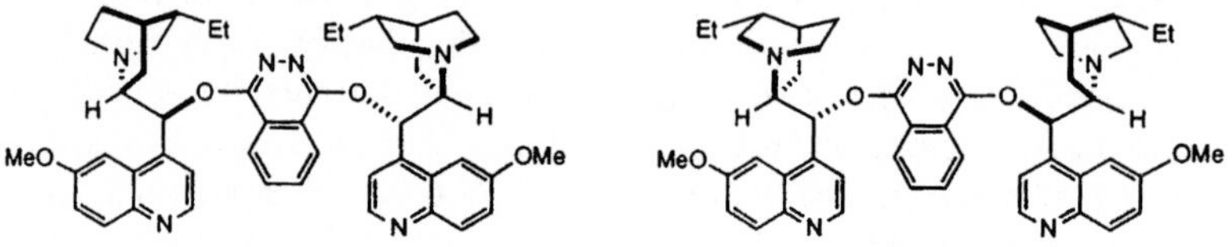

(DHQD)$_2$-PHAL,   (24)
Ligand used in AD-mix-β

(DHQ)$_2$-PHAL,   (25)
Ligand used in AD-mix-α

class of olefin	entry	olefin	AD-mix-β (DHQD)$_2$-PHAL % ee	config'n	AD-mix-α (DHQ)$_2$-PHAL % ee	config'n	CH$_3$SO$_2$NH$_2$
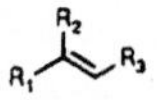	1		98	R	95	S	+
	2	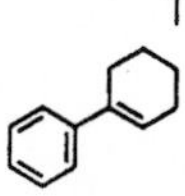	99	R,R	97	S,S	+
	3	 n-Bu ⌇ n-Bu	97	R,R	93	S,S	+
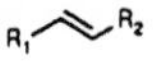	4	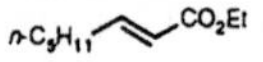 n-C$_5$H$_{11}$ ⌇ CO$_2$Et	99	2S,3R	96	2R,3S	+
	5	⌇ CO$_2$Et	97	2S,3R	95	2R,3S	+
	6	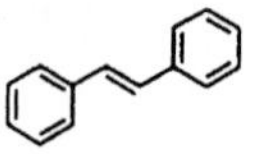	>99.5	R,R	>99.5	S,S	+
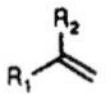	7	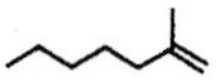	78	R	76	S	–
	8	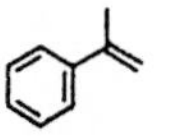	94	R	93	S	–
	9	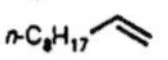 n-C$_8$H$_{17}$ ⌇	84	R	80	S	–
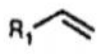	10	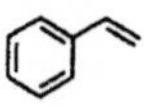	97	R	97	S	–
	11	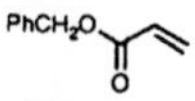 PhCH$_2$O ⌇	77	S	70	R	–
	12	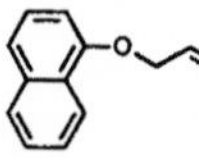	91	S	88	R	–

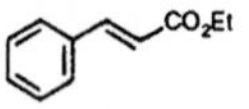

Scheme 6. Enantioselective *cis*-dihydroxylation of various olefins employing phthalazine-derived ligands according to Sharpless et al. [13d].

medium in the presence of NMO the hydrolysis is only slow and thereby provides an entrance into the second, less efficient catalytic cycle. Based on these optimized reaction conditions, a conveniently applicable and in the meantime also commercially available reagent mixture (AD-mix) was formulated in which all inorganic ingredients and the respective ligand (quinine ligand: AD-mix α, quinidine ligand: AD-mix β, see Scheme 6) are blended. This reagent mixture is dissolved in *tert*-butanol/H$_2$O and the olefin is simply added.

Finally, Sharpless et al. varied the substituents attached to the OH-group of the cinchona alkaloids in a wide range and thereby discovered specific ligands that give optimized enantioselectivity for the respective classes of olefins (Scheme 6). They found that four out of the six olefin classes can be converted to the *cis*-diols and with high enantiomeric excess if the phthalazine-derived ligands (DHQD)$_2$-PHAL *(24)* and (DHQ)$_2$-PHAL *(25)* (PHAL-class, see Scheme 7) are used (Schemes 6 and 7). [13d, e] For *cis*-disubstituted olefins the highest, albeit still moderate, *ee* values are recorded if indolinylcarbamoyl groups are attached to the alkaloids (IND-class, see

Scheme 7), [13f] and for tetrasubstituted alkenes the best results are obtained if pyrimidine-based ligands are employed (PYR-class, see Scheme 7). [13g] Finally, for terminal olefins, an improvement in the stereoselectivity was achieved if the OCH$_3$ group in the quinoline moieties of the PHAL ligands was replaced by OR groups carrying longer alkyl chains. [13h]

Under the reaction conditions given and employing the ligands mentioned above a variety of alkenes can be converted to the corresponding vicinal diols with high *ee* values. Typically these reactions are run at 0–25 °C using 1–5 mole % of the ligand and 0.2–1 mole % of the inorganic osmate. [13d, f, g] With the development of this methodology the enantioselective *cis*-dihydroxylation has reached a high level of practicality and efficiency. It has not only been applied to simple unfunctionalized olefins, but also, for instance, α, β-unsaturated carbonyl compounds [13i, k] and -acetals, [13l] enol ethers [13m] and α-hetero-substituted styrene derivatives [13n] have been investigated. Particularly remarkable is the kinetic resolution of the chiral fullerene C$_{76}$ [13o] and the respective asymmetric bisosmylation of C$_{60}$ [13p]

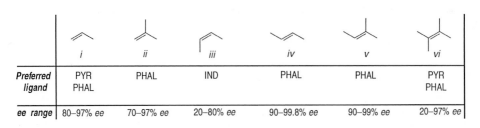

	i	*ii*	*iii*	*iv*	*v*	*vi*
Preferred ligand	PYR PHAL	PHAL	IND	PHAL	PHAL	PYR PHAL
ee range	80–97% ee	70–97% ee	20–80% ee	90–99.8% ee	90–99% ee	20–97% ee

PYR-class   PHAL-class   IND-class

Scheme 7. Ligand preference in enantioselective dihydroxylation as a function of olefin substitution pattern.

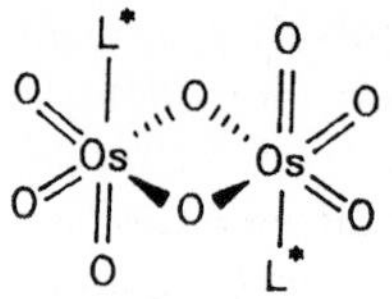

carried out recently. Despite intensive efforts Sharpless et al. could not devise a mechanistic model to rationalize the enantioselectivity observed in the cinchona alkaloid-mediated oxidations. They could only devise a mnemonic device that allows prediction of the sense and the approximate level of the enantioselectivity. [13c] This guide is based on the experience gained in the osmylation of ca. 90 different olefins. On the other hand, Corey et al. proposed to apply the model developed for the ligand (19) (see above) [11b] to the transformations in the presence of the ligands containing one cinchona alkaloid. In addition, they presented an analogous mechanistic scheme for the PHAL ligands carrying two alkaloid moieties. [13q] Corey et al. assume that in these cases μ-oxo-bridged bis-OsO$_4$ species (26a) and (26b) are formed from two OsO$_4$ molecules and the chiral ligands. Due to the electron donation of the nitrogen atoms

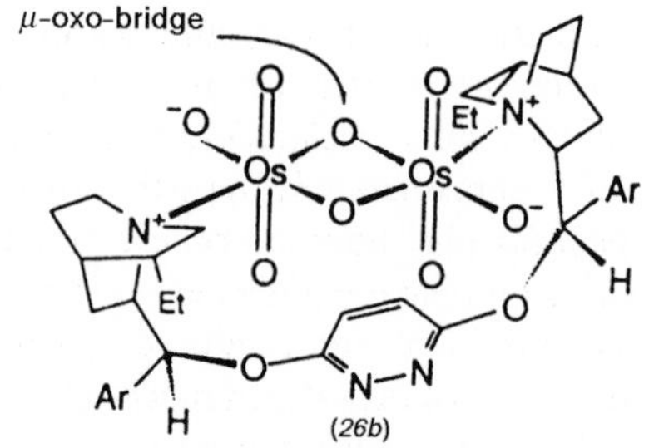

(26a)

(26b)

into the σ*-orbitals of the *trans*-oriented Os–O bonds the equatorial and the axial oxygen atoms once more become non-equivalent, so that the observed phenomenon of "ligand-accelerated catalysis" can be explained. How-

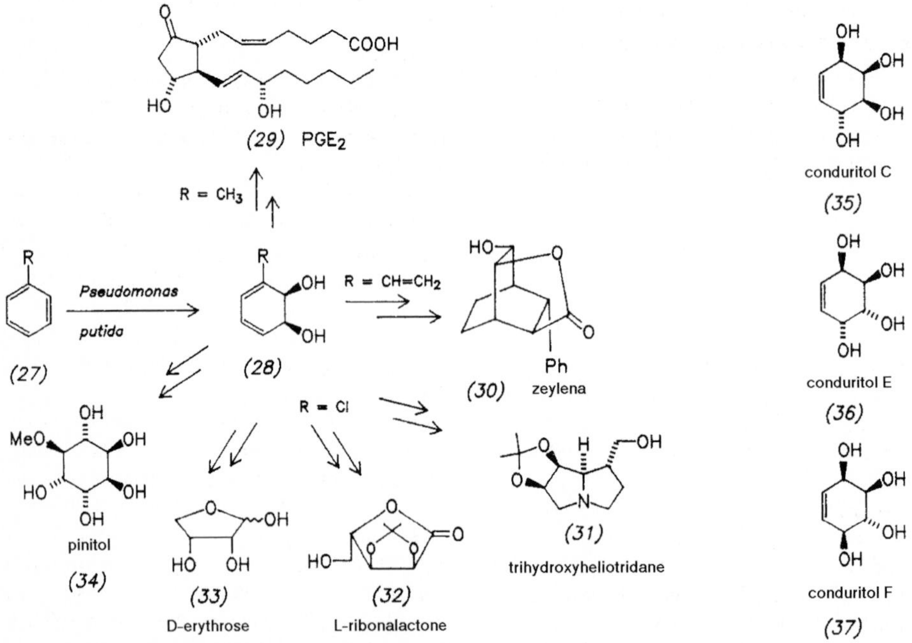

Scheme 8. Biocatalytic oxidation of aromatic compounds to the *cis*-dihydrodiols (28) and synthesis of natural products employing the chiral dienes (28) according to Hudlicky et al. [16].

ever, Sharpless et al. [13r] recently presented kinetic data and related observations that do not support this notion.

As for other classical chemical enantioselective reactions biocatalytic transformations have provided a competing methodology to the osmium mediated *cis*-dihydroxylation. Particularly interesting is the conversion of aromatic compounds *(27)* to *cis*-diols *(28)* carried out with high enantioselectivity by *Pseudomonas putida* (Scheme 8). [15] In the meantime the alcohols *(28)* have become commercially available and have been used by several groups as advantageous intermediates for the synthesis of various natural products. In particular, Hudlicky et al [16] demonstrated the viability of the dienes *(28)* as chiral building blocks. Thus, for instance *(28)* (R = CH$_3$) was transformed into prostaglandin PGE$_2$ *(29)*, [16a] and the analogous vinyl derivative *(28)* (R = CH = CH$_2$) served as starting material for the synthesis of the plant metabolite (–)-zeylena *(30)*. [16b] From the chloro-substituted diene *(28)* (R = Cl) Hudlicky et al. built up the alkaloid trihydroxyheliotridane *(31)*, [16c] the carbohydrates L-ribonolactone *(32)* [16d] and D-erythrose *(33)*, [16e] as well as the polyols (+)-pinitol *(34)*, [16f] conduritol C *(35)*, [16g] conduritol E *(36)* [16h] and conduritol F *(37)*. [16h]

References

[1] Reviews: a) D. Schinzer, *Nachr. Chem. Tech. Lab.* **1989**, *37*, 1294; b) R. A. Johnson and K. B. Sharpless in *Catalytic Asymmetric Synthesis* (Ed.: I. Ojima), VCH, Weinheim, **1993**, p. 103.

[2] a) Review: E. N. Jacobsen in *Catalytic Asymmetric Synthesis* (Ed.: I. Ojima), VCH, Weinheim, **1993**, p. 159; b) E. N. Jacobsen, W. Zhang, A. R. Muci, J. R. Ecker, L. Decker, *J. Am. Chem. Soc.* **1991**, *113*, 7063.

[3] a) R. Criegée, *Liebigs Ann. Chem.* **1936**, *522*, 75; b) Criegée, B. Marchand, H. Wannowius, *Liebigs Ann. Chem.* **1942**, *550*, 99.

[4] S. G. Hentges, K. B. Sharpless, *J. Am. Chem. Soc.* **1980**, *102*, 4263.

[5] R. Annunziata, M. Cinquini, F. Cozzi, L. Raimondi, S. Stefanelli, *Tetrahedron Lett.* **1987**, *28*, 3139.

[6] T. Yamada, K. Narasaka, *Chem. Lett.* **1986**, 131.

[7] M. Tokles, J. K. Snyder, *Tetrahedron Lett.* **1986**, *27*, 3951.

[8] Short review: H. Waldmann, *Nachr. Chem. Tech. Lab.* **1991**, *39*, 1124.

[9] a) K. Tomioka, M. Nakajima, K. Kogra, *J. Am. Chem. Soc.* **1987**, *109*, 6213; b) K. Tomioka, M. Nakajima, Y. Iitaka, K. Koga, *Tetrahedron Lett.* **1988**, *29*, 573, c) K. Tomioka, M. Nakajima, K. Koga, *J. Chem. Soc., Chem. Commun.* **1989**, 1921; d) K. Tomioka, M. Nakajima, K. Koga, *Tetrahedron Lett.* **1990**, *31*, 1741; e) M. Nakajima, K. Tomioka, Y. Iitaka, K. Koga, *Tetrahedron* **1993**, *49*, 10793; f) M. Nakajima, K. Tomioka, K. Koga, *Tetrahedron* **1993**, *49*, 10807.

[10] a) T. Oishi, M. Hirama, *J. Org. Chem.* **1989**, *54*, 5834; b) M. Hirama, T. Oishi and S. Ito, *J. Chem. Soc. Chem., Commun.* **1989**, 665; c) T. Oishi, K. Iida, M. Hirama, *Tetrahedron Lett.* **1993**, *34*, 3573; d) O. Sato, M. Hirama, *Synlett* **1992**, 705.

[11] a) E. J. Corey, P. DaSilva Jardine, S. Virgil, P.-W. Yuen, R. D. Connell, *J. Am. Chem. Soc.* **1989**, *111*, 9243; b) E. J. Corey, G. I. Lotto, *Tetrahedron Lett.* **1990**, *31*, 2665.

[12] a) E. N. Jacobsen, I. Markó, W. S. Mungall, G. Schröder, K. B. Sharpless, *J. Am. Chem. Soc.* **1988**, *110*, 1968, b) E. N. Jacobsen, I. Markó, M. B. France, J. S. Svendsen, K. B. Sharpless, *J. Am. Chem. Soc.* **1989**, *111*, 737; c) J. S. Wai, I. Markó, J. S. Svendsen, M. G. Finn, E. N. Jacobsen, K. B. Sharpless, *J. Am. Chem. Soc.* **1989**, *111*, 1123; d) J. S. Svendsen, I. Markó, E. N. Jacobsen, C. Pulla Rao, S. Bott, K. B. Sharpless, *J. Org. Chem.* **1989**, *54*, 2263; e) B. M. Kim, K. B. Sharpless, *Tetrahedron Lett.* **1990**, *31*, 3003; f) B. B. Lohray, T. H. Kalantar, B. M. Kim, C. Y. Park, T. Shibata, J. S. Wai, K. B. Sharpless, *Tetrahedron Lett.* **1989**, *30*, 2041. For recent reviews see: g) B. B. Lohray, *Tetrahedron: Asymmetry* **1992**, *3*, 1317; h) R. A. Johnson, K. B. Sharpless in *Catalytic Asymmetric Synthesis* (Ed.: I. Ojima), VCH, Weinheim, **1993**, p. 227.

[13] a) H. Kwong, C. Sorato, Y. Ogino, H. Chen,
K. B. Sharpless, *Tetrahedron Lett.* **1990**, *31*,
2999; b) Y. Ogino, H. Chen, H.-L. Kwong,
K. B. Sharpless, *Tetrahedron Lett.* **1991**, *32*,
3965; c) K. B. Sharpless, W. Amberg, M. Bel-
ler, H. Chen, J. Hartung, Y. Kawanami,
D. Lübben, E. Manoury, Y. Ogino, T. Shibata,
T. Ukita, *J. Org. Chem.* **1991**, *56*, 4585; d) K. B.
Sharpless, W. Amberg, Y. L. Bennani, G. A.
Crispino, J. Hartung, K.-S. Jeong, H.-
L. Kwong, K. Morikawa, Z.-M. Wang, D. Xu,
X.-L. Zhang, *J. Org. Chem.* **1992**, *57*, 2768; e)
W. Amberg, Y. L. Bennani, R. J. Chadha,
G. A. Crispino, W. D. Davis, J. Hartung, K.-
S. Jeong, Y. Ogino, T. Shibata, K. B. Sharpless,
J. Org. Chem. **1993**, *59*, 844; f) L. Wang, K. B.
Sharpless, *J. Am. Chem. Soc.* **1992**, *114*, 7568;
g) K. Morikawa, J. Park, P. G. Andersson,
T. Hashiyama, K. B. Sharpless, *J. Am. Chem.
Soc.* **1993**, *115*, 8463; h) M. P. Arrington, Y. L.
Bennani, T. Göbel, P. Walsh, S.-H. Zhao, K. B.
Sharpless, *Tetrahedron Lett.* **1993**, *34*, 7375; i)
K. Morikawa, K. B. Sharpless, *Tetrahedron
Lett.* **1993**, *34*, 5575; k) P. J. Walsh, K. B.
Sharpless, *Synlett* **1993**, 605; l) R. Oi, K. B.
Sharpless, *Tetrahedron Lett.* **1992**, *33*, 2095; m)
T. Hashiyama, K. Morikawa, K. B. Sharpless,
J. Org. Chem. **1992**, *57*, 5067; n) Z.-M. Wang,
K. B. Sharpless, *Synlett* **1993**, 603; o) J. M
Hawkins, A. Meyer, *Science* **1993**, *260*, 1918; p)
J. M. Hawkins, A. Meyer, M. Nambu, *J. Am.*

Chem. Soc. **1993**, *115*, 9844; q) E. J. Corey,
M. C. Noe, S. Sarshar, *J. Am. Chem. Soc.* **1993**,
115, 3828; r) H. C. Kolb, P. G. Andersson, Y. L.
Bennani, G. A. Crispino, K.-S. Jeong, H.-
L. Kwong, K. B. Sharpless, *J. Am. Chem. Soc.*
1993, *115*, 12226.

[14] a) T. Shibata, D. C. Gilheany, B. K. Blackburn,
K. B. Sharpless, *Tetrahedron Lett.* **1990**, *31*,
3817; b) Y. Ogino, H. Chen, E. Manoury, T. Shibata, M. Beller, D. Lübben, K. B. Sharpless,
Tetrahedron Lett. **1991**, *32*, 5761.

[15] a) D. T. Gibson, M. Hensley, H. Yoshika,
R. Mabry, *Biochemistry* **1970**, *9*, 1626, b) For a
review see: H. L. Holland: *Organic Synthesis
with Oxidative Enzymes*, VCH, Weinheim,
1992.

[16] a) T. Hudlicky, H. Luna, G. Barbieri, L. D.
Kwart, *J. Am. Chem. Soc.* **1988**, *110*, 4735; b)
T. Hudlicky, G. Seoane, T. Pettus, *J. Org.
Chem.* **1989**, *54*, 4239; c) T. Hudlicky, H. Luna,
J. D. Price, F. Rulin, *J. Org. Chem.* **1990**, *55*,
4683; d) T. Hudlicky, J. Price, *Synlett* **1990**, 159;
e) T. Hudlicky, H. Lunn, J. D. Price, F. Rulin,
Tetrahedron Lett. **1989**, 30, 4053; f) T. Hudlicky,
J. Price, F. Rulin, T. Tsunoda, *J. Am. Chem.
Soc.* **1990**, *112*, 9439; g) T. Hudlicky, J. D.
Price, H. Luna, C. M. Andersen, *Synlett*, **1990**,
309; h) T. Hudlicky, H. Luna, H. F. Olivo,
C. Andersen, T. Nugent, J. D. Price, *J. Chem.
Soc., Perkin Trans. 1* **1991**, 2907.

Enantioselective Deprotonation and Protonation

Herbert Waldmann

The development of asymmetric transformations belongs to the major goals of organic synthesis. In addition to diastereoselective syntheses that have already been developed to high efficiency, enantioselective methods are gaining increasing attention. In particular, the enantiodifferentiating transfer of protons is of special interest, [1] since this process may provide elegant routes for the conversion of racemic substances into pure enantiomers ("deracemization") or for the generation of optically active intermediates from prochiral *meso* compounds. Due to the low steric demand of the proton the realization of these processes with high stereoselectivity is considered to be particularly demanding.

Deprotonation of *meso*-Ketones

Koga et al. [2] and Simpkins et al. [3] independently used chiral lithium amides for the asymmetric deprotonation of cyclic *meso*-ketones and for the resolution of racemic 2-alkylcyclohexanones. [2c] For instance, the 4-alkylketones *(1)* were converted into the silyl-enol ethers *(3)* and *(6)* by means of the chiral bases *(2)*, [2a] *(4)* [2a] and *(5)* (Scheme 1), [3a, b] Whereas *(2)*, which contains an additional coordination site, induces the preferred formation of the *(R)*-enantiomer *(3)*, the *(S)* enantiomer *(6)* is formed predominantly if *(4)* and *(5)* are employed. The silylenol ethers

can, for instance be transformed to various functionalized cyclic ketones and dicarboxylic acids. [2, 3] Accordingly, cryptone *(7)* was synthesized with 67 % *ee* and used for the synthesis of the marine metabolite (+)-brasilenol *(8)*. [4] By application of this methodology also symmetric disubstituted ketones like *(9)* [3] and *(15)* [2b, 5] can be enolized enantioselectively. In the case of the 2,6-dimethylcyclohexanone *(9)* the highest stereoisomeric excess is achieved with the bornylamines *(10)* and *(11)*. For the monoacetal *(15)* of bicyclo[3.3.0]octene-3,17-dione the use of *(2)* and *(16)*, which differ from *(4)* only by the achiral *N*-substituent, is most advantageous. The steric outcome of these deprotonations is identical to the result obtained for the 4-alkylcyclohexanones. The silylenol ethers *(12)*, *(17)* and *(19)* are also versatile intermediates for the construction of natural products and physiologically active compounds.

For instance, the lactones aeginetolide *(13)* and dihydroactinidiolide *(14)* were obtained from *(12)*, [6] and *(17)* provides a rapid access to the carbacyclin precursor *(18)*. Although Koga et al. and Simpkins et al. studied the deprotonation of *meso*-ketones in detail, they could not offer a rationalization for the observed stereoselection. It is, however, striking that bases that carry additional heteroatoms, for examples *(2)*, induce high enantiomeric excess only after addition of HMPTA, whereas lithium amides of type *(4)*

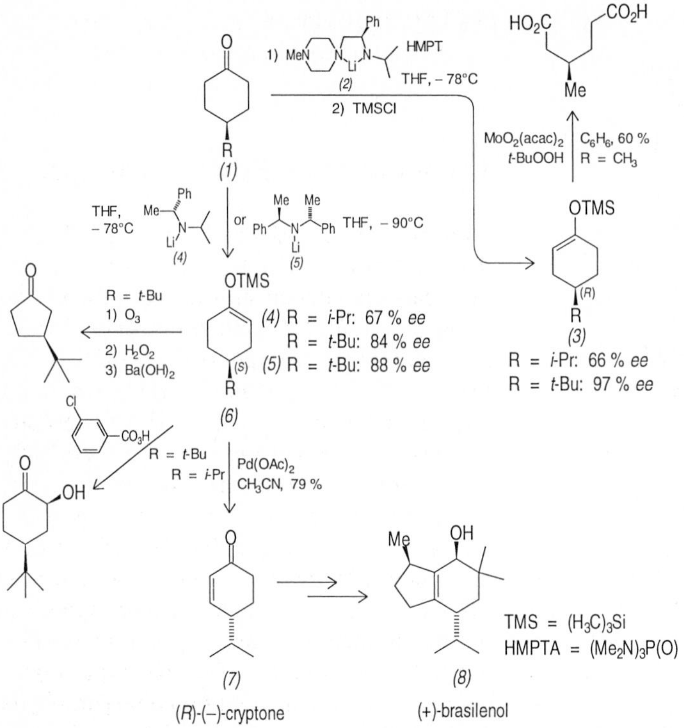

Scheme 1. Enantioselective deprotonation of 4-alkylcylohexanones according to Koga et al. [2] and Simpkins et al. [3].

are most effective in the absence of this strongly complexing reagent.

Opening of *meso*-Epoxides

For the highly selective rearrangement of epoxides *(20)* to allylic alcohols *(23)* and *(24)* N,N-disubstituted aminomethylpyrrolidines, in particular *(21)*, which were used first by Asami [7] for this purpose, show the most advantageous properties. The highest selectivity is recorded for cyclohexene oxide; other cyclic and acyclic oxiranes react with lower *ee*. In these cases the addition of bases like DBU results in a marked improvement of the enantiodifferentiation. Asami et al. assume that the *cis*-coordinated five-membered chelates

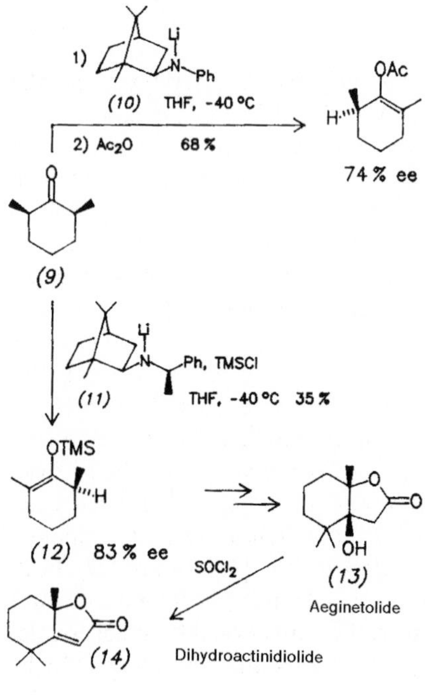

1) MeN(2)
Ph
OTMS

(2)
HMPTA, THF, -78 °C
2) TMSCl

(15) (19)

R = –(CH₂)₂– :
87 % ee

R = –CH₂C(Me)₂CH₂– :
94 % ee

1) Me
Ph
N
Li
(16)
THF, -78 °C

2) TMSCl

OTMS

1) MeLi
2) NC–COOEt
3) AcOH

COOEt

(17) (18)

R = –(CH₂)₂– : 83 % ee
R = –CH₂C(Me)₂CH₂– : 84 % ee

(22) are the decisive intermediates in the deprotonations. In these arrangements the second alkyl group does not undergo unfavorable steric interactions with the pyrrolidine substituent of the base. Compound *(21)* is also effective for more complex epoxides, for example, the *cis*-oxiranes *(26)* are converted with high selectivity to the functionalized cyclopentenediols *(27)*, [8] and the *exo*-epoxide *(28)* delivers the bicyclic allylic alcohol *(29)* with 58% enantiomeric excess. [9] The allylic alcohols, also, can be employed as central intermediates in further syntheses. Thus, from (S)-cyclohexene-1-ol *(24)* (*n* = 2) two diastereomers of the nor leukotriene D₄ analogue *(25)* were constructed. [10] From *(27)* both enantiomers of 4-hydroxycyclopentenol and of 2-oxabicyclo[3.3.0]oct-6-ene are accessible, [8] which are versatile building blocks

for the construction of prostaglandins and other cyclopentanoid natural products.

Elimination of Proton Acids HX

Duhamel et al. [11] and Sakai et al. [12] describe the elimination of HCl from the symmetrical β-halogen-substituted carboxylic acids *(30)* and *(31)* and the elimination of triflic acid from the *meso*-triflate *(33)*. In the first case the adamantylmethyl-substituted phenylethylamide *(32)* shows the highest stereoselectivity. [11] If the size of the sterically demanding alkyl group is reduced, lower *ee* values are obtained; however, the size of the 4-substituent of the carboxylic acid is not important. Interestingly, the prochiral *cis*-carboxylic acid *(30)* and the corresponding *trans*-compound *(31)* deliver opposite enantiomers on deprotonation. This observation proves that the amide in both prochiral carboxylic acids abstracts the same enantiotopic H-atom. For the elimination of triflic acid from *(33)* N,N-dimethylphenylethylamine *(34)* is clearly superior to other bases. [12] To achieve a high selectivity in this case, the carbonyl group must be converted into an acetal, which shields the concave face of the molecule. Bicyclo[3.3.0] oct-2-ene *(35)* is obtained with >90% *ee*. It can be employed as a valuable starting material for the construction of various natural products, for example, coriolin, loganin, etc.

Deprotonation of Alkylcarbamates

Hoppe et al. succeeded in the highly enantioselective deprotonation of achiral and racemic N,N-dialkyl-allyl- and -alkylcarbamates such as *(36)*, *(40)* and *(41)* [13] (Scheme 2). On treatment of the urethanes with *sec*-BuLi in

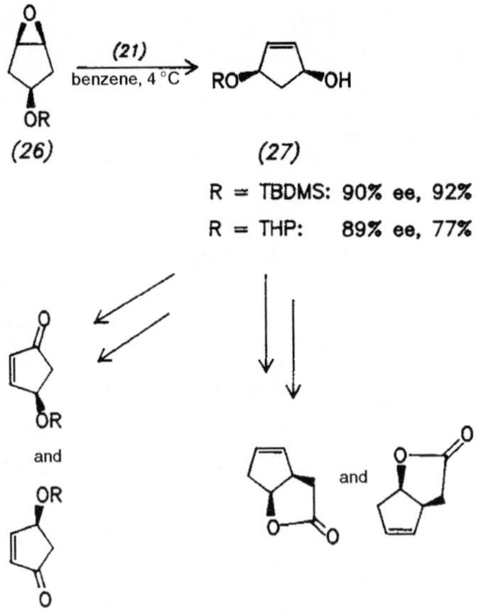

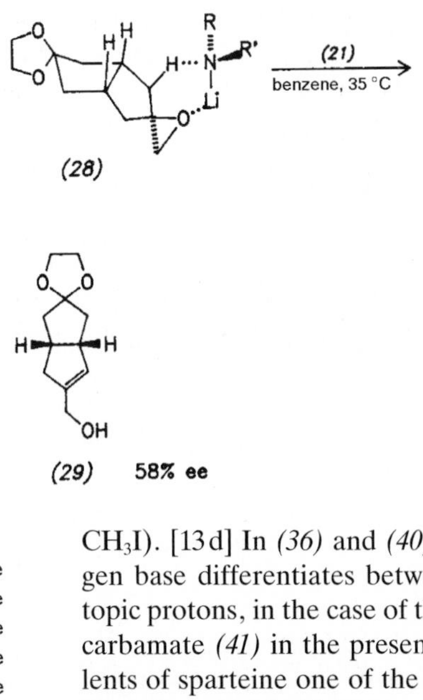

(26)

(21)
benzene, 4 °C

(27)

R = TBDMS: 90% ee, 92%
R = THP: 89% ee, 77%

and

and

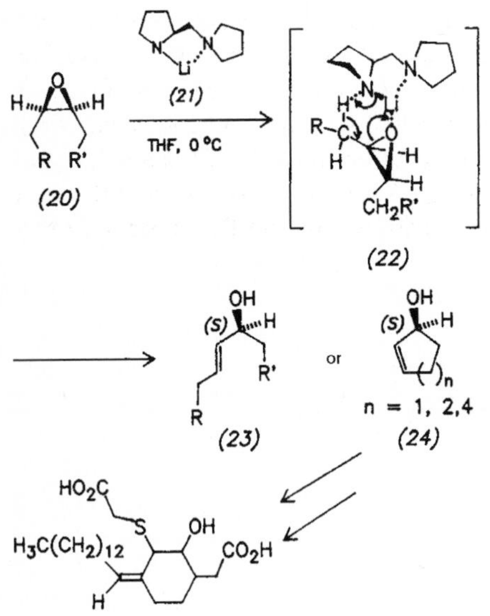

(21)
THF, 0 °C

(20)

(22)

(28)

(21)
benzene, 35 °C

OH
(S)   H

R

R'

or

OH
(S)   H

n = 1, 2, 4

(23)   (24)

(29)   58% ee

HO₂C

H₃C(CH₂)₁₂

OH

CO₂H

H

(25)

R, R' = CH₂:	41% ee
R, R' = (CH₂)₂:	92% ee
R, R' = (CH₂)₄:	58% ee
R = R' = H:	72% ee
R = R' = Et:	60% ee

the presence of the nearly C_2-symmetric alkaloid (–)-sparteine (37) the chiral lithium alkyls (38) and (42) are formed, which remain configurationally stable due to complexation with the urethane carbonyl oxygen. The anions can be trapped in high yield and with high selectivity by various electrophiles (ClSnMe₃, CO₂,

CH₃I). [13d] In (36) and (40) the chiral nitrogen base differentiates between two enantiotopic protons, in the case of the racemic allylic carbamate (41) in the presence of 0.5 equivalents of sparteine one of the two enantiomers is preferred with high selectivity. After stannylation at the 3-position of the allylic system and transmetallation with TiCl₄ the homoaldol (43) is obtained with 82% *ee*. [13c] The two possible anions formed by deprotonation of the allylic carbamate (40) are in rapid equilibrium, but fortunately only one of the two diastereomeric complexes (42) (R = H) crystal-

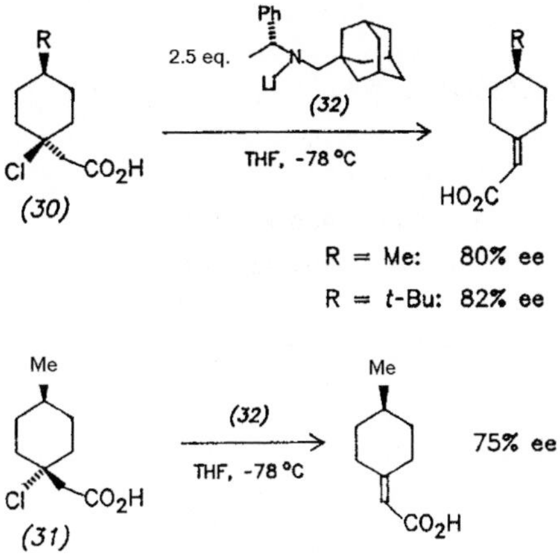
2.5 eq.
(32)
THF, -78 °C
(30)
R = Me: 80% ee
R = t-Bu: 82% ee
(32)
THF, -78 °C
75% ee
(31)

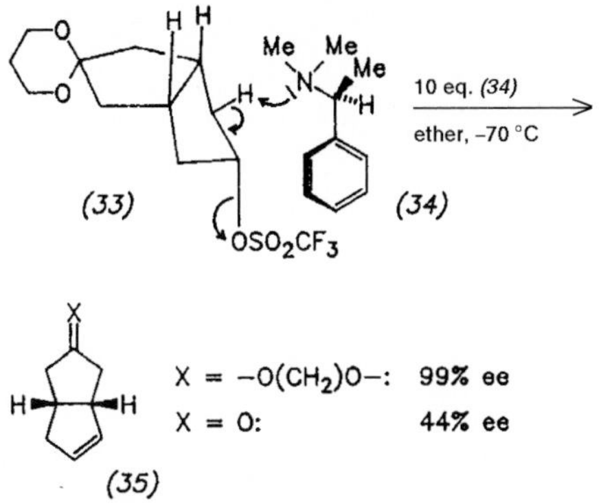
10 eq. (34)
ether, -70 °C
(33)
(34)
X = –O(CH2)O–: 99% ee
X = O: 44% ee
(35)

lizes from the solution under the particular reaction conditions. [13b] Here, too, transmetallation to the analogous titanium compound opens a route to highly enantioselective transformations. From the protected homoallylic alcohols natural products like the querus lactone A *(44)* are advantageously available.

Protonation of Enolates

Duhamel et al. studied in detail the protonation of the enolates *(45)* generated from aromatic imines of amino acid esters. [14] The chiral proton source of choice in particular is the (2*R*, 3*R*)-bis-pivaloyltartaric acid *(46)*. The size of the amino acid side chain does not show a significant influence on the efficiency of the stereoselection, but the electronic prop-

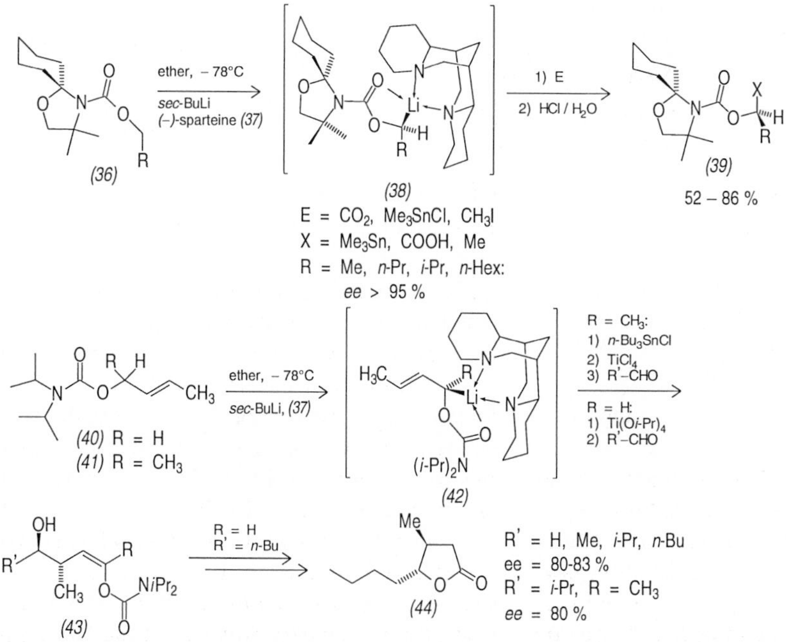
ether, – 78°C
sec-BuLi
(–)-sparteine (37)
(36)
(38)
1) E
2) HCl / H2O
(39)
52 – 86 %
E = CO2, Me3SnCl, CH3I
X = Me3Sn, COOH, Me
R = Me, n-Pr, i-Pr, n-Hex:
ee > 95 %
(40) R = H
(41) R = CH3
ether, – 78°C
sec-BuLi, (37)
(42)
R = CH3:
1) n-Bu3SnCl
2) TiCl4
3) R'–CHO
R = H:
1) Ti(Oi-Pr)4
2) R'–CHO
(43)
R = H
R' = n-Bu
(44)
R' = H, Me, i-Pr, n-Bu
ee = 80-83 %
R' = i-Pr, R = CH3
ee = 80 %

Scheme 2. Enantioselective deprotonation of *N,N*-dialkylcarbamates according to Hoppe et al. [13].

(45)

(46)

R = Ph, base = LDA, X = H, 50% ee
R = Ph, base = LDA, X = OMe, 70% ee
R = Ph, base = (47), X = H, 70% ee

(47)

(48) → (49)

(46), -70 °C, 16 h → (S):(R) = 90:10

The same model can also be applied if instead of the imines α-hydroxyketones like benzoin (48) are used. [15] After deprotonation with KH to give the enediolate (49) and protonation with (46) (S)-benzoin is obtained with an enantiomer ratio of 90:10. Of particular interest is the finding that the protonation of (49) to give the enediol is fast and that the stereoselectivity is established in the subsequent relatively slow keto-enol tautomerization.

(50)

HX* =

(51)   (52)

erties of the substituents in the aromatic ring of the Schiff's base and the structure of the base used for deprotonation are of particular importance. [14b–d] Suitable chiral lithium amides, for example, (47) induce a significant increase in the optical yield as compared with the use of lithium diisopropylamide (LDA). [14b] In all cases the (2R, 3R)-configured proton source (46) preferably attacks the *Re* face of the enolates (45), which are stabilized by complexation. In these cases, therefore, the sense of the stereoselection seems to be predictable with a high probability. [14c]

Hünig et al. used the conformationally fixed cyclic ester enolates (50) as model compounds to study enantiodifferentiating protonations. [16] α-Hydroxy carboxylic acid esters such as (51) and (52) proved to be efficient proton sources; O-acyl tartaric acids, for example, (45), C₂-symmetric diols and different carbohydrates induced only lower *ee* values. Whereas the secondary amine used for deprotonation does not seem to be involved in the protonation of (50), the cations play a decisive role. Thus, in the protonation of (50) by (51) the *ee* value drops to ca. 2% if NaHMDS or KHMDS is used. However, after generation of an ate complex by means of Ti(OiPr)₄ it rises to 61%. The authors conclude from these findings that a tight ion pair must be formed to achieve an efficient stereo-selection. This is attacked at the cation by the carbonyl group of the chiral α-hydroxy ester. Thereby the proton would be transferred per-pendicular to the double blond to the α-carbon of the enolate. Rebek et al. investi-gated the tricyclic lactams *(53)* and *(54)*, which can be obtained from Kemp's triacid and chiral alkylarylamines by diastereomer separation, as proton sources in this trans-formation. [17] For aromatic α-hydroxy-carboxylic acid esters only unreproducible results were obtained, but the alkyl-substituted intermediates *(50)* were proto-nated with high selectivity. To account for the observation that the enantiomeric excess increases with increasing steric demand of the "R" group, the authors propose the transition state model *(55)*. In the latter the lithium ion is complexed by two oxygen atoms and the

alkyl group remains outside the cage struc-ture. Fehr et al. described the stereoselective protonation of an acyclic ketoenolate without additional complexation site. [18] To construct (–)-damascone *(59)* from the ketene *(56)* the magnesium enolate *(57)* was generated by addition of allylmagnesium chloride. Com-pound *(57)* was then converted to a 1:1 Li-Mg complex by treatment with the lithium alk-oxide *(58)* (X = Li). Protonation of this inter-mediate with the respective alcohol *(58)* (X = H) proceeds with virtually complete enanti-oselectivity (*ee* >98%). (–)-Damascone is formed by isomerization of the double bond. By analogy its enantiomer may be obtained from the easily accessible *ent-(58)*.

(56)

(57)

(59): (–)-Damascone

(53)

(54)

(51), R = Ph: 42% ee *(S)*
(52), R = Ph: 53% ee *(S)*
(53), R = t-Bu: 72% ee *(R)*
(54), R = t-Bu: 91% ee *(S)*
(54), R = Ph: 0% ee

(55)

According to Pracejus et al., ketenes like *(60)* (R = H) can be protonated enantioselec-tively by achiral alcohols in the presence of catalytic amounts of chiral bases, for example, acylated cinchona alkaloids, to give esters like *(61)*. [19] Whereas the analogous transforma-tions employing exclusively chiral alcohols display only a low selectivity, [20] in these cases the addition of an achiral base causes a significant increase in the diastereomeric excess. Thus, Rüchardt et al. demonstrated that the ester *(63)* is formed from *(60)* and *(S)*-phenylethyl alcohol in the presence of pyri-dine with nearly 80% *de*. [21] The authors propose that in the reaction an intermediary

Ph, O
Me OMe   76% ee
(61)

toluene ↑ R = H
– 110 °C │ 1 mol-%
         │ acetyl-
         │ quinine

Me

(60)

R = H
0 °C
toluene
1.1 equiv.
pyridine

(62)

77% Me
     Ph   OH

R•–OH │ R = iBu
      │ 3 equiv. NMe₃
      │ toluene, – 78 °C

Me Ph
H   Ph
     O Me
    (63)

76% de

Me
        COOR•   H⁺ or OH⁻
R

Me
        OH
R   O
(65)

R•–OH = HO   (R)   90%, 99% de (R,R)
        (52)

R•–OH =   OH
         (S)   O   92%, 89% de (S,S)
              O
         (64)

enolbetaine (62) is protonated by the alcohol from below and is converted to the ester with simultaneous loss of pyridine. In similar transformations Larsen and Corley et al. [22] recorded a dramatic increase of the stereoselectivity by employing α-hydroxycarboxylic acid esters as chiral proton sources. Lactic acid esters like (64) induced ca. 90% de and with pantolactone (52) the (R)-enantiomer was formed even with 99% de. This process,

offers, for example, a very practical method for the highly stereoselective generation of optically active arylpropionic acids (65) from their racemates. Those compounds are of particular interest since they are widely used as non-steroidal anti-inflammatory drugs.

References

[1] a) Reviews: L. Duhamel, P. Duhamel, J.-C. Launay, J.-C. Plaquevent, *Bull. Soc. Chim. Fr.* **1984**, II-421; b) N. S. Simpkins, *Chem. Ind.* **1988**, 387; c) P. C. Coc, N. S. Simpkins, *Tetrahedron: Asymmetry* **1991**, 2, 1.

[2] a) R. Shirai, M. Tanaka, K. Koga, *J. Am. Chem. Soc.* **1986**, 108, 543; b) H. Izawa, R. Shirai, H. Kawasaki, H. Kim, K. Koga, *Tetrahedron Lett.* **1989**, 30, 7221; c) H. Kim, H. Kawasaki, M. Nakajima, K. Koga, *Tetrahedron Lett.* **1989**, 30, 6537.

[3] a) N. S. Simpkins, *Chem. Rev.* **1990**, 19, 335; b) C. M. Cain, R. P. Cousins, G. Coumbarides, N. S. Simpkins, *Tetrahedron* **1990**, 46, 523; c) N. S. Simpkins, *J. Chem. Soc., Chem. Commun.* **1986**, 88; see also 1c).

[4] A. E. Greene, A. A. Serra, E. J. Barreiro, P. R. Costa, *J. Am. Chem. Soc.* **1987**, 52, 1169.

[5] J. Leonard, J. D. Hewitt, D. Ouali, S. K. Rahman, S. J. Simpson, R. F. Newton, *Tetrahedron: Asymmetry* **1990** 1, 699.

[6] M. C. Cain, N. S. Simpkins, *Tetrahedron Lett.* **1987**, 28, 3723.

[7] a) M. Asami, *Chem. Lett.* 1984, 829, b) M. Asami, *Bull. Chem. Soc. Japan.* **1990**, 63, 721.

[8] M. Asami, *Tetrahedron Lett.* **1985**, 26, 5803.

[9] J. Leonard, J. D. Hewitt, D. Ouali, S. J. Simpson, R. F. Newton, *Tetrahedron Lett.* **1990**, 46, 6703.

[10] J. S. Sabol, R. J. Cregge, *Tetrahedron Lett.* **1989**, 30, 3377.

[11] a) L. Duhamel, A. Ravard, J.-C. Plaquevent, *Tetrahedron: Asymmetry* **1990**, 1, 347; b) L. Duhamel, A. Ravard, J.-C. Plaquevent, D. Davoust, *Tetrahedron. Lett.* **1987**, 28, 5517.

[12] H. Kashibara, H. Suemune, T. Kawahara, K. Sakai, *Tetrahedron Lett.* **1987**, 28, 6489.

[13] a) D. Hoppe, T. Krämer, J.-R. Schwark, O. Zschage, *Pure Appl. Chem.* **1990**, *62*, 1999; b) D. Hoppe, O. Zaschge, *Angew. Chem.* **1989**, *101*, 67; *Angew. Chem. Int. Ed. Engl.* **1989**, *28*, 69; c) O. Zschage, J.-R. Schwark, D. Hoppe, *Angew. Chem.* **1990**, *102*, 336; *Angew. Chem. Int. Ed. Engl.* **1990**, *29*, 296; d) D. Hoppe, F. Hintze, P. Tebben, *Angew. Chem.* **1990**, *102*, 1457; *Angew. Chem. Int. Ed. Engl.* **1990** 29, 1422. For experimental details and further references see: O. Zschage, J.-R. Schwark, T. Krämer, D. Hoppe, *Tetrahedron* **1992**, *48*, 8377 and O. Zschage, D. Hoppe, *Tetrahedron* **1992**, *48*, 8389.

[14] a) L. Duhamel, J.-C. Plaquevent, *J. Am. Chem. Soc.* **1978**, *100*, 7415; b) L. Duhamel, J.-C. Plaquevent, *Tetrahedron Lett.* **1980**, *21*, 2521; c) L. Duhamel, J.-C. Plaquevent, *Bull. Soc. Chim. Fr.* **1982**, II-75, d) L. Duhamel, P. Duhamel, S. Fouquay, J. J. Eddine, O. Peschard, J.-C. Plaquevent, A. Ravard, R. Solliard, J.-Y. Valnot, H. Vincens, *Tetrahedron* **1988**, *44*, 5495.

[15] L. Duhamel, J.-C. Launay, *Tetrahedron Lett.* **1983**, *24*, 4209.

[16] U. Gerlach, S. Hünig, *Angew. Chem.* **1987**, *99*, 1323; *Angew. Chem. Int. Ed. Engl.* **1987**, *26*, 1283.

[17] D. Potin, K. Williams, J. Rebek, Jr., *Angew. Chem.* **1990**, *102*, 1485; *Angew. Chem. Int. Ed. Engl.* **1990**, *29*, 1420.

[18] C. Fehr, J. Galindo, *J. Am. Chem. Soc.* **1988**, *110*, 6909.

[19] a) H. Pracejus, *Liebigs Ann. Chem.* **1960**, *634*, 9; b) H. Pracejus, G. Kohl, *Liebigs Ann. Chem.* **1969**, *722*, 1.

[20] E. Anders, E. Ruch, I. Ugi, *Angew. Chem.* **1973**, *85*, 16; *Angew. Chem. Int. Ed. Engl.* **1973**, *12*, 25.

[21] a) J. Jähne, C. Rüchardt, *Angew. Chem.* **1981**, *93*, 919; *Angew. Chem. Int. Ed. Engl.* **1981**, *20*, 885; b) U. Salz, C. Rüchardt, *Tetrahedron Lett.* **1982**, *23*, 4017.

[22] R. D. Larsen, E. G. Corley, P. Davis, P. J. Reider, E. J. Grabowski, *J. Am. Chem. Soc.* **1989**, *111*, 7650.

Carbohydrate Complexes in Enantioselective Carbon–Carbon Bond Formation

Karl Heinz Dötz

The remarkable progress that has been achieved in stereoselective carbon–carbon bond formation is mostly due to the development of novel organometallic reagents. An already classical example is provided by organotitanium compounds, the synthetic potential of which was recognized in the early 1980s. [1] In comparison with Grignard and organolithium compounds from which they are generated by a simple in-situ transmetalation, the titanium reagents exhibit an often dramatically improved chemo- and diastereoselectivity, as shown by the addition to aldehydes and ketones. [2] Whereas methyl lithium *(1a)* cannot differentiate between benzaldehyde and acetophenone the methyl titanium reagent *(1b)* is perfectly aldehyde selective (Scheme 1). Equally impressive is the *diastereoselectivity* that – in favor of the Cram product *(3a)* – is observed in the addition of *(1b)* to chiral aldehydes such as *(2)*. In contrast, powerful *enantioselective* methods have been restricted to a few examples for a long time (e.g. *(4)* → *(5)* [3]). Obviously, the incorporation of chelating ligands into compounds of type *(6a)* is not necessarily associated with a sufficient stability of configuration.

This goal, however, is achieved if a *pentahapto* cyclopentadienyl ligand is incorporated into the reagent. Coordination of chiral coligands (two unidentate coligands X or one bidentate coligand with C_2-symmetry to avoid the generation of an undesired metal stereo-center) leads to compounds of type *(6b)*. Carbohydrates turned out to be especially effective in reagents that allow high asymmetric induction both in the nucleophilic addition to aldehydes and in aldol reactions. [4] A potent allyl transfer reagent is complex *(10)* which can be prepared in a 100 g scale from the chloro precursor *(9)* following a two-step sequence from cyclopentadienyltitanium trichloride *(7)* and diacetone glucose (DAGOH) *(8)*. [5] Complex *10* prefers addition to the *re*-face of aldehydes, thereby affording homoallyl alcohols *(11)* in good chemical yields (50–90 %) and very good enantioselectivity (86–95 % *ee*). The selectivity depends only to a minor extent on the temperature, as shown in the addition of *(10)* to benzaldehyde: The *ee* value drops from 90 % to 80 % when the temperature is raised from –74 °C to 0 °C. Controlled hydrolysis of the reaction mixture affords undefined titanate species that are transformed to the trichloride *(7)*. Thus, both the metal component and the released chiral auxiliary can be recovered (Scheme 2).

The enantioselectivity of the allyl transfer depends markedly on the combination of the coligands: It is essential that *two* carbohydrate ligands are attached to the metal center together with *one* allyl group. The *ee* value of the allylation of benzaldehyde decreases to 46 % and 21 %, respectively, if the bis(diacetone glucose) complex CpTiCl(ODAG)₂ *(10)*

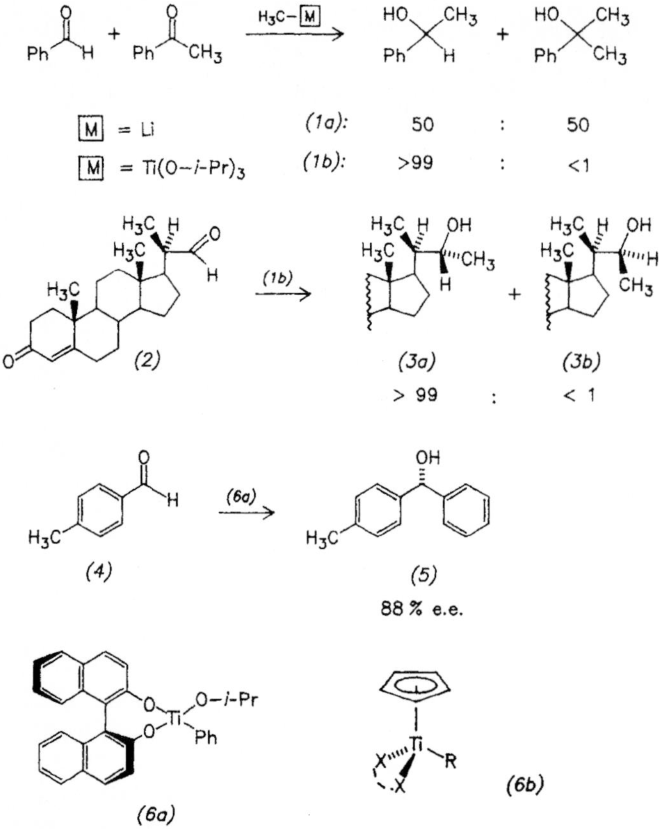

Scheme 1. Diastereoselective and enantioselective C–C bond formation using lithium and titanium organometallics.

is replaced by its diallyl(ODAG) and allyl chloro ODAG analogues. However, the enantioselectivity is unaffected (but only in this example) by increasing bulk and donor capacity of the cyclopentadienyl ligand, as shown by trimethylsilyl or pentamethyl substitution. [6]

Upon reaction with complex *(10)* chiral racemic aldehydes such as *(12)* afford the stereoisomeric homoallyl alcohols *(13)*–*(16)*. The pair of diastereomers *(13)/(14)* arises from *(R)-(12)*, while *(S)-(12)* leads to *(15)/(16)*. The only moderate enantiomeric differentiation (2.5:1) and the fact that the more reactive enantiomer reveals the better diastereoselectivity [97 % *de*, for *(R)-(12)*] may be rational-

ized in terms of a Cram selectivity that is overruled by the enantiofacial differentiation [71 % *de* for *(S)-(12)*]. Furthermore, allyl complex *(10)* exhibits the well-known chemoselectivity of titanium reagents: At –74 °C no reaction occurs with ketones; at 0 °C, however, aryl ketones give tertiary homoallyl alcohols (albeit with induction reduced to 50 % *ee*).

A complementary allyl transfer to the *si*-face of aldehydes occurs with complexes containing threitol ligands. The ligand of choice is the diphenyl derivative *(R,R)-(17)*, accessible from natural tartaric acid in 88 % yield. Upon reaction with CpTiCl₃ it forms a stable chloro chelate *(R,R)-(18)*, which serves as precursor

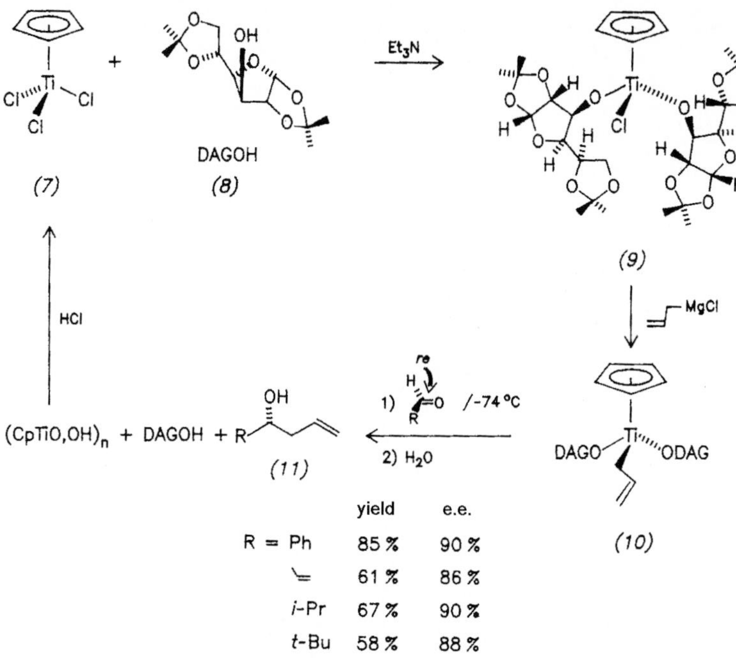

Scheme 2. Allyl transfer reactions with the bis(DAGO)titanium reagent *(10)*.

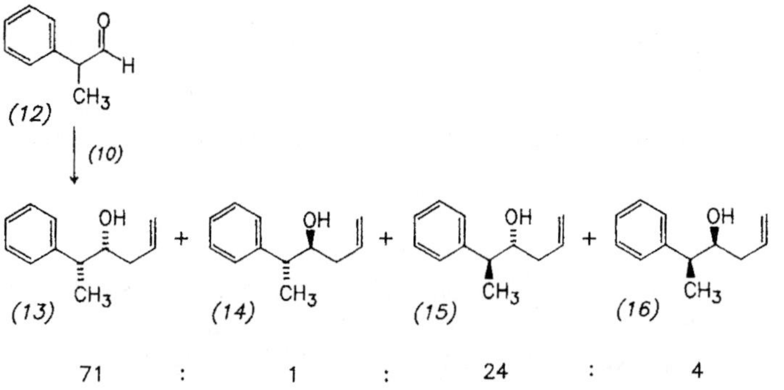

R = Ph	85 %	90 %
＝	61 %	86 %
i-Pr	67 %	90 %
t-Bu	58 %	88 %

71   :   1   :   24   :   4

for the allyl complex *(R,R)-(19)*. This compound adds to the *si*-face of aldehydes producing the homoallyl alcohols *(20)* with excellent enantioselectivity combined with very good chemical yields. Similarly successful is the allyl transfer to α-chiral aldehydes. Since reagent *(19)* is accessible both as *(R,R)*- and *(S,S)*-enantiomer, depending on the configuration of the tartaric acid precursor, the L-serinal derivative *(21)* [7] allows the formation

of both diastereomers *(22a)* and *(22b)*. The diastereoselective potential of the titanium complexes becomes evident in comparison with well-established boron reagents, for example, diisopinocampheny(allyl)borane, [8] as demonstrated for 2-phenyl-butanal *(23)*.

Complexes of type *(10)* and *(19)* are also well-suited for the transfer of substituted allyl groups. Independent on whether the titanium reagent is generated from the organometallic

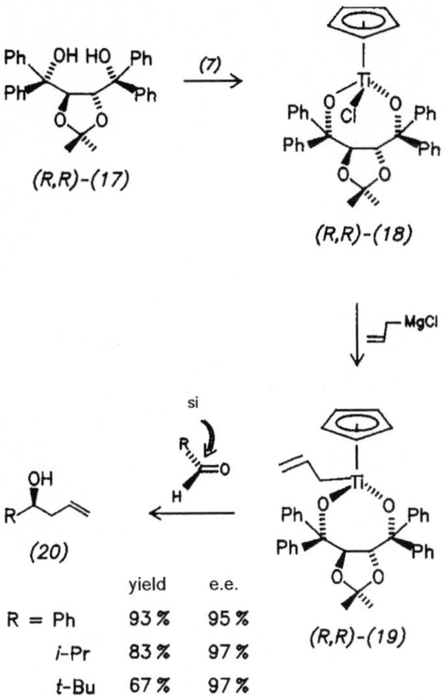

(R,R)-(17)
(7)
(R,R)-(18)
MgCl
si
(20)
(R,R)-(19)
yield e.e.
R = Ph 93 % 95 %
i-Pr 83 % 97 %
t-Bu 67 % 97 %

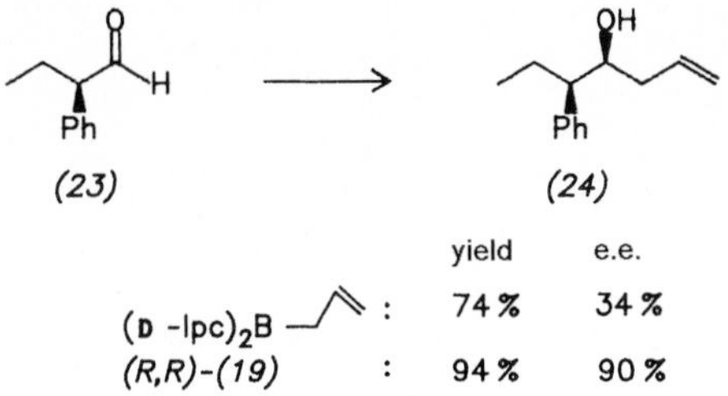

(23)
(24)
yield e.e.
(D -Ipc)2B : 74 % 34 %
(R,R)-(19) : 94 % 90 %

E-or *Z*-allyl precursor the *anti*-products *(26)* are formed exclusively. Obviously, the η^1-allyl titanium species *(25)* undergoes a rapid equilibration in which the *trans*-isomer containing the metal coordinated to the sterically unshielded allyl terminus prevails. A similar *trans*-preference was observed earlier for crotyl titanium reagents. [9] Using the boron methodology *syn*-crotyl adducts also have been obtained. [10]

Whereas the cyclopentadienyl complexes *(10)* and *(19)* are efficient *allyl* transfer reagents, their alkyl analogues are unreactive in *alkyl* transfer reactions. Taking advantage of the increasing Lewis acidity encountered in complexes containing the higher homologues of titanium this drawback can be overcome by switching to zirconium or hafnium. Thus, the pentamethylcyclopentadienyl complexes of zirconium(IV) and hafnium(IV), accessible from their chloro precursors *(27)* and methyl lithium or methyl magnesium bromide, undergo a highly enantioselective methyl transfer to aromatic aldehydes. [6, 11] In contrast to the trisalkoxy titanium compounds MeTi $(OR^*)_3$, the *ee* values are independent

Boc
(21)
(S,S)-(19)
95 %
(R,R)-(19)
93 %
OH
Boc
OH
Boc
(22b)
(22a)
>95 % d.e.
>95 % d.e.

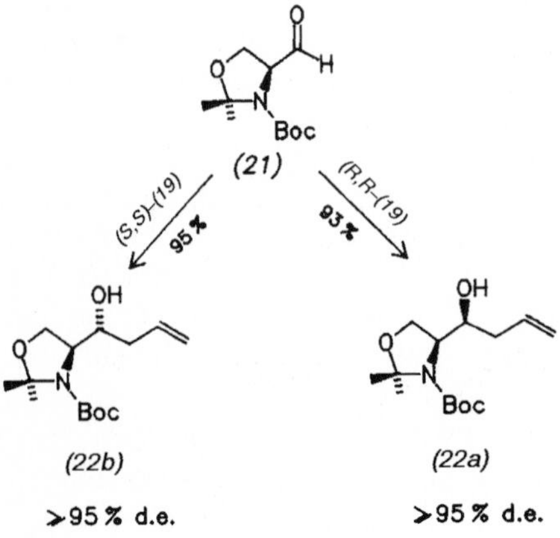

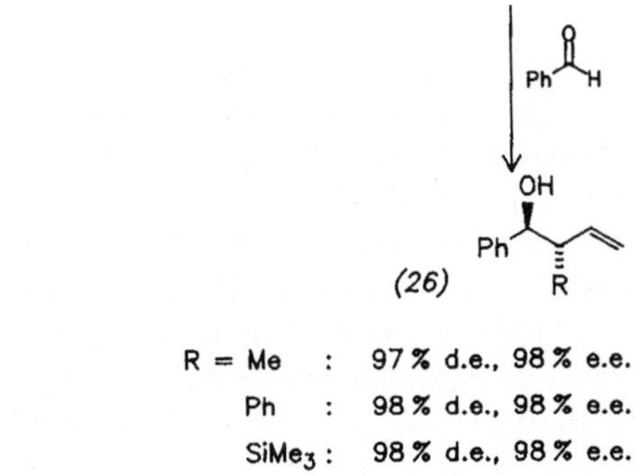

(R, R)-(18)
M = K, Li, MgCl
(25)
(26)
R = Me : 97 % d.e., 98 % e.e.
Ph : 98 % d.e., 98 % e.e.
SiMe3 : 98 % d.e., 98 % e.e.

1) R–[M]

2) Ph–CHO

−78 °C

(27)

Ph–CH(OH)–R

(28)

R–[M]	M	e.e.
MeMgBr	Zr	98 %
MeLi	Zr	97.5 %
MeMgBr	Hf	96.5 %
MeLi	Hf	97.5 %
BuLi	Zr	68 %
BuLi	Hf	65 %

(29)

(9)

(R,R)-(18)

DAGO–Ti–ODAG

(30)

(32)

−78 °C

R–CHO

R–CH(OH)–CH₂–CO–O–t-Bu

(31)

CH₃–CH(OH)–CH₂–CO–O–t-Bu

(33)

	yield	e.e.	temp.	yield	e.e.
R = n-Pr	51 %	94 %	−78 °C	80 %	78 %
i-Bu	81 %	94 %	0 °C		54 %
Ph	69 %	95 %			

of the metal used for transmetalation. However, the enantioselectivity distinctly decreases if benzaldehyde is replaced by non-aromatic aldehydes and if longer or branched alkyl groups are transferred. [4a]

Carbohydrate titanium complexes have also been successfully applied in aldol reactions. This is demonstrated for acetate enolates, which have caused problems in aldol reactions until recently. [12] The titanium enolate *(30)*, prepared in situ from lithium enolate *(29)* and the chloro complex *(9)*, reacts with aldehydes to give β-hydroxy esters *(31)*. [13] This route is particularly interesting for α,β-unsaturated aldehydes, since their aldols cannot be obtained from β-keto esters by enantioselective hydrogenation. [14] Unfortunately, the reaction of the threitol complex *(32)* obtained from *(R,R)-(18)* is not similarly selective, as demonstrated for isovaleraldehyde. [6] Another difference between the DAGO and threitol reagents arises from the temperature dependence of the enantioselectivity: The *ee* value observed for the DAGO complex *(30)* remains unchanged between −78 °C and ambient temperature, whereas for the threitol derivative *(32)* the induction decreases distinctly with increasing temperature.

A major advantage of these aldol reactions is that they can be easily scaled up. For instance, the insect pheromone *(−)-(S)-*ipsenol *(35)* has been synthesized on a 50 g scale. [15] Starting from *tert*-butyl acetate and CpTiCl₃ the β-hydroxy ester *(34)* was obtained via the titanium enolate *(30)*; subsequent elaboration (8 steps, mainly protection/deprotection procedures) afforded the diene alcohol *(35)* in 13 % overall yield.

CH₃–CO–O–t-Bu → (7) → (30) →

(34)

8 steps →

(35)

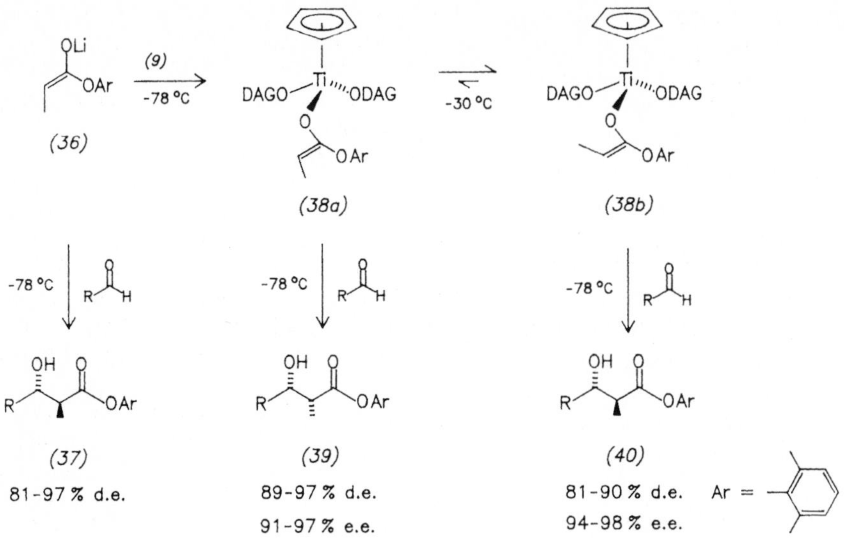

Scheme 3. Formation of *syn-* and *anti-*aldols from a common titanium enolate precursor.

An extension of these studies in propionate aldol reactions involves the lithium enolate (36) which has been used by Heathcook in a diastereoselective access to *anti-β*-hydroxy-*α*-methyl carboxylic esters (37). [16] Transme-

talation at –78 °C to give titanium enolate (38a) followed by addition of the aldehyde, however, affords the *syn*-aldol products (39) with comparable *de* values. [4, 6, 17] The reaction is temperature-dependent: At –30 °C the

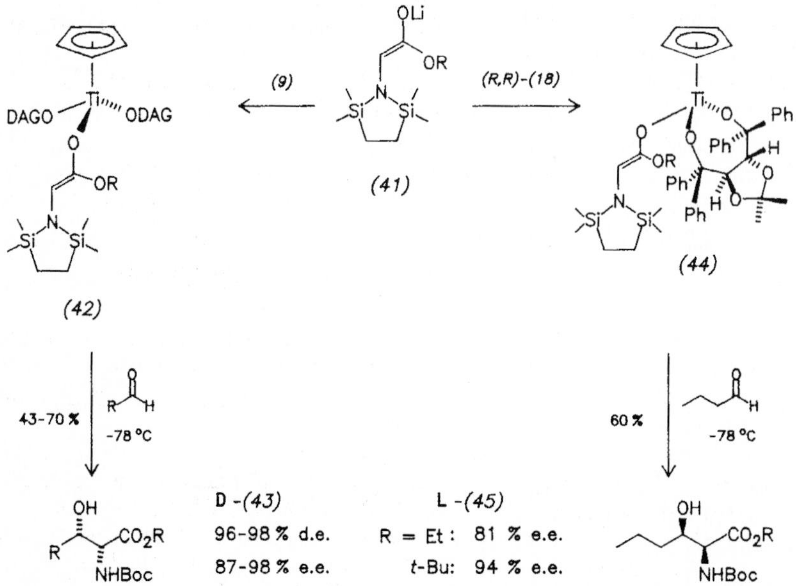

Scheme 4. Synthesis of D- and L-hydroxy amino acid esters from glycine ester enolates.

E-enolate undergoes a rearrangement (probably to give the *Z*-isomer *38b)*; as a consequence, the *anti*-esters are again obtained, as known for the lithium enolate series. Thus, both epimers *(39)* and *(40)* are accessible from the same precursor *(38a)* with good diastereoselectivity. In both cases the titanium enolate adds to the *re*-face of the aldehyde as previously observed for the allyl transfer using the DAGO reagent *(10)* (Scheme 3).

Finally, carbohydrate-modified titanium reagents can be used in the functionalization of amino acids. [18] Transmetalation of the lithium enolate of the *N*-protected glycine ester *(41)* generates the DAGO titanium enolate *(42)*, which attacks the *re*-face of an aldehyde and thus affords the D-*syn*-β-hydroxy-α-amino acid ester *(43)*. In a complementary fashion, the threitol complex *(44)*, obtained from the same lithium enolate by transmetalation using *(R,R)-(18)*, adds to the *si*-face of butyraldehyde and thus leads to the L-amino acid ester L-*(45)* (Scheme 4).

To rationalize the attack at complementary faces of the aldehyde by the DAGO and threitol complexes X-ray studies have been performed for the chloro complexes *(9)* and *(R,R)-(18)*. [6, 19] The major difference in the molecular structure of both compounds is a pair of Ti-O-C angles that differ by 7 and 10°, respectively, for the *re*- and *si*-side. It is tempting to make this enantiomeric distortion responsible for the efficient chiral induction that – in fact – is the opposite of that for reagents derived from *(9)* (*re*-side attack) and *(R,R)-(18)* (*si*-side attack).

References

[1] a) M.T. Reetz, R. Steinbach, J. Westermann, R. Peter, *Angew. Chem.* **1980**, *92*, 1044; *Angew. Chem. Int. Ed. Engl.* **1980**, *19*, 1011; b) B. Weidmann, D. Seebach, *Helv. Chim. Acta* **1980**, *63*, 2451.

[2] Reviews: a) D. Seebach, B. Weidmann, L. Widler in *Modern Synthetic Methods*, (Ed.: R. Scheffold), Vol. 3, Salle, Frankfurt/Main and Sauerländer, Aarau, **1983**; b) M.T. Reetz, *Organotitanium Reagents in Organic Synthesis*, Springer, Berlin, **1986**.

[3] a) A.G. Olivero, B. Weidmann, D. Seebach, *Helv. Chim. Acta* **1981**, *64*, 2485; b) D. Seebach, A.K. Beck, R. Imwinkelried, S. Roggo, A. Wonnacott, *Helv. Chim. Acta* **1987**, *70*, 954.

[4] Reviews: a) R.O. Duthaler, A. Hafner, *Chem. Rev.* **1992**, *92*, 807; b) R.O. Duthaler, A. Hafner, M. Riediker, *Pure Appl. Chem.* **1990**, *62*, 631.

[5] a) M. Riediker, R.O. Duthaler, *Angew. Chem.* **1989**, *101*, 488; *Angew. Chem. Int. Ed. Engl.* **1989**, *28*, 494; b) A. Hafner, R.O. Duthaler, R. Marti, G. Rihs, P. Rothe-Streit, F. Schwarzenbach, *J. Am. Chem. Soc.* **1992**, *114*, 2321.

[6] R.O. Duthaler, A. Hafner, M. Riediker in *Advances in Organic Synthesis via Organometallics*, (Eds.: R.W. Hoffmann and K.H. Dötz), Vieweg, Wiesbaden, **1991**, p. 285.

[7] P. Garner, J.M. Park. *J. Org. Chem.* **1987**, *52*, 2361.

[8] H.C. Brown, K.S. Bhat, R.S. Randad, *J. Org. Chem.* **1989**, *54*, 1570.

[9] M.T. Reetz, M. Sauerwald, *J. Org. Chem.* **1984**, *49*, 2292.

[10] R.W. Hoffmann, *Pure Appl. Chem.* **1988**, *60*, 123.

[11] R.O. Duthaler, A. Hafner, P.L. Alsters, R. Rothe-Streit, G. Rhis, *Pure Appl. Chem.* **1992**, *64*, 1897.

[12] a) M.T. Reetz, F. Kunisch, P. Heitmann, *Tetrahedron Lett.* **1986**, *27*, 4721; b) S. Masamune, T. Sato, B.-M. Kim, T.A. Wollmann, *J. Am. Chem. Soc.* **1986**, *108*, 8279; c) Review: M. Braun, *Angew. Chem.* **1987**, *99*, 24; *Angew. Chem. Int. Ed. Engl.* **1987**, *26*, 24.

[13] R.O. Duthaler, P. Herold, W. Lottenbach, K. Oertle, M. Riediker, *Angew. Chem.* **1989**, *101*, 490; *Angew. Chem. Int. Ed. Engl.* **1989**, *28*, 495.

[14] R. Noyori, T. Ohkuma, M. Kitamura, H. Takaya, N. Sayo, H. Kumobayashi, S. Akutagawa, *J. Am. Chem. Soc.* **1987**, *109*, 5856.

[15] K. Oertle, H. Beyeler, R.O. Duthaler, W. Lottenbach, M. Riediker, E. Steiner, *Helv. Chim. Acta* **1990**, *73*, 353.

[16] C.H. Heathcock, M.C. Pirrung, S.H. Montgomery, J.Lampe, *Tetrahedron* **1981**, *37*, 4087.

[17] R.O. Duthaler, P.Herold, S.Wyler-Helfer, M.Riediker, *Helv. Chim. Acta* **1990**, *73*, 659.

[18] G.Bold, R.O. Duthaler, M.Riediker, *Angew. Chem.* **1989**, *101*, 491; *Angew. Chem. Int. Ed. Engl.* **1989**, *28*, 497.

[19] M.Riediker, A.Hafner, U.Piantini, G.Rihs, A.Togni, *Angew. Chem.* **1989**, *101*, 493; *Angew. Chem. Int. Ed. Engl.* **1989**, *28*, 499.

Asymmetric Aza-Diels–Alder Reactions

Herbert Waldmann

Hetero-Diels–Alder reactions with aza-substituted dienes or dienophiles are powerful methods for the regio- and stereoselective synthesis of nitrogen heterocycles [1] and can play pivotal roles in the construction of many natural products and biologically active compounds. Therefore, the development of methods that allow these transformations to be carried out asymmetrically is of particular interest to organic synthesis. In contrast to the asymmetric carbo-Diels–Alder reaction the analogous cycloadditions with aza-substituted dienes and dienophiles have gained much less attention, and only recently have several efficient asymmetric aza-Diels–Alder processes been developed.

Imino Dienophiles and Dienes

Whereas simple, non-activated Schiff's bases display only a low reactivity in hetero-Diels–Alder reactions, *C*- or *N*-acylimines prove to be more efficient dienophiles. Thus, the imine *(2a)*, derived from *(R)*-phenylethylamine and glyoxylic acid ester *(1)* reacts even at room temperature with cyclic dienes in DMF as solvent and in the presence of equimolar amounts of trifluoroacetic acid as well as catalytic amounts of water to yield the bicyclic amino acid esters *(4)* and *(5)* with high diastereoselectivity [2a] (Scheme 1). The

analogous cycloadditions employing acyclic derivatives of butadiene give rise to substituted pipecolic acid esters *(6)*, from which the auxiliary can be easily removed by hydrogenolysis, and that are interesting intermediates for the construction of thrombin inhibitors such as *(7)*. The presence of small amounts of water is absolutely necessary to effect the cycloadditions, which do not proceed in dry DMF. The authors therefore propose the iminium ion *(3)* to be the reactive intermediate, which, for steric reasons, is attacked predominantly from the *si*-side. If the imine *(3)* is generated from racemic phenylethyl amine the respective cycloadditions proceed in dichloromethane and in the presence of $BF_3 \cdot OEt_2$ with complete facial selectivity. [2b] In contrast, to the *N*-phenylethyl-substituted Schiff's base *(2a)*, the *N*-sulfonyldienophiles *(2b–2f)*, carrying a chiral auxiliary in the ester function [2c, d] in purely thermal, uncatalyzed cycloadditions with various dienes induce a diastereomeric excess of only 20–30%. [2a, e] However, a significant improvement is recorded for *(2e)* and *(2f)* if the cycloadditions are promoted by a Lewis acid. [2d] *N*-Acyl imines are the reactive intermediates in the Lewis acid mediated reaction of the lactam *(8)* with the diene *(9)* [3a] and in the reaction of the enamine *(14)* with the α-bromo substituted amino acid ester *(12)* [3b] (Scheme 2). On treatment with $ZnCl_2$ *(8)* eliminates acetic acid and the *N*-acyl imine

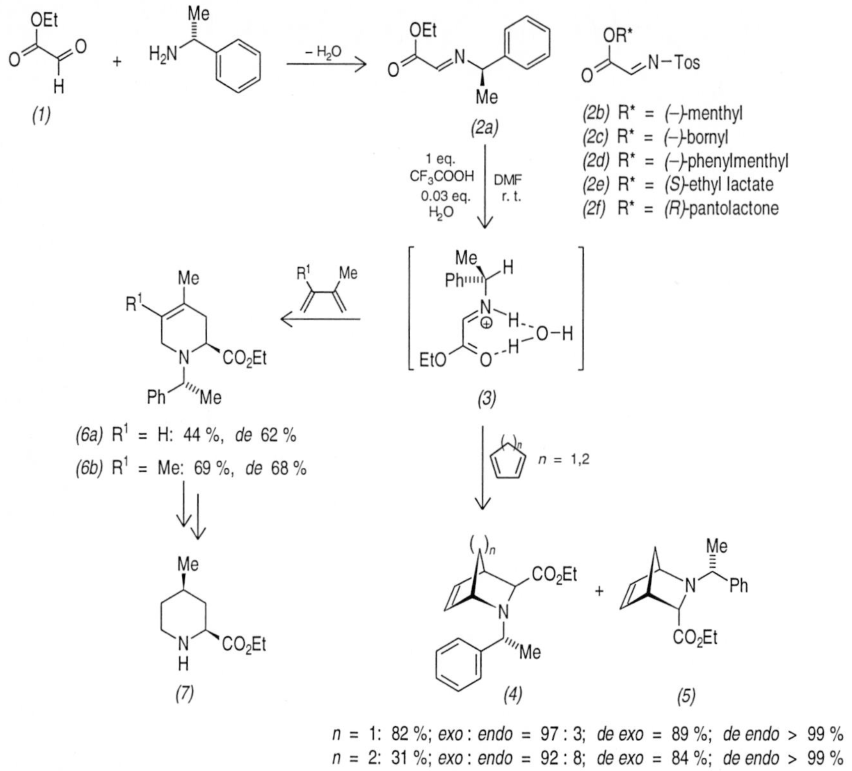

OEt

(1)

Me
H$_2$N

− H$_2$O

OEt
(2a)
Me

OR*
N−Tos

(2b) R* = (−)-menthyl
(2c) R* = (−)-bornyl
(2d) R* = (−)-phenylmenthyl
(2e) R* = (S)-ethyl lactate
(2f) R* = (R)-pantolactone

1 eq.
CF$_3$COOH   DMF
0.03 eq.   r. t.
H$_2$O

Me
R^1
CO$_2$Et
N
Ph   Me

(6a) R^1 = H: 44 %, de 62 %
(6b) R^1 = Me: 69 %, de 68 %

R^1 Me

Me
Ph
N
EtO   O   H–O–H
O

(3)

n = 1,2

Me
CO$_2$Et
N
H
(7)

( )$_n$
CO$_2$Et
N
Me
(4)

Me
N   Ph
CO$_2$Et
(5)

n = 1: 82 %; exo : endo = 97 : 3; de exo = 89 %; de endo > 99 %
n = 2: 31 %; exo : endo = 92 : 8; de exo = 84 %; de endo > 99 %

Scheme 1. Asymmetric aza-Diels–Alder reactions with the chiral C-acyl imine (2a).

thereby formed delivers the carbacephem (10) from which the carbapenem (11) can be formed. In the presence of NEt$_3$ the ester (12) eliminates HBr to generate the chiral hetero diene (13) which reacts with the chiral enamine (14) to give the cycloadduct (15). After hydrolysis, the amino acid ester (16) is obtained from (15) with complete stereoselectivity and in quantitative yield. [3b] Recently, a highly efficient asymmetric intramolecular cycloaddition with an azo compound generated in situ by oxidation of a proline derived hydrazide has been reported. [3c] In the cycloadditions mentioned above the C=N group is activated by an electron-withdrawing C- or N-acyl substituent. This goal can also be reached by protonation of the heteroanalogous carbonyl function or, in general, by its complexation with a Lewis acid. For instance, the imine (17) reacts with the electron-rich Brassards diene (18) in the presence of an aluminum Lewis acid in a highly stereocontrolled manner to give the cycloadduct (19) which is converted on hydrolysis to the α,β-unsaturated amine (20) (Scheme 3). [4a]

For such Lewis acid-mediated aza-Diels–Alder reactions with the Brassard diene valine tert-butyl ester has proven to be an efficient mediator of selectivity. [4b, c] The imines (21), obtained from this primary amine and aliphatic, aromatic and further functionalized aldehydes, react with diene (18) to give the addition products (23) with high diastereomeric excess (Scheme 3). In these cases, the primary adducts (23), which can be addressed as orthoesters, can not be isolated. Rather, on

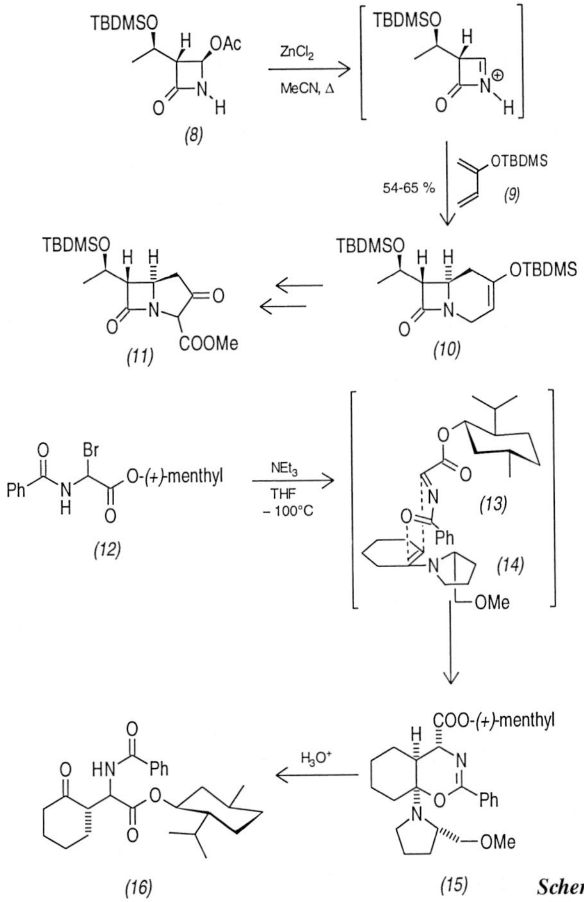

Scheme 2. Asymmetric aza-Diels–Alder reactions with chiral *N*-acyl imines.

hydrolytic workup they are converted into the amides *(24)* and the esters *(25)*, which, however, can easily be subsequently cyclized to the lactams *(24)*. For the removal of the chiral auxiliary from *(24)* the amino acid α-carbon is converted into an acetal center by means of a Curtius rearrangement. Cleavage of the intermediate urethanes thereby generated leads to the desired heterocycles, for example *(26)*, in good yield (Scheme 3). This reaction sequence is most advantageously carried out as a one-pot reaction. To explain the steric course of the reactions between the imines *(21)* and the diene *(18)*, the Diels–Alder

type transition state *(22)* was proposed, in which the diene attacks the *si*-side of the imine double bond. In this arrangement the amino acid ester adopts an *anti*-Felkin–Anh type conformation to minimize unfavourable steric interactions between the voluminous substituents on C-1 of the diene and the sterically demanding amino acid side chain. However, a stepwise reaction consisting of the formation of the esters *(25)* and their subsequent cyclization to the lactams *(24)* could not be rigorously ruled out.

Such a pathway may also be involved in the reactions of the electron rich silyloxydienes

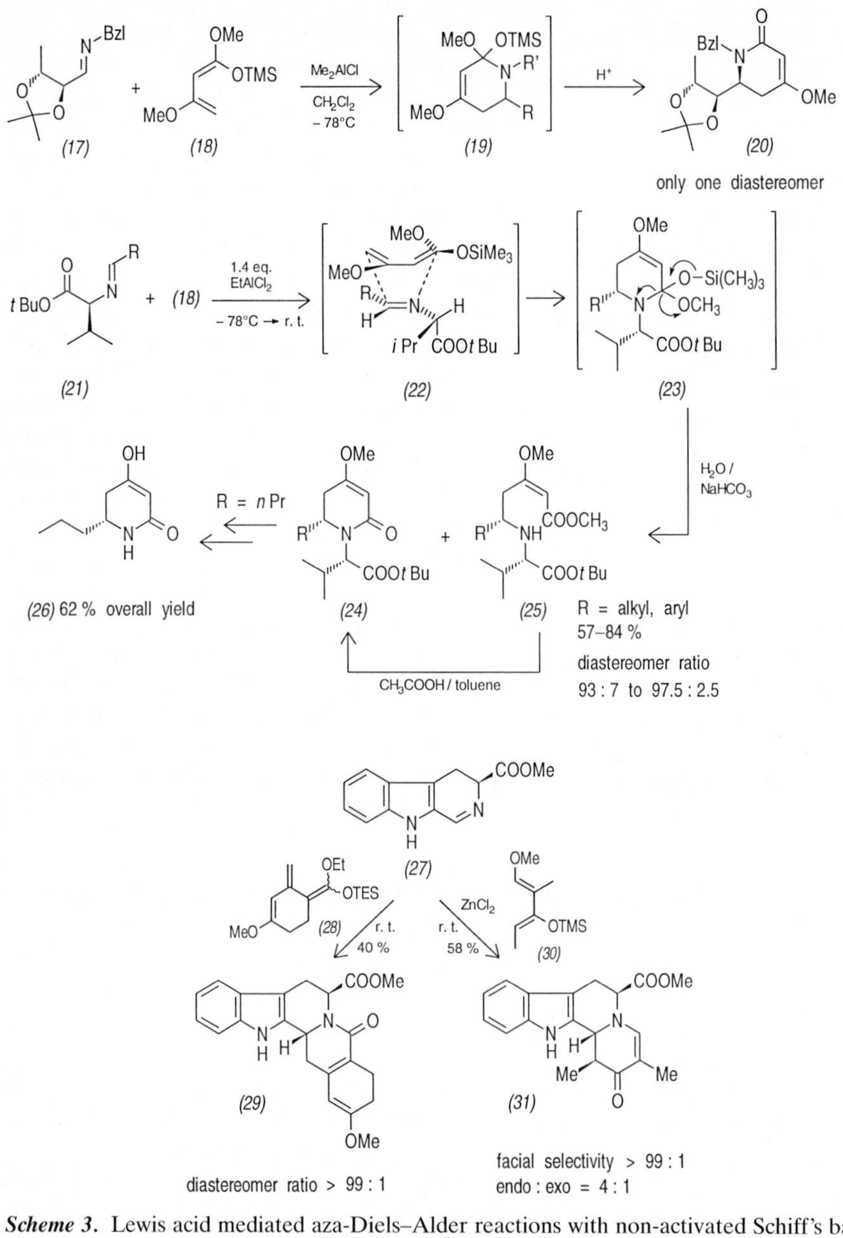

Scheme 3. Lewis acid mediated aza-Diels–Alder reactions with non-activated Schiff's bases.

(28) and *(30)* with the imine *(27)* derived from tryptophan. [4d] These cyclizations proceed with high stereoselectivity to give the polycyclic compounds *(29)* and *(31)* (Scheme 3), which may be interesting congeners to alkaloids. However, a cycloaddition and not a stepwise process clearly occurs if the carbohydrate-derived imines *(32)* are treated with alkyl-substituted butadienes *(33)* in the presence of ZnCl$_2$ to give the dehydropiperidi-

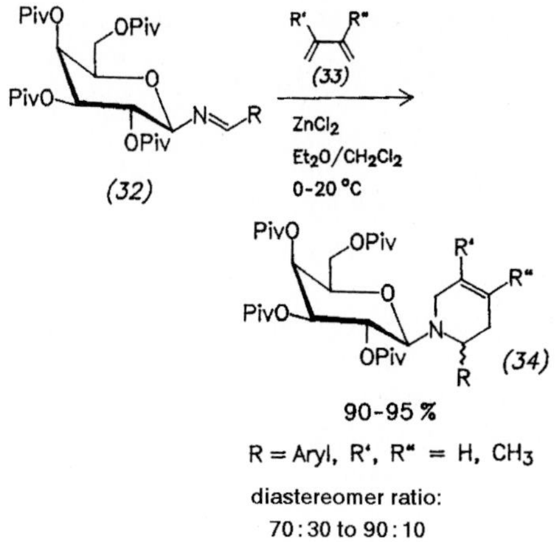
PivO
OPiv
O
PivO
N
R
OPiv
(32)
R'
R"
(33)
ZnCl2
Et2O/CH2Cl2
0-20 °C
PivO
OPiv
R'
R"
O
PivO
N
OPiv
R
(34)
90-95 %
R = Aryl, R', R" = H, CH3
diastereomer ratio:
70 : 30 to 90 : 10

nes *(34)* with appreciable diastereomeric ratios. [5]

Recently, highly diastereoselective Lewis acid mediated aza-Diels–Alder cycloadditions with chiral imines and dienes carrying a chiral auxiliary have also been reported. [6] In these processes 2-amino-1,3-butadienes embodying 2-(methoxymethyl)pyrrolidine as mediator of chirality were employed with great success.

Protonated imines are the reactive heterodienophiles in aza-Diels–Alder reactions in aqueous solution. [7] Under conditions that are otherwise typical for Mannich reactions, ammonium salts *(35)* and aldehydes, in particular formaldehyde, form iminium intermediates *(36)* in situ that undergo cycloadditions with various dienes (Scheme 4). If *(S)*-phenylethylamine is used as mediator of chirality in the Diels-Alder reaction with cyclopentadiene, the diastereomers *(37)* and *(38)* are formed in a ratio of 80:20. [7a] However, in this reaction the stereoselectivity is rather sensitive to the steric bulk and the electronic properties of the chiral auxiliary. Thus, with the respective 4-bromophenyl or the 4-NO_2-phenyl analogue and *(S)*-1-phenylpropylamine the diastereomers are formed in a ratio of ca. 1:1. [7g] If amino acid esters are employed as mediators of selectivity isomer ratios up to 93:7 are reached (Scheme 4). [7b–e] This enhanced level of stereodiscrimination can be explained by the assumption that in these cases in the aqueous medium highly ordered transition states *(40)*, with min-

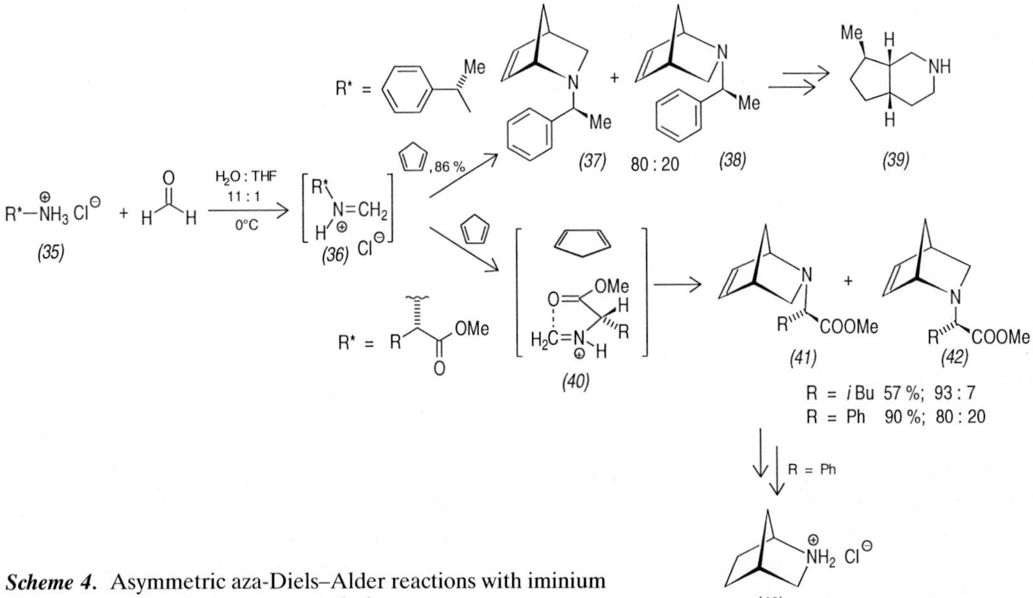
R* =
Me
N
Me
+
N
Me
Me
H
NH
H
, 86 %
(37)
80 : 20
(38)
(39)
O
H2O : THF
11 : 1
0°C
R*–NH3 Cl
H
H
(35)
R*
N=CH2
H
(36) Cl
O
OMe
H
H2C=N
R
H
(40)
R* = R
OMe
O
N
+
N
R
COOMe
R
COOMe
(41)
(42)
R = i Bu 57 %; 93 : 7
R = Ph 90 %; 80 : 20
R = Ph
NH2 Cl
(43)

Scheme 4. Asymmetric aza-Diels–Alder reactions with iminium salts generated in situ in aqueous solution.

imized hydrophobic surfaces, are passed in which the diene attacks the iminium intermediates from the direction opposite to the amino acid side chain. The chiral auxiliary is easily removed from the phenylethylamine and the phenylglycine derivatives obtained from these transformations by hydrogenation. Thereby enantiomerically pure nitrogen heterocycles, for example (43), become available, which can be advantageously employed in the construction of alkaloids. For instance, the terpene alkaloid (−)-δ-N-normethylskantine (39) was constructed from (38) making use of a [3,3]-sigmatropic rearrangement as the key step. [7f]

A 2-aza-1,3-butadiene that is activated via N-arylation is present in the isoquinolinium salt (44). [8] This heterodiene undergoes asymmetric cycloadditions with chiral enol ethers. The cycloadducts (45) formed initially are further converted by the solvent methanol into the tetralines (46) and (47). It is striking that the most advantageous chiral auxiliaries for this process contain aromatic groups but a rationalization of the observed direction of stereoselection is not at hand. If tetrabenzyl glucose is employed as source of chirality, its removal can readily be affected by a three-step sequence.

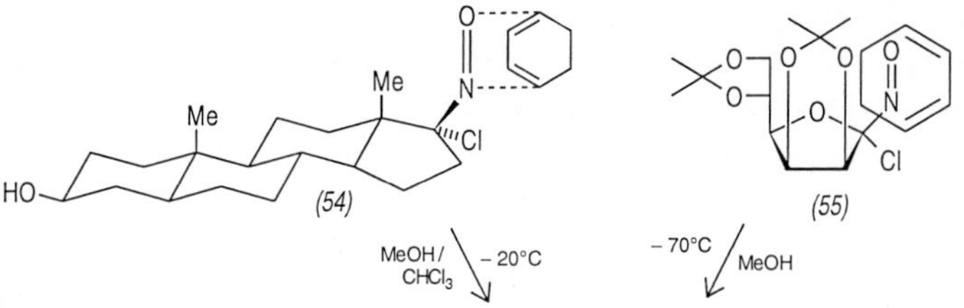

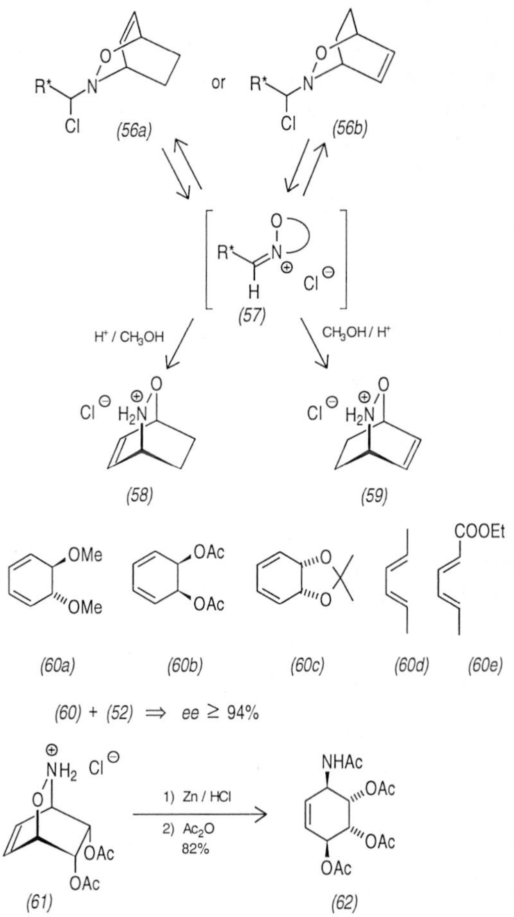

N-Sulfinyl Dienophiles

N-Sulfinyl compounds are heterodienophiles that already under mild conditions react with various dienes to give dehydrothiazine oxides. The N-sulfinyl carbamates (48) and (51) carrying 8-phenylmenthol or a camphor derivative react with open-chain butadiene derivatives and cyclohexadiene in the presence of SnCl$_4$ and TiCl$_4$, respectively, to give, for instance, the cycloadducts (50) and (53) in a ratio ≥ 98:2. [9] The heterocycles obtained in these transformations can be transformed into various synthetically useful derivatives, for example, homoallylamines, by well established techniques. In both cases the steric course of the cycloaddition is rationalized by assuming that *endo* transition states (49) and (52) are involved in which one diastereotopic face of the dienophile is efficiently shielded by the chiral auxiliary.

Nitroso Dienophiles and Dienes

N-Acyl- and α-chloronitroso compounds are very reactive heterodienophiles that undergo hetero-Diels–Alder reactions with various dienes at low temperature. Thus, the 17-chloro-17-nitrosoepiandrosterone derivative (54) and the 1-chloro-1-nitrosomannose (55) in alcoholic solution react with cyclohexadiene to give the expected cycloadducts (56a) and (56b) (Scheme 5). The latter are in equilibrium with the iminium salts (57) that immediately release the chiral auxiliary on solvolysis. By this procedure the dihydrooxazines (58) and (59) are obtained with an enantiomeric excess of ≥ 96 % [10] (the absolute configura-

Scheme 5. Asymmetric hetero-Diels–Alder reactions with chiral α-chloro-α-nitroso compounds.

tion initially [10a–c] assigned to (58) and (59) was subsequently corrected [10d]). The observed sense of asymmetric induction is explained by the involvement of an *exo*-attack of the diene on the sterically more accessible face of the nitroso dienophiles (Scheme 5).

The mannosyl dienophile *(55)* also undergoes highly enantioselective cycloadditions with the substituted cyclohexadienes *(60a–c)* [10d–f] and the butadiene derivatives *(60d)* and *(60e)*. [10c] The dihydrooxazines are valuable precursors for the construction of inosamines. For instance, the cycloadduct *(61)* obtained from *(60b)* can be readily converted to the tetraacetate *(62)* of conduramine A 1 by reductive cleavage of the N–O bond and subsequent acetylation (Scheme 5). [10e] The use of chiral *N*-acylnitroso compounds in asymmetric hetero-Diels–Alder reactions has been investigated thoroughly. [11] These studies revealed that simple derivatives of mandelic acid and proline [11a–i] are only ineffective auxiliary groups in this process. However, high levels of asymmetric induction are attained if C_2-symmetric amines [12] are employed as mediators of chirality. [11k–n] In general, the chiral hetero dienophiles *(64)* are generated in situ from the corresponding hydroxamic acids *(63)* by oxidation with periodate. They subsequently react with various dienes, for example, cyclohexadiene, to deliver the desired heterocycles *(65)* with high yield and virtually complete stereoselectivity. [11e, m] If 2-azabutadienes are employed in this process, an interesting and highly stereoselective route to amino acids is opened up. [11n] The stereochemical course of these hetero Diels–Alder cycloadditions can be understood by assuming an *endo*-attack of the respective diene on the *S-cis*-conformer *(64)* of the dienophile. In addition to the pyrrolidine derivatives, the nitroso compounds *(66)* [11o] and *(67)* [11p] which are derived from camphor, proved to be highly stereodirecting chiral heterodienophiles that render the desired cycloadducts available with high isomer ratios.

(63)

(64)

(65)

R^1 = CH$_3$, R^2 = H: 81 %, de 98 %

R^1 = CH$_2$OMe, R^2 = H: 88 %, de > 99 %

R^1, R^2 = –Ph 90 %, de > 99 %

(66)

n = 1: 91 %, de > 98 % de

n = 2: 94 %, de > 98 % de

(67)

62–94 %, de ⩾ 95 %

dienes:

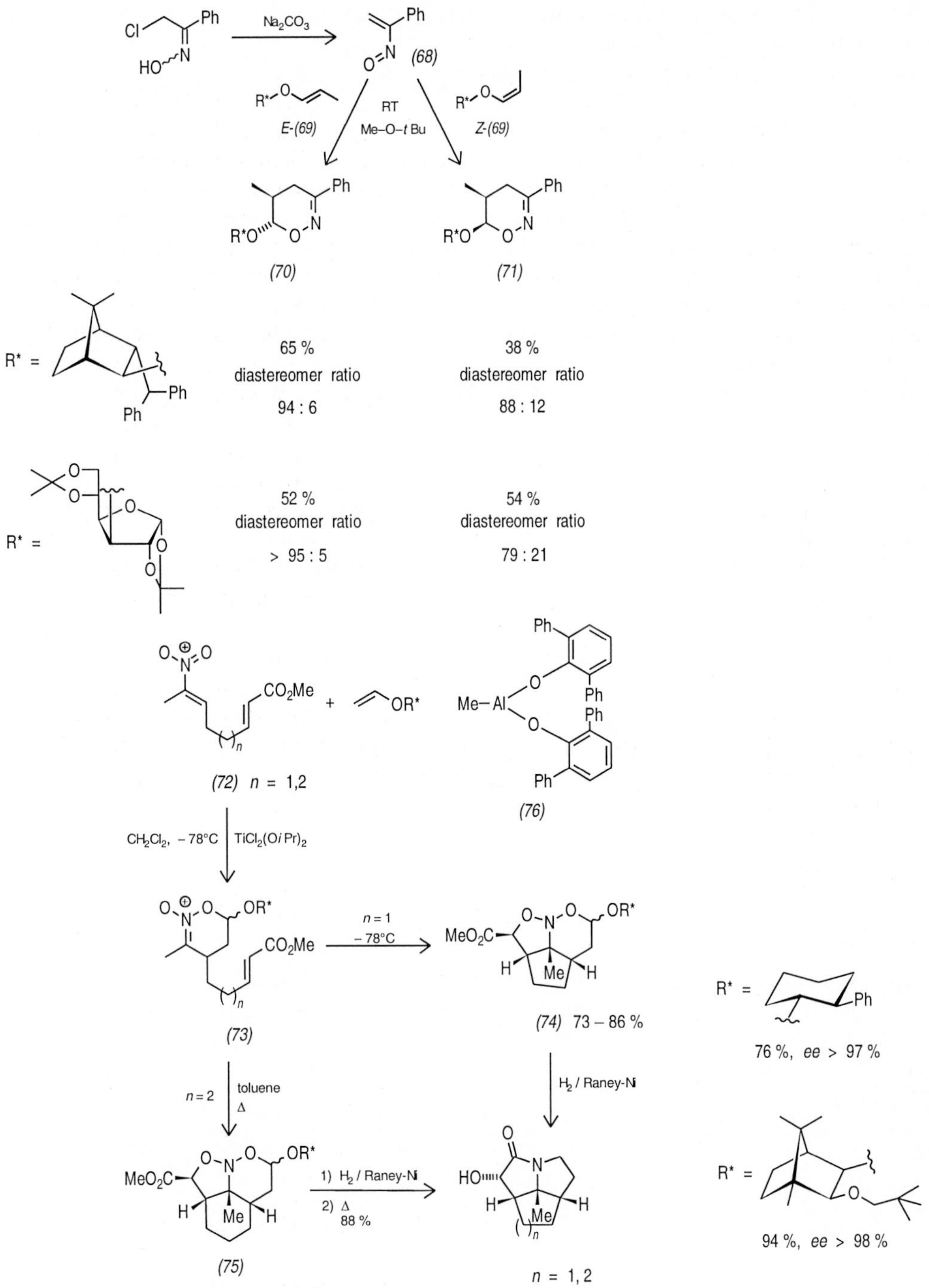

Scheme 6. Asymmetric hetero-Diels–Alder reactions with nitroso- and nitroalkenes.

Hetero-Diels–Alder reactions with inverse electron demand in which the nitroso group is part of the hetero diene system are observed if chiral enol ethers react with nitrosoalkenes, for example, *(68)*, generated in situ (Scheme 6). In these transformations a camphor derivative and diacetone glucose have been identified as advantageous chiral auxiliaries. [13] If the heterodiene *(68)* is treated with *E/Z*-mixtures of enol ethers *(69)* derived from these mediators of chirality, the *E*-isomers react not only faster but also with higher selectivity to give the *trans*-1,2-oxazines *(70)* with high diastereomeric excess. However, the *cis*-configured heterocycles *(71)* are also obtained from the *Z*-enol ethers with appreciable results.

Similar to nitrosoalkenes nitroalkenes also undergo hetero-Diels–Alder reactions with inverse electron demand. Thus, *(72)* reacts with chiral enol ethers in the presence of $TiCl_2(OiPr)_2$ already at low temperature to give nitronates *(73)*, which, on warming to ambient temperature ($n = 1$) or after heating ($n = 2$) undergo a spontaneous [3+2]-cycloaddition to the nitroso acetals *(74)* and *(75)*. [14] Both steps proceed in a highly stereoselective manner. The sense of the diastereoselection can be reversed by using the aluminum catalyst *(76)* instead of a titanium Lewis acid.

In conclusion, despite the great potential that aza-Diels–Alder reactions hold for the synthesis of heterocycles, in particular physiologically active compounds, natural products and analogs thereof, only a very limited number of efficient chiral auxiliaries has been introduced for these processes. Therefore, research directed at the development of new methods for the highly stereoselective execution of these cycloadditions is worthwile and rewarding.

References

[1] Review on asymmetric hetero Diels-Alder reactions: H. Waldmann, *Synthesis* **1994**, 535; general reviews on hetero Diels-Alder reactions: a) D. L. Boger, S. M. Weinreb, *Hetero Diels-Alder Methodology in Organic Synthesis*, Academic Press, San Diego, **1987**; b) S. M. Weinreb in *Comprehensive Organic Synthesis*, Vol. 5 (Ed.: L. A. Paquette), Pergamon Press, Oxford **1991**, p. 401; c) D. L. Boger in *Comprehensive Organic Synthesis*, Vol. 5 (Ed.: L. A. Paquette), Pergamon Press, Oxford, **1991**, p. 451; d) S. M. Weinreb, R. R. Staib, *Tetrahedron* **1982**, 38, 3087; e) D. Boger, Chem. Rev. **1986**, *86*, 781.

[2] a) P. D. Bailey, G. R. Brown, F. Korber, A. Reed, R. D. Wilson, *Tetrahedron: Asymmetry* **1991**, *2*, 1263; b) L. Stella, H. Abraham, J. Feneau-Dupont, B. Tinant, J. P. Declercq, *Tetrahedron Lett.* **1990**, 31, 2603; c) M. Maggini, M. Prato, G. Scorrano, *Tetrahedron Lett.* **1990**, 31, 6243; d) P. Hamley, G. Helmchen, A. B. Holmes, D. R. Marshall, J. W. M. MacKinnon, D. F. Smith, J. W. Ziller, *J. Chem. Soc., Chem. Commun.* **1992**, 786.

[3] a) A. I. Meyers, T. J. Sowin, S. Scholz, Y. Ueda, *Tetrahedron Lett.* **1987**, *28*, 5103; b) R. Kober, K. Papadopoulos, W. Miltz, D. Enders, W. Steglich, *Tetrahedron* **1985**, *41*, 1693; c) T. Sheradsky, J. Milvitskaya, I. E. Pollak, *Tetrahedron Lett.* **1991**, *32*, 133.

[4] a) M. Midland, J. McLoughlin, *Tetrahedron Lett.* **1988**, *29*, 4653; b) H. Waldmann, M. Braun, M. Dräger, *Angew. Chem.* **1990**, *102*, 1445; *Angew. Chem. Int. Ed. Engl.* **1990**, 29, 1468; c) H. Waldmann, M. Braun, M. Dräger, *Tetrahedron: Asymmetry*, **1991**, *2*, 1231; d) S. Danishefsky, M. Langer, C. Vogel, *Tetrahedron Lett.* **1985**, *26*, 5983.

[5] W. Pfrengle, H. Kunz, *J. Org. Chem.* **1989**, *54*, 4261.

[6] J. Barluenga, F. Aznar, C. Valdés, A. Martín, S. García-Granda, E. Martín, *J. Am. Chem. Soc.* **1993**, *115*, 4403.

[7] a) S. D. Larsen, P. A. Grieco, *J. Am. Chem. Soc.* **1985**, *107*, 1768; b) P. A. Grieco, A. Bahsas, *J. Org. Chem.* **1987**, *52*, 5746; c) H. Waldmann, *Angew. Chem.* **1988**, *100*, 307; *Angew. Chem. Int. Ed. Engl.* **1988**, 27, 307; d)

H. Waldmann, *Liebigs Ann. Chem.* **1989**, 231;
e) H. Waldmann, M. Braun, *Liebigs Ann. Chem.* **1991**, 1045; f) M. M. Cid, U. Eggenauer, H. P. Weber, E. Pombo-Villar, *Tetrahedron Lett.* **1991**, *32*, 7233; g) E. Pombo-Villar, J. Boelsterli, M. M. Cid, J. France, B. Fuchs, M. Walkinshaw, H.-P. Weber, *Helv. Chim. Acta* **1993**, *76*, 1203.

[8] A. Choudhury, R. W. Franck, R. B. Gupta, *Tetrahedron Lett.* **1989**, *30*, 4921.

[9] a) J. K. Whitesell, D. James, J. F. Carpenter, *J. Chem. Soc., Chem. Commun.* **1985**, 1449; b) S. W. Remiszewski, J. Yang, S. M. Weinreb, *Tetrahedron Lett.* **1986**, *27*, 1853.

[10] a) M. Sabuni, G. Kresze, H. Braun, *Tetrahedron Lett.* **1984**, *25*, 5377; b) H. Felber, G. Kresze, H. Braun, A. Vasella, *Tetrahedron Lett.* **1984**, 25, 5381; c) H. Felber, G. Kresze, R. Prewo, A. Vasella, *Helv. Chim. Acta* **1986**, *69*, 1137; d) H. Braun, R. Charles, G. Kresze, M. Sabuni, J. Winkler, *Liebigs Ann. Chem.* **1987**, 1129; e) O. Werbitzky, K. Klier, H. Felber, *Liebigs Ann. Chem.* **1990**, 267; f) K. Schürrle, B. Baier, W. Piepersberg, *J. Chem. Soc., Chem. Commun.* **1991**, 2407; g) Very recently, an efficient nitrosodienophile that is easily accessible from ribose has been developed: H. Braun, H. Felber, G. Kresze, F. P. Schmidtchen, R. Prewo, A. Vasella, *Liebigs Ann. Chem.* **1993**, 261.

[11] a) A. Miller, T. M. Paterson, G. Procter, *Synlett* **1989**, 32; b) A. Miller, G. Procter, *Tetrahedron Lett.* **1990**, *31*, 1041; c) A. Miller, G. Procter, *Tetrahedron Lett.* **1990**, *31*, 1043; d) A. D. Morley, D. M. Hollinshead, G. Procter, *Tetrahedron Lett.* **1990**, *31*, 1047; e) G. W. Kirby, M. Nazeer, *Tetrahedron Lett.* **1988**, *29*, 6173; f) A. Defoin, H. Fritz, C. Schmidlin, J. Streith, *Helv. Chim. Acta* **1987**, *70*, 554; g) A. Brouillard-Poichet, A. Defoin, J. Streith, *Tetrahedron Lett.* **1990**, *30*, 7061; h) A. Defoin, J. Pires, J. Streith, *Synlett* **1991**, 417; i) A. Defoin, A. Brouillard-Poichet, J. Streith, *Helv. Chim. Acta* **1992**, *75*, 109; k) A. Defoin, J. Pires, I. Tissot, T. Tschamber, D. Bur, M. Zehnder, J. Streith, *Tetrahedron: Asymmetry* **1991**, *2*, 1209; l) A. Defoin, A. Brouillard-Poichet, J. Streith, *Helv. Chim. Acta* **1991**, *74*, 103; m) V. Gouverneur, L. Ghosez, *Tetrahedron: Asymmetry* **1990**, *1*, 363; n) V. Gouverneur, L. Ghosez, *Tetrahedron Lett.* **1991**, *32*, 5349; o) V. Gouverneur, L. Ghosez, *Tetrahedron: Asymmetry* **1991**, *2*, 1172; p) S. F. Martin, M. Hartmann, J. A. Josey, *Tetrahedron Lett.* **1992**, *33*, 3583.

[12] Short review: H. Waldmann, *Nachr. Chem. Tech. Lab.* **1991**, *39*, 1142.

[13] a) T. Arnold, H.-U. Reissig, *Synlett* **1990**, 514; b) T. Arnold, B. Orschel, H.-U. Reissig, *Angew. Chem.* **1992**, *104*, 1084; *Angew. Chem. Int. Ed. Engl.* **1992**, *31*, 1033.

[14] a) S. E. Denmark, C. Senananyake, G.-D. Ho, *Tetrahedron* **1990**, *46*, 4857; b) S. E. Denmark, M. E. Schnute, *J. Org. Chem.* **1991**, *58*, 1859.

C_2-Symmetric Amines as Chiral Auxiliaries

Herbert Waldmann

The use of chiral auxiliary groups derived from natural products like terpenes, amino acids and carbohydrates to direct the steric course of asymmetric transformations belongs to the most powerful methods of stereoselective synthesis. The majority of the most efficient auxiliaries that are often employed for this purpose do not display striking symmetry elements. However, it has recently been demonstrated that the concept of using C_2-symmetric chiral auxiliary groups like the amines *(1)–(5)* and their enantiomers, as mediators of the stereochemical information, is particularly viable. [1] The introduction of this symmetry element may have three consequences: (1) both faces of the auxiliary become equivalent, (2) the number of the competing diastereomeric transition states is reduced significantly, (3) the atoms which are located on the C_2-axis become chirotopic but not stereogenic, [1, 2] i.e., no new stereocenters are formed that would additionally complicate the situation. In addition to the use of diols [3] the application of the amines and diamines *(1)–(5)* turned out to be particularly rewarding. In a strict sense, due to the sp³ hybridization of the N-atoms in the ground state, *(1)–(3)* are not C_2-symmetrical. However, reactions carried out on derivatives of these chiral auxiliary groups proceed as if this condition was fulfilled. To characterize this, the term "functional C_2-symmetry" was coined. [4]

[To achieve a uniform and thereby better understandable presentation, in the following paragraphs the results in some cases are presented for the enantiomer of *(1)–(5)* which actually was not used by the authors in their original work. This is marked in the text with an (*).]

The pyrrolidines *(2)* prove to be particularly effective chiral auxiliary groups. Thus, the alkylation of the enamine *(6)(*)* with different alkyl halides delivers the cyclohexanones *(7)*

(1)

(2a): R = CH₃
(2b): R = CH₂OCH₂OCH₃
(CH₂OMOM)
(2c): R = CH₂OCH₃

(2d)

(3)

Bzl = –CH₂–Ph

(4a): R = CH₃
(4b): R = SO₂CF₃

(5): R = SO₂R'

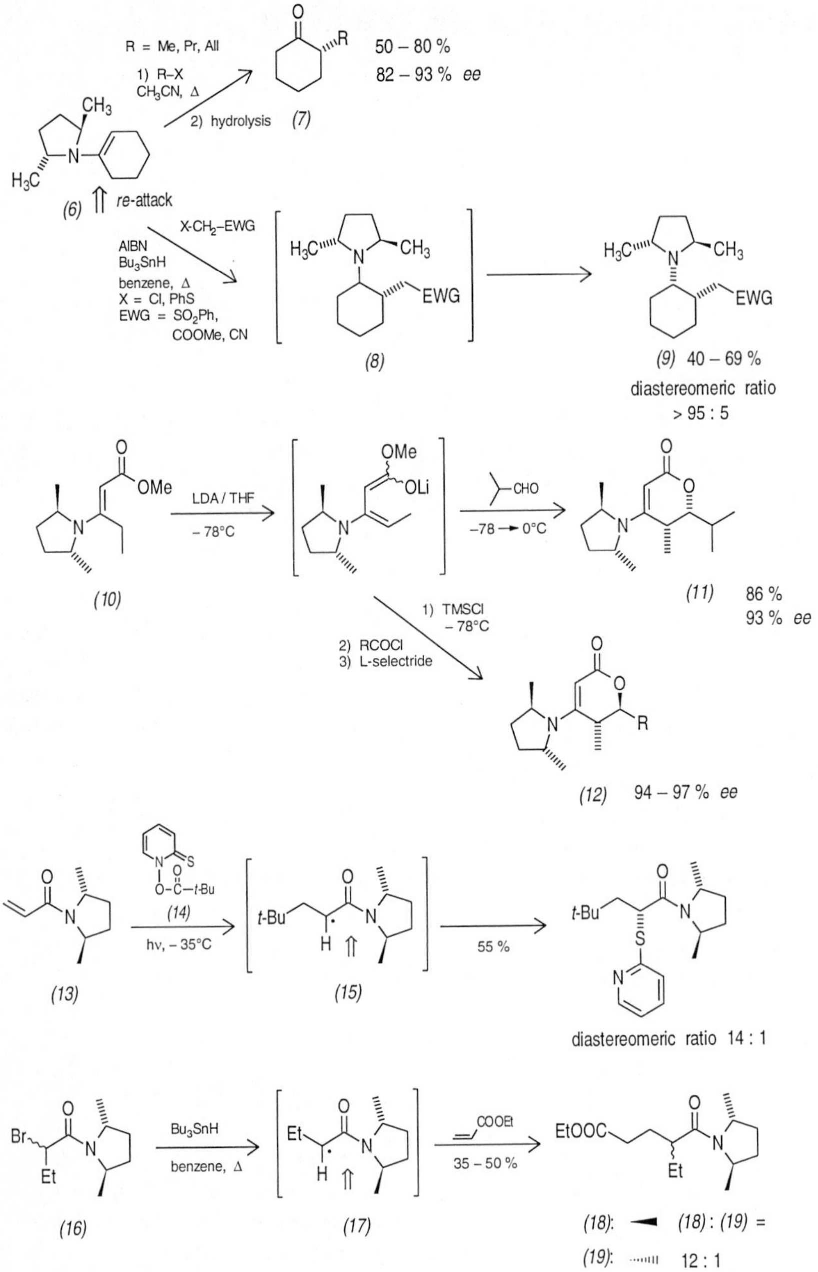

Scheme 1. Use of the 2,5-dimethylpyrrolidine (2a) as chiral auxiliary.

with optical purities of 82–95 % [5], and after deprotonation the vinylogous urethane *(10)* reacts with aldehydes [6a] and acid chlorides [6b] (*) with high diastereoselectivity to give the adducts *(11)* and *(12)* (Scheme 1). The (2*R*, 5*R*)-dimethylpyrrolidine *(2a)* furthermore turned out to be an efficient auxiliary group in asymmetric intermolecular radical reactions. In a series of papers [7] Giese et al. and Porter et al. demonstrated that reactions of amide radicals that embody *(2a)* proceed with high stereoselectivity (Scheme 1). Thus, on irradiation the acrylamide *(13)* and the *tert*-butyl radical donor *(14)* form the chiral radical *(15)*, which reacts further with the thiocarbonyl group of *(14)*. [7c] Alternatively, for instance, from the bromoamide *(16)* the chiral radical *(17)* can be generated, which even at 80 °C reacts with acrylic acid ethyl ester to give the products *(18)* and *(19)* in a ratio of 12:1. [7d] In a similar but photochemically-induced reaction this value rises even to 36:1. [7d] In earlier studies [7a, b, e] (*) the authors had already demonstrated that the addition of chiral radicals to amides of fumaric acid and alkylidene malonic acids derived from *(2a)* proceeds with high selectivity. Finally, the radical transformations of the enamine *(6)* to give the *cis*-disubstituted cyclohexanes *(9)* display *cis/trans* and diastereomer ratios of >95:5, respectively. [8] In the reactions compiled in Scheme 1 the C_2-symmetric auxiliary causes the educts to preferably adopt one out of a multitude of possible conformations. Consequently, the steric course of the nucleophilic and radical transformations carried out with *(6)* is explained by the fact that the conformation shown for *(6)* in Scheme 1 is particularly low in energy end directs the attack to the *re*-side. The amide radicals *(15)* and *(17)* most probably are present as *Z*-conformers and are attacked predominantly *anti* with respect to the neighbouring methyl group of the dimethylpyrrolidine.

The doubly alkoxy-substituted pyrrolidine *(2b)* was used by Katsuki et al. in a variety of organometallic syntheses (Scheme 2). The enolates generated from *(20)* undergo alkylations [9 a, c, f] (*), acylations [9b] and aldol reactions [9e] with excellent stereoselectivity. The β-ketoamides *(21)* are converted to the β-hydroxyamides *(22)* or *(23)* by employing suitable reducing reagents. [9b, d] Alternatively, *(22)* and *(23)* would have to be formed by *syn*- or *anti*-selective aldol reactions. Furthermore, *(2b)* proves to be efficient in Wittig rearrangements [9g] [Scheme 2, *(24)* → *(25)*] and reductions of α-keto amides [9h] (*). The use of the MOM protecting group on the one hand causes a supporting chelation of the involved metal ions. On the other hand this acetal is easily hydrolyzed and thereby liberates OH groups, which support the hydrolytic removal of the amide auxiliary in refluxing 1N HCl.

The enantiomer of the methoxy-substituted auxiliary *(2c)* induces high stereoselectivity in Diels–Alder reactions, hetero Diels–Alder reactions and iodolactonizations. For instance, after activation by TBDMS triflate the acrylamide *(26)* reacts at low temperature with the bis-(silyloxy)-substituted cyclohexadiene *(27)* to give the *endo*- and *exo* isomers *(28)* and *(29)* with high *de* values [10a] (Scheme 3). In the presence of the europium salt Eu(fod)₃ at 80 °C in toluene the *exo* compound *(29)* is formed with an isomer ratio 95:5. Similarly, the carbamoynitroso dienophile *(30)* is converted to the corresponding cycloadducts with a selectivity of 87–98 %. [10b] If the tetrasubstituted pyrrolidine *(2d)* is used as chiral auxiliary in the hetero Diels-Alder reaction with cyclohexadiene the diastereomeric excess even exceeds 99 %. [10c] As already pointed out above for the radical reactions with *(2a)*, the functional C_2-symmetry of the auxiliaries in these latter cases, too, causes the dienophiles to adopt preferred conformations that are attacked from the more accessible sides (Scheme 3). In the course of the enantioselective iodolactonization of *(31)* the *trans*-lactone *(33)* is pro-

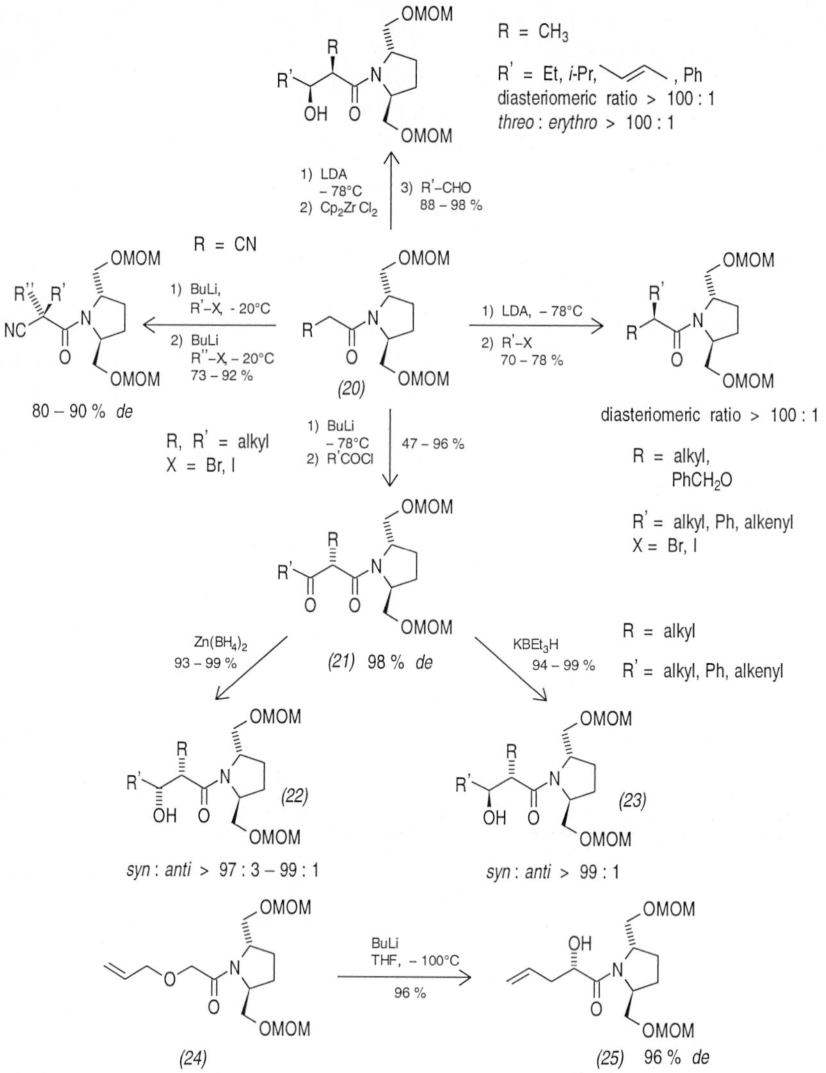

Scheme 2. Use of the 2,5-dialkoxymethylpyrrolidine *(2b)* as chiral auxiliary.

duced with high enantiomeric excess and a *cis*-isomer is not isolated. [10d] Here the presence of the C_2-symmetric auxiliary makes a differentiation between the *E*- and the *Z*-conformer unnecessary. Its influence is responsible for the fact that the hypothetical transition state *(32)* is energetically lower than three competing arrangements. If the enantio-

mer of *(2a)* [11a] or the piperidine derivative *(3)* [4] are employed similar observations are made. Furthermore *(3)* [4] and *(1)* [11b] prove to be efficient stereodirecting groups in aldol reactions. Effenberger et al. [12] could carry out an intermolecular electrophilic addition that proceeds with practically complete diastereoselectivity. In the reaction of the amide

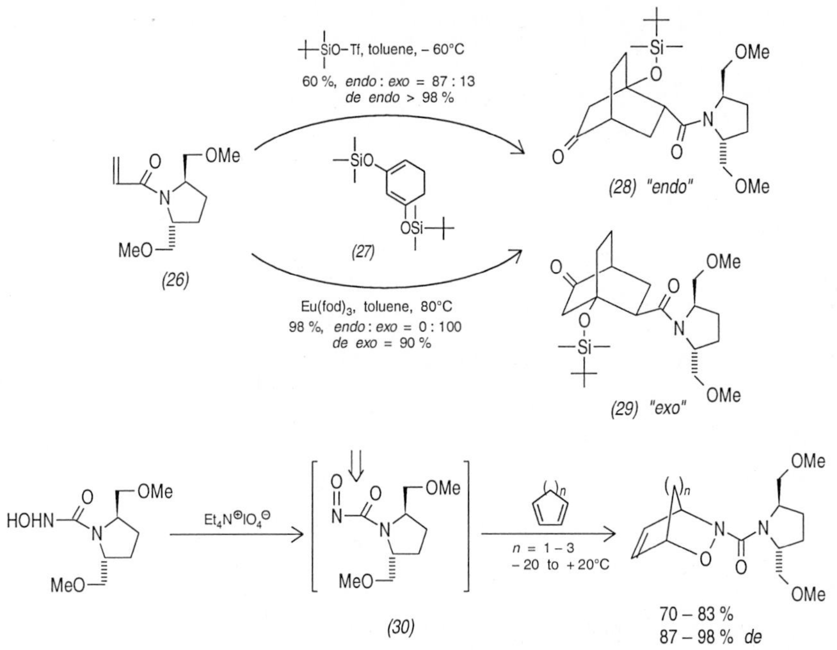

Scheme 3. Use of the auxiliary *(2c)* in Diels–Alder reactions.

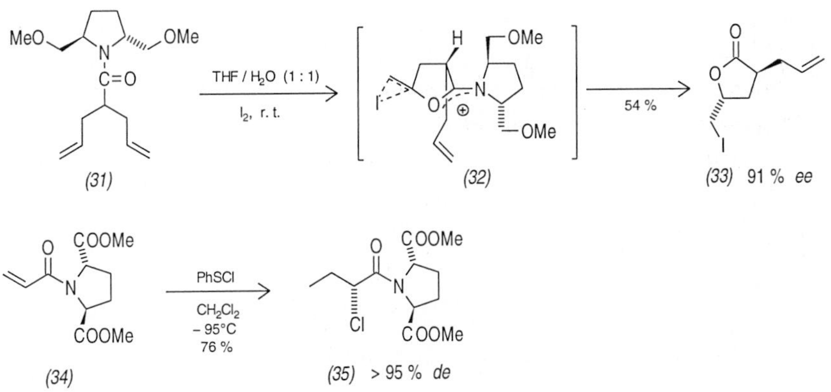

Scheme 4. Diastereoselective iodolactonizations and electrophilic additions to amides of *C2*-symmetric pyrrolidines.

(34) with a sulfur electrophile only the *α*-chloroamide *(35)* was formed (Scheme 4).

The examples highlighted above demonstrate that by employing the amines *(1)–(3)* very high diastereomeric excesses can be achieved in various reactions. However, these auxiliaries are often only available by multistep syntheses and, in particular, their removal from the reaction products in the majority of the cases can be effected only under drastic conditions.

The difficulty mentioned last does not have

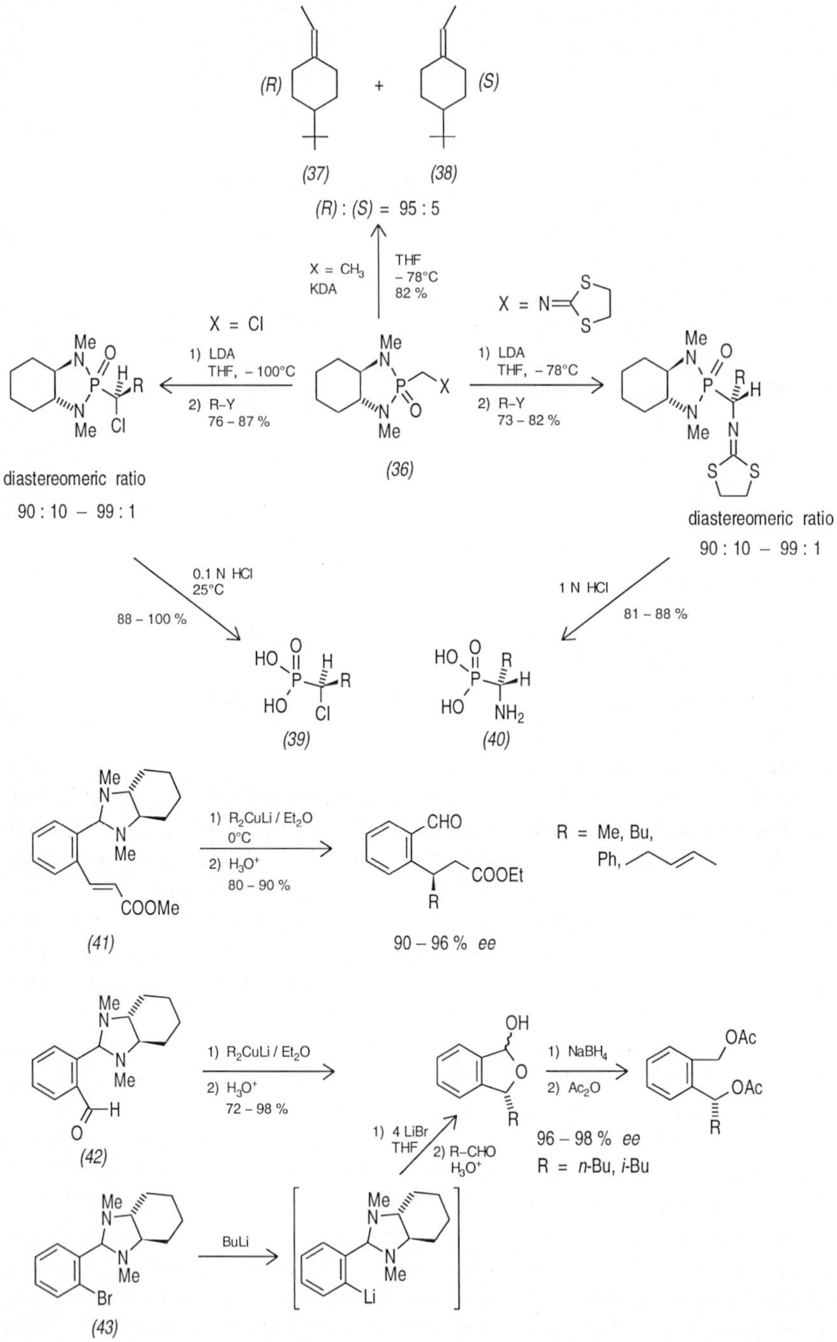

Scheme 5. Use of the cyclohexanediamine *(4a)* as chiral auxiliary group in diastereoselective syntheses.

to be met in the applications of the diamine *(4a)* shown in Scheme 5. Hanessian et al. [13a] used this cyclohexane derivative and its enantiomer for the synthesis of axially dissymmetric and optically active olefins via Horner olefination. The potassium salt of the phosphonamide *(36)* obtained from *(4)*, for instance, reacts with 4-*tert*-butylcyclohexanone to give the olefins *(37)* and *(38)* in a 95:5 ratio. Furthermore, the chiral auxiliary proves its efficiency in the alkylation of the respective α-nitrogen-substituted [13c] phosphonic acid amide *(36)*. After mildy acidic hydrolysis the α-chloro- and α-amino-phosphonic acids *(39)* and *(40)* are obtained in high yield. Such compounds are, for instance, of interest as building blocks for peptidomimetics. Alexakis et al. used the enantiomer of *(4a)* and converted aromatic aldehydes into the aminals *(41)–(43)*. The addition of cuprates to the cinnamic acid derivative *(41)* [14a] and to the carbonyl compound *(42)* [14b] in both cases proceeds with high diastereomer differentiation. Also the reaction of the aryl lithium compound, which is generated from *(43)*, to aliphatic aldehydes displays a marked selectivity [14c] (*). The possibility to reach high isomer ratios in the transformations carried out with the aminals *(41)–(43)* can be attributed to a large extent to the fact that the acetalic C-atoms due to the C_2-symmetry of the chiral auxiliary *(4a)* do not become new stereocenters (see above).

The enantioselective addition of zinc alkyls to carbonyl compounds can be steered efficiently with the sulfonamide *(4b)*. Yoshioka et al. [15] demonstrated that catalytic amounts of *(4b)* together with titanium tetraisopropylate catalyze the highly enantiodifferentiating addition of these organometallic reagents to aldehydes. The authors propose that complexes *(44)* are formed as intermediates. In *(44)* the sulfonamide groups enhance the Lewis acidity of the metal and thereby set up an efficient catalysis, which proceeds with turnover numbers up to 2000. Corey et al. also

used the activating influence of sulfonic acid moieties in the development of the complexes *(45)*, *(49)* and *(53)*, which are derived from the stilbene diamine derivative *(5)*, as reagents for enantioselective syntheses (Scheme 6). Already at low temperature *(45)* catalyzes the Diels–Alder reaction between the oxazolidinone *(46)* and the dienes *(47)*, in which the cycloadducts *(48)* are formed with 95% ee. [16a]. Highly enantiodifferentiating aldol reactions can be carried out by means of boron enolates, which are generated from the bromoboranes *(49)* and *(53)* [16 a, d, e] Compound *(49)*, for instance, proves to be efficient in the synthesis of the pheromone sitophilur *(50)*, whereas *(53)*, is a particularly advantageous reagent for transformations involving ester enolates. If phenylthio esters are used in the respective reactions, *syn* aldols like *(54)* are the major products, but on the other hand from *tert.*-butyl esters the *anti* isomers *(55)* are

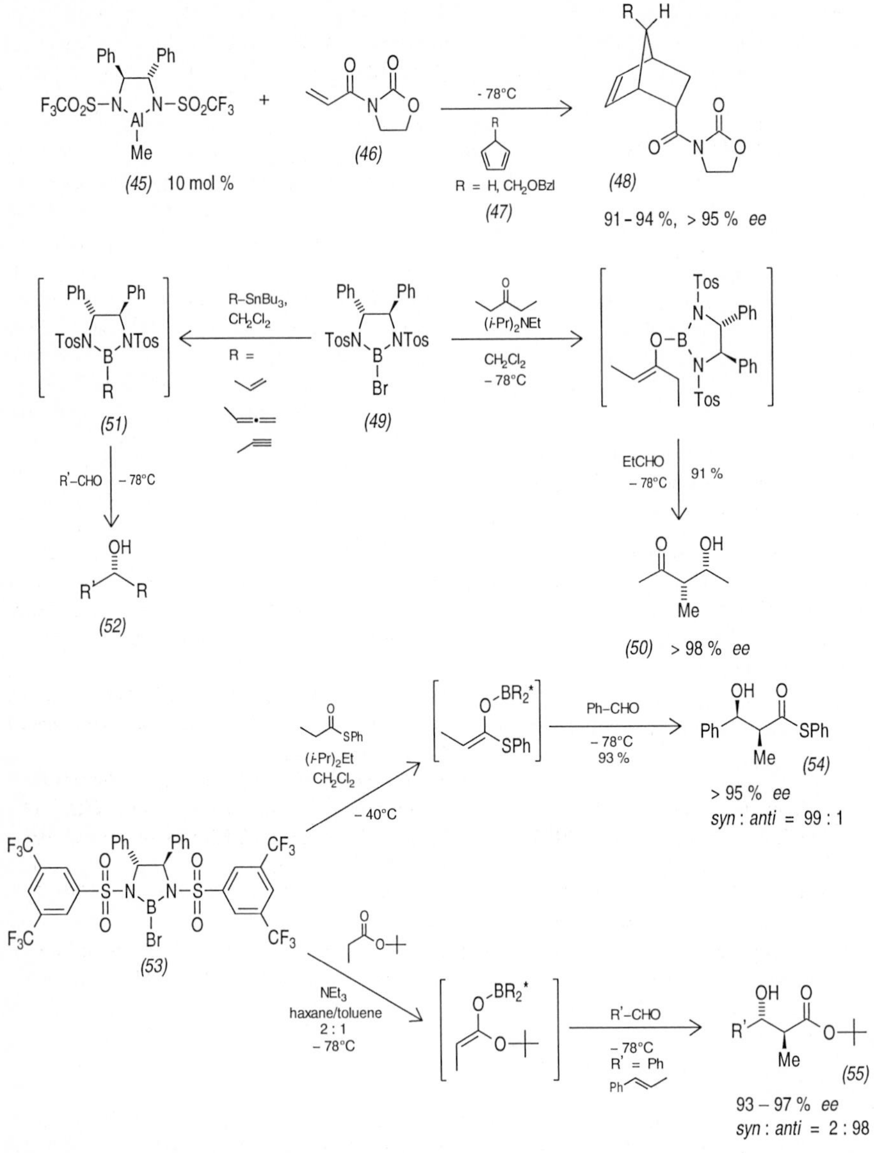

Scheme 6. Use of the stilbenediamine *(5)* in enantioselective transformations.

obtained with high isomer ratios. Furthermore, ester enolates derived from *(53)* allow highly stereoselective Mannich reactions [16g] and Ireland-Claisen rearrangements [16h] to be carried out. From *(49)* and allyl-, allenyl- or propargylstannanes in situ excellent reagents for nucleophilic additions to aldehydes are generated. [16b, c] They make the carbinols *(52)* available with an enantiomeric excess of 95–99 % and in high yield.

$R^1, R^2 =$ alkyl, aryl, COOMe

75–90%
92–98% ee

(57)

In addition to these transformations, the *N,N'*-dialkyl stilbenediamine *(56)* was found to be a highly stereodirecting ligand in OsO₄-mediated *cis*-hydroxylations of different olefins. [16 f] In these processes the diols *(57)* are formed with uniformly high *ee* values exceeding 92 %.

References

[1] J.K. Whitesell, *Chem. Rev.* **1989**, *89*, 1581.

[2] R. Breslow, J. Siegel, *J. Am. Chem. Soc.* **1984**, *106*, 3319.

[3] a) H.-J. Altenbach, *Nachr. Chem. Tech. Lab.* **1988**, *36*, 1212; b) A. Alexakis, P. Mangeney, *Tetrahedron:Asymmetry* **1990**, 477.

[4] a) S. Najdi, M. Kurth, *Tetrahedron Lett.* **1990**, *31*, 3279; b) S. Najdi, D. Reichlin, M.J. Kurth, *J. Org. Chem.* **1990**, *55*, 6241.

[5] J.K. Whitesell, S.W. Felman, *J. Org. Chem.* **1977**, *42*, 1663.

[6] a) R.H. Schlessinger, E.J. Iwanowitz, J.K. Springer, *J. Org. Chem.* **1986**, *51*, 3070; b) R. Schlessinger, J.R. Tata, J.P. Springer, *J. Org. Chem.* **1987**, *52*, 708.

[7] a) N.A. Porter, D.M. Scott, B. Lacher, B. Giese, H.G. Zeitz, H.J. Lindner, *J. Am. Chem. Soc.* **1989**, *111*, 8311; b) D.M. Scott, A.T. McPhail, N.A. Porter, *Tetrahedron Lett.* **1990**, *31*, 1679; c) B. Giese, M. Zehnder, M. Roth. H.G. Zeitz, *J. Am. Chem. Soc.* **1990**, *112*, 6741; d) N.A, Porter, E. Swann, J. Nally,

A. McPhail, *J. Am. Chem. Soc.* **1990**, *112*, 6740; e) N.A. Porter, D.M. Scott, I.J. Rosenstein, B. Giese, A. Veit, H.G. Zeitz, *J. Am. Chem. Soc.* **1991**, *113*, 1791.

[8] P. Renaud, S. Schubert, *Synlett* **1990**, 624.

[9] a) Y. Kawanamaki, Y. Ito, T. Kitagawa, Y. Taniguchi, T. Katsuki, M. Yamaguchi, *Tetrahedron Lett.* **1984**, *25*, 857; b) Y. Ito, T. Katsuki, M. Yamaguchi, *Tetrahedron Lett.* **1984**, *25*, 6015; c) M. Enomoto, Y. Ito, T. Katsuki, M. Yamaguchi, *Tetrahedron Lett.* **1985**, *26*, 1343; d) Y. Ito, T. Katsuki, M. Yamaguchi, *Tetrahedron Lett.* **1985**, *26*, 4643; e) T. Katsuki, M. Yamaguchi, *Tetrahedron Lett.* **1985**, *26*, 5807; f) T. Hanamozo, T. Katsuki and M. Yamaguchi, *Tetrahedron Lett.* **1986**, *27*, 2463; g) M. Uchikawa, T. Hanamoto, T. Katsuki, M. Yamaguchi, *Tetrahedron Lett.* **1986**, *27*, 4577; h) Y. Kawanami, I. Fujita, S. Asahara, T. Katsuki, M. Yamaguchi, *Bull. Chem. Soc. Japan* **1989**, *62*, 3598.

[10] a) H. Lamy-Schelkens, L. Ghosez, *Tetrahedron Lett.* **1989**, *30*, 5891; b) V. Gouverneur, L. Ghosez, *Tetrahedron: Asymmetry* **1990**, *1*, 363; c) A. Defoin, A. Brouillard-Poichet, J. Streith, *Helv. Chem. Acta.* **1991**, *74*, 103; d) K. Fuji, M. Node, Y. Naniwa, T. Kawabata, *Tetrahedron Lett.* **1990**, *31*, 3175.

[11] a) D.J. Hart, H.C. Huang, R. Krishnamurthy, T. Schwartz, *J. Am. Chem. Soc.* **1989**, *111*, 7507; b) D. Tanner, C. Birgersson, *Tetrahedron Lett.* **1991**, *32*, 2533.

[12] F. Effenberger, H. Isak, *Chem. Ber.* **1989**, *122*, 545.

[13] a) S. Hanessian, D. Delorme, S. Beaudoin, Y. Leblanc, *J. Am. Chem. Soc.* **1984**, *106*, 5754; b) S. Hanessian, Y. L. Bennai and D. Delorme, *Tetrahedron Lett.* **1990**, *31*, 6461; c) S. Hanessian, Y. L. Bennai, *Tetrahedron Lett.* **1990**, *31*, 6465.

[14] a) A. Alexakis, R. Sedrani, P. Mangeney, J. F. Nornant, *Tetrahedron Lett.* **1988**, *29*, 4411; b) A. Alexakis, R. Sedrani, J. F. Normant, P. Mangeney, *Tetrahedron: Asymmetry* **1990**, *1*, 283; c) M. Commercon, P. Mangeney, T. Tejero, A. Alexakis, *Tetrahedron: Asymmetry* **1990**, *1*, 287.

[15] a) M. Yoshioka, T. Kawakita, M. Ohno, *Tetrahedron Lett.* **1989**, *30*, 1657; b) H. Takahashi, T. Kawakita, M. Yoshioka, S. Kobayashi, M. Ohno, *Tetrahedron Lett.* **1989**, *30*, 7095.

[16] a) E. J. Corey, R. Imwinkelried, S. Pikul, Y. B. Xiang, *J. Am. Chem. Soc.* **1989**, *111*, 5493; b) E. J. Corey, C.-M. Yu, S. S. Kim, *J. Am. Chem. Soc.* **1989**, *111*, 5495; c) E. J. Corey, C.-M. Yu, D.-H. Lee, *J. Am. Chem. Soc.* **1990**, *112*, 878; d) E. J. Corey, S. S. Kim, *Tetrahedron Lett.*, **1990**, *31*, 3715; e) E. J. Corey, S. S. Kim, *J. Am. Chem. Soc.* **1990**, *112*, 4976; f) E. J. Corey, P. DaSilva Jardine, S. Virgil, P.-W. Yuen, R. D. Connel, *J. Am. Chem. Soc.* **1989**, *111*, 9243; g) E. J. Corey, C. P. Decicco, R. C. Newbold, *Tetrahedron Lett.* **1991**, *32*, 5287; h) E. J. Corey, D.-H. Lee, *J. Am. Chem. Soc.* **1991**, *113*, 4026; i) Review: E. J. Corey, *Pure Appl. Chem.* **1990**, *62*, 1209.

B. Organometallic Reagents in Organic Synthesis

Iron η^5-Complexes in Organic Synthesis

Dieter Schinzer

Metal-mediated C–C bond formations are becoming more and more important in organic synthesis. The following chapter will draw attention to a small but growing part of the field: C–C bond formation with iron η^5-complexes.

As early as 1960 E. O. Fischer et al. carried out pioneering work and complexed 1,3-cyclohexadiene with iron pentacarbonyl to obtain the desired η^4-complex *(2)*. [1] Hydride abstraction of *(2)* in the presence of trityl cation yielded the η^5-complex and triphenyl methane. But it took almost 8 years before the potential was recognized in organic synthesis (Scheme 1).

Most of the starting materials for this type of chemistry can be synthesized from aromatic precursors using the Birch reduction. [2] Therefore it was evident that this particular research group started projects with η^5-iron complexes. The interesting question of the regioselectivity of the hydride abstraction was also studied first by Birch et al., who showed that hydride abstraction from *(1)* (R = OMe) is completely regioselective in favor of *(4)*.

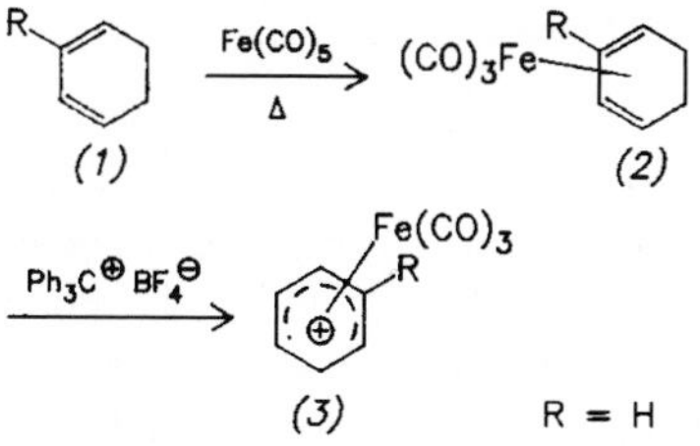

Scheme 1

The latter is formally the more unstable compound. The effect can be understood by HOMO–LUMO interactions of the cation and orbitals of the iron. The orbitals of the cation *(4)* are energetically closer to those of the iron atom, which favors overlap and therefore stabilization of this particular cation. Cation *(5)* cannot have this stabilization because the energy of the allyl cation is too low relative to that of the metal (Scheme 2). [3]

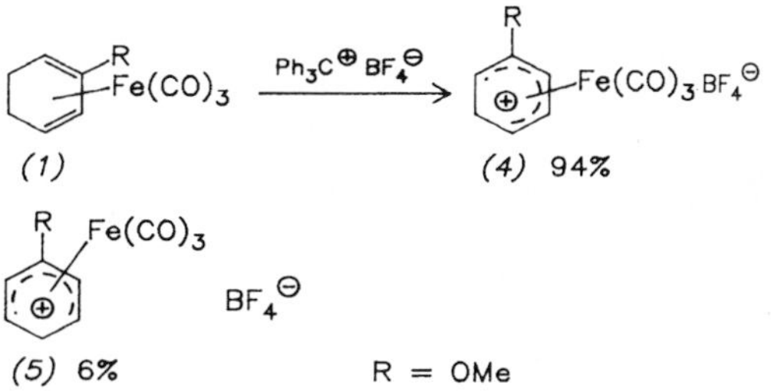

Scheme 2

The real potential for these compounds in synthesis is reactions with nucleophiles. This preparatively very important area was been started by Pearson's research group. [4] In general, soft nucleophiles, like alcohols, amines, enolates, and other stabilized anions, can be applied. The stereochemical course of these reactions mostly leads diastereoselectively to the so-called *exo*-products *(6)*, relative to the metal (Scheme 3).

The metal will be removed oxidatively under very mild conditions with various reagents to give *(7)*. In particular, metal salts

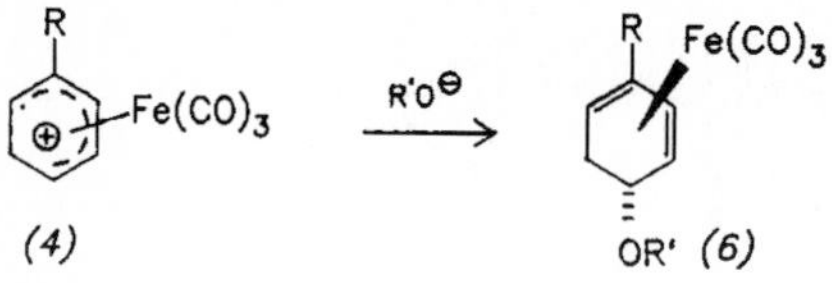

(4)

R = OMe   *Scheme 3*

(Fe^{3+}, Cu^{2+}) and trimethylamine *N*-oxide have been used most successfully (Scheme 4).

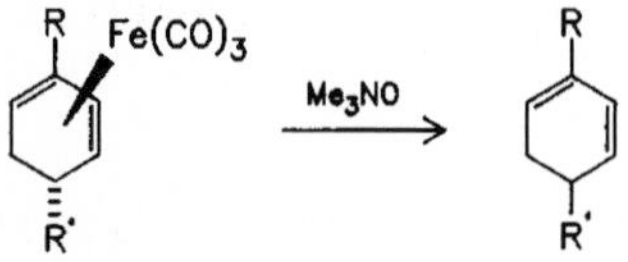

(6)   (7)   *Scheme 4*

A smooth reaction is obtained if alkoxides and amines are used as nucleophiles (Scheme 5). [5]

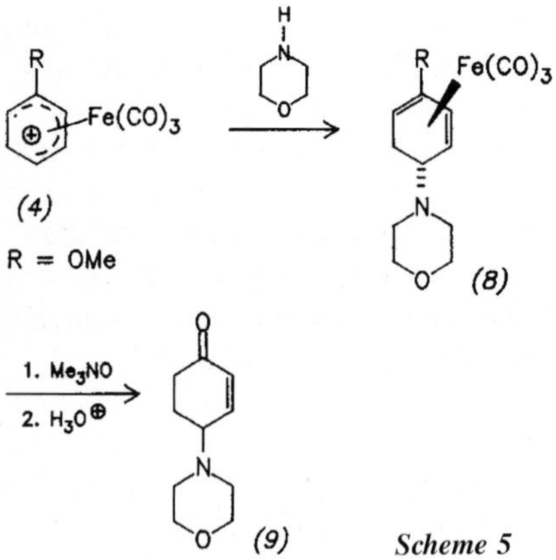

(4)

R = OMe

(9)   *Scheme 5*

Reactions with enolates [6], enamines [7] and allylic silanes [8] in particular should be pointed out, because all these transformations can be done with high chemical yield. No additional Lewis acid is required for reactions with allyl silanes, which demonstrates the high reactivity of these complexes (Scheme 6).

More important for synthesis are combined processes in which two bonds are closed. The first example of this type was published by Birch et al. [9] An oxidative cyclization is followed a C–C bond formation with an activated 1,3-dicarbonyl compound yielding a

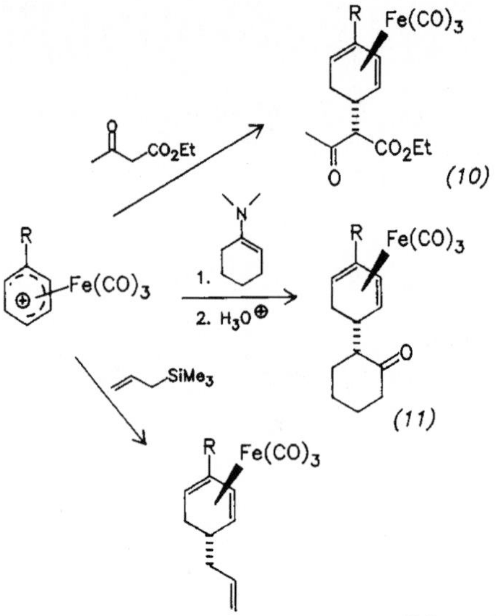

R = OMe   (12)   *Scheme 6*

dihydro benzofuran derivative (16). As reagent for the oxidative cyclization MnO_2 was used (Scheme 7).

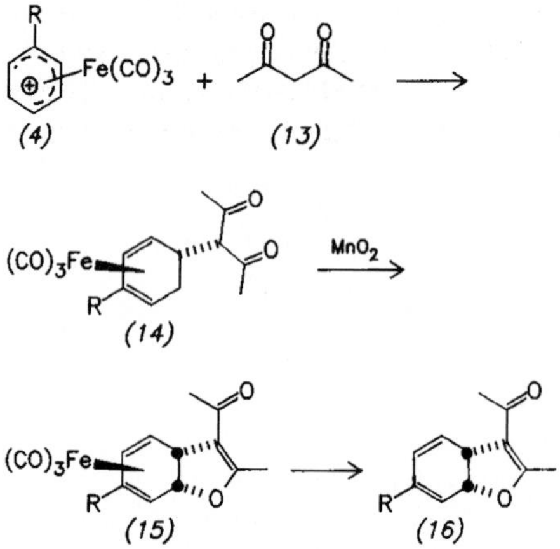

(4)   (13)

(14)

R = H   *Scheme 7*

A useful sequence of reactions can be realized by the use of acetone, as shown by Pearson et al. [10] The authors obtained a cyclic ether (18)→(19), which was cleaved under acidic conditions to yield a second η^5-complex, which reacted with an external nucleophile. The overall result was the stereoselective introduction of two substituents *"exo"* to the iron (Scheme 8).

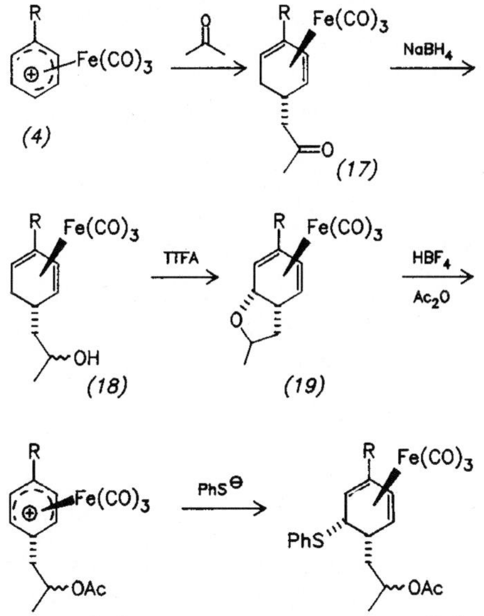

(4) ———→ (17)

(18) —TTFA→ (19) —HBF₄/Ac₂O→

(20) *Scheme 8* (21)

Another important task in organic synthesis is the introduction of quaternary centers, which is always a good testing ground for new synthetic methods. Again, Pearson et al. were

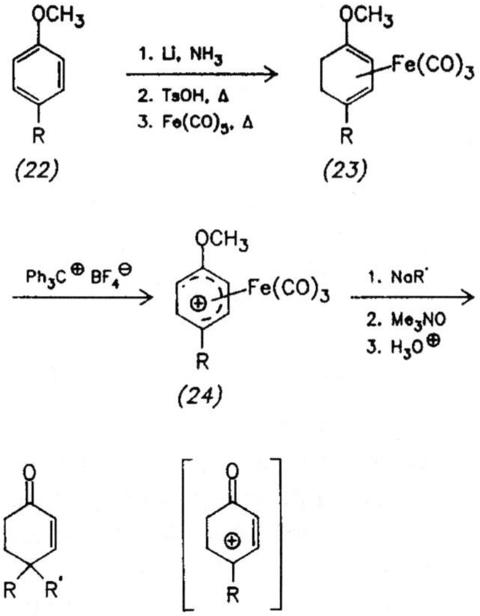

(22) 1. Li, NH₃ 2. TsOH, Δ 3. Fe(CO)₅, Δ → (23)

(23) —Ph₃C⊕ BF₄⊖→ (24) 1. NaR' 2. Me₃NO 3. H₃O⊕ →

(25) (25a) *Scheme 9*

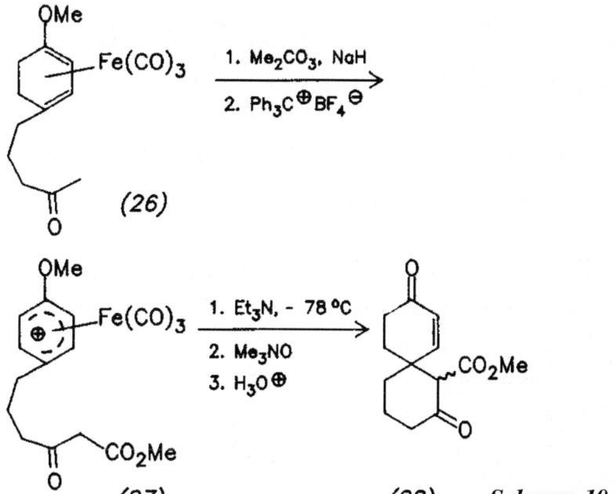

(26) 1. Me₂CO₃, NaH 2. Ph₃C⊕BF₄⊖ →

(27) 1. Et₃N, – 78 °C 2. Me₃NO 3. H₃O⊕ → (28) *Scheme 10*

successful in establishing such a method. [11] Substituted aromatic compounds of type *(22)* can be transformed into iron complexes after Birch reduction. Cation *(24)* can be obtained by the use of trityl cation, and subsequent reaction with a soft nucleophile yields compounds of type *(25)*. The cation *(24)* represents a synthon for compound *(25a)* (Scheme 9).

A spiro-cyclization was realized by this technique in two steps. In addition to the problems involved in the synthesis of quaternary centers, now a spiro ring had to be constructed. The only drawback in this sequence was the use of only "soft" nucleophiles, which limits the general scope of transformations with these complex cations. The method was demonstrated by the synthesis of spiro-undecane *(28)* (Scheme 10). [12]

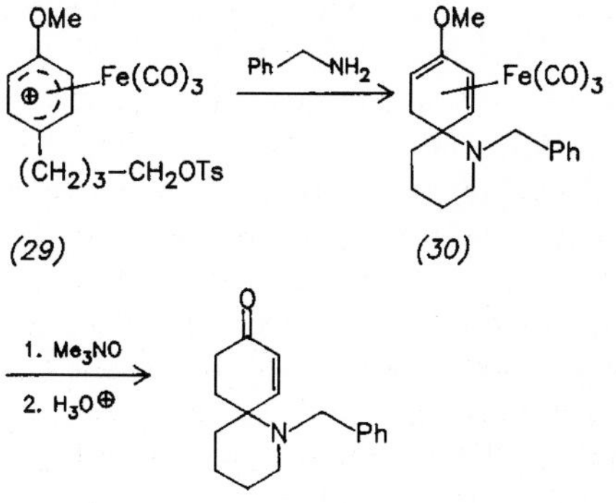

(29) —Ph⌒NH₂→ (30)

(30) 1. Me₃NO 2. H₃O⊕ → (31) *Scheme 11*

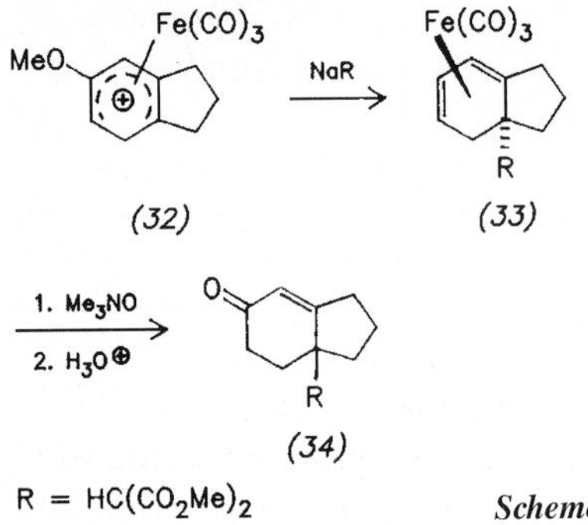

(32)   (33)

(34)

R = HC(CO₂Me)₂

R = $HC(CO_2Me)_2$ *Scheme 12*

A useful extension of this concept was the synthesis of aza-systems. Again Pearson has synthesized several spiro-piperidines and -pyrrolidines. [13] The complex cation *(29)* was transformed with benzylamine to *(30)* via an in situ substitution with the tosylate and subsequent ring closure to *(30)*. The spiro-piperidines in particular are important precursors for the biologically active histrionicotoxine [14] framework (Scheme 11).

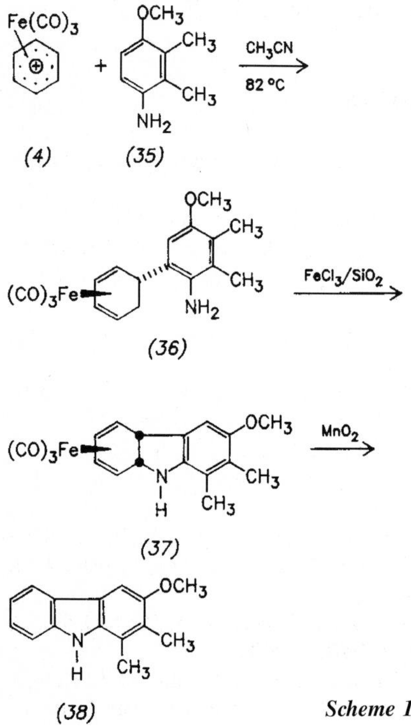

(4)   (35)

(36)

(37)

(38)   *Scheme 13*

Starting from bicyclic compounds like *(32)* angular substituents can be introduced. The complex cation *(32)* can be treated with nucleophiles to obtain – after oxidative demetalation – enones of type *(34)* (Scheme 12). [15]

Very elegant combinations of the concepts presented have been done by Knölker et al. These authors have already synthesized all the known and biologically active carbazomycines. [16, 17] Knölker used a combination of aromatic substitution and oxidative cyclization to synthesize the carbazomycine skeleton. Starting with complex cation *(4)*, addition of the aromatic amine *(35)* yielded *(36)*. Oxidative cyclization to *(37)* followed by demetalation with MnO_2 gave directly the natural product *(38)* (Scheme 13).

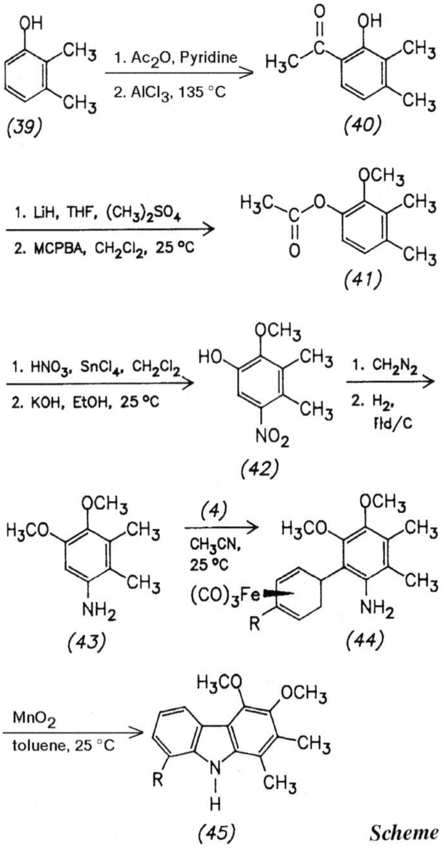

(39)   (40)

(41)

(42)

(43)   (44)

(45)   *Scheme 14*

By the use of the same strategy carbazomycine A and B could be synthesized starting from aromatic precursors. Knölker started with the commercially available phenol *(39)*, which was transformed in a short sequence of steps into the required amine *(43)*: acylation, followed by methylation of the phenol; the missing oxygen atom was introduced by a Baeyer-Villiger reaction. Finally, regioselective nitration and subsequent reduction gave *(43)*. Key reaction in this synthesis was an electrophilic aromatic substitution with *(4)* yielding hexasubstituted aromatic compound *(44)*! This demonstrates again the high reactivity of these complexes. The synthesis was fin-

ished by oxidative cyclization and final demetalation to give the natural product *(45)* (Scheme 14).

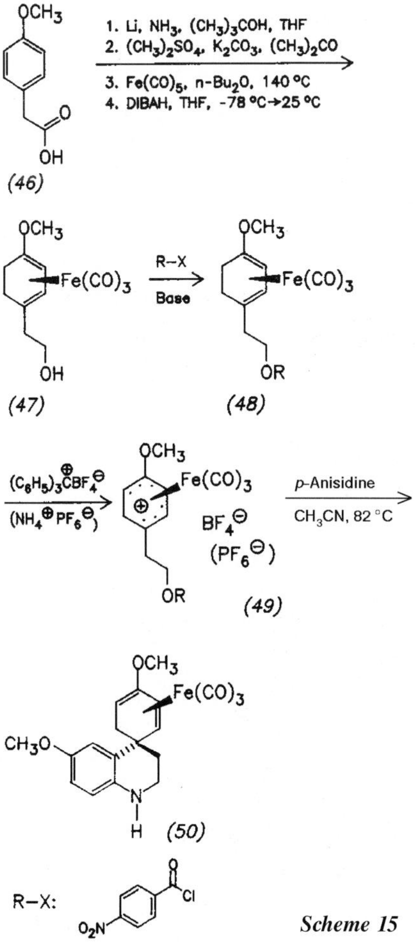

Finally, an iron-mediated diastereoselective spiro-annulation will be presented. In contrast to the method described by Pearson et al. [13] this process, reported by Knölker et al., is a one-pot procedure in which both the C–C and the C–N bonds are formed. [18]

The desired iron complex was synthesized in four steps. After oxygen-functionalization and hydride abstraction complex cation *(49)* was obtained. Addition of *p*-anisidine gave directly the spiro-cycle *(50)*, via aromatic substitution reaction followed by in situ substitution of the benzoate (Scheme 15).

The last examples presented in particular show the high potential of this technique in constructing various ring skeletons. The whole concept is very much extended by recent findings from Knölker's laboratories in which the complexation of the required dienes with 1-aza-1,3-butadiene can be carried out as a high yield catalytic process. [19, 20]

Scheme 15

References

[1] E.O. Fischer, R.D. Fischer, *Angew. Chem.* **1960**, *72*, 919.

[2] See textbooks on organic chemistry.

[3] A.J. Pearson, in: *Comprehensive Organometallic Chemistry*. Vol 8, 939, Pergamon Press, **1982**.

[4] A.J. Pearson: *Metallo-Organic Chemistry*, Wiley, **1985**.

[5] A.J. Birch, B.E. Cross, J. Lewis, D.A. White, S.B. Wild, *J. Chem. Soc. (A)* **1968**, 332.

[6] A.J. Birch, K.B. Chamberlain, M.A. Haas, D.J. Thompson, *J. Chem. Soc. Perkin Trans. I* **1973**, 1882.

[7] R.E. Ireland, G.G. Brown, Jr., R.H. Stanford, Jr., T.C. McKenzie, *J. Org. Chem.* **1974**, *39*, 51.

[8] L.F. Kelly, A.S. Narula, A.J. Birch, *Tetrahedron Lett.* **1980**, *21*, 871.

[9] A.J. Birch, K.B. Chamberlain, D.J. Thompson, *J. Chem. Soc. Perkin Trans. I* **1973**, 1900.

[10] A.J. Pearson, *J. Chem. Soc. Chem. Commun.* **1980**, 488.

[11] A.J. Pearson, *J. Chem. Soc. Perkin Trans. I* **1979**, 1255.

[12] A.J. Pearson, *J. Chem. Soc. Perkin Trans. I* **1980**, 400.

[13] A.J. Pearson, P. Ham, D.C. Rees, *Tetrahedron Lett.* **1980**, *21*, 4637.

[14] E.J. Corey, J.F. Arnelt, G.N. Widiger, *J. Am. Chem. Soc.* **1975**, *97*, 430.

[15] A.J. Pearson, *J. Chem. Soc. Perkin Trans. I* **1978**, 495.

[16] H.-J. Knölker, M. Bauermeister, D. Bläser, R. Boese, J.-B. Pannek, *Angew. Chem.* **1989**, *101*, 225.
Angew. Chem. Int. Ed. Engl. **1989**, *28*, 223.

[17] H.-J. Knölker, M. Bauermeister, *J. Chem. Soc. Chem. Commun.*, **1989**, 1468.

[18] H.-J. Knölker, R. Boese, K. Hartmann, *Angew. Chem.* **1989**, *101*, 1745.
Angew. Chem. Int. Ed. Engl. **1989**, *28*, 1678.

[19] H.-J. Knölker, P. Gonser, *Synlett*, **1992**, 517.

[20] H.-J. Knölker, P. Gonser, P.G. Jones, *Synlett* **1994**, 405.

Rhodium-Catalyzed Carbenoid Cyclizations

Karl Heinz Dötz

Diazo reagents serve as an example for a class of compounds on which interest has focussed again after a rather quiet period of some decades. The copper-catalyzed decomposition of diazo ketones and diazo acetate represents a long-known standard method for carbene transfer reactions. [1] In these processes carbenoids containing a metal-coordinated electrophilic carbene are supposed to be the reactive intermediates. [2] The second generation of catalysts, represented by dinuclear rhodium carboxylates, has led to a significantly improved methodology. [3] These compounds are characterized by a dimetal unit strongly hold together by four bridging carboxylate ligands; this results in a compact and stable skeleton that resists substitution in the equatorial positions. The axial position, however, is readily accessible for carbene coordination. Rhodium (II) acetate is often used as catalyst, generating a carbenoid of type *(1)*. Since rhodium carboxylates are inert towards redox reactions with diazo compounds, only low concentrations of the catalysts are required, for example, 0.05 mol % for the cyclopropanation of methoxycyclohexene *(2)*.

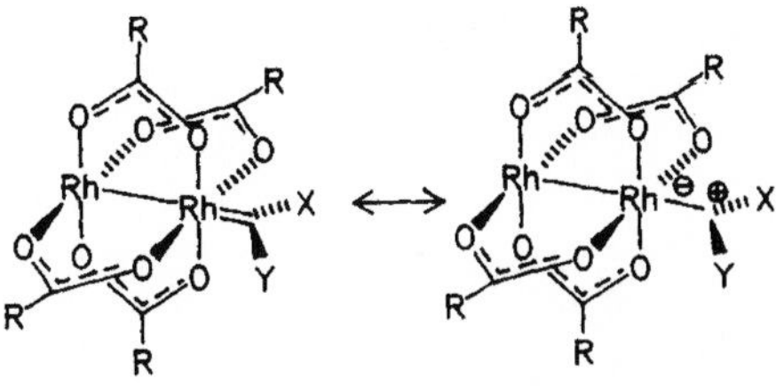

The catalytic cycle (Scheme 1) suggests that the activity of the catalyst A depends on the degree of coordinative unsaturation of the metal center. The catalyst acts as an electrophile and adds the diazo compound to give complex B, which undergoes elimination of N₂ affording the carbenoid C. The final steps

(1a)   *(1b)*

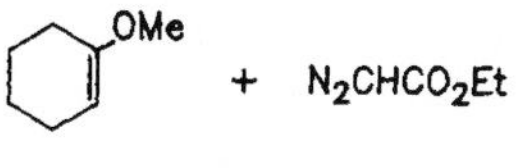

(2)

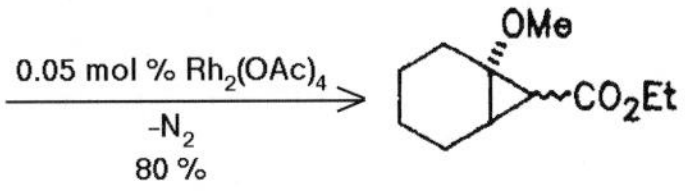

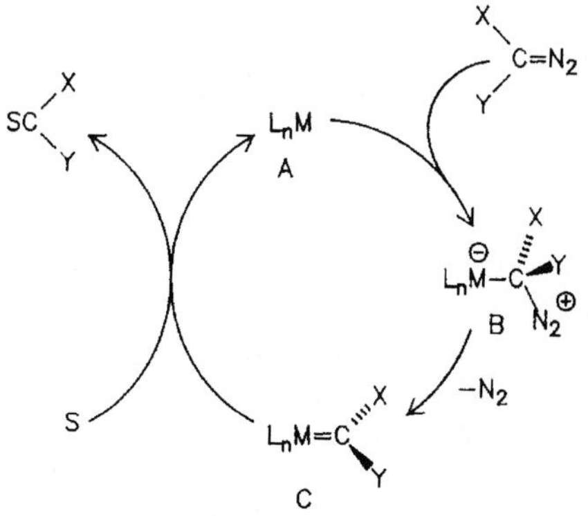

Scheme 1. Catalytic cycle for the cyclopropanation of an olefinic substrate S with the carbenoid C using rhodium carboxylate A as catalyst.

of the catalytic cycle involve carbene transfer to the substrate S combined with the regeneration of the catalyst. In catalytic reactions no direct evidence for carbenoid C has been provided so far. However, support for the involvement of a carbenoid intermediate comes from the observation that alkenes undergo enantioselective cyclopropanation in the presence of optically active catalysts; [2, 4] moreover, a linear reactivity/selectivity correlation exists for a wide range of olefins between the $Rh_2(OAc)_4$-catalyzed cyclopropanation with phenyldiazomethane and the stoichiometric cyclopropanation using the isolable and well-characterized benzylidene tungsten complex $(CO)_5W=CHPh$. [5]

Synthetically useful applications involve cycloaddition of carbenoids to aromatic C–C bonds, as demonstrated by furyl diazo ketones, which are modified to give acyl vinyl cycloalkenones. Previously, copper sulfate in boiling cyclohexane has served as catalyst; [6] using rhodium acetate in dichloromethane, however, the transformation (3)→(4) occurs with improved yield even at ambient temperature. [7] Whereas the reaction is compatible with a variety of substitution patterns in the furan ring, the length of the diazo ketone chain is crucial: C_4 and C_5 side chains readily undergo ring-closure to give five- and six-membered enones. In contrast, the furfuryl diazo ketone (3c) leads to a complex mixture of products under identical conditions. If benzene is used as a solvent instead of dichloromethane, the intermolecular cycloaddition prevails to give cycloheptatriene (5) via a norcaradiene intermediate. This primary cycloaddition product can be isolated in the benzofuran series: the tetracyclic product (6) is obtained, which allows subsequent acid-catalyzed modification to give ketone (7) or a thermal rearrangement to isomer (8).

The cycloaddition/cycloreversion sequence has been exploited in a novel access to compounds of the β-ionone series. [8] The diazo ketone (9), accessible in several steps from 2-

(3)

(4)

	R	n	yield
(3a), (4a):	H	1	86 %
(3b), (4b):	H	2	87 %

(3c)

(5)

(6)

(7)
88 %

(8)
82 %

methylfuran, undergoes concomitant de-amination and cyclization in the presence of rhodium acetate to give *(Z)-oxo-β-ionone (10)*, which has been modified to the *(E)-*isomer *(11)* under iodine catalysis. Cycloaddition to the benzene ring affords a direct and efficient access to the hydroazulene skeleton. This strategy has been exploited in the key steps of a confertin synthesis: Under rhodium mandelate catalysis diazo ketone *(12)* is transformed quantitatively to hydroazulenone *(13b)*, which exists in a rapid equilibrium with its valence tautomer *(13a)*. [9]

for *intra*molecular reactions: Starting from acyclic diazo compounds such as *(14)*, five-membered rings are the preferred products; *trans*-cyclopentanones *(15)* are formed diastereoselectively suggesting a six-membered chair-like transition state. [11] Cyclic diazo precursors, for example, *(16)*, however, lead to the formation of six-membered rings, for example *(17)*. [12] Given a favorable geometry of the transition state, C–H insertion can compete successfully with cycloaddition; for instance, the unsaturated diazo ketoester *(18)* affords the *trans*-cyclopentanone *(19)*. [13]

(9)

(10)   *(11)*

(12)

(13a)   *(13b)*

(14)   *(15)*

(16)   *(17)*

(18)   *(19)*

The development of the rhodium catalysts had a significant impact on the carbene insertion in carbon hydrogen bonds. Based on the electrophilic character of the carbenoid, the selectivity increases in the order of primary, secondary and tertiary C–H bonds. [10] Whereas in ethers α-C–H insertion is favored, phenyl, vinyl and ester groups lead to a deactivation of α-C–H bonds. In addition to electronic factors the conformation plays an important role. This is particularly important

The selectivity of the C–H insertion depends on the catalyst used. The diazo acetate *(20)* is not able to discriminate between primary and tertiary C–H bonds in the presence of rhodium acetate; if rhodium acetamide is used as catalyst, however, only the tertiary C–H insertion product *(21)* is formed. [14] Selective insertion in N-α-C–H bonds of diazo acetoacetamides occurs even though they are deactivated by an N-carboalkoxy-ethyl substituent, as demonstrated for *(23)*. [15] Apparently, in this example the influence of electronic factors is over-compensated by a favored conformation of the carbenoid inter-mediate *(24)*, adopted as a consequence of the bulky *N-tert*-butyl group, and the *trans-β*-lactam *(25)* is formed exclusively. Evidence for this argument comes from the observation that the reaction becomes unselective if the *tert*-butyl substituent is replaced by an *n*-butyl group, which – concerning its steric demand – is comparable with the *β*-propionate side chain. The *α*-C–H bond of the carboalkoxy group and the *β*-C–H bond of the *n*-butyl sub-stituent compete for the insertion; the *β*-lactam *(27)* and the *γ*-lactam *(28)* are formed in a 1:9 ratio along with the seven-membered ring *(29)*, which arises from the carbonyl ylide intermediate *(30)*. The influence of the con-formation exerted on the stereoselectivity is obvious from the cyclization of amidoacetate *(31)*. Using rhodium perfluorobutyrate Rh$_2$(pfb)$_4$ as catalyst the *β*-lactam *(33)* is obtained in a diastereoselective cyclization, as anticipated from the preferred conformation of intermediate *(32)*.

A breakthrough in the synthesis of bicyclic *β*-lactams has been achieved by the carbene insertion in N–H bonds of 2-acetidinones. This strategy has become the standard method for access to derivatives of the carbapenem and the carbacephem series. An elegant example is provided by the synthesis of thiena-mycin *(37)* [16]. The key step of this sequence is the rhodium acetate-catalyzed cyclization of *α*-diazo-*β*-ketoester *(35)*, which is readily accessible from *β*-ketoester *(34)*. The cycliza-tion is remarkable in two aspects: First, it pro-ceeds quantitatively via N1–C2 ring closure, and, second, is leads exclusively to the more stable *exo*-carbapenam isomer *(36)*, which may be rationalized in terms of a facile epi-merization at C2 via an enol intermediate. A similar reaction path – starting from a simple *β*-lactam precursor *(38)* via diazo ketoester *(39)* – leads to homothienamycin *(41)*, a com-pound with a carbacephem skeleton. [17]

The insertion of carbenes into N–H bonds can be further exploited for the synthesis of *β*-lactam analogues. Standard diazo transfer to hydrazides using tosyl azide affords the diazo derivatives *(42)* in variable yields; subsequent rhodium carbenoid induced cyclization in boiling benzene leads to 1,2-diazetidinones *(43)*. [18] Similarly, the diazo precursors *(44)* are well-suited for the construction of 1,3-bridged "anti-Bredt"-*β*-lactams *(45)*. The methyl derivative *(45a)* decomposes in solu-tion (decomposition in CHCl$_3$: T$_{1/2}$ ≈ 1 h); the isopropyl analogue *(45b)*, however, is stable under these conditions. The reason for this is that sterically demanding substituents in the 3-position block the addition of nucleophiles,

(20)

(21)   *(22)*

Rh$_2$L$_4$	yield	*(21)*		*(22)*
Rh$_2$ (OAc)$_4$	81 %	53	:	47
Rh$_2$ (acam)$_4$	96 %	>99	:	<1

(23)   *(24)*   *(25)*

96 %

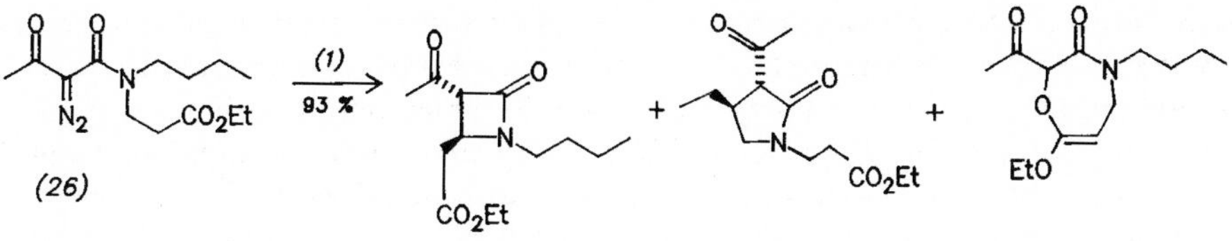

(26) (27) (28) (29)

7 : 62 : 31

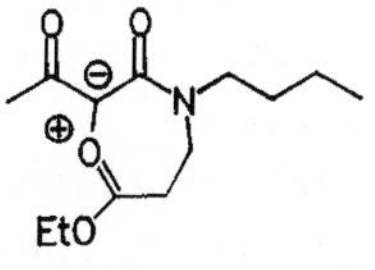

(30)

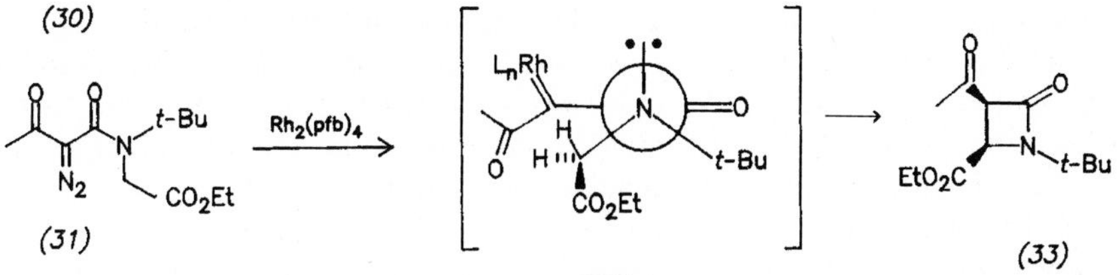

(31) (32) (33)

89 %

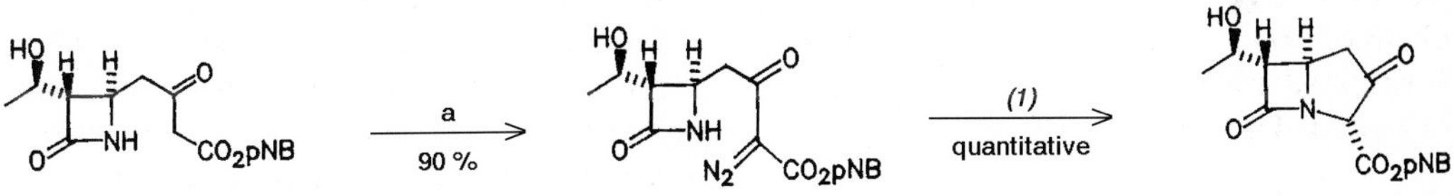

(34) (35) (36)

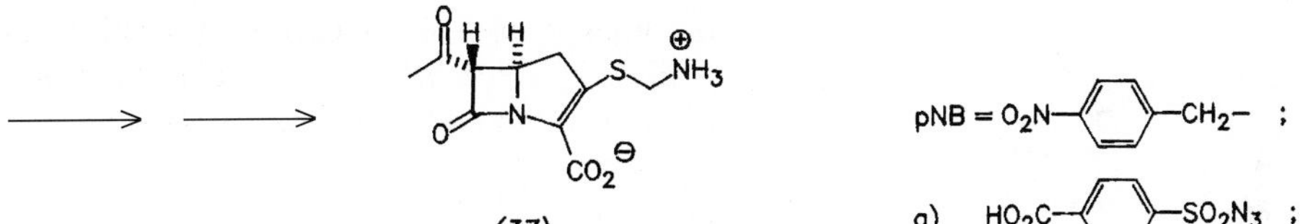

(37)

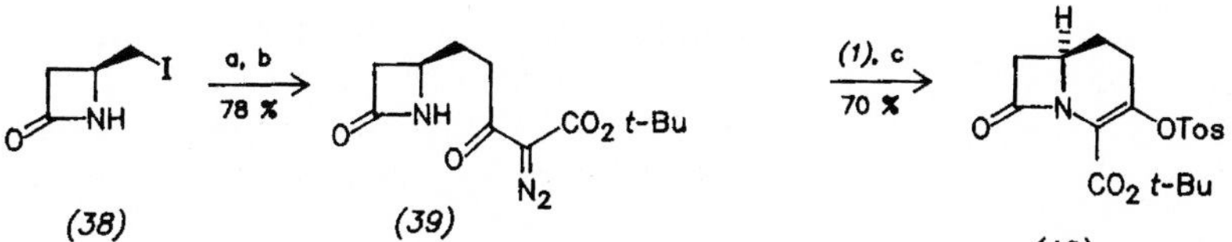

(38) (39) (40)

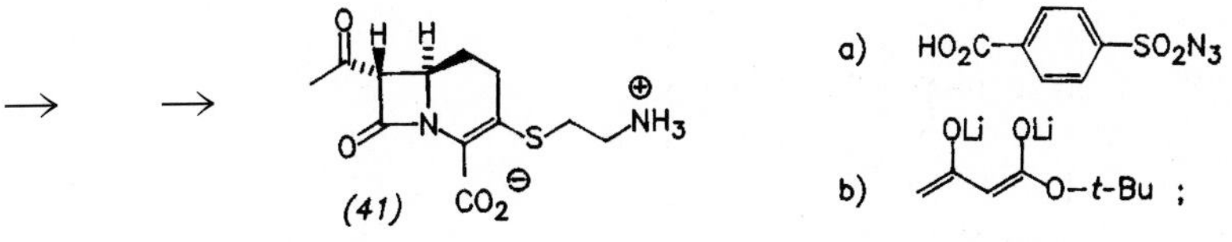

(41)

c) Tos_2O, NEt_3

which can occur only from the β-lactam face opposite the bridge and which is responsible for the decomposition. [19]

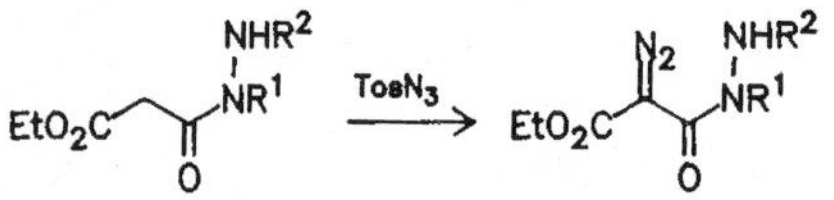

(42)

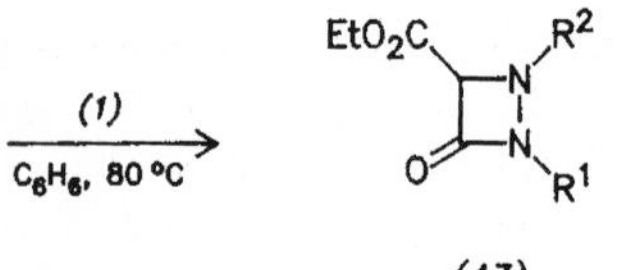

(43)

R¹	R²	yield
Ph	–CH₂Ph	87 %
–CH₂CO₂Et	–CO₂CH₂Ph	93 %

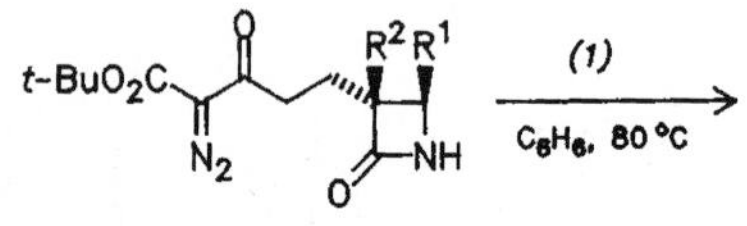

(44)

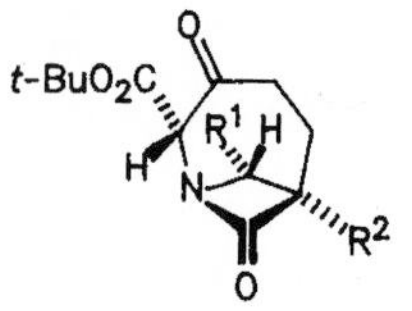

(45)

	R¹	R²	yield
(45a) :	Me	H	50 %
(45b):	i-Pr	H	20 %

The rhodium carbenoid route offers an attractive access to ylides, which provides a useful alternative to the base-induced standard methodology. Moreover, in comparison with methods based on the thermal or photochemical generation of carbenes, it is significantly more selective and – due to its scope and its mild conditions – it is generally also

superior to the copper-catalyzed decomposition of diazo carbonyl compounds. [20] In the presence of rhodium acetate (1) diazo sulfide (47), which is accessible from the thiolactone (46) and lithiodiazoacetate with subsequent in-situ alkylation, is modified into the cyclic sulfonium ylide (48). This type of intermediate can be isolated in some cases; it may undergo a 1,2-Stevens rearrangement, for example, to give (49), or a 2,3-sigmatropic rearrangement, for example, to afford (50). [21] β-Elimination has been only observed if exo-cyclic β-hydrogen atoms are present; in this case, a ready fragmentation leads to derivatives of type (51). In contrast, endo-cyclic hydrogen atoms do not take part in β-elimination.

This methodology can be extended to the generation of stable five- and six-membered sulfoxonium ylides such as (52) or (53), although side reactions involving deoxygenation may occur. [22] An interesting application is provided by a synthetic approach to pyrroli-

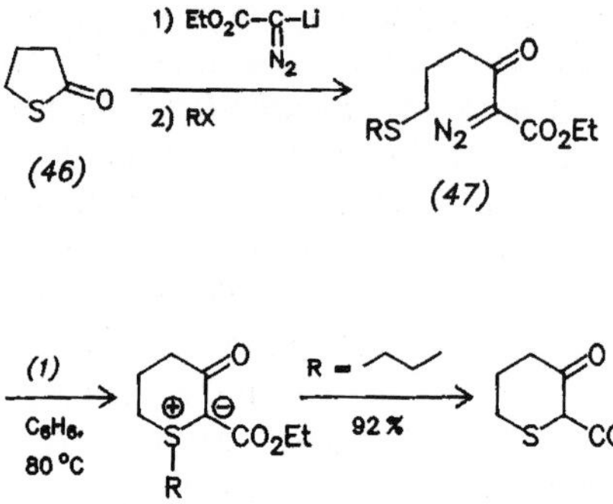

(46)

(47)

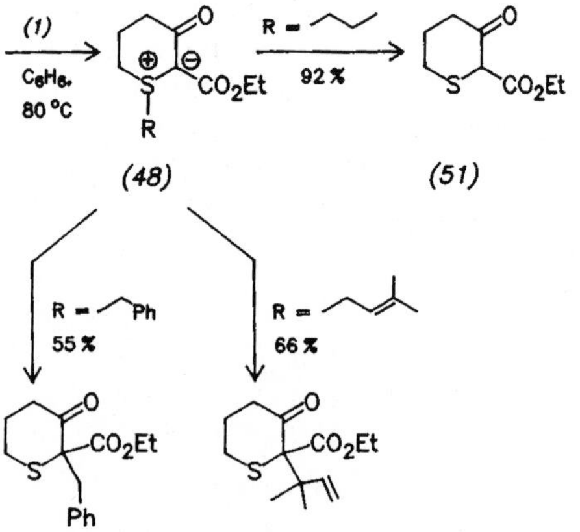

(48)

(51)

(49)

(50)

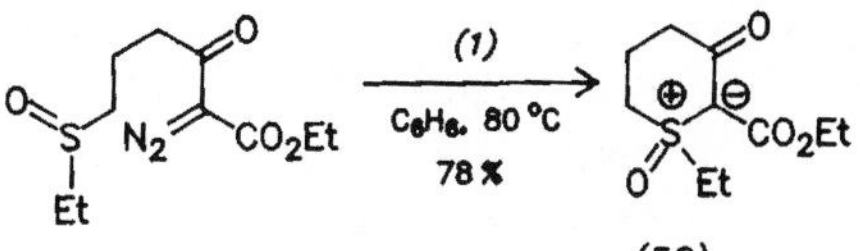

(52)

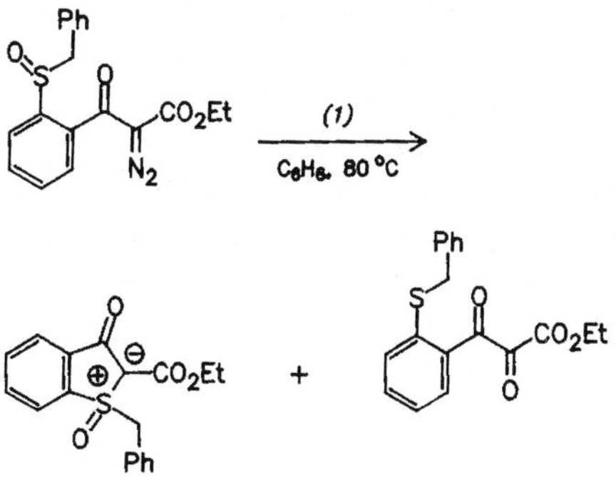

(53)   (54)

58 %   6 %

zidine alkaloids, which involves cyclic sulfonium ylides in the key step. [23]

The carbene transfer to carbonyl groups offers a simple direct route to 1,3-dipoles that can be subsequently either trapped in cycloaddition reactions or that may undergo rearrangement leading to modified 1,3-dipoles (Scheme 2). Often, the diazo carbonyl precursors are easily accessible. For example, the ring-opening of phthalic acid anhydride by 3-methyl-3-buten-1-ol followed by diazotation affords ester (55), which under $Rh_2(OAc)_4$-catalysis undergoes ring closure to give the carbonyl ylide (56). The 1,3-dipole can be either trapped intramolecularly leading to the tetracyclic skeleton (57) or it can be exploited in an intermolecular [3 + 2]cycloaddition with

dimethyl acetylenedicarboxylate (DMAD). [24] The cycloaddition is regioselective: The carbonyl ylide intermediate generated from 1-diazo-2,5-pentandione (59) adds to benzaldehyde with exclusive formation of acetal (60), while addition to methyl propiolate affords the regioisomer (61a) in a 15:1 preference over (61b). [25]

Starting from allyl aryl diazopentandione (62) no intermolecular cycloaddition occurs even in the presence of an activated dipolarophile such as DMAD. In this example – in contrast to the oxygen-stabilized carbonyl ylide (56) – no efficient resonance stabiliza-

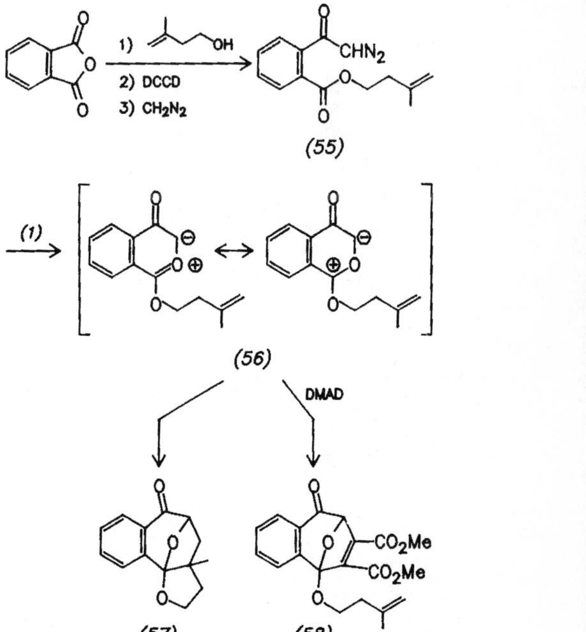

(55)

(56)

(57)   (58)

80 %   65 %

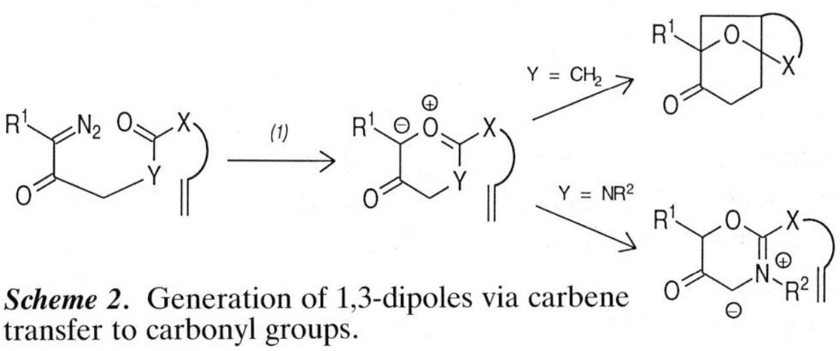

Scheme 2. Generation of 1,3-dipoles via carbene transfer to carbonyl groups.

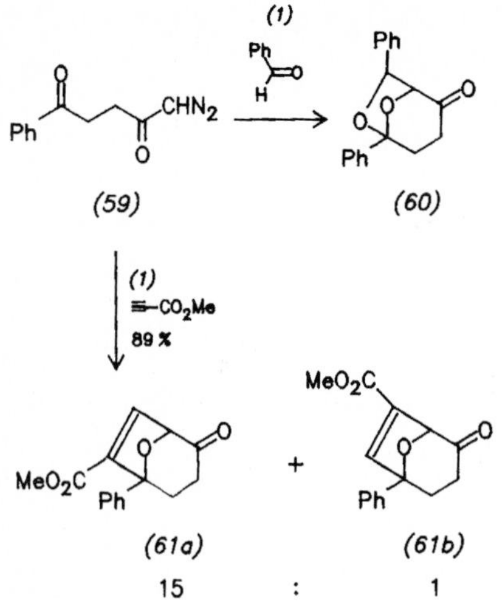
(1)
Ph
H
O
CHN2
Ph
O
Ph
O
O
Ph
O
(59)
(60)
(1)
≡-CO2Me
89 %
MeO2C
MeO2C
O
O
Ph
+
O
O
Ph
(61a)
(61b)
15 : 1

tion of the intermediate 1,3-dipole by an α-heteroatom can be provided. As a consequence, the resulting carbonyl ylide is short-lived and the bimolecular reaction can no longer successfully compete with the intramolecular cycloaddition leading to the formation of *(63)*.

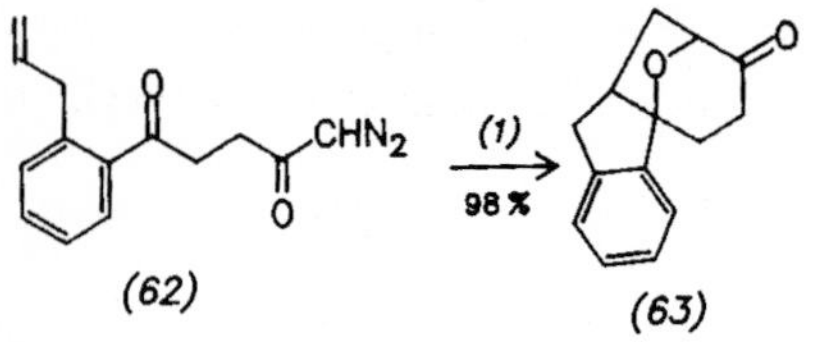
CHN2
(1)
98 %
(62)
(63)

The scope of the formation of carbonyl ylide intermediates is further extended by a "dipole cascade" sequence. The reaction of *N*-acetylpyrrolidine *(64)* with DMAD affords only small amounts of the carbonyl ylide cycloaddition product *(65)*. Instead the dihydropyrrolizine *(67)* is obtained as the major product, which is obviously formed via a carbonyl ylide–azomethine ylide isomerization followed by addition to the dipolarophile to give *(66)* and by a final 1,3-alkoxy migration. [26] Calculations indicate that the carbonyl ylide *(69)* derived from *(68)* is more stable than the azomethine ylide intermediate *(70)*. Experimental evidence for this result has been provided

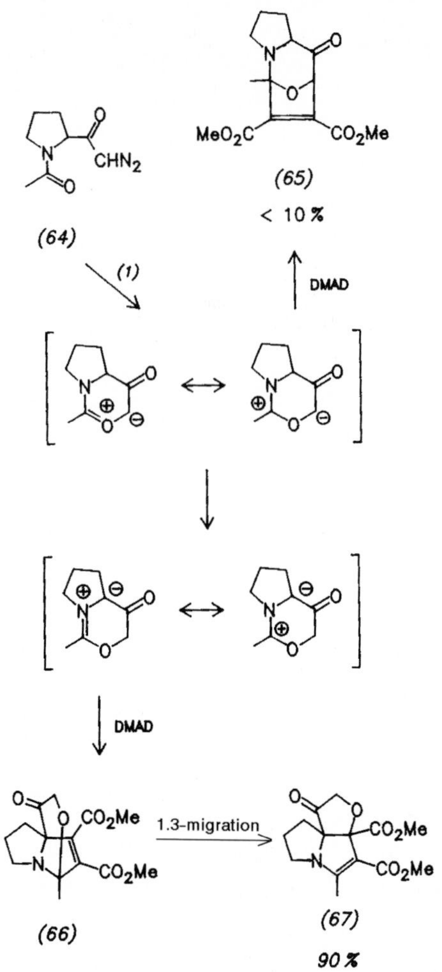
N
CHN2
(64)
(1)
MeO2C
CO2Me
(65)
< 10 %
DMAD
DMAD
CO2Me
CO2Me
1.3-migration
CO2Me
CO2Me
(66)
(67)
90 %

by the exclusive formation of the *N,O*-acetal *(71)* from the reaction of *(68)* with DMAD. The scope of synthetic applications of diazo ketoamides is demonstrated by the aliphatic analogue *(72)*. In this example, instead of a direct cycloaddition, the carbonyl ylide *(73)* prefers isomerization to give the *N,O*-acetal *(74)*, which then adds to the dipolarophile and finally affords the spiro compound *(75)*. [27]

In summary, it should be pointed out that the carbenoid intermediate faces a general competition of ylide formation and insertion reaction. The example *(76)* demonstrates that the formation of the ylide is especially favored

Rhodium-Catalyzed Carbenoid Cyclizations   73

(68)   (69)

(71)
90 %

(70)

(72)   (73)

(74)

(75)
60 %

(76)

(77)   (78)

X = O: 1   :   1,2
X = S: 1   :   9

in the presence of an easily polarizable group such as a sulfur atom. [28]

Based on the broad synthetic scope of catalyzed carbene reactions and the mild reaction conditions required rhodium carbenoids have become well-established. So far, rhodium acetate has been used as catalyst almost exclusively. It can be anticipated that both scope and selectivity can be further improved by a more elaborated "fine-tuning" of the rhodium ligand sphere.

References

[1] W. Kirmse Carbene Chemistry, Academic Press, New York, 1971.
[2] H. Nozaki, S. Moriuti, H. Takaya, R. Noyori, Tetrahedron Lett. 1966, 5239.
[3] Reviews: a) A. J. Hubert, A. F. Noels, A. J. Anciaux, P. Teyssié, Synthesis 1976, 600; b) G. Maas, Top. Curr. Chem., 1987 137, 75; c) M. P. Doyle, Chem. Rev. 1989, 86, 919; d) J. Adams, D. M. Spero, Tetrahedron 1991, 47, 1765; e) A. Padwa, K. E. Krumpe, Tetrahedron 1992, 48, 5385.
[4] T. Aratani, Y. Yoneyoshi, T. Nagase, Tetrahedron Lett. 1982, 23, 685.
[5] M. P. Doyle, J. H. Griffin, V. Bagheri, R. L. Dorow, Organometallics 1984, 3, 53.

[6] M.N. Nwaji, O.O. Onyiriuka, *Tetrahedron Lett.* **1974**, 2255.

[7] A.Padwa, T.J. Wisnieff, E.J. Walsh, *J. Org. Chem.* **1989**, *54*, 299.

[8] E.Wenkert, R.Decorzant, F.Näf, *Helv. Chim. Acta* **1989**, *72*, 756.

[9] M.Kennedy, M.A. McKervey, *J. Chem. Soc. Chem. Commun.* **1988**, 1028.

[10] A.Demonceaux, A.F. Noels, A.J. Hubert, P.Teyssié, *Bull. Soc. Chim. Belg.* **1984**, *93*, 945.

[11] D.F. Taber, R.E. Ruckle, Jr., *J. Am. Chem. Soc.* **1986**, *108*, 7686.

[12] D.E. Cane, P.J. Thomas, *J. Am. Chem. Soc.* **1984**, *106*, 5295.

[13] P.Ceccherelli, M.Curini, M.C. Marcotullio, O.Rosati, E.Wenkert, *J. Org. Chem.* **1990**, *55*, 311.

[14] M.P. Doyle, V.Bagheri, M.M. Pearson, J.D. Edwards, *Tetrahedron Lett.* **1989**, *30*, 7001.

[15] M.P. Doyle, J.Taunton, H.Q. Pho, *Tetrahedron Lett.* **1989**, *30*, 5397.

[16] T.N. Salzmann, R.W. Ratcliffe, B.G. Christensen, F.A. Bouffard, *J. Am. Chem. Soc.* **1980**, *102*, 6161.

[17] T.N. Salzmann, R.W. Ratcliffe, and B.G. Christensen, *Tetrahedron Lett.* **1980**, *21*, 1193.

[18] G. Lawton, C.J. Moody, C.J. Pearson, *J. Chem. Soc. Perkin I* **1987**, 899.

[19] R.M. Williams, B.H. Lee, M.M. Miller, O.P. Anderson, *J. Am. Chem. Soc.* **1989**, *111*, 1073.

[20] E.Vedejs, *Acc. Chem. Res.* **1984**, *17*, 358.

[21] C.J. Moody, R.J. Taylor, *Tetrahedron Lett.* **1988**, *29*, 6005.

[22] C.J. Moody, A.M.Z. Slawin, R.J. Taylor, D.J. Williams, *Tetrahedron Lett.* **1988**, *29*, 6009.

[23] T.Kametani, H.Yukawa, T.Honda, *J. Chem. Soc. Perkin I* **1988**, 833.

[24] A.Padwa, S.P. Carter, H.Nimmesgern, P.D. Stull, *J. Am. Chem. Soc.* **1988**, *110*, 2894.

[25] A.Padwa, G.E. Fryxell, L.Zhi, *J. Org. Chem.* **1988**, *53*, 2875.

[26] A.Padwa, D.C. Dean, L.Zhi, *J. Am. Chem. Soc.* **1989**, *111*, 6451.

[27] A.Padwa, L.Zhi, *J. Am. Chem. Soc.* **1990**, *112*, 2037.

[28] A.Padwa, S.F. Hornbuckle, G.E. Fryxell, P.D. Stull, *J. Org. Chem.* **1989**, *54*, 817.

Nickel-Activated C₁-Synthons

Karl Heinz Dötz

Transition metals can be used to activate substrates by coordination and couple them under stereoelectronic control by the coordination sphere of the metal. A classical example of this type of reaction is provided by a "naked" nickel template, which mediates the cyclooligomerization of unsaturated hydrocarbons. [1] This methodology is similarly attractive for carbon–carbon bond formation using C₁-synthons. In this respect, interest has focused almost exclusively on carbon monoxide, as demonstrated in industrial processes such as hydroformylation and the Monsanto acetic acid process. The use of carbon dioxide, however, has been limited so far to less spectacular applications – in spite of its ubiquitous natural occurrence. Nevertheless, the synthetic potential of carbon dioxide [2] and its heterocumulene analogues, the isocyanates, [3] can be widely extended in combination with an appropriate transition metal.

Carbon dioxide combines the properties of a Lewis acid (at carbon) and a Lewis base (at oxygen). These properties result in various modes of coordination that – in general – can be traced back to complexes *(1)* and *(2)*, which have both been structurally characterized by X-ray analysis. [4, 5] Compounds of type *(1)* are particularly interesting for the synthetic organic chemist: They are supposed to be key intermediates in the transition metal-assisted coupling of CO_2 with unsaturated substrates such as alkenes and alkynes,

although direct experimental support is still missing. Instead, metalalactones *(4)* have been observed, the formation of which may be rationalized in terms of an oxidative coupling of coordinated CO_2 with the alkene ligand via intermediate *(3)*. Low-valent "late" transition metals turned out to be efficient templates: Among them, nickel(0) stabilized by amine or phosphine ligands plays a key role. [2, 3, 6]

(1)

$Cy = c\text{-}C_6H_{11}$

(2)

(3)

(4)

L = amine, phosphine

Stability and reactivity of the metalacycles depend on the nature of the coligands. Well-balanced properties are observed for the 1,8-diazabicyclo[5.4.0]undec-7-ene (DBU) complex (5), [7] which is accessible from bis(cyclooctadiene)nickel (Scheme 1). It is stable at ambient temperature and its structure is well-characterized by X-ray analysis. Its synthetic potential is focused on the nickel carbon bond, which is prone to insertion by π-systems. This reaction is particularly useful for monofunctionalized alkenes or allene, which afford monocarboxylic acids (6) and (7) after insertion followed by hydrolysis; butadiene, however, gives only disappointingly low yields of alkenoic acids (8a) and (8b). The selectivity is controlled both by the metal and the coligands. This fact is demonstrated by the ferra-

lactone (9), an iron analogue of (5), which undergoes insertion of excess CO_2 into the Fe–C bond. [8] In the presence of the sterically demanding bis(dicyclohexylphosphino)ethane (DCPE) dimethylsuccinate (10) is obtained as the main product after hydrolysis and esterification. On the other hand, the methylmalonic acid derivative (11) is formed nearly exclusively, if DCPE is substituted for trimethylphosphine.

Metalalactones (4) serve as potential C_3-carboxyl synthons and can be exploited in the homologization of alkyl halides, as exemplified in the modification of the D-ring side chain in steroids. Usually, cross-coupling of the bispyridyl nickel complex (12) with the steroid halide (13) is hampered by undesired side reactions. For instance, the oxidative

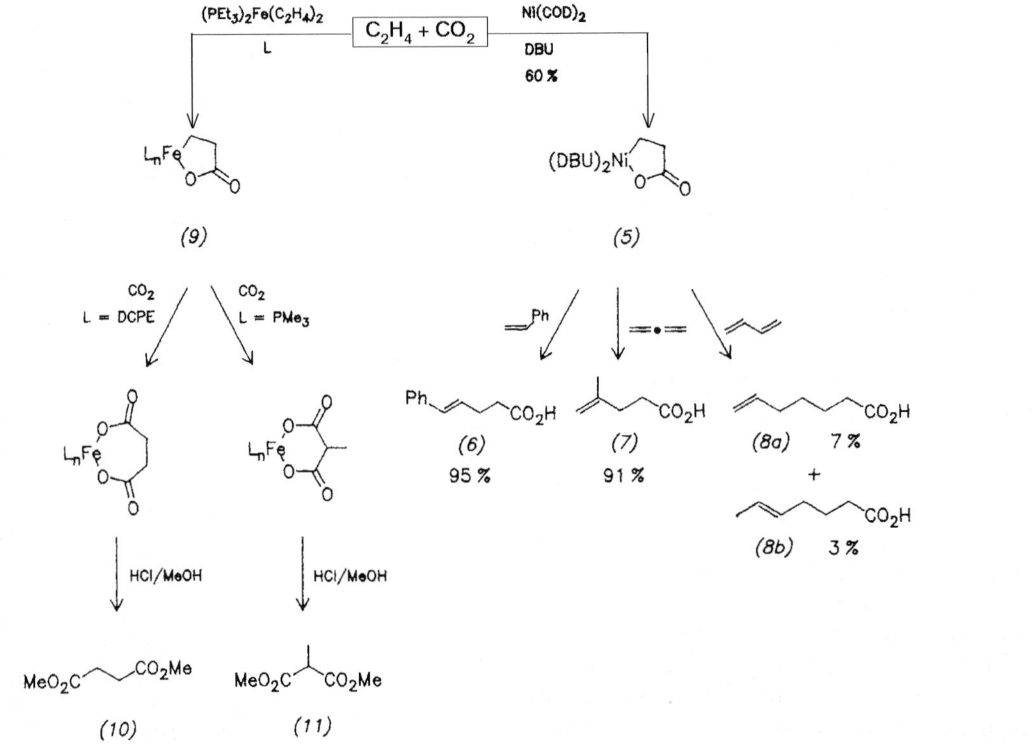

	yield	(10) : (11)
L = DCPE:	49 %	10 : 1
L = PMe₃:	60 %	1 : 400

Scheme 1. Influence of coligands on the reactivity of metalalactones.

addition product of *(13)* undergoes β-hydride elimination and finally gives the propenyl steroid *(14)* in low yield. Using MnI$_2$ and ultrasound conditions, however, C–C coupling occurs readily and the carboxylic acid *(15)* is formed in 80 % yield. [9] The cross-coupling is chemoselective: Enone functionalities in ring A remain unchanged.

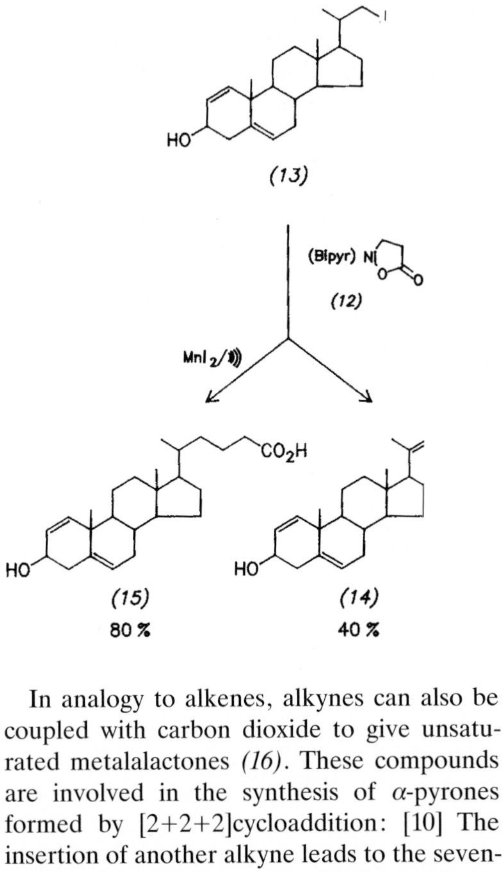

and which shows a α-pyrone selectivity of 96 %. [9b, 10b] More recent studies demonstrate that this type of reaction can be extended to functionalized alkynes (e.g., ethoxyethyne [11]) and diynes. [12] Again, the choice of coligands attached to the nickel center is crucial: Terminally disubstituted diynes require unidentate phosphine ligands whereas *P,N*-chelating systems should be used if unsubstituted diynes are involved. Furthermore, the coligands control the regioselectivity of the CO$_2$/diyne coupling: In the presence of the *P,N*-chelating ligand *(21)* the silyldiyne *(22)* is predominantly cyclocarboxylated to give the 7-silyl derivative *(23);* if, however, tri-*n*-octylphosphine is used as coligand, the regi-

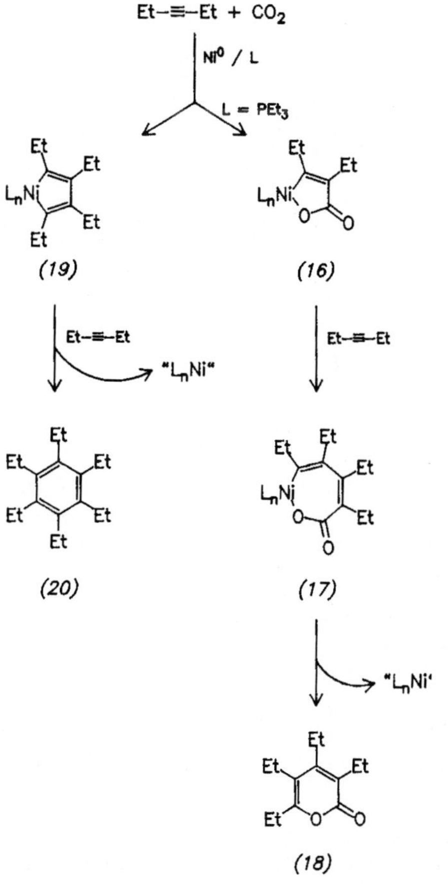

In analogy to alkenes, alkynes can also be coupled with carbon dioxide to give unsaturated metalalactones *(16)*. These compounds are involved in the synthesis of α-pyrones formed by [2+2+2]cycloaddition: [10] The insertion of another alkyne leads to the seven-membered metalacycle *(17)*, which undergoes reductive elimination to α-pyrone *(18)*. It is important for synthetic applications that competing reaction paths such as alkyne cyclotrimerization to give arene *(20)* via metalacyclopentadiene *(19)* can be slowed down efficiently by appropriate catalysts. An efficient system is Ni(COD)$_2$/PEt$_3$ (1:2), which is used in acetonitrile in amounts of 0.1 equivalents

(13)

(12)

(15) 80 %

(14) 40 %

Et—≡—Et + CO$_2$

Ni^0 / L

L = PEt$_3$

(19)

(16)

(20)

(17)

(18)

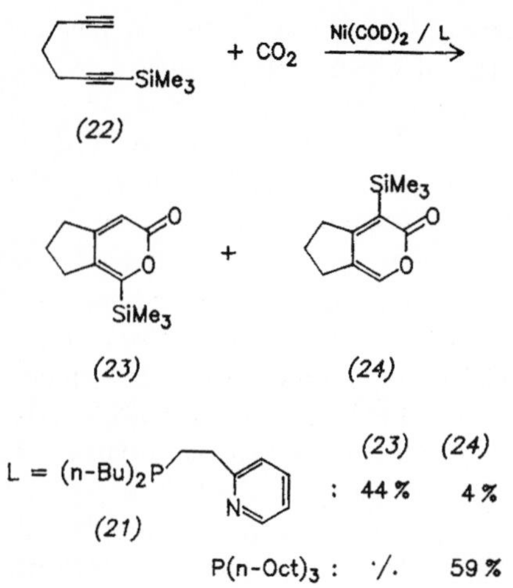

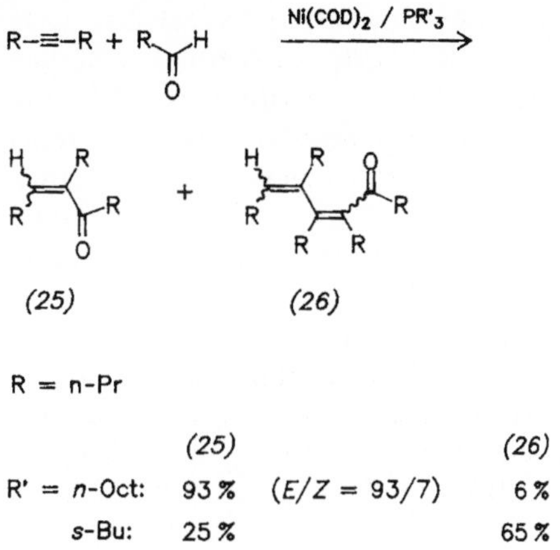

+ CO2
Ni(COD)2 / L
SiMe3
(22)
(23)
(24)
L = (n-Bu)2P
N
(21)
(23) (24)
: 44 % 4 %
P(n-Oct)3 : ·/· 59 %

R—≡—R + R—CHO
Ni(COD)2 / PR'3
(25)
(26)
R = n-Pr
(25) (26)
R' = n-Oct: 93 % (E/Z = 93/7) 6 %
s-Bu: 25 % 65 %

ochemistry is reversed and the 3-silyl isomer *(24)* is obtained as the only product.

Similar catalytic systems allow the addition of aldehydes to alkynes. Application of appropriate coligands, for example tri(*n*-octyl)phosphine, leads in a "hydroacylation reaction" predominantly to *E*-α,β-enones *(25);* [13] more bulky phosphines increase the amount of dienone *(26)*. The mechanism of this reaction is unclear and two alternatives must be discussed: The initial step might involve either oxidative addition of the aldehyde to the zero-valent nickel center (path a) or the formation of a nickelaoxacyclopentene intermediate (path b). So far, the hydroacylation has been applied mainly to symmetric dialkylalkynes. The addition of aldehydes to unsymmetric alkynes suffers from low regioselectivity; a synthetically useful selectivity *(27)*/*(28)* requires α-branched alkyne side chains, which, in turn, results in decreased yields.

The activation of CO_2 can be combined with the oligomerization of 1,3-dienes. [1] This strategy leads to the formation of C_{4n+1}-carboxylic acids, the chain lengths of which depend again on the coligands present in the nickel catalyst. If a bulky basic diphosphine, for example DCPE, is used, only one diene unit is coupled to carbon dioxide. [14] Unsym-

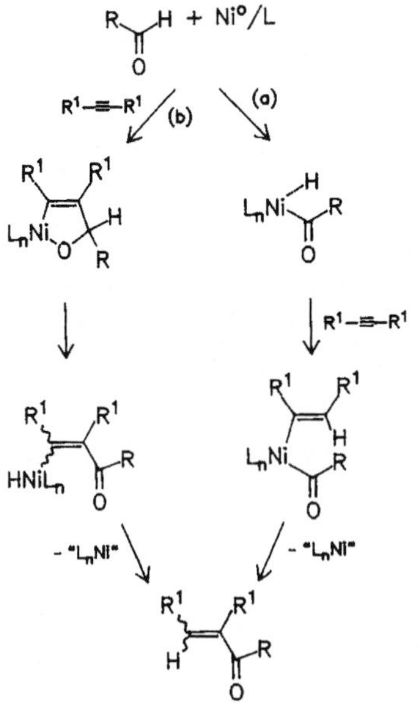

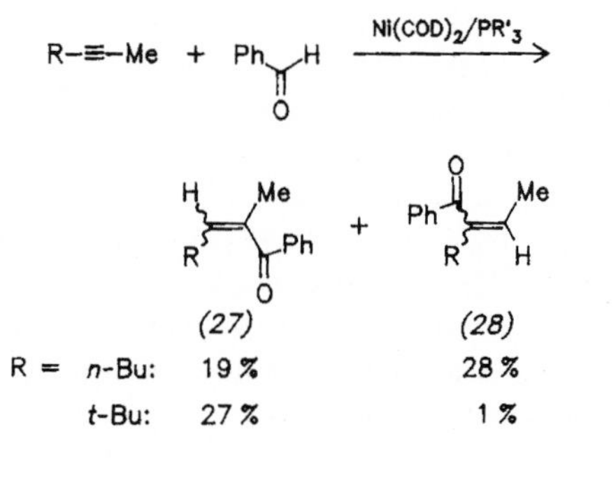

+ Ni°/L
R1—≡—R1
(b)
(a)
LnNi
HNiLn
R1—≡—R1
- "LnNi"
- "LnNi"

R—≡—Me + Ph—CHO
Ni(COD)2/PR'3
(27)
(28)
R = n-Bu: 19 % 28 %
t-Bu: 27 % 1 %

metrical 1,3-dienes such as piperylene afford, after treatment with methanolic hydrogen chloride, two pairs of unsaturated methyl carboxylate isomers *(31a, b)* and *(32a, b)*, which are derived from the π-allyl complex intermediates *(29)* and *(30)*. A pyridine-modified nickel catalyst allows the coupling of CO_2 with two diene moieties to give the C_9-ester *(33)* or, at elevated temperature, its C_{18}-dimer *(34)*. [15] A combined diene-trimerization and carboxylation to linear C_{13}-triene and -tetraene carboxylic esters is effected if pentafluoropyridine (PFP) is used as a coligand (Scheme 2). [16]

The key role of the coligand sphere is further obvious from coupling reactions involving a dimerization of butadiene. In the presence of carbon dioxide a *stoichiometric* reaction mediated by the nickel/triphenylphosphine-system affords a 9:1 mixture of cyclopentane carboxylic acids *(36)* and *(37)*. If, however, the less basic tri(isopropyl)-phosphite acts as a coligand, exclusively the diene *(37)* is formed in a *catalytic* reaction (30 cycles). [17] Similar catalytic telomerization reactions have been earlier observed with palladium systems. [18] A more recent example demonstrates the efficiency of cationic palladium(II)/triphenylphosphine catalysts. In comparison with other telomers lactone *(38)* is formed in 96% chemoselectivity. [19]

Similar to carbon dioxide, isocyanates can be incorporated into nickelacycles together with C_2-π-systems. [20] On reaction with alkynes these heterocumulenes give five-membered unsaturated metalacycles *(39)*

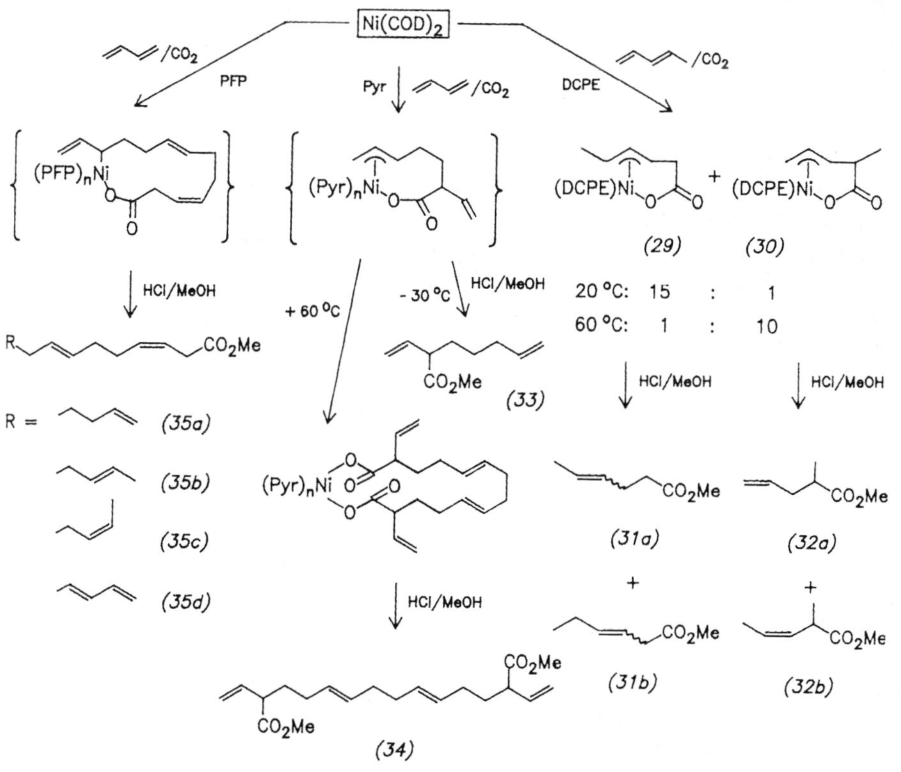

Scheme 2. Activation of CO_2 combined with the oligomerization of 1,3-dienes: Influence of coligands on the formation of carboxylic acids.

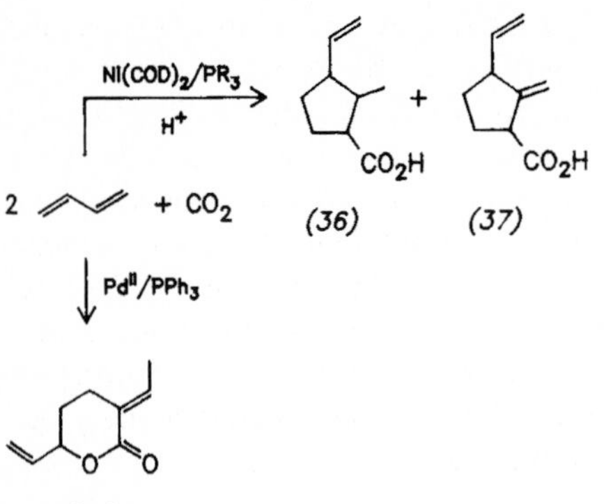

$$2 \quad \diagup\diagdown\diagup + CO_2 \xrightarrow[\text{H}^+]{\text{Ni(COD)}_2/\text{PR}_3}$$

(36)   (37)

(with CO_2H groups)

↓ Pd^{II}/PPh_3

(38)

which subsequently undergo reactions analogous to those already described for their oxa congeners such as metalalactone (16). The cross-coupling with alkyl halides affords acrylic amides (40). Insertion of electron-deficient alkynes into the nickel–carbon bond

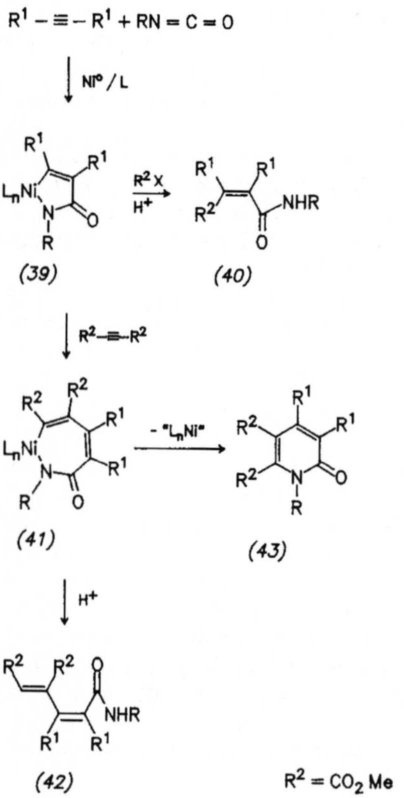

$$R^1 - \equiv - R^1 + RN = C = O$$

↓ Ni^0/L

(39)

$$\xrightarrow[\text{H}^+]{R^2X}$$

(40)

↓ $R^2 - \equiv - R^2$

(41)

$$\xrightarrow{- \text{"L}_n\text{Ni"}}$$

(43)

↓ H^+

(42)

$R^2 = CO_2 Me$

followed by hydrolysis leads to diene amides (42) while reductive elimination from the seven-membered nickelacycle intermediate (41) gives the 2-pyridone (43). [21] These types of nickelacycles have been further used in oxidative degradation reactions. The metal-assisted coupling of allene and phenylisocyanate generates a (C,N)-chelate ligand that undergoes dimerization to give the 1,5-diene diamide (45) on oxidation of complex (44) with $FeCl_3$. [22]

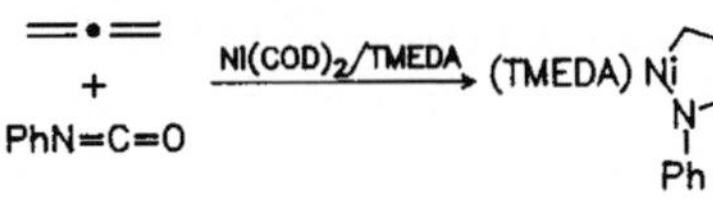

$$\equiv \cdot \equiv \; + \; PhN = C = O \xrightarrow{\text{Ni(COD)}_2/\text{TMEDA}} \text{(TMEDA) Ni}$$

(44)

$$\xrightarrow{Fe^{III}}$$

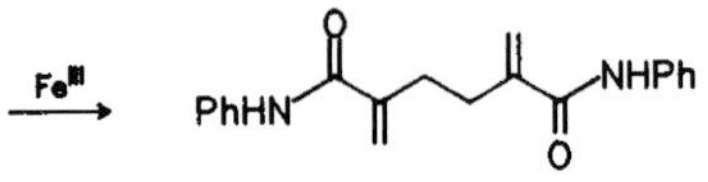

(45)

To carry out this type of reaction catalytically, a profound knowledge of those steps is required that effect the cleavage of the metal–carbon bond. A key step is the β-hydride elimination, which has been studied in the coupling of vinylcyclohexane and phe-

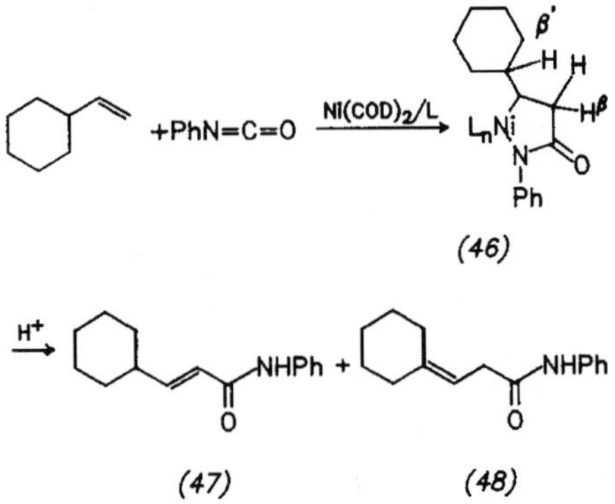

$$+ PhN = C = O \xrightarrow{\text{Ni(COD)}_2/L} L_n Ni$$

(46)

$$\xrightarrow{H^+}$$

(47) + (48)

	(47)		(48)
L = PEt$_3$	: 50	:	50
P (O-i-Pr)$_3$	: 95	:	5
P (O-Ph-o-Ph)$_3$	: 14	:	86

nylisocyanate. [23] A bis(cyclooctadiene)-nickel/triethylphosphine catalyst leads to a 1:1 mixture of amides *(47)* and *(48)*; apparently, in the metalacycle *(46)*, the rates of elimination of the β- and the β'-H-atoms are equal. Less basic phosphine ligands, however, allow differentiation: Tri(isopropyl)phosphite favors the β-H-elimination [*(47)*/*(48)* = 95/5] whereas the even bulkier tri(*ortho*-phenylphenyl)phosphite preferentially activates the β'-position [*(47)*/*(48)* = 14/86].

The potential of isocyanate C$_1$-synthons in organic synthesis is not restricted to nickel, as demonstrated by two examples referring to the higher homologue palladium. The epoxide *(49)*, accessible from the diene *via* enantioselective Sharpless epoxidation, reacts with tosylisocyanate in a palladium-catalyzed hydroxyamination with retention of configuration to give oxazolidinone *(50)*, an intermediate in the synthesis of (–)-acosamine *(51)*. [24] It is an attractive feature of this route that – independent of the configuration of the epoxide [e.g., *(52)*] – the *cis*-oxazolidinone [e.g., *(53)*] is obtained.

(49)

(50)   *(51)*

(52)

(53)

from   *cis*-*(52)*   :   100%
          trans-*(52)*   :   80%

[Pd] = (DBA)$_3$Pd$_2$·CHCl$_3$/P(O^{i}Pr)$_3$; R =

DBA = dibenzalacetone

References

[1] P. W. Jolly, G. Wilke, *The Organic Chemistry of Nickel*, Vol. II, Academic Press, New York, **1975**.

[2] Reviews: a) P. Braunstein, D. Matt, D. Nobel, *Chem. Rev.* **1988**, *88*, 747; b) A. Behr, *Angew. Chem.* **1988**, *100*, 681; *Angew. Chem. Int. Ed. Engl.* **1988**, *27*, 661.

[3] Review: P. Braunstein, D. Nobel, *Chem. Rev.* **1989**, *89*, 1927.

[4] M. Aresta, C. F. Nobile, *J. Chem. Soc. Dalton Trans.* **1977**, 708.

[5] S. Gambarotta, F. Arena, C. Floriani, P. F. Zanazzi, *J. Am. Chem. Soc.* **1982**, *104*, 5082.

[6] H. Hoberg, in *Carbon Dioxide as a Source of Carbon*, (Eds.: M. Aresta, G. Forti), Reidel, Dordrecht, **1987**.

[7] H. Hoberg, Y. Peres, C. Krüger, Y. H. Tsay, *Angew. Chem.* **1987**, *99*, 799; *Angew. Chem. Int. Ed. Engl.* **1987**, *26*, 771.

[8] H. Hoberg, K. Jenni, K. Angermund, C. Krüger, *Angew. Chem.* **1987**, *99*, 141; *Angew. Chem. Int. Ed. Engl.* **1987**, *26*, 153.

[9] a) G. Bräunlich, R. Fischer, B. Nestler, D. Walther, *Z. Chem.* **1989**, *29*, 417; b) D. Walther in *Advances in Organic Synthesis via Organometallics*, (Eds.: R. W. Hoffmann, K. H. Dötz), Vieweg, Wiesbaden, **1991**, p. 77.

[10] a) Y. Inoue, Y. Itoh, H. Kazama, H. Hashimoto, *Bull. Chem. Soc. Japan* **1980**, *53*, 3329; b) D. Walther, H. Schönberg, E. Dinjus, J. Sieler, *J. Organomet. Chem.* **1987**, *334*, 377; c) H. Hoberg, D. Schäfer, G. Burkhart, C. Krüger, M. R. Romao, *J. Organomet. Chem.* **1984**, *266*, 203.

[11] T. Tsuda, K. Kunisada, N. Nagahama, S. Morikawa, T. Saegusa, *Synth. Commun.* **1990**, *20*, 313.

[12] T. Tsuda, S. Morikawa, N. Hasegawa, T. Saegusa, *J. Org. Chem.* **1990**, *55*, 2978.

[13] T. Tsuda, T. Kiyoi, T. Saegusa, *J. Org. Chem.* **1990**, *55*, 2554.

[14] H. Hoberg, D. Schaefer, B. W. Oster, *J. Organomet. Chem.* **1984**, *266*, 313.

[15] H. Hoberg, Y. Peres, A. Milchereit, S. Gross, *J. Organomet. Chem.* **1988**, *345*, C 17.

[16] H. Hoberg, D. Bärhausen, *J. Organomet. Chem.* **1989**, *379*, C 7.

[17] H. Hoberg, S. Gross, A. Milchereit, *Angew. Chem.* **1987**, *99*, 567; *Angew. Chem. Int. Ed. Engl.* **1987**, *26*, 571.

[18] a) Y. Inoue, Y. Sasaki, H. Hashimoto, *Bull. Chem. Soc. Japan* **1978**, *51*, 2375; b) A. Behr, K.-H. Juszak, W. Keim, *Synthesis* **1983**, 574.

[19] P. Braunstein, D. Matt, D. Nobel, *J. Am. Chem. Soc.* **1988**, *110*, 3207.

[20] H. Hoberg, *J. Organomet. Chem.* **1988**, *358*, 507.

[21] H. Hoberg, B. W. Oster, *J. Organomet. Chem.* **1983**, *252*, 359.

[22] H. Hoberg, E. Hernandez, K. Sümmermann, *J. Organomet. Chem.* **1985**, *295*, C 21.

[23] H. Hoberg, D. Guhl, *Angew. Chem.* **1989**, *101*, 1091; *Angew. Chem. Int. Ed. Engl.* **1989**, *28*, 1035.

[24] B. M. Trost, A. R. Sudhakar, *J. Am. Chem. Soc.* **1987**, *109*, 3792; *J. Am. Chem. Soc.* **1988**, *110*, 7933.

Aminocarbene Complexes in Ligand- and Metal-Centered Carbon–Carbon Bond Formation

Karl Heinz Dötz

Transition metals have become indispensable tools in organic synthesis. Typical examples involve transmetalation reactions starting from classical organometallic materials such as Grignard or organolithium compounds. These routes allow in situ generation of reagents that – like organotitanium compounds [1] – often exhibit a dramatically improved chemo- and stereoselectivity. In addition, increasing attention has focussed on stable and unambiguously characterizable metal complexes provided that they are accessible by a reasonable synthetic effort in medium scale batches. These requirements are met with transition metal carbene complexes: [2] Compounds containing a high-valent metal center (Schrock-type metal carbenes) are used in carbonyl olefination processes – with the Tebbe–Grubbs reagent $Cp_2Ti=CH_2$ [3] generated in situ as the most prominent example – or as catalysts in olefin metathesis such as $(R_FO)_2(NR)W=CHtBu$. [4] On the other hand, the organic-synthetic interest in low-valent metal center derived complexes (Fischer type metal carbenes), [5] for example, $(CO)_5Cr=C(OR)R'$, has concentrated on stoichiometric cycloaddition reactions.

Initial interest has focussed on Fischer carbene complexes containing *alkoxy*carbene ligands; [6, 7] more recently, remarkably stereoselective reactions have been reported for *amino*carbene complexes. These compounds can be easily prepared from metal carbonyls either via alkoxy- and acyloxycarbene precursors *(1)* and *(2)*, respectively, or via carbonyl metalates *(3)* [8, 9] (Scheme 1).

The synthetic potential of metal aminocarbenes arises from the conjugate carbene anions as well as from cycloaddition reactions. The intrinsic α-CH-acidity of alkylcarbene complexes $((CO)_5Cr=C(OMe)Me$: $pK_a = 8$ (PPN^+ salt in THF; [10] $(CO)_5Cr=C(NMe_2)Me$: $pK_a = 20.4$ (K^+ salt in DMSO) [11] has been exploited in aldol reactions. The anions derived from alkoxycarbene complexes *(1)* are more stable and give aldol adducts in synthetically useful yields only if the carbonyl compound is complexed first with a Lewis acid. [12] The deprotonation of aminocarbene complex *(4)*, however, leads to a more reactive conjugate base *(5)*, which rapidly and cleanly adds to aldehydes and ketones without the assistance of a Lewis acid. These reactions proceed with a remarkable diastereofacial selectivity, as shown by the addition of *(5)* to chiral aldehydes: [13] *d,l*-2-phenylpropionaldehyde affords an 80% yield of the aldol adduct as a *(6a):(6b)* = >40:1 (l:u) mixture of diastereomers. A similar stereoselection has been restricted so far to Lewis acid-promoted reactions with silyl enol ethers. [14] With *d,l*-2-methyl-3-phenylpropionaldehyde the stereoselectivity is reduced to *(7a):(7b)* = 4.1:1, but is still competitive with that observed with customary α-unsubstituted enolates. The chromium carbonyl fragment

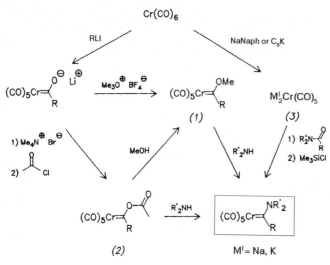

Scheme 1. Synthesis of aminocarbene complexes from alkoxy- or acyloxycarbene precursors and from carbonyl metalates.

can be cleaved by oxidation (e.g., with DMSO or dimethyldioxirane) [15, 16] or by UV-irradiation (via carbene CO coupling to give a coordinated ketene intermediate [17]) and formation of lactone *(8)* (Scheme 2).

The facial selectivity of carbene anions is further expressed in the Michael addition to γ-chiral enones. Presently, the method of choice for such diastereoselective 1,4-additions is the Lewis acid-mediated addition of enol silanes, [18] which affords a selectivity of up to 30:1. The lowest preference in this series (5:1) was reported for the TiCl$_4$-assisted addition of the silyl enol ether of 3,3-dimethylbutanone to 1,4-diphenyl-2-penten-1-one. This enone, however, reacts readily at –78 °C with the conjugate base of aminocarbene complex *(9)* to give ketoaldehyde *(11)* in a diastereomeric ratio *l:u* = 21:1 after acidic cleavage of the primary Michael addition product *(10)*. [11] Asymmetric Michael reactions of optically active proline derived carbene ligands with five- and six-membered cyclic enones generally proceed with only moderate enantiomeric excess (60–76 % *ee*). As expected,

steric bulk at the reactive center of the enone increases the selectivity to 95 % *ee*.

The characteristic feature of Fischer type metal carbenes is the electrophilicity of the carbene carbon atom. Since the pentacarbonyl metal fragment is isolobal with an oxygen atom, [19] the reactivity of carbene complexes is expected to correlate with that of their organic carbonyl congeners. In intermolecular Diels-Alder reactions methoxy(vinyl)carbene complexes react 2×10^4 times faster and more regioselectively than their acrylate analogues. [20] Starting materials for the intramolecular version (IMDA) are accessible in a two-step procedure based on subsequent addition of a nucleophile and an electrophile to a carbonyl ligand followed by nucleophilic substitution at the carbene center. As demonstrated in the intermolecular reaction, the metal carbonyl fragment is a strongly electron-withdrawing functional group as powerful as a Lewis acid-coordinated carbonyl oxygen atom. An illustrative example is provided by the following IMDA reaction with inverse electron demand: The (diallylamino)furylcarbene complex *(12)*

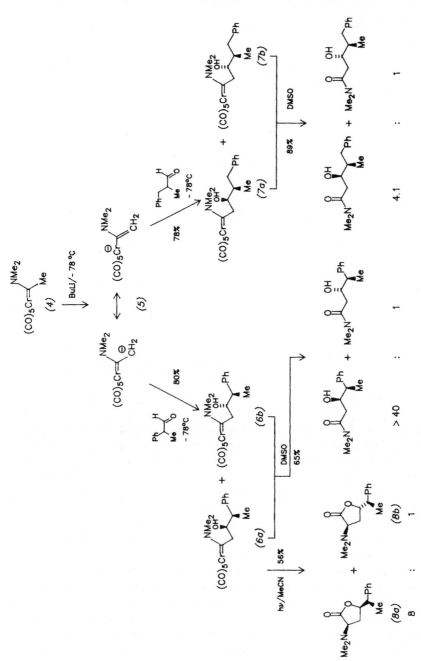

Scheme 2. Aminocarbene complex based diastereoselective aldol reactions.

1) BuLi / -78°C

2)

(CO)₅Cr= N Ph
         CH₃

(9)

(CO)₅Cr= N Ph
              O
         Ph
    Ph

(10)

CF₃COOH

H O   O Ph
      Ph

(11)
76%
l : u = 21 : 1

W(CO)₆

1) O Li

2) Cl

(CO)₅W= O   NH
         O

-40 °C

(CO)₅W= N
              O

(12)

-40 °C

O= N   CₒN   (CO)₅W= N
    O   O

(14)   *(13)*

(CO)₅Cr= N OMe
         CH₃

1) BuLi / -78°C

2) R   O
    R

(CO)₅Cr= N OMe
         R R
         O

CF₃COOH

H O R R
      O

R = H : 41% , 64% ee (S)
R = Me : 51% , 95% ee (S)

obtained from hexacarbonyl tungsten undergoes spontaneous cyclization even under the conditions of its formation at –40 °C. A single diastereomer *(13)* is observed in which the newly formed five-membered rings are *trans*-fused. [21] In comparison, the cyclization of analogous furan carboxylic amides requires reflux conditions in benzene and toluene, respectively, for several hours. Finally, the metal carbene bond can be modified by oxidation to give the lactam *(14)*.

Aminocarbene ligands are better donors than their alkoxy counterparts, which results in a more efficient back-bonding from the metal to the carbonyl ligands and has an important consequence for metal-mediated cyclization reactions, for example with alk-

ynes (Scheme 3). While *alkoxy*carbene ligands in *(15)* (D = OR) undergo annulation by the alkyne *and* by a carbonyl ligand via ketene intermediate *(16)* to give hydroquinone *(17)*, in general, no CO-incorporation is observed with *amino*carbene complexes which generate alkyne insertion products in which the resonance form *(B)* prevails. Instead, a five-membered carbene annulation product *(18)*, in which the less crowded alkyne carbon atom is connected to the carbene carbon atom, is obtained after acidic work-up. [22] Ring formation is regio- and diastereoselective, as demonstrated by the intramolecular example *(19)*→*(20)*. [23] Only one diastereomer is observed in which the former alkyne substituents in the central five-membered ring are *trans* and – more remarkably – the benzylic substituent R and the metal carbonyl fragment are bonded to the same face of the tricyclic ring system. A plausible mechanism suggests that the stereochemistry in the central ring is determined by a suprafacial 1,5-sigmatropic hydrogen migration *(21)* → *(22)*. Thus, the carbene annulation methodology allows the formation of coordinated substi-

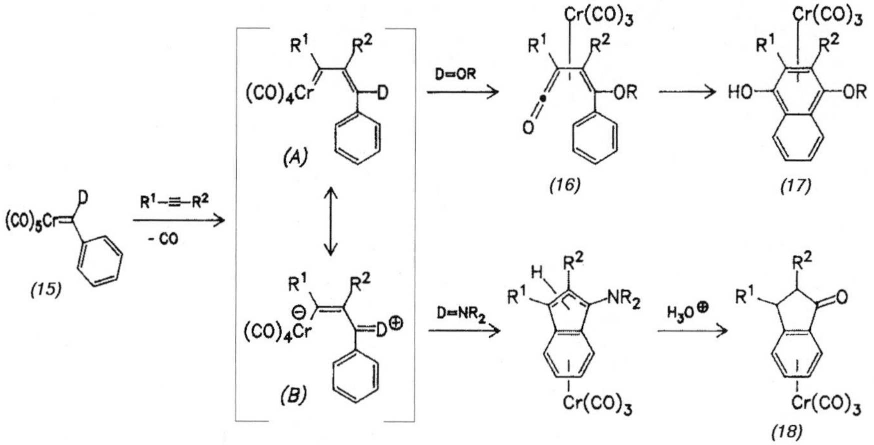

Scheme 3. Metal-induced cyclization reactions using alkoxycarbene or aminocarbene complexes.

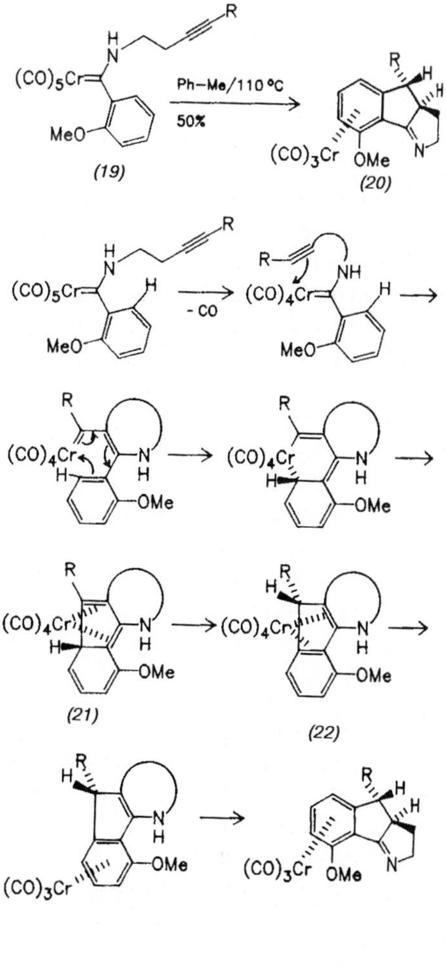

tuted fused ring systems that can be further functionalized under the stereocontrol exerted by the bulky Cr(CO)$_3$ fragment. [24]

Moreover, the versatility of the carbene annulation allows the formation of heterocycles. The iminocarbene ligand in *(23)*, formed by iminolysis from the methoxy precursor, adds to alkynes with the same regioselectivity as discussed above to give pyrrole *(24)*. [25] If the donor ability of the aminocarbene ligand is reduced, the formation of six-membered annulation products is favoured. [26] *N*-Acylation of the benzylaminocarbene complex by (BOC)$_2$O occurs with concomitant decarbonylation and affords directly the activated aminocarbene chelate *(25)*. Its annula-

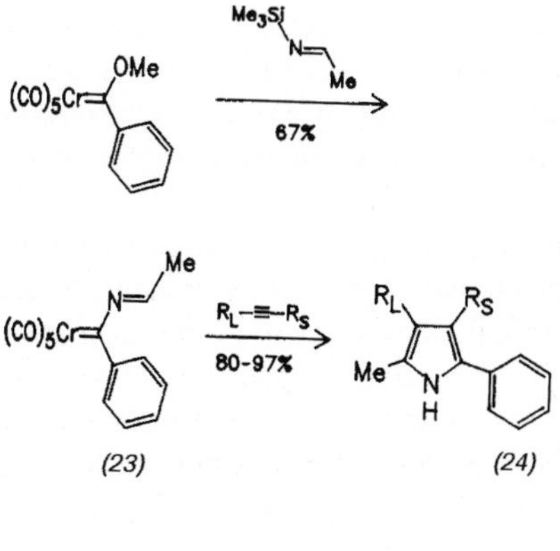

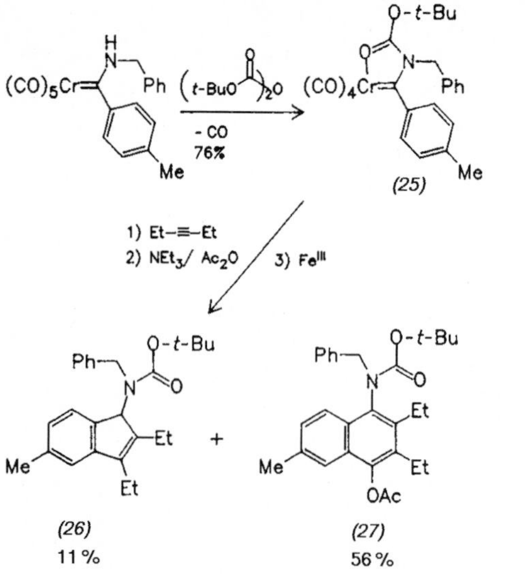

(25)

(26)   (27)
11%   56%

tion by 3-hexyne leads to a 5:1 preference of the aminonaphthol (27) over the aminoindene (26).

The carbene annulation is a *thermally* induced sequence; it can be explained in terms of a primary decarbonylation of the carbonyl carbene complex followed by coordination of the alkyne and subsequent coupling of the alkyne and carbene ligands. [6] On the other hand, it has been suggested that a direct carbonyl carbene coupling occurs under *photochemical* conditions. Assisted by metal ligand charge transfer a ketene is formed in the coordination sphere of the metal and can be subsequently trapped by protic nucleophiles or by [2+2] cycloaddition. An interesting application involves the addition of imines which opens up a new diastereoselective route to β-lactams. [27] Since aminocarbene complexes containing α-hydrogen substituents are accessible via the carbonyl metalate/formamide route and since chiral information can be readily incorporated into the amino group, this strategy is suited for the synthesis of biologically relevant optically active compounds. For instance, the (R)-aminocarbene complex (29), accessible from (R)-phenylglycinol via

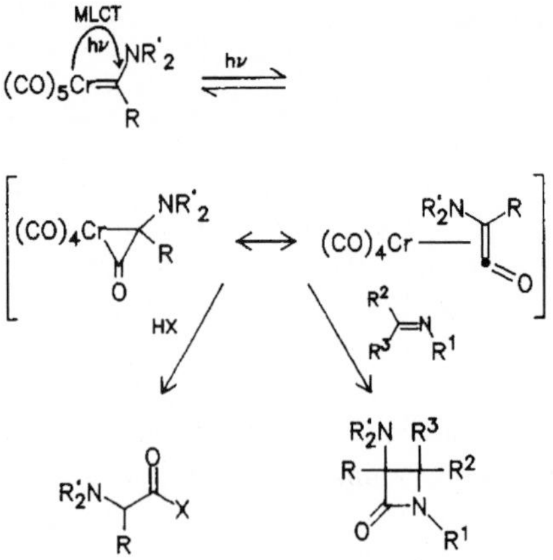

the formamide (28), reacts with a series of imines to give β-lactams (30) in good to excellent diastereoselectivity. It is important that the new chiral center formed next to the carbonyl group has the same absolute configuration as the asymmetric center in the chiral auxiliary. This correlation also holds for the (S)-series: The (S)-carbene complex (32) leads to the (S)-lactam (33). The chiral auxiliary is fi-

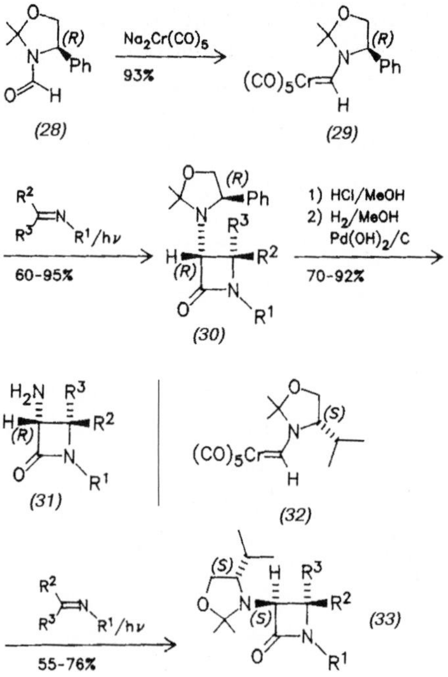

(28)   (29)

(30)

(31)   (32)

(33)

nally cleaved and recovered in a two-step hydrolysis/hydrogenolysis process, for example *(30→(31)*.

The photochemically generated amino ketene intermediates can be further exploited in the synthesis of amino acid derivatives. Irradiation in the presence of water did not afford synthetically useful results. However, if the reaction is carried out in alcoholic solution, the amino acid esters *(35)* are formed in good yields and interesting diastereoselectivity. [28] It is surprising that, in this case, the newly formed stereocenter has the *opposite* absolute configuration compared with that of the chiral auxiliary used. Unfortunately, no high-yield synthetic route to the methyl(amino)carbene complexes *(34)* is available so far. If this problem might be solved, aminocarbene complexes could provide an attractive access to optically natural and unnatural amino acids. An additional avantage arises from the α-CH-acidity in the carbene side chain, which may be used in homologization, as outlined in the formation of *(36)*.

The asymmetric protonation leading to the amino acid ester offers an efficient approach to "chiral glycine" [2-²H₁]glycine, an important substance for mechanistic studies in biochemistry. Photolysis of *(R)* or *(S)* carbene complex *(37)* in methanol-d/acetonitrile produced the glycine precursor *(38)* in *(R, R)* or

(37) (S) or (R)

(38) (S,S) or (R,R)
93%, >97% D, 86%de

(38) (S,S)

(39)
(S), 92%, 84%ee

(S, S) absolute configuration: Cleavage of the oxazolidine ring afforded the optically active glycine *(39)* with >97% monodeuteration, 84% *ee*, and 74% overall chemical yield. [29]

A final remark concerns the handling of the carbene complexes described. Their preparation does not require a more sophisticated technique than that used for Grignard and organolithium reagents. Aminocarbene complexes are air-stable in the solid state and can be stored for months. In solution, however, an inert gas atmosphere (nitrogen or argon) should be provided. The uncomplicated handling of metal carbenes has clearly assisted their development from exotic organometallics to customary synthetic reagents.

References

[1] M.T. Reetz, *Organotitanium Reagents in Organic Synthesis*, Springer, Berlin **1986**.

[2] a) K.H. Dötz, H. Fischer, P. Hofmann, F.R. Kreißl, U. Schubert, K. Weiss, *Transition Metal Carbene Complexes*, VCH, Weinheim **1983**; b) R.R. Schrock in *Reactions of Coordinated Ligands*, (Ed.: P.S. Braterman), Vol. 1, Plenum Press, New York, **1986**, p. 221 ff.; c) K.H. Dötz, ibid. p. 285 ff.

[3] K.A. Brown-Wensley, S.L. Buchwald, L. Cannizzo, L. Clawson, S. Ho, D. Meinhardt, J.R. Stille, D. Straus, R.H. Grubbs, *Pure Appl. Chem.* **1983**, 55, 1733.

(34)   hν/MeOH   96%   *(35)*

≥ 93% d.e.

1) BuLi
2) Br
3) hν/MeOH
5 bar CO
72%

(36)

[4] R.R. Schrock, *J. Organomet. Chem.* **1986**, *300*, 249.

[5] Reviews: a) E.O. Fischer, *Angew. Chem.* **1974**, *86*, 651; b) E.O. Fischer, *Adv. Organomet. Chem.* **1976**, *14*, 1.

[6] Reviews: a) K.H. Dötz, *Angew. Chem.* **1984**, *96*, 573; *Angew. Chem. Int. Ed. Engl.* **1975**, *14*, 644; b) W.D. Wulff in *Comprehensive Organic Synthesis*, Vol.5 (Eds.: B.M. Trost, I.Fleming), Pergamon Press, New York, **1991**, p.1065–1113.

[7] H.U. Reißig, in J.Mulzer, H.-J. Altenbach, M.Braun, K.Krohn, H.-U. Reißig *Organic Synthesis Highlights*, VCH, Weinheim, **1991**, p.186ff.

[8] E.O. Fischer, J.A. Connor, *J. Chem. Soc. A* **1967**, 578.

[9] R.Imwinkelried, L.S. Hegedus, *Organometallics* **1988**, 7, 702.

[10] C.P. Casey, R.L. Anderson, *J. Am. Chem. Soc.* **1974**, *96*, 1230.

[11] B.A. Anderson, W.D. Wulff, A. Rahm, *J. Am. Chem. Soc.* **1993**, *115*, 4602.

[12] W.D. Wulff, S.R. Gilbertson, *J. Am. Chem. Soc.* **1985**, *107*, 503.

[13] W.D. Wulff, B.A. Anderson, A.J. Toole, *J. Am. Chem. Soc.* **1989**, *111*, 5485.

[14] C.H. Heathcock, L.A. Flippin, *J. Am. Chem. Soc.* **1983**, *105*, 1667.

[15] C.P. Casey, T.J. Burkhardt, C.A. Bunnell, J.C. Calabrese, *J. Am. Chem. Soc.* **1977**, *99*, 2117.

[16] A.-M. Lluch, L.Jordi, F.Sanchez-Baeza, S.Richart, F.Camps, A.Messeguer, J.M. Moreto, *Tetrahedron Lett.* **1992**, *33*, 3021.

[17] L.S. Hegedus, G. de Weck, S. D'Andrea, *J. Am. Chem. Soc.* **1988**, *110*, 2122.

[18] C.H. Heathcock, D.E. Uehling, *J. Org. Chem.* **1986**, *51*, 279.

[19] a) R.Hoffmann, *Angew. Chem.* **1982**, *94*, 725; *Angew. Chem. Int. Ed. Engl.* **1982**, *21*, 711. b) F.G.A. Stone, *Angew. Chem.* **1984**, *96*, 85; *Angew. Chem. Int. Ed. Engl.* **1984**, *23*, 89.

[20] a) W.D. Wulff, W.E. Bauta, R.W. Kaesler, P.J. Lankford, R.A. Miller, C.K. Murray, D.C. Yang, *J. Am. Chem. Soc.* **1990**, *112*, 3642; b) W.D. Wulff, T.S. Powers, *J. Org. Chem.* **1993**, *58*, 2381.

[21] K.H. Dötz, R.Noack, K.Harms, G.Müller, *Tetrahedron* **1990**, *46*, 1235.

[22] a) K.H. Dötz, H.G. Erben, K.Harms, *J. Chem. Soc. Chem. Commun.* **1989**, 692; b) C.Alvarez, A.Parlier, H.Rudler, R.Yefsah, J.C. Daran, C.Knobler, *Organometallics* **1989**, 8, 2253.

[23] K.H. Dötz, T.Schäfer, K.Harms, *Synthesis* **1992**, 146.

[24] a) E.P. Kündig, *Pure Appl. Chem.* **1985**, *57*, 1855; b) P.J. Dickens, J.P. Gilday, J.T. Negri, D.A. Widdowson, *Pure Appl. Chem.* **1990**, *62*, 575.

[25] V.Dragisich, C.K. Murray, B.P. Warner, W.D. Wulff, D.C. Yang, *J. Am. Chem. Soc.* **1990**, *112*, 1251.

[26] D.B. Grotjahn, F.E.K. Kroll, T.Schäfer, K.Harms, K.H. Dötz, *Organometallics* **1992**, *11*, 298.

[27] a) L.S. Hegedus, R.Imwinkelried, M.Alarid-Sargent, D.Dvorak, Y.Satoh, *J. Am. Chem. Soc.* **1990**, *112*, 1109; b) L.S. Hegedus, L.M. Schultze, J.Toro, C.Yijun, *Tetrahedron* **1985**, *41*, 5833.

[28] L.S. Hegedus, M.A. Schwindt, S. De Lombaert, R.Imwinkelried, *J. Am. Chem. Soc.* **1990**, *112*, 2264.

[29] L.S. Hegedus, E.Lastra, Y.Narukawa, D.C. Snustad, *J. Am. Chem. Soc.* **1992**, *114*, 2991.

Organolanthanides in Reduction and Nucleophilic Addition Methodology

Karl Heinz Dötz

The organometallic chemistry of the lanthanides has been a sleeping beauty for quite a long period. This might be due, in part, to the term "rare earth metals", which is misleading to some extent. These elements are produced in quantities of about 40 000 t per year (mainly as oxides), and, for instance, the natural abundance of samarium is twice as high as that of boron. Based on the development of efficient protocols for their preparation and on their structural characterization [1] organolanthanides have witnessed an explosive growth from NMR shift reagents to versatile reagents in organic chemistry during the past decade, as documented in recent reviews. [2] Lanthanide cations favor high coordination numbers and are potent oxophilic "hard" Lewis acids; these properties can be exploited in the activation of carbonyl compounds. This report deals with compounds containing divalent samarium and trivalent cerium; these elements are among the less expensive in the lanthanide series.

The impetus of organolanthanides in organic synthesis came from compounds containing the metal in oxidation state II. [3] Due to its electron donor properties (E_o (Sm(II)/Sm(III) = –1.55 V) samarium was particularly attractive. Starting from the element and 1,2-diiodoethane Kagan developed an efficient route to samarium diiodide; [4] this reagent can be stored for months under an inert gas atmosphere and is now commercially available

as a solution in THF. Its reactivity resembles that of a Grignard reagent and it is now well-established as a soluble one-electron donor in Barbier reactions with alkyl halides. [4, 5] The reaction is catalyzed by $FeCl_3$ and is most attractive for ketones whereas with aldehydes subsequent Meerwein–Ponndorf–Verley–Oppenauer oxidation is often observed. Illustrative examples are provided by a short route to frontalin *(1)* [6] and by the annulation of five- (but not six-!) membered rings, which occurs with high stereoselectivity $[(2a):(2b) = 99.5:0.5]$. [7] Alkyl halides can be replaced by acyl chlorides (e.g. α-alkoxy(-acyl)chlorides), which also lead to Barbier-type products *(3)* via decarbonylation [8].

The reduction potential of samarium diiodide has been widely used for the deoxygenation of epoxides, for pinacol formation from aldehydes and for the coupling of acyl chlorides. [5, 9] More recently, this reagent has been shown to induce stereoselective cyclizations. The reduction of carbonyl com-

(1)

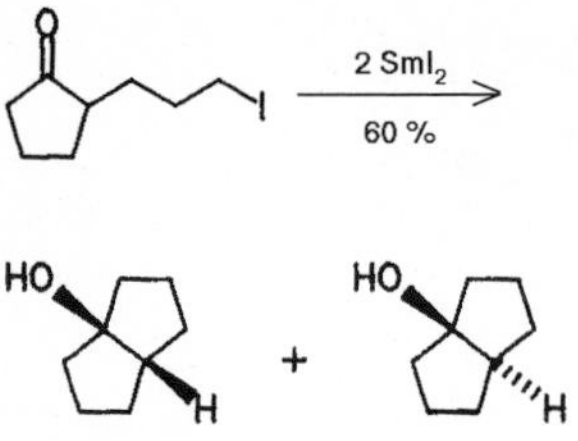

(2a)   +   (2b)

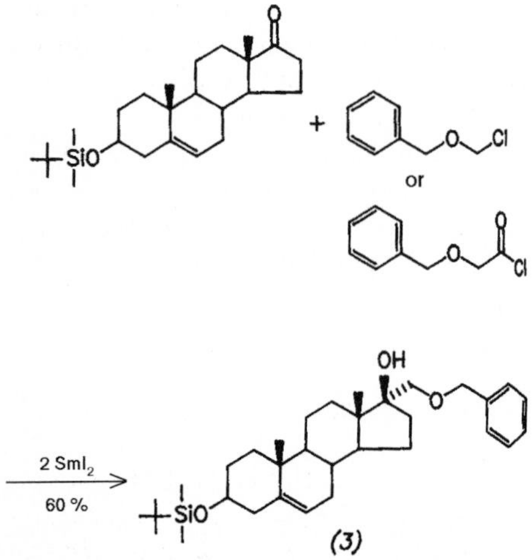

or

$\xrightarrow[60\%]{2\,SmI_2}$

(3)

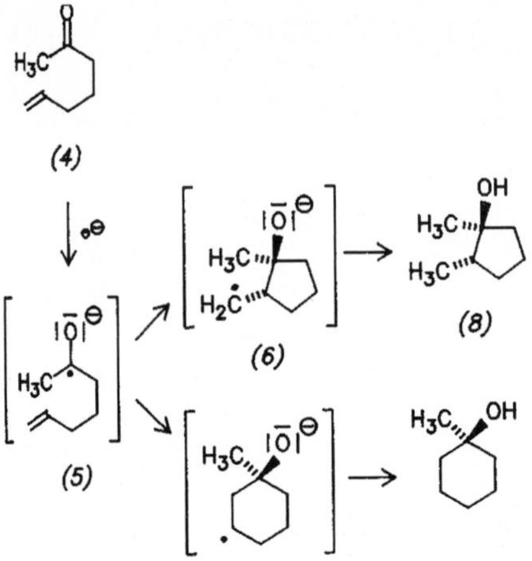

(4)

(5)

(6)

(7)

(8)

pounds by SmI₂ is expected to generate ketyl intermediates, which can be exploited in the construction of up to 3 adjacent stereocenters. Studies of the reduction of 6-hepten-2-ones indicate that 5-heptenyl radicals *(5)* are formed that prefer *exo* cyclization to give the primary methylcyclopentyl radical *(6)* over *endo* ring closure leading to the cyclohexanol skeleton via the secondary radical *(7)*. [10] The reaction is not only regioselective, but also diastereoselective, as demonstrated by the formation of *cis*-1,2-dimethylcyclopentanol *(8)* as the main product.

Besides its reduction potential another interesting aspect of SmI₂ is its coordination properties. Its propensity to chelate formation can be used in control over the relative configuration of a third adjacent stereocenter. [11] An illustrative example is provided by the cyclization of the unsaturated β-ketoesters

(9). The scope of the reaction includes the synthesis of spiro compounds *(10)*; the diastereoselectivity is high to excellent but drops if enolizable substrates *(9b)* are used. Under similar conditions – and with a diastereoselec-

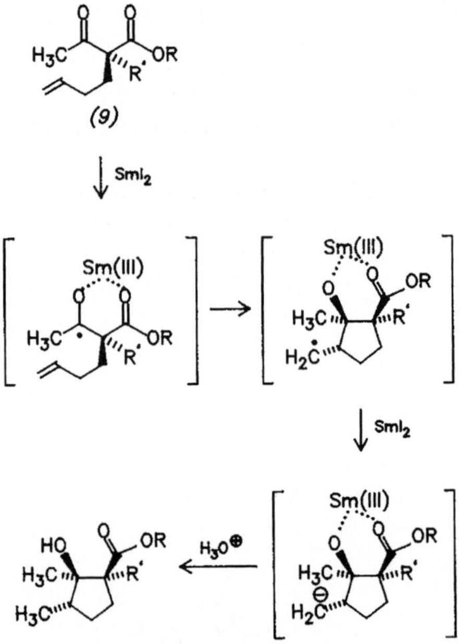

(9)

$\downarrow SmI_2$

$\downarrow SmI_2$

	R	R'	diastereo-selectivity
(a)	Et	Et	200:1
(b)	Me	H	20:1

tivity (220:1) distinctly superior to that reported for customary procedures [12] – the intramolecular pinacol coupling proceeds [(11)→(12)]. Sometimes problems arise from undesired reduction of carbonyl groups to alcohols [e.g. (13)→(14)] as encountered in a total synthesis of (+)-coriolin; [13] this side reaction can be suppressed in the presence of chelating additives such as HMPA, which favor the aldehyde/alkene coupling to give (15).

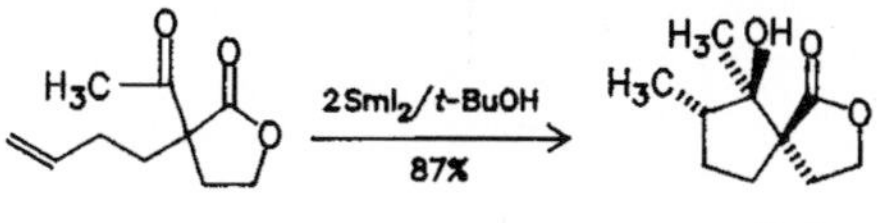

(10)

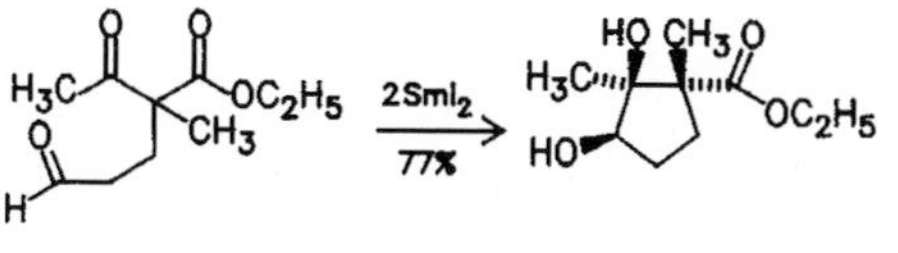

(11)   (12)

The electron donor properties of samarium diiodide facilitate not only radical carbon–carbon bond formation, but have been also used in selective fragmentation reactions, as demonstrated by two examples aiming at steroid and anthracycline synthesis. Cleavage of the xanthate (16) using reductants such as Bu₃SnH/azaisobutyronitrile leads to complex mixtures, whereas SmI₂ affords the 9,10-*seco*-

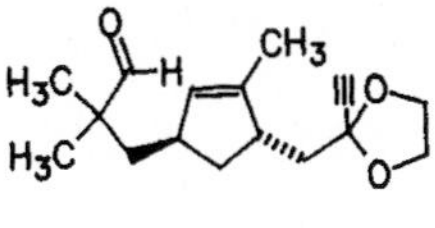

steroid (17) in high yield under mild conditions. [14] For the reduction of α-heterosubstituted ketones this reagent is even the method of choice. [15] The compatibility of SmI₂ with a wide range of substituents is demonstrated by the lactone (18), which serves as precursor for the A-ring (19) in an enantioselective synthesis of (+)-pillaromycinone. [16] The scope of this reagent in reductive processes is explained by its ability to undergo two rapid electron transfer steps and thus to avoid an undesirable high stationary concentration of radical intermediates.

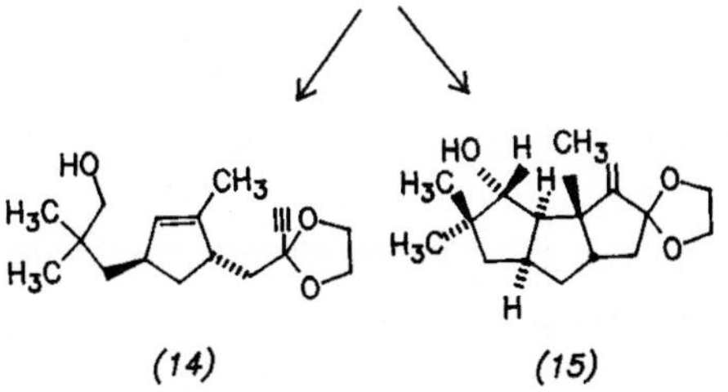

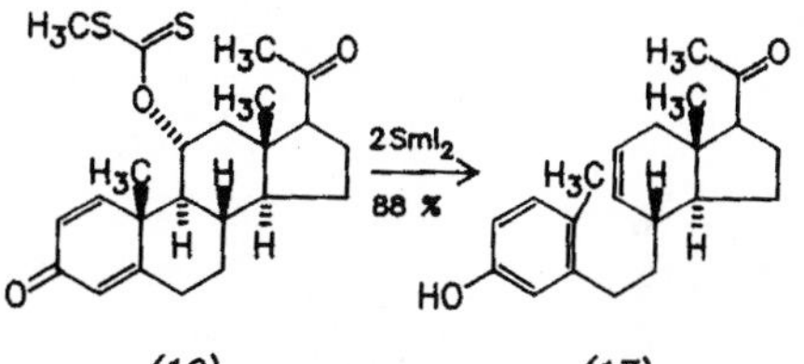

(16)   (17)

Kagan's synthesis of SmI₂ involves a carbenoid intermediate – similar to the Simmons–Smith reagent [17] – which looks promising for olefin cyclopropanation. Studies in which the reagent was generated from samarium activated by HgCl₂ have led to interesting results. [18] The cycloaddition proceeds with good to excellent yield and, in comparison with the customary Simmons–Smith method (which, in

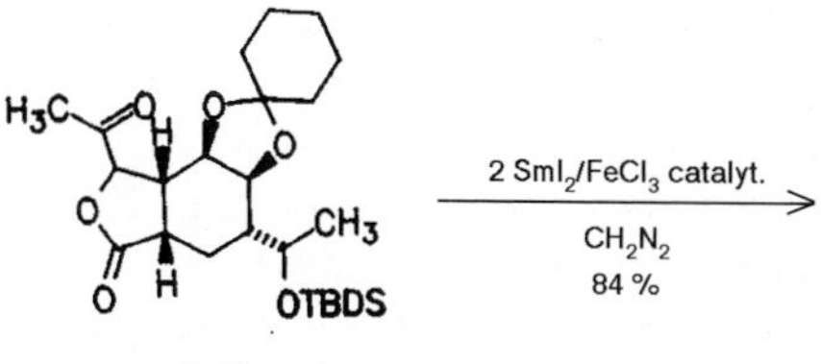

(13)

(18)

(14)   (15)

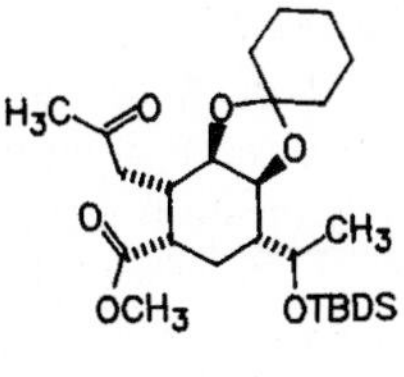

(19)

general, requires harsher conditions), with improved stereoselectivity. The major advantage is that only *allyl* alcohols but not homoallyl alcohols and non-conjugated or functionalized C–C bonds undergo cyclopropanation. In geraniol *(20)* and nerol *(21)* the C6–C7 bond is unaffected. Thus, the SmI_2-method is complementary to the triethylaluminium/diiodomethane route, which results in methylene transfer to the remote C–C bond. [19]

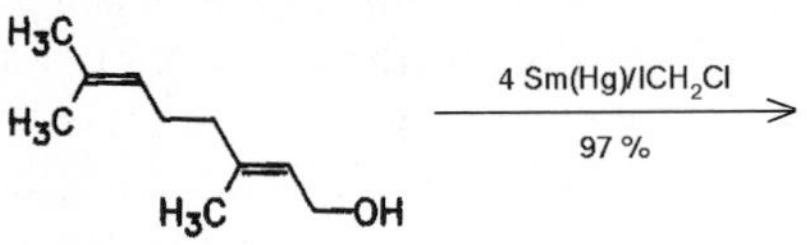

4 Sm(Hg)/ICH₂Cl
97 %

(20)

4 Sm(Hg)/ICH₂Cl
97 %

(21)

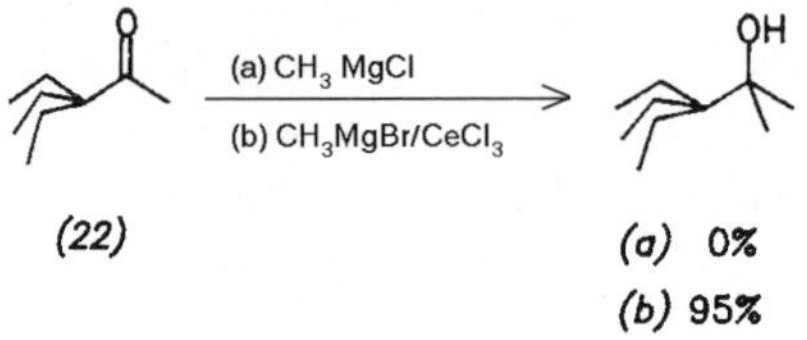

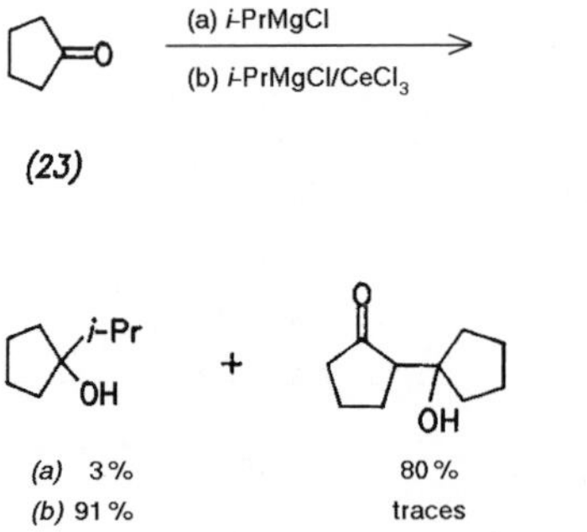

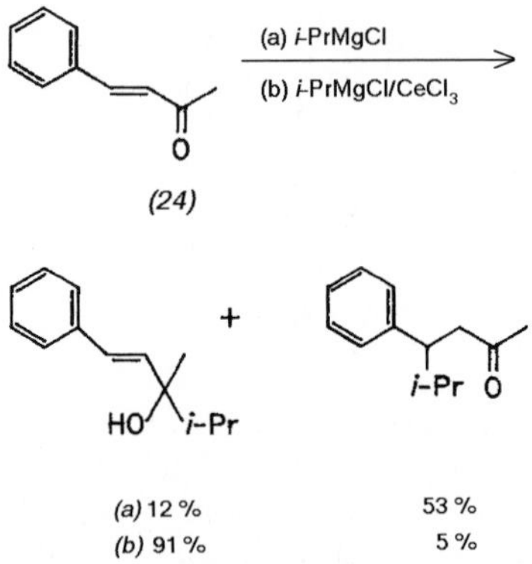

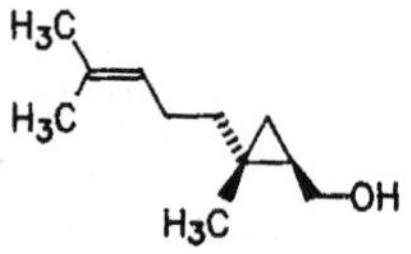

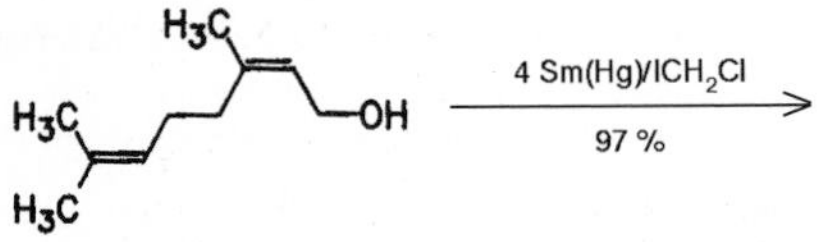

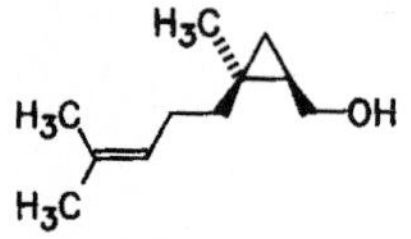

Another major application of lanthanides in organic chemistry involves transmetalation of Grignard and organolithium compounds. Transmetalation often improves chemoselectivity and concomitantly enhances stereocontrol. Among the d-block elements, organotitanium [20] and organocopper reagents [21] are well-established in synthetic methodology. More recently, attention has focused on organocerium(III) compounds, [22] which are generated *in situ* from $CeCl_3$ or CeI_3 at 0 °C in THF. [23] Their structure, both in the solid state and in solution, is still a matter of speculation. Nevertheless, these reagents "$RCeX_2$" are less basic than their organolithium or Grignard precursors and, as a consequence, they add to enolizable substrates, such as *(22)*, whereas deprotonation and subsequent aldol reaction can be suppressed, as demonstrated for cyclopentanone *(23)*. Moreover, with enones *(24)* 1,4-addition is preferred over 1,2-addition. [24]

(a) CH₃ MgCl
(b) CH₃MgBr/CeCl₃
(22)
(a) 0%
(b) 95%

(a) i-PrMgCl
(b) i-PrMgCl/CeCl₃
(23)
i-Pr
OH
+
OH
(a) 3%
(b) 91%
80%
traces

(a) i-PrMgCl
(b) i-PrMgCl/CeCl₃
(24)
+
HO
i-Pr
i-Pr
(a) 12%
(b) 91%
53%
5%

A remarkable observation was made along a reaction sequence aimed at the synthesis of the DEF substructure of the antitumor antibiotic nogalamycin. [25] An attempt to add the metalated arene *(25)* to the chiral ketone *(26)* established that the aryl*lithium* compound *(25a)* produced the desired alcohol *(27a)* while the lithium-derived organo*cerium* reagent *(25b)* led to the unnatural epimer *(27b)*. This result might reflect the different coordination abilities of both metal ions: Whereas in *(25b)* the cerium cation is supposed to prefer chelation by the adjacent benzyloxy group, coordination of the lithium ion by the solvent THF is discussed for *(25a)*. Further studies should aim at the generality of the unexpected finding that metal tuning of a carbanion is a possibility to perform carbonyl addition reactions with opposite stereochemistry.

dride cyclopentenol *(29)* is obtained with a selectivity of 97:3, while in the absence of $CeCl_3$ the saturated alcohol *(30)* is formed exclusively. The reaction proceeds smoothly at ambient temperature, requires neither exclusion of air nor moisture and is compatible with a range of functional groups such as esters in *(31)* and tosylates. [27] Chiral enones are reduced stereoselectively, especially at low temperature: Dihydropyrone *(33)* affords exclusively the *all-cis* allyl alcohol *(34)*. In the hydride reduction a complimentary stereocontrol similar to the carbon–carbon bond formation mentioned above is observed depending on whether $CeCl_3$ is present or not. The saturated ketone *(35)*, in which the phosphine oxide moiety allows chelation of cerium (III), is reduced by $NaBH_4$ to give *threo-(36)* as the main product. In the presence of $CeCl_3$, however, the formation of *erythro-(36)* prevails [28].

The synthetic application of lanthanides is not restricted to stoichiometric reactions. Due to their Lewis acid properties LnX_3-type com-

(26)

(25a) : M = Li
(25b) : M = CeCl$_2$

(27a) : R = OH, R' = Me
(27b) : R = Me, R' = OH

Bn = Benzyl

Another application of organometals containing trivalent lanthanides refers to carbonyl reduction. An established reagent is $NaBH_4$/$CeCl_3$ in methanol (Luche reagent), which selectively modifies enones to allyl alcohols. [26] The role of the lanthanide component is obvious from the reduction of cyclopentenone *(28)*: In combination with sodium borohy-

(28)

(a) NaBH$_4$ / CeCl$_3$
(b) NaBH$_4$

(29)   *(30)*
(a): 97 : 3
(b): 0 : 100

(31)

NaBH$_4$ / CeCl$_3$
91 %

(32)

(33)

NaBH$_4$ / CeCl$_3$
92 %

(34)

Bn = Benzyl

(35)

(36)

	threo	:	*erythro*
(a):	15	:	85
(b):	85	:	15

pounds are used as catalysts in aldehyde-selective acetalization, [29] in Friedel–Crafts [30] and, most impressively, in hetero Diels–Alder reactions. The latter process tolerates a wide range of dienophiles: Highly electrophilic aldehydes as well as unactivated ketones can be employed using Eu(fod)$_3$ or analogous commercially available NMR shift reagents. [31] Somewhat surprisingly, 1,3-dialkoxybutadienes *(37)* provide dihydropyran cycloadducts *(38)* resulting from *endo*-addition of the dienophile even though there are no obvious secondary orbital interactions. A possible explanation is based on steric arguments and arises from a preferred coordination of the lanthanide ion *anti* to the alkyl group of the aldehyde. On these grounds, the *endo*-selectivity might be rationalized in terms of the effective size of the alkyl group versus the catalyst–solvent ensemble. [32]

(37)   *(38)*

In summary, lanthanides exhibit a versatile chemistry and we should be prepared for some more surprises.

References

[1] a) H. Schumann, *Angew. Chem.* **1984**, *96*, 475; *Angew. Chem. Int. Ed. Engl.* **1984**, *23*, 474; b) W. J. Evans, *Polyhedron* **1987**, *6*, 603; c) *Organometallic Chemistry of the f-Elements* (Eds.: T. Marks, R. D. Fischer), Reidel, Dordrecht, **1979**.

[2] a) H. B. Kagan, J. L. Namy, *Tetrahedron* **1986**, *42*, 6573; b) H. B. Kagan, *New J. Chem.* **1990** *14*, 453; c) J. A. Soderquist, *Aldrichimica Acta* **1991**, *24*, 15; d) D. P. Curran, T. L. Fevig, C. P. Jasperse, M. J. Totleben, *Synlett*, **1992**, 943; e) G. A. Molander, *Chem. Rev.* **1992**, *92*, 29.

[3] D. F. Evans, G. V. Fazakerley, R. F. Philips, *J. Chem. Soc. (A)* **1971**, 1931.

[4] J. L. Namy, P. Girard, H. B. Kagan, *Nouv. J. Chim.* **1977**, *1*, 5.

[5] P. Girard, J. L. Namy, H. B. Kagan, *J. Am. Chem. Soc.* **1980**, *102*, 2693.

[6] T. Imamoto, T. Takeyama, M. Yokoyama, *Tetrahedron Lett.* **1984**, *25*, 3225.

[7] G. A. Molander, J. B. Etter, *J. Org. Chem.* **1986**, *51*, 1778.

[8] M. Sasaki, J. Collin, H. B. Kagan, *Tetrahedron Lett.* **1988**, *29*, 4847.

[9] J. L. Namy, J. Souppe, H. B. Kagan, *Tetrahedron Lett.* **1983**, *24*, 765.

[10] B. Giese, *Radicals in Organic Synthesis: Formation of Carbon-Carbon Bonds*, Pergamon Press, New York, **1986**.

[11] G. M. Molander, C. Kenny, *J. Am. Chem. Soc.* **1989**, *111*, 8236.

[12] E. J. Corey, R. Danheiser, S. Chandrasekaran, *J. Org. Chem.* **1976**, *41*, 260.

[13] T. L. Fevig, R. L. Elliott, D. P. Curran, *J. Am. Chem. Soc.* **1988**, *110*, 5064.

[14] T. P. Anathanarayan, T. Gallagher, P. Magnus, *J. Chem. Soc. Chem. Commun.* **1982**, 709.

[15] G. A. Molander, G. Hahn, *J. Org. Chem.* **1986**, *51*, 1135.

[16] J. D. White, E. G. Nolen, Jr., C. H. Miller, *J. Org. Chem.* **1986**, *51*, 1150.

[17] H. E. Simmons, R. D. Smith, *J. Am. Chem. Soc.* **1959**, *81*, 4256.

[18] G. A. Molander, L. S. Harring, *J. Org. Chem.* **1989**, *54*, 3525.

[19] K. Maruoka, Y. Fukutani, and H. Yamamoto, *J. Org. Chem.* **1985**, *50*, 4412.

[20] M. T. Reetz, *Organotitanium Reagents in Organic Synthesis*, Springer, Berlin, **1986.**

[21] G. H. Posner, *An Introduction to Synthesis Using Organocopper Reagents*, Wiley, New York, **1980.**

[22] T. Imamoto, *Pure Appl. Chem.* **1990**, *62*, 747.

[23] T. Imamoto, T. Kusumoto, Y. Tawarayama, Y. Sugiura, T. Mita, Y. Hatanaka, M. Yokoyama, *J. Org. Chem.* **1984**, *49*, 3904.

[24] T. Imamoto, N. Takiyama, K. Nakamura, T. Hatajima, Y. Kamiya, *J. Am. Chem. Soc.*, **1989**, *111*, 4392.

[25] M. Kawasaki, F. Matsuda, S. Terashima, *Tetrahedron Lett.* **1985**, *26*, 2693.

[26] J.-L. Luche, *J. Am. Chem. Soc.* **1978**, *100*, 2226.

[27] R. A. Raphael, S. J. Telfer, *Tetrahedron Lett.* **1985**, *26*, 489.

[28] J. Elliott, S. Warren, *Tetrahedron Lett.* **1986**, 27, 645.

[29] R. Baudouy, P. Prince, *Tetrahedron* **1989**, *45*, 2067.

[30] N. Mine, Y. Fujiwara, H. Taniguchi, *Chem. Lett.* **1986**, 357.

[31] M. M. Midland, R. S. Graham, *J. Am. Chem. Soc.* **1984**, *106*, 4294.

[32] M. Bednarski, S. Danishefsky, *J. Am. Chem. Soc.* **1983**, *105*, 3716.

Carbon–Carbon Bond Formation with Group Four Metallocenes

Martin Maier

In many areas of synthetic organic chemistry, methodological progress and new reactions are based on the use of transition metals. Transition metals have become almost indispensable, particularly for the formation of C–C and C–H bonds. [1] Because they easily form complexes with unsaturated fragments or undergo insertion reactions, transition metals are able to make molecules in their coordination sphere react with each other which would normally not be able to do so. Moreover, transition-metal induced reactions can proceed with high selectivity. Finally, with the use of multidentate and chiral ligands, it is possible to prepare catalysts that allow one to perform enantioselective reactions. [2, 3] Besides being able to activate molecules towards certain reactions, transition metals are also able to stabilize highly reactive species, thereby making them available for selective transformations. In this connection, the chemistry of group four metallocenes has been investigated very intensively, whereby some interesting results have come to light that are the subject of this review.

Insertion of Unsaturated Species into Zirconium–Carbon Bonds

What might be the reason for the renaissance of the group four metallocenes? [4] Primarily, this is due to the fact that very simple and practical methods could be found to generate species of the type "Cp$_2$Zr" or their equivalents. This fragment represents a 14-electron compound with a d^2-configuration. It possesses a free electron pair and, in addition, two unoccupied valence orbitals. With regard to reactivity, this zirconium species might best be compared with a singlet carbene (e.g. Cl$_2$C:); although the zirconium moiety contains one more empty orbital. With the combination of empty orbitals and an occupied orbital, it is understandable that "Cp$_2$Zr" reacts readily with many types of unsaturated compounds with concomitant formation of the corresponding metallacycles. For these complexes, one can formulate two mesomeric resonance forms *(1a)* and *(1b)*, whereby a large amount

(1a)   *(1b)*

cf.: C=C   134 pm
C≡C   120 pm

128.6 pm

(2)

of π-back bonding from the metal to the organic fragment causes the resulting complexes to behave chemically like metallacyclopropanes (= alkene complexes) or metallacyclopropenes (= alkyne complexes). This is also expressed in the bond lengths found for the alkyne complex (2). [5] Thus, the bond length of 128.6 pm for the original triple bond shows that in the complexed form this bond has a high degree of double bond character. A preparatively very important reaction of the complexes between "Cp₂Zr" and unsaturated fragments is the insertion of polar (carbonyl compounds, nitriles) and even unpopular, unsaturated molecules (alkenes, alkynes) into the Zr–C bond with the formation of five-membered metallacycles, which in turn may be functionalized in many ways (reaction with CO, RNC, H⁺, I₂ etc.). Originally, zirconocene–alkene and alkyne complexes were pre-

pared by reduction of zirconocene dichloride with metals (Mg/HgCl₂; Na/Hg) in the presence of an alkene or alkyne, although a significant amount of by products could be formed. A new method by Buchwald et al. relies on the concerted elimination of methane from monoorgano-methyl-zirconocene by β-elimination. [4e] It was also discovered by Buchwald that such complexes can be stabilized with trimethyl phosphine to give 18-electron complexes. For example, he used this method for the regiospecific functionalization of naphthalenes [6] (Scheme 1). In this case, naphtyl bromides such as (3) served as educts, which, after metalation (cf. (4)) react with zirconocene methyl chloride under substitution of chloride to give compounds of type (5). On warming aryne complexes (6) are formed as intermediates, which react with nitriles to give the azametallacycles (7). From these, the acyl-

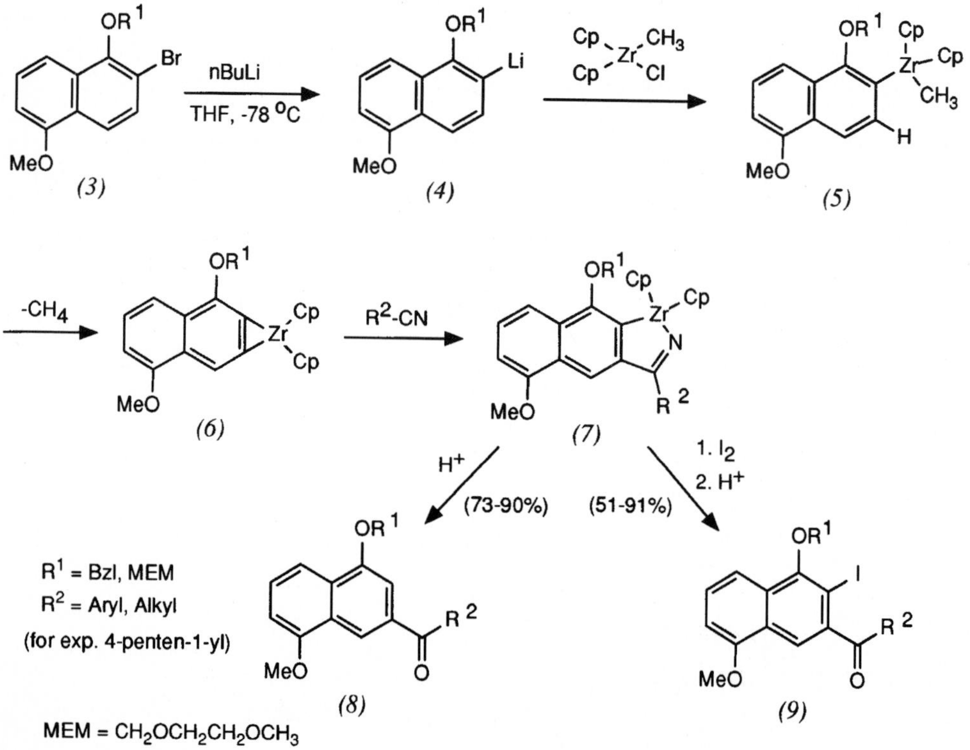

Scheme 1. Insertion of nitriles into aryne-zirconium complexes.

naphthalenes *(8)* can be generated by acid hydrolysis. If the azametallacycles *(7)* are treated with iodine prior to the hydrolysis, iodoacylnaphthalenes *(9)* are formed. As can be seen, the acyl substituent is in the *meta*-position to the OR^1-group. That means that by using this method an *anti*-Friedel–Crafts acylation of aromatic compounds is realized. It should be noted that fundamental studies of aryne–metal complexes were carried out by Erker et al., who succeeded in the first preparation of an aryne-zirconocene complex by elimination of benzene from diphenylzirconocene. [7]

Clearly, the simplest procedure for the generation of the fragment "Cp_2Zr" consists of the reaction of zirconocene dichloride with alkylmetal compounds (Scheme 2). Thus, first the corresponding dialkylzirconocenes *(10)* are formed, which are converted to alkene-zirconocene complexes *by* β-hydrogen abstraction and elimination of one molecule of alkane. [8] For example, with *n*-butyllithium, the 1-butene-zirconocene complex is obtained whereas treatment of zirconocene dichloride with ethylmagnesium bromide or ethyllithium produces the ethene complex. On reaction of these alkene complexes with other olefins, five-membered metallacycles are formed by insertion (e.g. *(12)* and *(16)*), which, for example, can be cleaved by hydrolysis. For steric reasons, the alkyl residues are directed

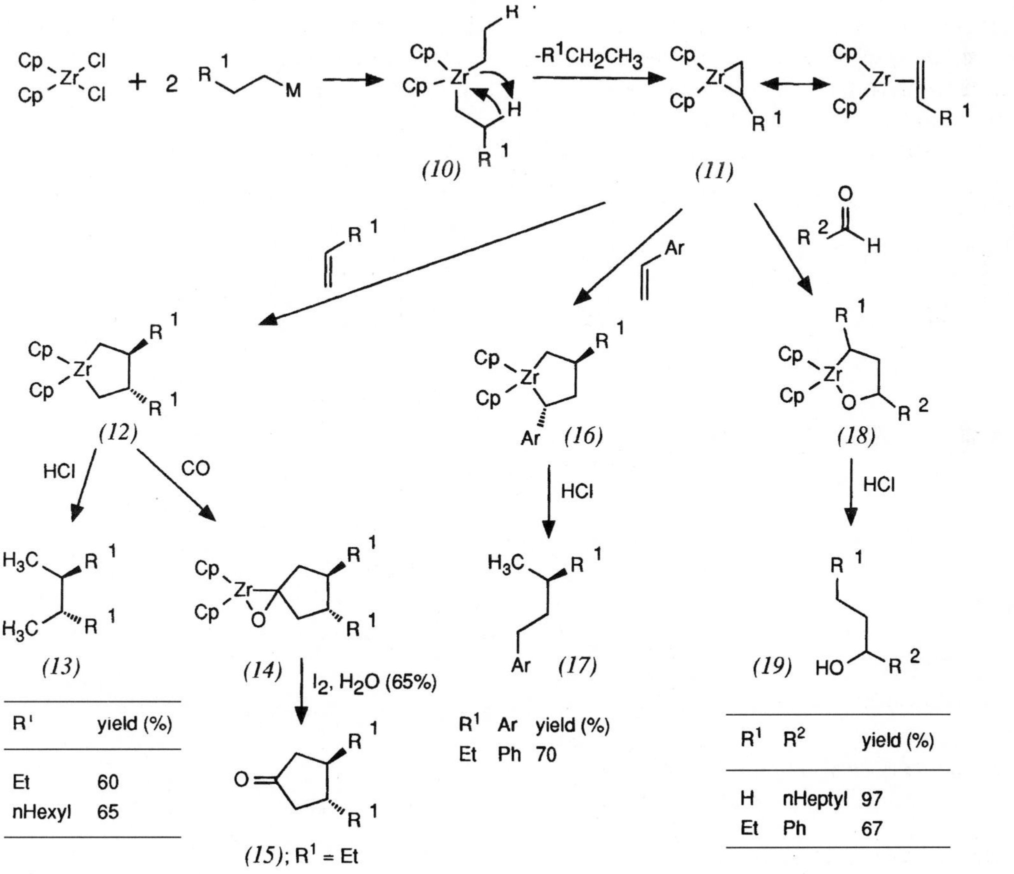

R¹	yield (%)
Et	60
nHexyl	65

R¹	Ar	yield (%)
Et	Ph	70

R¹	R²	yield (%)
H	nHeptyl	97
Et	Ph	67

Scheme 2. Zirconocene-induced coupling of alkenes with alkenes and aldehydes.

to the 3- and 4-positions, providing almost exclusively the *trans*-isomer *(12)*. [9] Some problems arise if one attempts to couple two different olefins. For example, treatment of Cp$_2$Zr(nBu)$_2$ with one equivalent of 1-octene gives a statistical mixture of homo- and heterocoupling products. If styrene derivatives are used for the second step, then the insertion takes place so that the aryl group occupies the α-position (cf. *(16)*). In this manner additional stabilization of the partial negative charge next to the zirconium is secured. In both cases these reductive coupling reactions are practically regiospecific. Instead of the second olefin, one can also use aldehydes, whereby hydrolysis of the metallacycle *(18)* furnishes alcohols of type *(19)*. [10] In contrast to the coupling of olefins, insertion of the aldehyde takes place into the less substituted Zr–C bond. If one wants to complex alkenes of which a polar organometal compound is not easily available, a variation is offered that

relies on an exchange reaction with complexed isobutene. [11] Similar studies with mixed dialkyl compounds of zirconocene revealed the relative reactivity of various alkylgroups as β-hydrogen donors. [12] In general, this reactivity correlates with the degree of substitution at the β-center, whereby the following order is valid: β-methyl > β-methylene > β-methine. The Negishi procedure is not only applicable to the preparation of alkene complexes, but can also be exploited in an elegant manner – at least in a formal sense – for the synthesis of intermediate zirconocene–alkyne complexes. Insertion of alkynes into the ethylene zirconocene complex leads first to zirconacyclopentenes *(20)*. From these, in turn, the ethylene can be displaced by other unsaturated fragments; a

(1a)   *(20)*

R^1	R^2	yield (%)
Et	Et	87
Ph	Me	84
Me$_3$Si	H	40

(20)

(21)
R^1 = R^2 = nPr
R^3 = R^4 = Et (81%)

(22)
R^1 = R^2 = nPr
R^3 = Me (65%)

(23)
R^1 = R^2 = nPr
R^3 = Ph (65%)
R^3 = nHex (52%)

(20)

(21)

(22)

(23)

(24)

R	yield (%)
H	80
Et	85
Ph	85
cC$_5$H$_{11}$	75

result that also indicates a relative weak C3–C4 bond. Mechanistically, alkyne complexes, formed by elimination of ethylene, are most likely passed through. Thus, reductive coupling of alkynes with aldehydes, nitriles, or other alkynes can be performed by application of this method (cf. *(21)–(23)*). [13]

Cyclizations with Group Four Metallocene Catalysts

Although zirconium compounds are toxicologically relatively unobjectionable, zirconocene-induced C–C bond formations that take place with catalytic amounts of zirconium would be even more attractive. That this is

indeed the case was also demonstrated by Negishi et al. [14] Coupling of the Grignard reagent with the olefin is induced by treatment of catalytic amounts of Cp_2ZrCl_2 with two equivalents of ethyl magnesium bromide and an olefin (in this case 1-decene). The reaction proceeds through the ethene complex *(1a)*, into which the olefin inserts. An additional advantage is that the reaction proceeds with pair selectivity, that is to say no homo-coupling products are formed. The Grignard reagent then cleaves the metallacycle *(24)* regiospecifically via an addition-elimination mechanism to give compound *(25)* from which the reactive complex *(1a)* is regenerated by reductive elimination.

Using secondary allylic alcohols as coupling partners the zirconocene-catalyzed addition of ehtylmagnesium bromide occurs with remarkable stereoselectivity. [15] While the use of protected allylic alcohols furnishes the *anti*-products, on the other hand the free allylic alcohols yield preferentially the *syn*-isomers. The Grignard reagents *(30)* formed in this manner can be functionalized in numerous

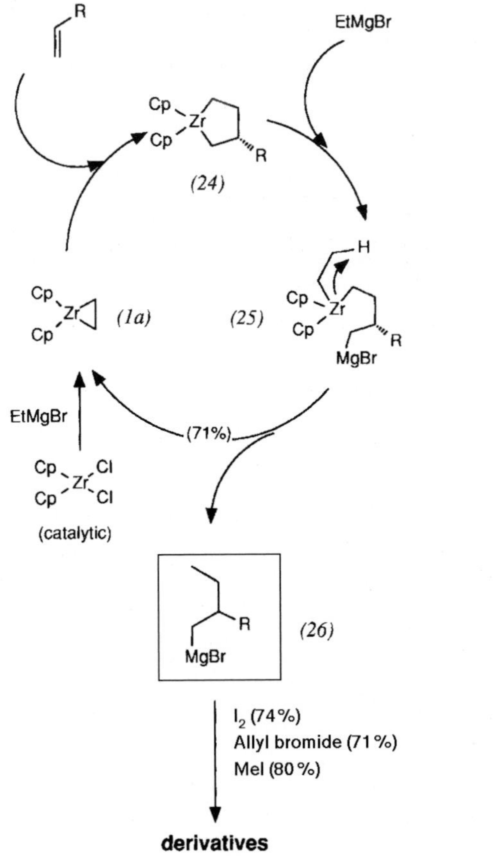

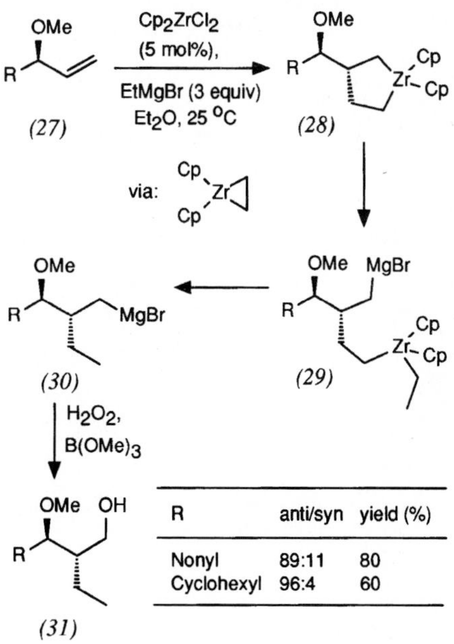

R	anti/syn	yield (%)
Nonyl	89:11	80
Cyclohexyl	96:4	60

(31)

derivatives

ways; one example is the oxidative formation of 1,3-diols *(31)*. Because of the above mentioned reaction mechanism, this reaction is not, unfortunately, suitable for the direct transfer of methyl groups.

With a view to the synthesis of cyclic or polycyclic natural products, intramolecular variants of these reductive coupling reactions are of particular interest. Indeed, zirconocene-induced cyclization of substrates that contain two unsaturated groups have proven to be very powerful. Under the influence of the species "Cp₂Zr", the cyclization of diynes, [16] eneynes, [17] and dienes [18] to zirconabicycles can be performed. Normally, these are not isolated, rather transformed into subse-

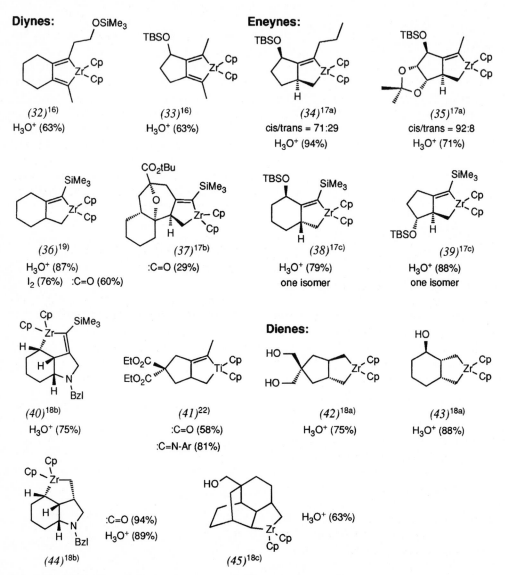

Diynes:

(32)[16]
H₃O⁺ (63%)

(33)[16]
H₃O⁺ (63%)

Eneynes:

(34)[17a]
cis/trans = 71:29
H₃O⁺ (94%)

(35)[17a]
cis/trans = 92:8
H₃O⁺ (71%)

(36)[19]
H₃O⁺ (87%)
I₂ (76%) :C=O (60%)

(37)[17b]
:C=O (29%)

(38)[17c]
H₃O⁺ (79%)
one isomer

(39)[17c]
H₃O⁺ (88%)
one isomer

(40)[18b]
H₃O⁺ (75%)

(41)[22]
:C=O (58%)
:C=N-Ar (81%)

Dienes:

(42)[18a]
H₃O⁺ (75%)

(43)[18a]
H₃O⁺ (88%)

(44)[18b]
:C=O (94%)
H₃O⁺ (89%)

(45)[18c]
H₃O⁺ (63%)

Scheme 3. Cyclization of diynes, eneynes, and dienes to give metallacycles.

quent products. Thus, the treatment of zirconacycles with the vinylcarbene analogues carbon monoxide and isonitriles provides five-membered ketones or imines, respectively. Work-up with acid or iodine gives products in which the Zr–C bonds are replaced by C–H or C–I bonds. Mechanistic studies suggest that in the case of eneynes a zirconacyclopropene-intermediate is formed first, in which the alkene is inserted in a second step. Scheme 3 depicts some representative zirconabicycles, that were generated by this method. The methods for the work-up and the corresponding yields are also given in this Scheme. In the cyclization of 1,6-dienes the two annulated rings usually turn out to be *trans*-connected, whereas from 1,7-dienes *cis*-annulated rings are formed preferentially (cf. *(42)* and *(43)*). With chiral substrates such reductive cyclizations proceed with high asymmetric induction. It should be noted that such cyclizations can be performed with catalytic amounts of "Cp₂Zr" if one uses stochiometric amounts of the Grignard-reagent. [19] However, the method does have its limitations. Thus, cyclizations to small rings (four-membered rings) and larger rings (seven-membered rings) fail in most cases. The cyclization also does not work with substrates that contain allylic ethers. In this case cleavage of the allylic ether with the formation of an allylic zirconium species is observed. Nevertheless, this represents a new method for the cleavage of allylic protecting groups. [20] Moreover, it was shown by Taguchi et al. that for allyl ethers in which the oxygen atom is part of a ring and simultaneously part of an acetal a zirconium-mediated ring contraction can be performed to yield carbocyclic compounds. [21] A case in point is the synthesis of an intermediate for carbocyclic oxetanocin. [22] Thus, treatment of *(46)* with Cp₂Zr followed by BF₃·Et₂O provided the chiral cyclobutanone *(48)*. Although the Negishi method for the generation of "Cp₂Zr", that is, Cp₂ZrCl₂ + *n*BuLi, is operationally very sim-

(46)

(47)   *(48)*

ple, the disadvantage is the lack of tolerance for functional groups. In a variant Buchwald et al. use instead the corresponding titanium reagent. (Cp₂TiCl₂ + 2 EtMgBr), which is characterized by lower oxophilicity and therefore does not, for example, attack ester functions (cf. *(41)*). [23] In the presence of isonitriles even catalytic amounts of titanocene are sufficient for the cyclization. [24]

Reactions of Hetero Olefin Complexes

The group four metallocenes not only complex molecules with CC-multiple bonds, but they are also able to bind hetero olefins. As an additional method for the synthesis of complexes of a zirconocene unit and a carbonyl

(49)   *(50)*

(52)   *(51)*

group, insertion of carbon monoxide in a Zr–C bond followed by migration of a second residue from the zirconium to the acyl-carbon can be used (cf. *(49)→(51)*). [25] As an aside, reaction of the above-mentioned zirconacycles with carbon monoxide and isonitriles also takes place according to this scheme. If carbonyl complexes with a α-H atoms are intermediates, enolates may be formed; a fact that restricts the use of carbonyl – and imine complexes for reductive coupling with unsaturated fragments. In addition, metallocene–heteroolefin complexes can also be prepared by direct synthesis from Cp$_2$Zr- and Cp$_2$Ti-equivalents (e.g. Cp$_2$Zr(CO)$_2$, Cp$_2$Ti(PMe$_3$)$_2$) and hetero olefins. [26] If such complexes are not stabilized by a further ligand, they usually associate to give dimers. For the preparation of heteroolefin complexes, the Buchwald procedure in particular has broad application. This method relies on the elimination of methane in the presence of trimethyl phosphine. By this route a large number of zirconocene thioaldehydes could be generated. [27] Thus, if dimethyl zirconocene is treated with one equivalent of a thiol, (alkylthio)methylzirconocenecomplexes are formed. At a temperature of around 80 °C the second molecule of methane is lost with the formation of the desired complexes, which can be stabilized with trimethyl phosphine. According to the same principle, zirconocene–imine complexes can be generated. However, in this case one starts from zirconocene methyl chloride, which is treated with *N*-silylated lithium amides. Subsequent elimination of methane is facilitated enormously by the *N*-silyl group and succeeds even at –10 °C. Imine complexes such as *(54)* react with a large number of unsaturated molecules with concomitant insertion in the Zr–C bond. Thus, with alkynes allyl amines are obtained after insertion and hydrolysis. [28]

By reaction of "Cp$_2$Zr" with the chiral imine *(57)*, Taguchi et al. generated the metallacycle *(58)*, which reacts with aldehydes ste-

(53)

THF,
-10 °C, - CH$_4$

R^{2}CHO

(55)

R^1	R^2	yield (%)
Ph	tBu	91

(54)

R^2≡≡R^3

R^1	R^2	R^3	yield (%)
H	Ph	Ph	75

(56)

reoselectively to give amino alcohols. [29] The major products are always the *threo*-compounds *(60)* and *(61)* (*threo*/*erythro* 2:1–10:1). The ratio of *(60)* to *(61)* is quite temperature dependent. At 0 °C and at room temperature mostly *(60)* is formed, whereas at higher temperature the ratio is reversed in favor of *(61)*. As this example shows, the Negishi procedure for the generation of "Cp$_2$Zr" is applicable to the preparation of imine complexes without undesired side-reactions, although the possible substrates are limited to imine derivatives of aromatic aldehydes.

If, however, aldehydes with α-H atoms are converted to the corresponding hydrazones, then it is possible to form complexes of them with zirconocene according to the Negishi method. As Livinghouse et al. reported, unsaturated hydrazones cyclize in high yield to give the metallacycles *(63)*. [30] These in turn

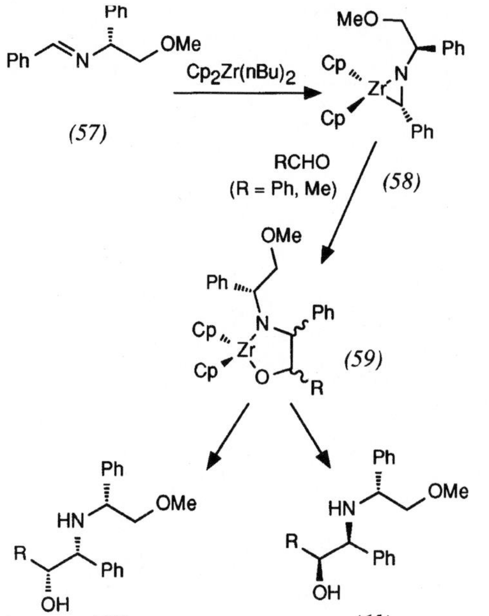

(57) → *(58)* → *(59)*

(60) R = Ph, Me

(61) e.g. R = Ph:

temp. (°C)	yield (%)	**60/61**
0	75	94:6
64	84	5:95

can be derivatized by hydrolysis and acetalization. If olefinic hydrazones are used, *cis*-cycloalkylhydrazides are formed with high selectivity. It should also be mentioned that the cyclization of unsaturated ketones is possible under the action of the "Cp₂Ti"-equivalent Cp₂Ti(PMe₃)₂. [26]

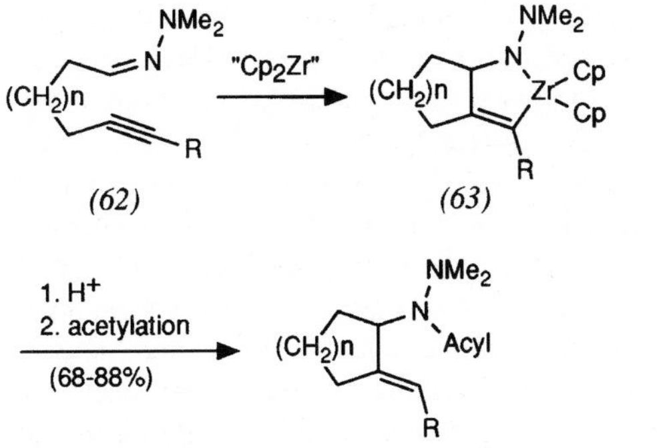

A completely different concept of using transition metals for C–C bond formation is to use carbene complexes that react with unsaturated fragments (olefins, alkynes, dienes, etc.) in the sense of pericyclic reactions. Accordingly, with complexes of metal–heteroatom double bonds one should be able to form carbon–heteroatom bonds in an analogous manner. In this regard, intramolecular variants have proven particularly elegant and effective. Livinghouse et al. describe the application of intramolecular [2+2]cycloadditions of metal–imido complexes with alkynes for the synthesis of nitrogen heterocycles. [31] If one treats alkynyl amines, for example *(65)* with CpTiCl₃ the titanium–imido complex *(66)* is formed first, which then undergoes a [2+2]cycloaddition to give *(67)*. Protonation cleaves the metallacycle to the enamine, which tautomerizes to the pyrroline *(68)*. Because formation of the titanium-imido complex is accompanied by the generation of HCl, catalytic amounts of CpTiCl₃ (0.3 equiv.) are sufficient. If equimolar quantities of CpTi(CH₃)₂Cl or CpZr(CH₃)₂Cl are used – whereby methane is eliminated – the metallacycles are not hydrolyzed and can be functionalized in various ways. It should be noted that Cp₂Zr=NR complexes practically do not undergo cycloaddition.

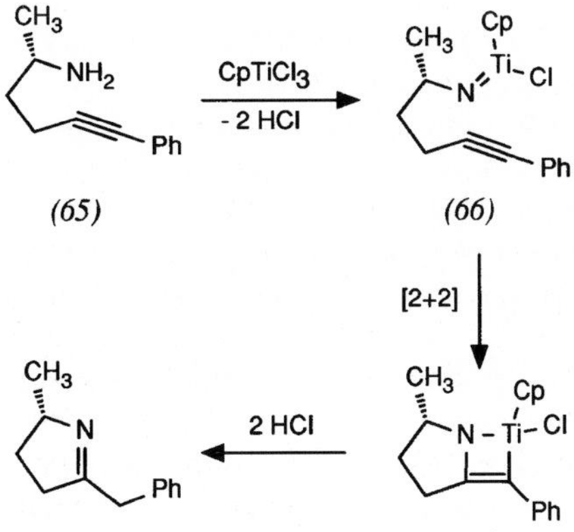

(62) "Cp₂Zr" → *(63)*

1. H⁺
2. acetylation
(68-88%)

(64)

R = Ph, nBu, SiMe₃

(65) CpTiCl₃ – 2 HCl → *(66)*

[2+2]

(67) 2 HCl → *(68)*

Hydrozirconation Reactions

Interesting developments are also seen for one of the most useful organometallic reactions: the hydrozirconation of olefins and alkynes. This reaction was discovered a couple of years ago by Schwartz et al., whereby Cp$_2$Zr(H)Cl serves as the reagent. [32] Another reaction of similar significance is the carboalumination reaction of alkynes catalyzed by zirconocene dichloride. [4d] Recently, some improved procedures for the preparation of Cp$_2$Zr(H)Cl became known. In addition, new valuable transformations of vinyl- and alkylzirconocenes were found, for example the above-mentioned synthesis of alkyne complexes according to Buchwald, the preparation of butenolides from propargyl alcohols, [33] and the hydrocyanation of olefins. [34] Moreover, by the hydrozirconation of highly functionalized, in part even metallated, substrates, it became possible to open up new applications for this reaction. For the preparation of Cp$_2$(H)Cl according to Buchwald, Cp$_2$ZrCl$_2$ is reduced with LiAlH$_4$, which produces a mixture of Cp$_2$Zr(H)Cl and over-reduced Cp$_2$ZrH$_2$. On treatment with dichloromethane the latter is then converted to Cp$_2$Zr(H)Cl, thus enabling a very efficient and cheap synthesis. [35] In some cases it is possible to use Cp$_2$Zr(H)Cl which has been generated in situ. This can be performed by reduction of Cp$_2$ZrCl$_2$ with LiEt$_3$BH, a procedure which works well for the hydrozirconation of many alkynes. [36]

As serveral groups have shown, it is possible to activate vinyl- and alkylzirconocenes with other metals (transmetalation) thereby making it possible to adjust the reactivity to the other reaction partner and the reaction. Thus, one succeeds in the conjugate addition of vinylzirconocenes to enones under the catalytic influence of [Ni(acac)$_2$]. [37] If the chlorine atom in compounds of type *(69)* is exchanged for methyl, a subsequent facile conversion to organocuprates *(71)* is possible.

(69)

(70)   *(71)*

(72)

(73); 92%

These in turn may be added effectively in a 1,4-mode to enones. The synthesis of the prostaglandin derivative *(73)* by this strategy may serve as an example. [38] In a similar manner alkyl groups can be transferred to enones by hydrozirconation of alkenes and transmetalation with catalytic amounts of Cu(*I*) salts. [39] If acid chlorides are used instead of enones this represents a flexible synthesis of ketones. [40] Clean transmetalation of vinylzirconocenes to vinylboranes is possible by reaction with haloboranes. [41] The advantage as compared with the direct hydroboration lies in higher regioselectivity.

The chloride in alkenyl- and alkylzirconocene chlorides can be replaced not only by nucleophiles, but it may also be removed with appropriate Lewis-acids, whereby cationic zirconocene species are generated that themselves possess Lewis-acid character and are therefore able to increase the electrophilicity of carbonyl groups. This fact was exploited by

(69)

(74)

$$Cp_2Zr^+\!-\!X \quad (75)$$

X = Cl or

examples:

(76); 90%

(77); 94%

(diastereomeric mixture 1:1)

(78)

(79)

(80)

examples:

R¹	R²	R³	R⁴	yield (%)
Hept	H	cHex	H	83
CH₂Ph	H	Pent	H	89
(CH₂)₄CN	H	cHex	H	55
-(CH₂)₄-		cHex	H	76

Suzuki et al. for the transfer of organic substituents to aldehydes. Whereas compound (69) reacts only slowly with aldehydes without added Lewis-acid, the corresponding additions are complete within minutes in the presence of catalytic amounts of AgClO₄. [42] It is assumed that with AgClO₄ zirconocene cations of type (75) are formed that are able to form a complex with the carbonyl group and thereby facilitate transfer of the alkenyl group from the zirconium to the aldehyde. Moreover, such cationic zirconium species are also important for glycosylation reactions. [43]

Particular interesting reagents are formed by the hydrozirconation of functionalized alkenes and alkynes. Thus Knochel et al. found that by hydrozirconation of alkenylzinc halides in dichloromethane 1,1-bimetallic reagents of type (79) are formed regiospecifically. [44] Although these reagents are not all to stable, they can be trapped with carbonyl groups to give olefins (80) in good yields. Aldehydes react selectively to (E)-disubstituted olefins. In an analogous manner the hydrozirconation of alkynylzinc halides yields 1,1-bimetallic alkenes which react with aldehydes to give allenes.

For the stereoselective preparation of synthetically useful (Z)-vinyltin compounds, Lip-

shutz et al. made use of the hydrozirconation of alkynyltin compounds. Initially, the 1,1-bimetallic alkenes are formed, which then are converted to the desired compounds by protonolysis. [45]

If the allylstannane (81) is treated with Cp₂Zr(H)Cl, the 1,1-bimetallic reagent is not formed but rather the 1,3-bimetallic reagent (82), which represents at the same time an allylzirconocene derivative. This reacts with carbonyl groups via the six-membered transition state (83) to give the intermediate (84). With Brönsted- or Lewis-acids spontaneous β-elimination of the tin and oxygen-functions takes place and (E)-1,3-dienes (85) are formed with high selectivity. [46]

Wittig-Analogous Reactions

The prototype example of a 1,1-bimetallic compound that contains a group four metallocene unit is the Tebbe reagent (86), which is

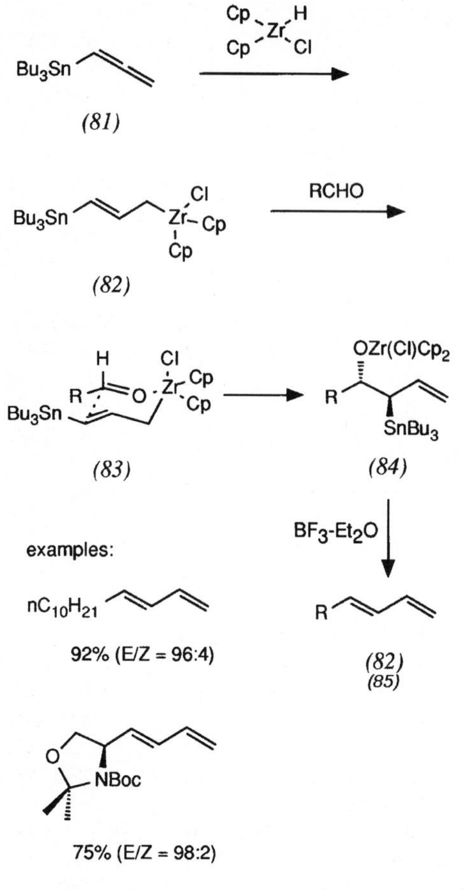

(81)

(82)

(83)

(84)

examples:

92% (E/Z = 96:4)

(82)
(85)

75% (E/Z = 98:2)

(86)

(87)   X = H, Alkyl, Aryl
                                          OR

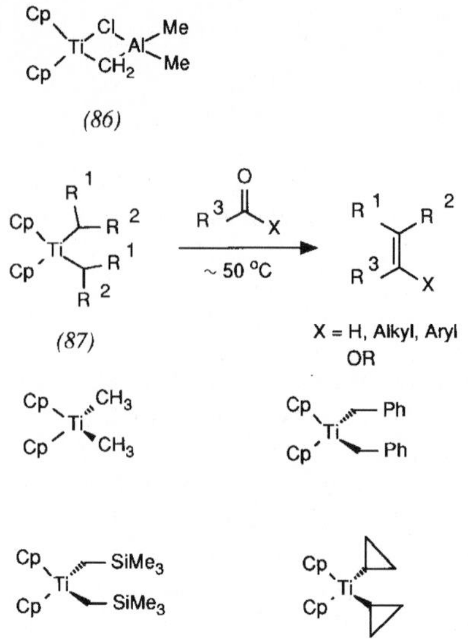

follows from the fact that compound *(87)* reacts with alkynes to form titanacyclobutenes. [48 d] These in turn can react with ketones and aldehydes to form six-membered metallacycles, whereby insertion is observed in the Ti–alkyl as well as the Ti–vinyl bond. [49]

formed from Cp_2TiCl_2 and two equivalents of $Al(CH_3)_3$ with the formation of methane and $Al(CH_3)_2Cl$. This complex is able to olefinate carbonyl groups in a Wittig-type reaction, whereby the advantage lies in the fact that even ester groups react with the formation of enol ethers. [47] Such olefination reactions are also possible with dialkyl titanocenes *(87)* although in this case a higher temperature is necessary. Despite that, this reaction represents an effective method for transferring methylene groups and for converting carbonyl compounds to styrene derivatives, vinyl silanes, or alkylidene cyclopropanes. [48] As reactive species are titanocene alkylidene complexes most probably intermediates. This

Outlook

As this summary demonstrates, group four metallocenes are versatile reagents for performing C–C bond formation. [50] Particularly with a view to asymmetric and enantioselective reactions these metallocenes conceal an enormous potential. Finally, it should be noted that the chemistry of the group four metallocenes carries many other facets, such as the stabilization of planar tetracoordinate carbon centers [51] or the use as catalysts for polymerization reactions of olefins. [52] Due to space limitations, these and many other interesting topics could not be considered.

References

[1] D. Seebach *Angew. Chem.* **1990**, *102*, 1363–1409; Angew. Chem. Int. Ed. Engl. **1990**, *29*, 1320.

[2] a) S.L. Blystone, *Chem. Rev.* **1989**, *89*, 1663–1679; b) *Chem. Rev.* **1992**, 5, issue number 5.

[3] Chiral group four metallocenes are also available from cyclopentadienyl ligands that are chiral in the E2-space: a) A. Schäfer, E. Karl, L. Zsolnai, G. Huttner, H.-H. Brintzinger, *J. Organomet. Chem.* **1987**, *328*, 87; b) S. Collins, B. A. Kuntz, Y. Hong, *J. Org. Chem.* **1989**, *54*, 4154–4158; c) R. B. Grossman, R. A. Doyle, S. L. Buchwald, *Organometallics* **1991**, *10*, 1501–1505.

[4] For reviews, see: a) G. Erker, *Acc. Chem. Res.* **1984**, *17*, 103–109; b) H. Yasuda, K. Tatsumi, A. Nakamura, *Acc. Chem. Res.* **1985**, *18*, 120–126; c) G. Erker, *Angew. Chem.* **1989**, *101*, 411–426; Angew. Chem. Int. Ed. Engl. **1989**, *29*, 397 d) E. Negishi, T. Takahashi, *Synthesis* **1988**, 1–19; e) S. L. Buchwald, R. B. Nielsen, *Chem. Rev.* **1988**, 1047–1058; f) E. I. Negishi in *Comprehensive Organic Synthesis* (Eds.: B. M. Trost, I. Fleming), Vol. 5, Pergamon, Oxford **1991**, 1163–1184; g) R. D. Broene, S. L. Buchwaldt, *Science* **1993**, *261*, 1696–1701.

[5] S. L. Buchwald, B. T. Watson, J. C. Huffman, *J. Am. Chem. Soc.* **1987**, *109*, 2544–2566.

[6] a) S. L. Buchwald, S. M. King, *J. Am. Chem. Soc.* **1991**, *113*, 258–265; b) see also: G. D. Cuny, A. Gutierrez, S. L. Buchwald, *Organometallics* **1991**, *10*, 537–539.

[7] G. Erker, K. Kropp, *J. Am. Chem. Soc.* **1979**, *101*, 3659–3660.

[8] a) E. Negishi, F. E. Cederbaum, T. Takahashi, *Tetrahedron Lett.* **1986**, *27*, 2829–2832; b) E. Neghishi, D. R. Swanson, T. Takahashi, *J. Chem. Soc., Chem. Commun.* **1990**, 1254–1255.

[9] a) E. Neghishi, F. E. Cederbaum, T. Takahashi, *Tetrahedron Lett.* **1986**, *27*, 2829–2832; b) E. Neghishi, D. R. Swanson, T. Takahashi, *J. Chem. Soc., Chem. Commun.* **1990**, 1254–1255.

[10] a) T. Takahashi, N. Suzuki, M. Hasegawa, Y. Nitto, K. Aoyagi, M. Saburi, *Chem. Lett.* **1992**, 331–334; b) for reactions of alkene-zirconocene-complexes with 1,3-dienes, see: E. Neghishi, S. R. Miller, *J. Org. Chem.* **1989**, *54*, 6014–6016.

[11] a) D. R. Swanson, E. Neghishi, *Organometallics* **1991**, *10*, 825–826; b) See also: P. Binger, P. Müller, R. Benn, A. Rufinska, B. Gabor, C. Krüger, P. Betz, *Chem. Ber.* **1989**, *122*, 1035–1042.

[12] E. Neghishi, T. Nguyen, J. P. Maye, D. Choueiri, N. Suzuki, T. Takahashi, *Chem. Lett.* **1992**, 2367–2370.

[13] T. Takahashi, M. Kageyama, V. Denisov, R. Hara, E. Neghishi, *Tetrahedron Lett.* **1993**, *34*, 687–690.

[14] T. Takahashi, T. Seki, Y. Nitto, M. Saburi, C. J. Rousset, E. Neghishi, *J. Am. Chem. Soc.* **1991**, *113*, 6266–6268.

[15] a) A. H. Hoveyda, Z. Xu, *J. Am. Chem. Soc.* **1991**, *113*, 5079–5080; b) A. F. Houri, M. T. Didiuk, Z. Xu, N. R. Horan, A. H. Hoveyda, *J. Am. Chem. Soc.* **1993**, *115*, 6614–6624.

[16] W. A. Nugent, D. L. Thorn, R. L. Harlow, *J. Am. Chem. Soc.* **1987**, *109*, 2788–2796.

[17] a) T. V. RajanBabu, W. A. Nugent, D. F. Taber, P. J. Fagan, *J. Am. Chem. Soc.* **1988**, *110*, 7128–7135; b) P. A. Wender, F. E. McDonald, *Tetrahedron Lett.* **1990**, *31*, 3691–3694; c) E. C. Lund, T. Livinghouse, *J. Org. Chem.* **1989**, *54*, 4487–4488.

[18] a) W. A. Nugent, D. F. Taber, *J. Am. Chem. Soc.* **1989**, *111*, 6435–6437; b) M. Mori, N. Uesaka, M. Shibasaki, *J. Org. Chem.* **1992**, *57*, 3519–3521; c) D. F. Taber, J. P. Louey, J. A. Lim, *Tetrahedron Lett.* **1993**, *34*, 2243–2246.

[19] K. S. Knight, R. M. Waymouth, *J. Am. Chem. Soc.* **1991**, *113*, 6268–6270.

[20] H. Ito, T. Taguchi, Y. Hanzawa, *J. Org. Chem.* **1993**, *58*, 774–775.

[21] H. Ito, Y. Motoki, T. Taguchi, Y. Hanzawa, *J. Am. Chem. Soc.* **1993**, *115*, 8835–8836.

[22] H. Ito, T. Taguchi, Y. Hanzawa, *Tetrahedron Lett.* **1993**, *34*, 7639–7642.

[23] R. B. Grossman, S. L. Buchwald, *J. Org. Chem.* **1992**, *57*, 5803–5805.

[24] S. C. Berk, R. B. Grossman, S. L. Buchwald, *J. Am. Chem. Soc.* **1993**, *115*, 4912–4913.

[25] a) G. Erker, U. Dorf, P. Czisch, J. L. Peterson, *Organometallics* **1986**, *5*, 668; b) R. M. Waymouth, K. R. Clausen, R. H. Grubbs, *J. Am. Chem. Soc.* **1986**, *108*, 6385; c) G. Erker,

M. Mena, C. Krüger, R. Noe, *Organometallics* **1991**, *10*, 1201–1204, and references cited therein.

[26] D. F. Hewlett, R. J. Whitby, *J. Chem. Soc., Chem. Commun.* **1990**, 1684–1686, and references cited therein.

[27] a) S. L. Buchwald, R. B. Nielsen, *J. Am. Chem. Soc.* **1988**, *110*, 3171–3175; b) W. Ando, T. Ohtaki, T. Suzuki, Y. Kabe, *J. Am. Chem. Soc.* **1991**, *113*, 7782–7784.

[28] a) S. L. Buchwald, B. T. Watson, M. W. Wannamaker, J. C. Dewan, *J. Am. Chem. Soc.* **1989**, *111*, 4486–4494; b) J. M. Davis, R. J. Whitby, A. Jaxa-Chamiec, *J. Chem. Soc., Chem. Commun.* **1991**, 1743–1745; c) N. Coles, R. J. Whitby, J. Blagg, *Synlett* **1992**, 143–145; d) see also: A. S. Guram, R. F. Jordan, *J. Org. Chem.* **1992**, *57*, 5994–5999; e) enantioselective example: R. G. Grossman, W. M. Davis, S. L. Buchwald, *J. Am. Chem. Soc.* **1991**, *113*, 2321–2322.

[29] H. Ito, T. Taguchi, Y. Hanzawa, *Tetrahedron Lett.* **1992**, *31*, 4469–4472.

[30] M. Jensen, T. Livinghouse, *J. Am. Chem. Soc.* **1989**, *111*, 4495–4496.

[31] a) P. L. McGrane, M. Jensen, T. Livinghouse, *J. Am. Chem. Soc.* **1992**, *114*, 5459–5460; b) P. L. McGrane, T. Livinghouse, *J. Org. Chem.* **1992**, *57*, 1323–1324.

[32] Review: a) J. Schwartz, J. A. Labinger, *Angew. Chem.* **1976**, *88*, 402; Angew. Chem. Int. Ed. Engl. **1976**, *15*, 333 b) see also ref. 4d); c) E. Negishi, T. Takahashi, *Aldrichimica Acta* **1985**, *18*, 31.

[33] S. L. Buchwald, Q. Fang, S. M. King, *Tetrahedron Lett.* **1988**, *29*, 3445–3448.

[34] S. L. Buchwald, S. J. LaMaire, *Tetrahedron Lett.* **1987**, *28*, 295–298.

[35] a) S. L. Buchwald, S. J. LaMaire, R. B. Nielsen, B. T. Watson, S. M. King, *Tetrahedron Lett.* **1987**, *34*, 3895–3898; b) S. M. Buchwald, S. J. LaMaire, R. B. Nielsen, B. T. Watson, S. M. King, *Org. Synth.* **1992**, *71*, 77–82.

[36] B. H. Lipshutz, R. Keil, E. L. Ellsworth, *Tetrahedron Lett.* **1990**, *31*, 7257–7260.

[37] R. C. Sun, M. Okabe, D. L. Coffen, J. Schwartz, *Org. Synth.* **1992**, *71*, 83–88.

[38] a) B. H. Lipshutz, E. L. Ellsworth, *J. Am. Chem. Soc.* **1990**, *112*, 7440–7441; b) K. A. Babiak, J. R. Behling, J. H. Dygos, K. T. McLaughlin, J. S. Ng, V. J. Kalish, S. W. Kramer, R. L. Shone, *J. Am. Chem. Soc.* **1990**, *112*, 7441–7442.

[39] P. Wipf, J. H. Smitrovich, *J. Org. Chem.* **1991**, *56*, 6494–6496.

[40] P. Wipf, W. Xu, *Synlett* **1992**, 718–720.

[41] T. E. Cole, R. Quintanilla, *J. Org. Chem.* **1992**, *57*, 7366–7370.

[42] H. Maeta, T. Hashimoto, T. Hasegawa, K. Suzuki, *Tetrahedron Lett.* **1992**, *33*, 5965–5968; see also: P. Wipf, W. Xu, *J. Org. Chem.* **1993**, *58*, 825–826.

[43] K. Suzuki, H. Maeta, T. Suzuki, T. Matsumoto, *Tetrahedron Lett.* **1989**, *30*, 6879–6882.

[44] C. E. Tucker, P. Knochel, *J. Am. Chem. Soc.* **1991**, *113*, 9888–9890; see also: R. D. Dennehy, R. J. Whitby, *J. Chem. Soc., Chem. Commun.* **1992**, 35–36.

[45] B. H. Lipshutz, R. Keil, J. C. Barton, *Tetrahedron Lett.* **1992**, *33*, 5861–5864.

[46] M. Maeta, T. Hasegawa, K. Suzuki, *Synlett* **1993**, 341–345.

[47] a) F. N. Tebbe, G. W. Parshall, G. S. Reddy, *J. Am. Chem. Soc.* **1978**, *100*, 3611–3613; b) S. H. Pine, G. Kim, V. Lee, *Org. Synth.* **1990**, *69*, 72–79.

[48] a) N. Petasis, E. I. Bzowej, *J. Am. Chem. Soc.* **1990**, *112*, 6392–6394; b) N. Petasis, E. I. Bzowej, *J. Org. Chem.* **1992**, *57*, 1327–1330; c) N. Petasis, I. Akritopoulou, *Synlett* **1992**, 665; d) N. Petasis, E. I. Bzowej, *Tetrahedron Lett.* **1993**, *34*, 943–946, and references cited therein.

[49] a) K. M. Doxsee, J. K. M. Mouser, *Tetrahedron Lett.* **1991**, *32*, 1687–1690; b) K. M. Doxsee, J. K. M. Mouser, J. B. Farahi, *Synlett* **1992**, 13–21, and references cited therein.

[50] Zirconium-catalyzed allylation reactions of allylic ethers: a) G. D. Cuny, S. L. Buchwald, *Organometallics*, **1991**, *10*, 363–365; b) N. Suzuki, D. Y. Kondakov, T. Takahashi, *J. Am. Chem. Soc.* **1993**, *115*, 8485–8486; c) J. P. Morken, M. T. Didiuk, A. H. Hoveyda, *J. Am. Chem. Soc.* **1993**, *115*, 6997–6998; 3-Oxacyclohexyne-zirconocene complex: M. C. J. Harris, R. J. Whitby, J. Blagg, *Synlett* **1993**, 705–707. Insertion of alkynes and alkenes into a cationic iminoacyl-zirconocene-complex: A. S. Guram, R. F. Jordan, *J. Org. Chem.* **1993**, *58*, 5595–5597. Intermolecular pinacol cross-coupling of anionic zirconaoxiranes with

aldehydes: F. R. Askham, K. M. Carroll, *J. Org. Chem.* **1993**, *58*, 7328–7329.

[51] a) G. Erker, *Nachr. Chem. Tech. Lab.* **1992**, *40*, 1099–1104; b) G. Erker, M. Albrecht, C. Krüger, S. Werner, *J. Am. Chem. Soc.* **1992**, *114*, 8531–8536; c) M. Albrecht, G. Erker, C. Krüger, *Synlett* **1993**, 441–448.

[52] a) W. Kaminsky, K. Külper, H. H. Brintzinger, F. W. R. P. Wild, *Angew. Chem.* **1985**, *97*, 507–508; Angew. Chem. Int. Ed. Engl. **1985**, *24*, 507. b) J. W. Röll, H.-H. Brintzinger, B. Rieger, R. Zalk, *Angew. Chem.* **1990**, *102*, 339; Angew. Chem. Int. Ed. Engl. **1990**, *29*, 279. c) G. Erker, *Pure Appl. Chem.* **1992**, *64*, 393–401; d) R. F. Jordan, P. K. Bradley, R. E. LaPointe, D. F. Taylor, *New J. Chem.* **1990**, *14*, 505–511; e) R. F. Jordan, *Adv. Organomet. Chem.* **1991**, *32*, 325; f) S. Collins, D. G. Ward, *J. Am. Chem. Soc.* **1992**, *114*, 5460–5462; g) G. Erker, M. Aulbach, M. Knickmeier, D. Wingbergmühle, C. Krüger, M. Nolte, S. Werner, *J. Am. Chem. Soc.* **1993**, *115*, 4590–4601, and references cited therein.

Aluminum Enolates

Herbert Waldmann

For the directed and highly selective formation of C–C bonds metal enolates are efficient reagents. Their reactivity is determined decisively by the nature of the respective metal ions. Thus, enolates embodying counterions of the first and the second main group, tin or zirconium and the respective titanium [1] and boron [2] reagents have been applied in numerous cases in organic synthesis. In addition, aluminum enolates have recently been added to the tools of current synthetic methodology.

Aluminum enolates can be generated by various techniques, for example, by nickel-catalyzed [3a, b] or photochemically initiated [3c, 9] conjugate addition of aluminum alkyls to α,β-unsaturated carbonyl compounds, by conjugate addition of the group "X" embodied in Lewis acids R_2AlX (X = SPh, [3d] SeMe, [3d] C≡CR, [3e] H [3f, g, 5, 6])

or the alkyl moiety of R_2AlCl [3c] to these electrophiles, and by direct deprotonation of a carbonyl compound and trapping of the enolate with an aluminum amide or phenolate. [3h, i] However, in the majority of cases a transmetalation of an analogous lithium enolate is carried out (see below). The aluminum alcoholates and the corresponding alkali metal derivatives often display distinctly different reactivities. Thus, cyclohexene oxide *(1)* reacts with lithium *tert*-butyl acetate to give *(2)* in only 8 % yield, whereas the diethylaluminum intermediate makes this alcohol available in 68 % yield. [4a] If enantiomeric-

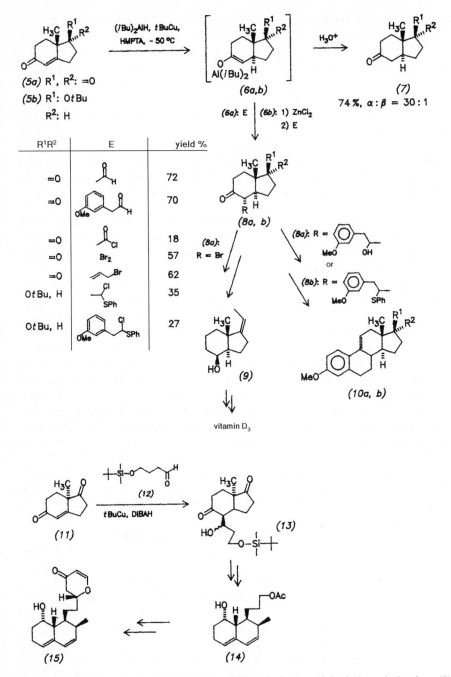

Scheme 1. Conjugate reduction and electrophilic substitution of the indane derivatives *(5)* and *(11)* via aluminum enolates.

ally homogenous *(S)*-propylene oxide is used the respective transformation proceeds with a stereoselectivity of 94:6. [4b] The selective conjugate reduction of α,β-unsaturated aldehydes, ketones and esters (e.g., *(3)* and *(4)*) can be carried out by protonation of the aluminum enolates that are formed by addition of diisobutylaluminum hydride (DIBAH) to the vinylogous carbonyl compounds. [3f, 5, 6] However, the desired selectivity and a sufficient reactivity are only achieved if a copper alkyl and HMPTA are added. Under these conditions the competing 1,2-addition does not occur and additional carbonyl groups present also remain unaffected. If the enolates are quenched with alkyl halides instead of acids, the corresponding α-alkylated esters, ketones and aldehydes are formed in high yield. [3g, 6] These transformations have proven their efficiency, for instance, in the construction of steroids [3f, 5, 6] (Scheme 1). By this means the chemoselective reduction of the tetrahydroindanedione *(5a)*, which is available in enantiomerically pure form by the Hajos–Eder-Sauer reaction, to give the *trans*-diketone *(7)* with a *trans/cis* selectivity of 30:1 can be carried out. [5b] In the course of the alternative catalytic reduction, on the other hand, the *cis*-configured product is predominantly generated. If the intermediate *(6a)* is trapped with aldehydes, allyl bromides, acyl chlorides or bromine, the *trans*-hydrindanedions *(8a)* are obtained. [6a] After transmetallation with $ZnCl_2$, *(6b)* was thioalkylated successfully with α-chloro-2-phenylthioalkanes. [6d] Compound *(8a)* (R = Br) can be converted in a few steps to the alcohol *(9)*, a well known congener to vitamin D3 and its hydroxylated analogs. From the aldol adduct *(8a)* (R = *m*-MeOPh-CH_2-CH(OH)-) estrogens like 9-(11)-dehydroestrone-3-methylether *(10a)* can easily be generated. [6c] However, for this purpose the use of the thioalkylation product *(8b)* seems to be more advantageous. [6d] Finally, this synthetic strategy opens a route to compactin *(15)* (Scheme 1). [6e] The (–)-enantiomer of the ketone *(5a)* reacts with the protected γ-hydroxyaldehyde *(12)* to give the diastereomeric mixture *(13)* (isomer ratio 2:1). After chromotagraphic separation the main component of *(13)* was transformed into the alcohol *(14)*, a known precursor of compactin. The application of aluminum enolates may also be beneficial in aldol reactions [3b, d, i, h, 7] and nucleophilic additions to Schiff's bases. In a particularly impressive manner this is demonstrated by the reactions of aldehydes with the chiral iron complexes *(16)* (Scheme 2). [7] Whereas the lithium enolates of the acetyl complexes *rac-(16a)* and the propionyl complex *rac-(16b)* (R = CH_3) react with aldehydes with virtually no stereocontrol, the aluminum compounds *(17a)* and *(17b)*, which were generated by transmetallation with Et_2AlCl, deliver the β-hydroxyacyl complexes *(18a)* and *(18b)* at –100 °C with high isomer ratios. [7b, c, d] In these transformations *(16a)* proves to be a highly diastereoselecting acetate aldol equivalent, whereas the propionyl derivative *(16b)* after oxidative decomplexation, makes predominantly the *anti*-aldols *(19b)* available. With other established auxiliaries both classes of aldol adducts are accessible with high diastereomeric excess in only a few cases. If the corresponding enantiomerically pure iron complexes are introduced instead of *rac-(16)*, asymmetric syntheses can be carried out. The reaction of *(S)-(16a)* with Boc-prolinal *(20)* gives the aldol adduct *(21)* with a selectivity of 300:1 (Scheme 2). [7e] After removal of the *N*-protecting group and oxidative decomplexation from *(21)* the pyrrolizidinone *(22)* is formed. Analogously, to assign the absolute configuration of some marine peroxides, *(S)-(16b)* was converted to the acyloxy iron complex *(23)* (*anti:syn* = 15:1). [7f] If *(17a)* is treated with imines *(24)* or nitrones β-aminoacyl iron complexes *(25)* are formed with high diastereoselectivity. [7b] On treatment with bromine in CS_2 they cyclize to give

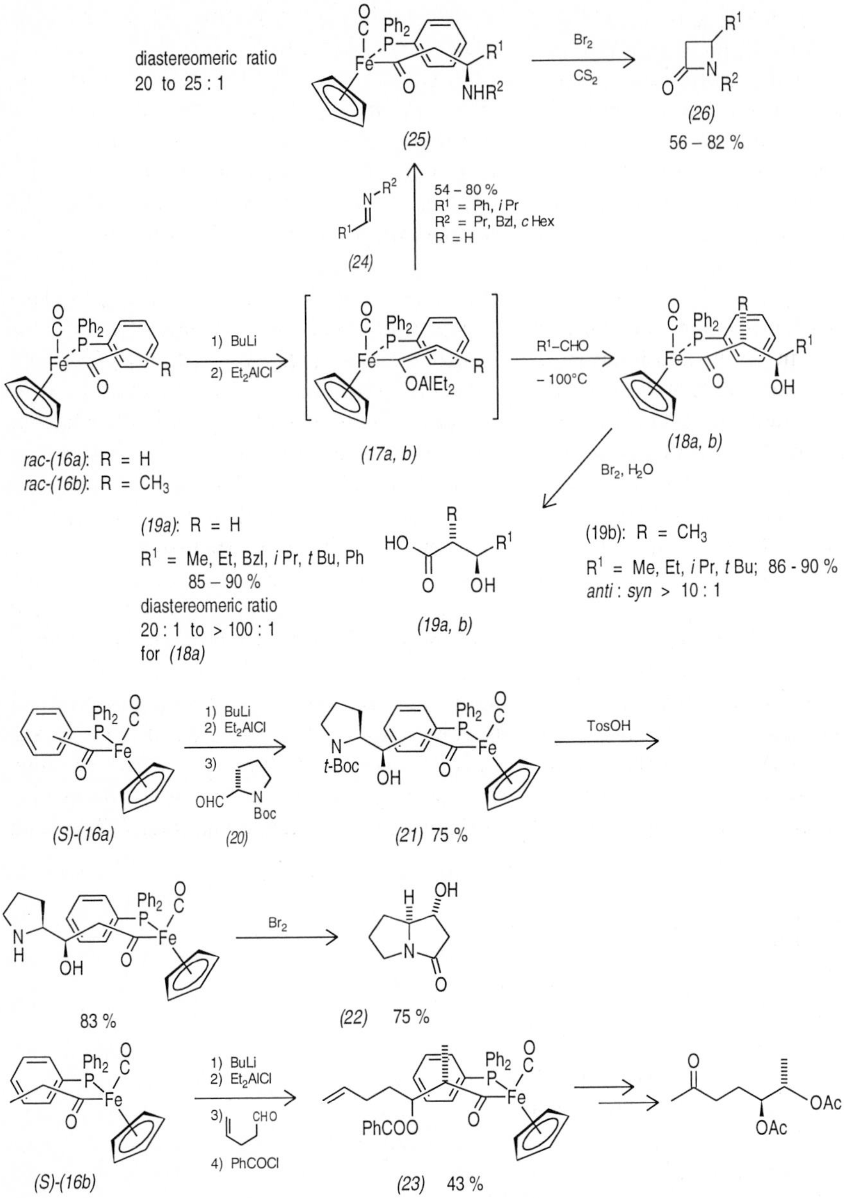

Scheme 2. Stereoselective aldol reactions employing aluminum enolates of the chiral iron complexes *rac-(16a, b), (S)-(16a)* and *(S)-(16b)*.

the lactams *(26)*. By analogy, the aluminum enolates of simple carboxylic acid esters and amino acid esters directly give β-lactams. [8]

Aluminum enolates can also be advantageously employed in asymmetric syntheses with α,β-unsaturated *N*-acyl urethanes, for example, the oxazolidinone *(27)*. [3c] Thus, Et₂AlCl reacts with this electrophile in an ionic reaction to form *(28a)*, which, after work up, delivers the β-branched amide *(29a)* in high yield and with a diastereomer ratio of 93:7. On the other hand, after photochemical initiation, the conjugate addition of Me₂AlCl proceeds by a radical mechanism. Such a pathway is unexpected for the combination of a strong Lewis acid with the polar acceptor *(27)*. Furthermore, the intermediates *(28)* can be oxidized stereoselectively with triplet oxygen (if Me₂AlCl is used) or with sulfoxaziridine *(30)* (if Et₂AlCl is used) to give β-branched α-hydroxy acid derivatives like *(31)*.

Aluminum enolates play a particularly interesting role in the development of a new method for the photochemically induced fixation of CO₂ (Scheme 3). [9] The transformations employed for this purpose are based on the observation that the nucleophilicity of axial ligands in (tetraphenylporphyrinato)aluminum complexes [(TPP)Al-X], for example, *(32)*, can be enhanced by irradiation with visible light. If, for instance, the enolates *(32a)* (obtained from (TPP)Al-NEt₂ and aromatic ketones) are irradiated in the presence of *N*-methylimidazole and CO₂, they add to the dioxide and the β-ketocarboxylate complexes *(33)* are formed, from which the β-keto acids *(34)* can be obtained in high yield. [9a] This reaction sequence can be regarded as an analogy to the biological assimilation of CO₂ during which pyruvate is carboxylated to oxaloacetate via phosphoenolpyruvate. Malonic acid derivatives are obtained if (TPP)aluminum alkyl complexes *(32b)* are irradiated in the presence of α,β-unsaturated esters like *(35)* and CO₂ (Scheme 3). [9b] Under these conditions, first a conjugate addition of the aluminum alkyls to the vinylogous carbonyl system occurs, and the enolates *(36)* formed thereby then react with CO₂ to give the malonate complexes *(37)*. On addition of diethylzinc this process proceeds even catalytically. If CO₂ is excluded in these experiments, *(36)* undergoes conjugate addition to surplus α,β-unsaturated ester to generate *(38)*, which

(29a): R = Et: 98 %, (βS):(βR) = 93:7
(29b): R = Me: 56 %, (βS):(βR) = 92:8

anti:syn >95:5   (31)
(βR):(βS) ≈ 90:10

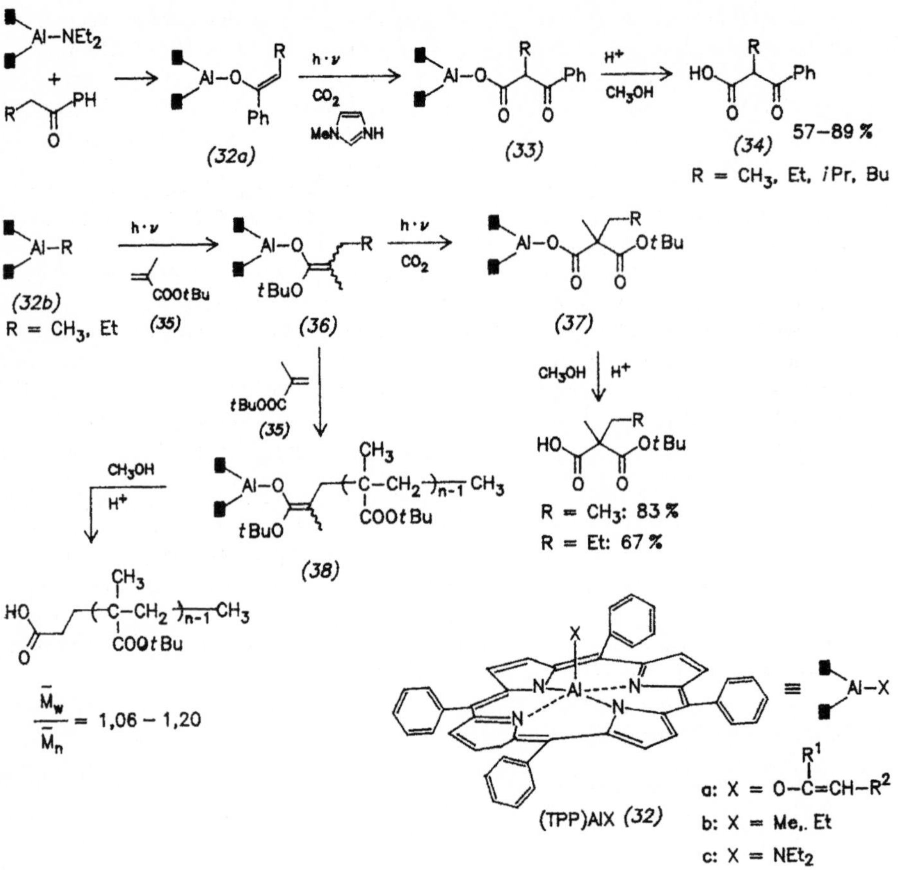

Scheme 3. Photochemically initiated fixation of CO_2 using (tetraphenylporphyrinato)-aluminum enolates.

can then undergo the same transformation again. [9c, d] Overall, the continuation of this reaction sequence allows for an anionic polymerization of the electron-deficient olefin in a process that has strong similarities to the well known group transfer polymerization. By this method living polymers with a narrow molecular weight distribution are built up, which can also be employed for the construction of block copolymers.

In the light of the examples highlighted above it may be expected that the variation of the metal in enolate chemistry also in future research will provide surprising and unexpected results. In particular the aluminum

enolates that have not been studied intensively so far may offer new opportunities for the development of new synthetic methodologies.

References

[1] K. H. Dötz, *Nachr. Chem. Tech. Lab.* **1990**, *38*, 1244.

[2] M. Braun, *Nachr. Chem. Tech. Lab.* **1990**, *38*, 1244.

[3] a) E. A. Jeffrey, A. Meisters, T. Mole, *J. Organomet. Chem.* **1974**, *74*, 365; b) E. A. Jeffrey, A. Meisters, T. Mole, *J. Organomet. Chem.* **1974**, *74*, 373; c) K. Rück, H. Kunz, *Angew.*

Chem. **1991**, *103*, 712; *Angew. Chem. Int. Ed. Engl.* **1991**, *30*, 694; d) A. Itoh, S. Ozawa, K. Oshima, H. Nozaki, *Tetrahedron Lett.* **1980**, 361; e) J. Schwartz, D. B. Carr, R. T. Hansen, F. M. Dayrit, *J. Org. Chem.* **1980**, *45*, 3053; f) T. Tsuda, T. Hayashi, H. Satomi, T. Kawamoto, T. Saegusa, *J. Org. Chem.* **1986**, *51*, 537; g) T. Tsuda, H. Satomi, T. Hayashi, T. Saegusa, *J. Org. Chem.* **1987**, *52*, 439; h) H. Nozaki, K. Oshima, K. Takai, S. Ozawa, *Chem. Lett.* **1979**, 379; i) J. Tsuji, T. Yamada, M. Kaito, T. Mandai, *Tetrahedron Lett.* **1979**, 2257.

[4] S. Danishefsky, T. Kitahara, M. Tsai, J. Dynak, *J. Org. Chem.* **1976**, *41*, 1669; b) T.-J. Sturm, A. E. Marolewski, D. S. Rezenka, S. K. Taylor, *J. Org. Chem.* **1989**, *54*, 2039.

[5] a) T. Tsuda, T. Kawamoto, Y. Kuwamoto, T. Saegusa, *Synth. Commun.* **1986**, 639; b) A. R. Daniewski, J. Kiegiel, *Synth. Commun.* **1988**, 115.

[6] a) A. R. Daniewski, J. Kiegiel, E. Piotrowska, T. Warchol, W. Wojciechowska, *Liebigs Ann. Chem.* **1988**, 593; b) A. R. Daniewski, J. Kiegiel, *J. Org. Chem.* **1988**, *53*, 5534; c) A. R. Daniewski, J. Kiegiel, *J. Org. Chem.* **1988**, *53*, 5535; d) U. Groth, T. Köhler, T. Taapken, *Tetrahedron* **1991**, *36*, 7583; e) A. R. Daniewski, M. R. Uskokovic, *Tetrahedron Lett.* **1990**, *31*, 5599.

[7] a) Review: S. G. Davies, *Aldrichimica Acta* **1990**, *23*, 31, b) L. S. Liebeskind, M. E. Welker, R. W. Fengl, *J. Am. Chem. Soc.* **1986**, 108, 6328; c) S. G. Davies, I. M. Dordor-Hedgecock, P. Warner, R. H. Jones, K. Prout, *J. Organomet. Chem.* **1985**, *285*, 213; d) S. G. Davies, I. M. Dordor-Hedgecock, J. C. Walker, P. Warner, *Tetrahedron Lett.* **1985**, 26, 2125; e) R. P. Beckett, S. G. Davies, *J. Chem. Soc., Chem. Commun.* **1988**, 160; f) R. J. Capon, J. K. MacLeod, S. J. Coote, S. G. Davies, G. L. Gravatt, I. M. Dordor-Hedgecock, M. Whittaker, *Tetrahedron* **1988**, *44*, 1637.

[8] F. H. van der Steen, G. P. M. van Mier, A. L. Spek, J. Kroon, G. van Koten, *J. Am. Chem. Soc.* **1991**, *113*, 5742–5750 and references given therein.

[9] a) Y. Hirai, T. Aida, S. Inoue, *J. Am. Chem. Soc.* **1989**, *111*, 3062; b) M. Komatsu, T. Aida, S. Inoue, *J. Am. Chem. Soc.* **1991**, *113*, 8492; c) M. Kuroki, T. Aida, S. Inoue, *J. Am. Chem. Soc.* **1987**, *109*, 4737; d) M. Kuroki, T. Watanabe, T. Aida, S. Inoue, *J. Am. Chem. Soc.* **1991**, *113*, 5903–5904.

C. Silicon in Organic Synthesis

Selective Transformations with Pentacoordinate Silicon Compounds

Dieter Schinzer

Selective reactions with acyclic systems are still a challenge in organic synthesis. [1] Besides selectivity the reagents used should be easily accessable, easy to handle, and the reactions should be simple to operate. Especially in organo metallic chemistry very useful reagents have been introduced during the last decade. [2–5] In particular silicon was of interest, because many transformations of vinyl- and allyl silanes with acceptor groups have been reported. [6–9] In the following paragraph a permanent growing part in organosilicon chemistry is presented: reactions with pentacoordinate silicon compounds.

Silicon atoms can be penta- and hexacoordinated besides the normal four bonds. Compact and electronegative ligands like fluorine are required to increase the degree of coordination. [10] The driving force in such reactions is the relative high Lewis acidity of silicon (Scheme 1). [11]

$$R-F + SiX_4 \rightleftharpoons R-F-SiX_4 \rightleftharpoons R^+ + FSiX_4^-$$

Scheme 1

This type of pentacoordinated compounds is a very mild Lewis acid, because the negative charge is stabilized by electronegative ligands. This combination of properties makes these reagents quite promising in reactions with acceptor groups.

Commercially available hydrosilanes of type *(2)* reduce carbonyl groups in the presence of fluoride ions to the desired alcohols (Scheme 2). [12]

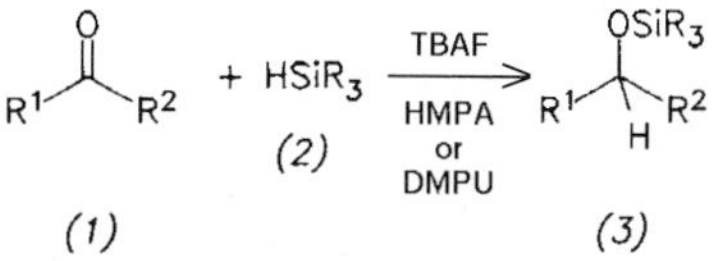

Scheme 2

The advantage of this new method is that no expensive metal catalyst is required because the intermediate formation of the pentacoordinated hydridosilicates activates the carbonyl group like a Lewis acid and transfers the hydride ion to yield the alcohol (Scheme 3).

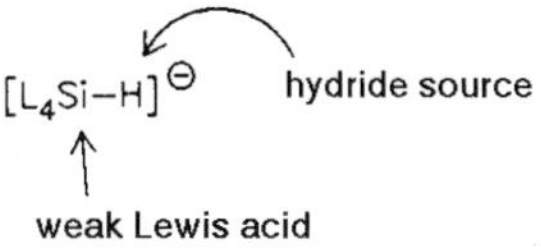

Scheme 3

The chemical yields in reductions of carbonyl groups are in general high. 2-Amino- and 2-hydroxyketones cannot be reduced at all or only with low diastereoselectivity using classical reagents. On the other hand hydridosilicates generated in situ reduce these compounds with excellent chemical and stereochemical yield (Scheme 4). [12]

Besides the regio- and chemoselectivity perfect control of stereoselectivity is obtained, as shown in the reduction of 2-methylcyclo-

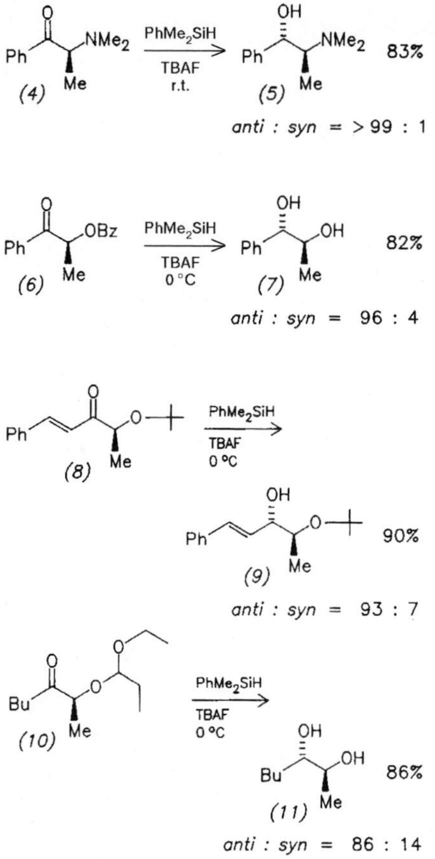

(4) → (5) 83%

anti : syn = > 99 : 1

(6) → (7) 82%

anti : syn = 96 : 4

(8) + → (9) 90%

anti : syn = 93 : 7

(10) → (11) 86%

anti : syn = 86 : 14

Scheme 4

hexanone *(12)* to give *trans*-2-methyl-cyclohexanol *(13)* in 94 % stereoselectivity (Scheme 5).

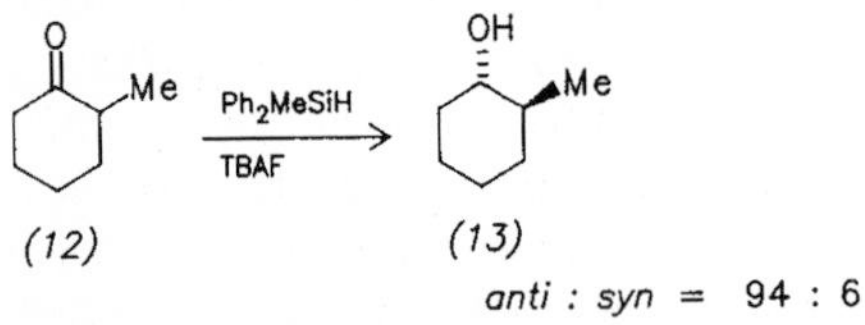

(12) → (13)

anti : syn = 94 : 6

Scheme 5

The selectivity obtained can be directly correlated with the bulkiness of the ligand used at silicon: If a phenyl group is replaced by a methyl group the selectivity drops to 76 %, as shown in the reaction *(12)→(13)* (Scheme 6).

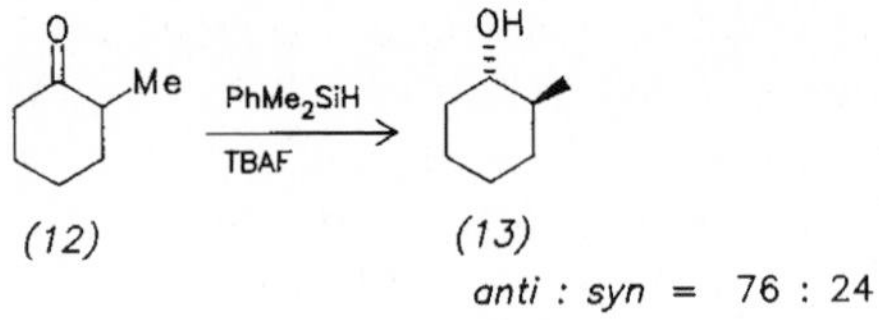

(12) → (13)

anti : syn = 76 : 24

Scheme 6

The *anti*-selectivity observed in acyclic systems can be explained by the Felkin-Anh model for transition states, which is based on steric interactions. [13] It should be mentioned that in no case has racemization been observed. For example, *(S)*-2-acetoxy-1-phenylpropanone *(14)* (88 % *ee*) was reduced to give (1*S*, 2*S*)-phenyl-1,2-propanediol, also in 88 % *ee* (Scheme 7).

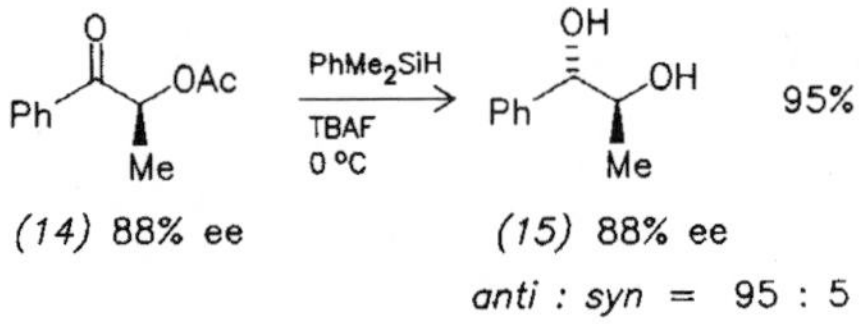

(14) 88% ee → (15) 88% ee 95%

anti : syn = 95 : 5

Scheme 7

The reduction described can also be run under acidic conditions but with the opposite stereochemical result: with excellent stereochemical yields the *syn*-products are obtained. These results can be explained by the use of the cyclic Cram model for transition states. [14] Again, the operation is very simple: instead of fluoride ion, trifluoroacetic acid (TFA) is added (Scheme 8).

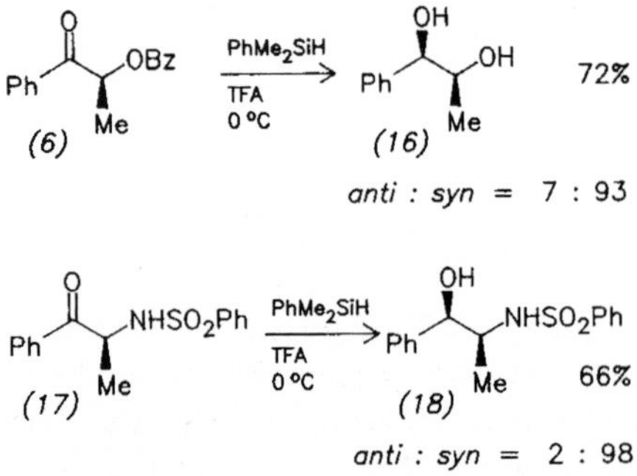

(6) → (16) 72%

anti : syn = 7 : 93

(17) → (18) 66%

anti : syn = 2 : 98

TFA = trifluoroacetic acid

Scheme 8

In order to demonstrate the usefulness of this stereoselective reduction in asymmetric synthesis two syntheses of simple, but pharmacologially active, compounds will be presented. [12] Compounds *(19)* and *(22)* are reduced diastereoselectively to *(20)* and *(23)* using the reducing system diphenyl methyl silane in the presence of TFA. Compound *(20)* is reduced again with LAH to give L-ephedrine *(21)* in 80% yield, and *(23)* is hydrolyzed with KOH to give L-methoxamine *(24)* in 84% yield (Scheme 9).

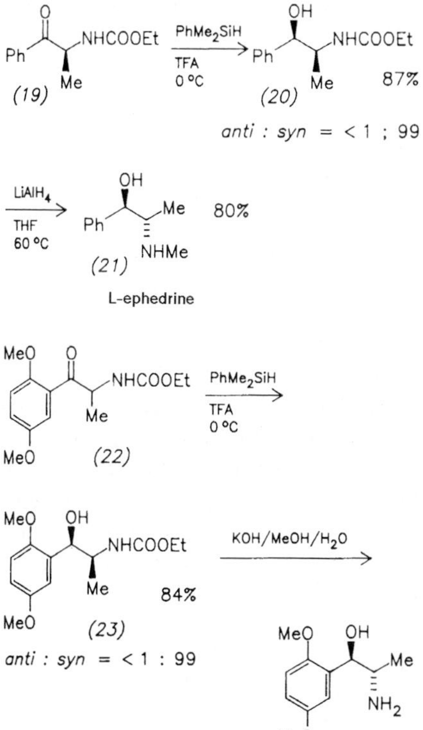

Scheme 9

Sakurai et al. have reported another entry to these new reducing reagents starting from trichlorosilane and dilithio catecholate. The silicate *(27)* obtained in situ, is stable in solution, reduces aldehydes and ketones, but does not reduce esters like *(34)* (Scheme 10). [15]

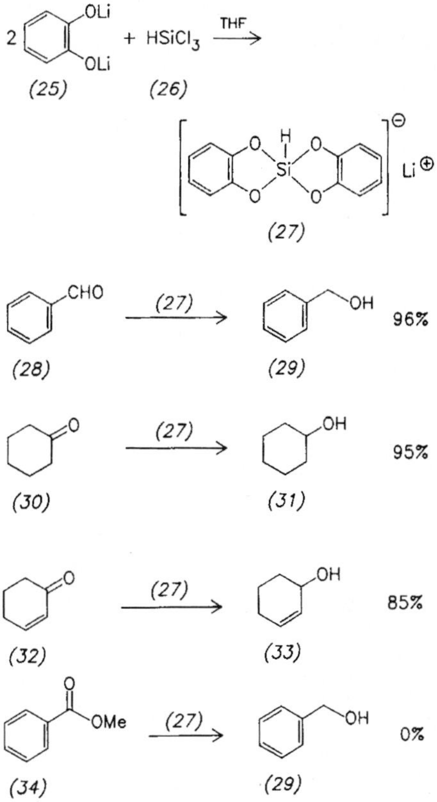

Scheme 10

The same group has isolated for the first time a compound of type *(27)*. [16]

C–C bond formation with allyl metal reagents is a very important reaction in organic synthesis. Reagents based on boron, [2] silicon, [17] tin, [5] and titanium [18] have been studied extensively. Even more important in this context are transformations using crotyl metal compounds in 1,2-additions to aldehydes. This type of transformation creates homo allyl alcohols containing two new asymmetric centers. In addition, these compounds can be transformed into β-hydroxy carbonyl compounds by oxidative cleavage, so the overall transformation is equivalent to the aldol reaction. [19]

Mixtures of *Z*- and *E*-crotyltrimethylsilanes react with aldehydes in a stereoconvergent way to yield *syn*-homo allylic alcohols. The reaction is promoted by titanium tetrachloride (Scheme 11). [17]

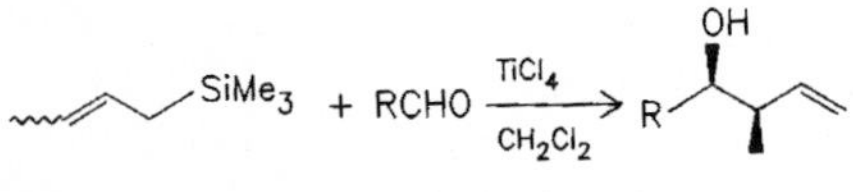

Scheme 11

The stereochemical outcome can be explained by an acyclic open transition state. [2] It is therefore impossible to synthesize *anti*-homo allylic alcohols by the use of crotyl silanes. The problem can be solved by the use of pentacoordinate silicates as shown by Sakurai et al. [20, 21] Reaction of allyltrichlorosilane in the presence of dilithio catecholate generated *(36)*, which adds diastereoselectivity to aldehydes without additional Lewis acids (Scheme 12).

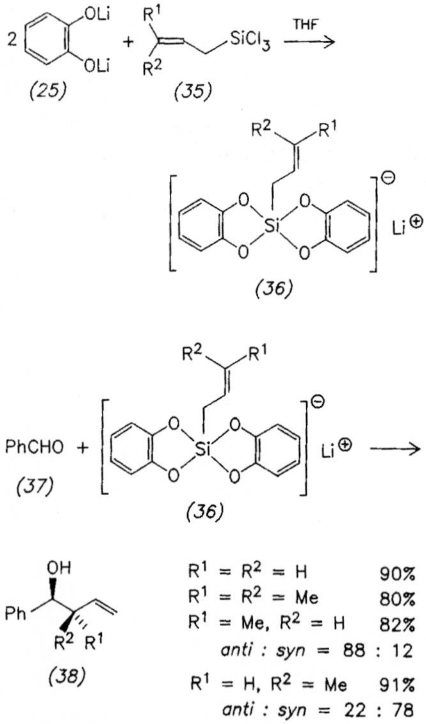

R^1 = R^2 = H   90%
R^1 = R^2 = Me   80%
R^1 = Me, R^2 = H   82%
  anti : syn = 88 : 12
R^1 = H, R^2 = Me   91%
  anti : syn = 22 : 78

Scheme 12

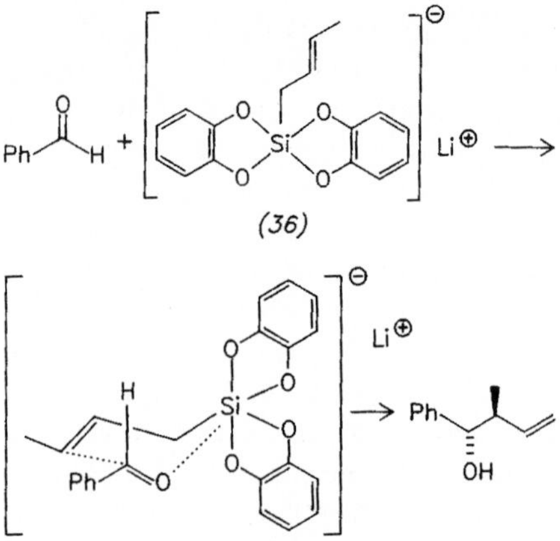

Scheme 13

(39) + CsF $\xrightarrow{\text{THF}}$ (40)

(40) + RCHO $\longrightarrow$

(41) (42)

R = Ph   92%; *anti : syn* = 99 : 1
R = n-C$_8$H$_{17}$ 96%; *anti : syn* = 99 : 1
R = CH$_3$(CH$_2$)$_3$CH(C$_2$H$_5$) 68%; *anti : syn* = 99 : 1
R = PhCH=CH$_2$ 77%; *anti : syn* = 98 : 2

(43) + CsF $\xrightarrow{\text{THF}}$ (44)

(44) + RCHO $\longrightarrow$

(45) (46)

R = Ph   96%; *anti : syn* = 1 : 99
R = n-C$_8$H$_{17}$ 89%; *anti : syn* = 2 : 98
R = CH$_3$(CH$_2$)$_3$CH(C$_2$H$_5$) 89%; *anti : syn* = 2 : 98
R = PhCH=CH$_2$ 77%; *anti : syn* = 2 : 98

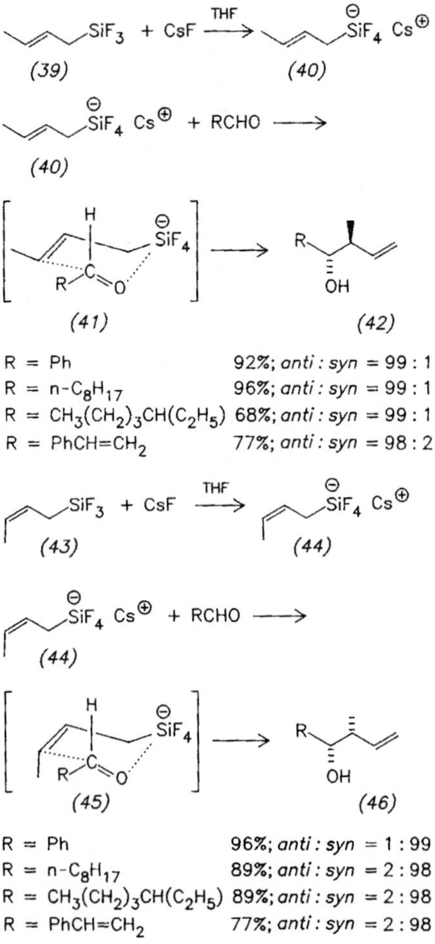

Scheme 14

The high degree of diastereoselectivity is consistent with a six-membered cyclic transition state, [2] in which, because of the electrophilic properties of silicon, a hexacoordinate silicon species is involved (Scheme 13).

Therefore both *syn-* and *anti-*homoallylic alcohols can be obtained with this new class of reagents, dependent whether *Z-* or *E-*crotyltrichlorosilanes are used as starting materials. The simplest way to carry out this transformation is to use of crotyltrifluorosilanes in the presence of CsF in THF. [20] Stereochemically homogeneous synthesis of either *(42)* or *(46)* is obtained by using pure *Z-* or *E-*trifluorocrotylsilanes, respectively (Scheme 14).

The results presented in this chapter clearly show that reductions and C–C bond forming reactions with pentacoorinate silicates are powerful means to construct various types of alicyclic compounds diastereoselectively.

References

[1] P. A. Bartlett, *Tetrahedron* **1980**, *36*, 3.

[2] R. W. Hoffmann, *Angew. Chem.* **1982**, *94*, 569. Angew. Chem. Int. Ed. Engl. **1982**, *21*, 555.

[3] M. T. Reetz, *Top. Curr. Chem.* **1982**, *106*, 1.

[4] B. Weidemann, D. Seebach, *Angew. Chem.* Angew. Chem. Int. Ed. Engl. **1983**, *22*, 474.**1983**, *95*, 12.

[5] Y. Yamamoto, K. Maruyama, *Heterocycles* **1982**, *18*, 357.

[6] T. A. Blumenkopf, L. E. Overman, *Chem. Rev.* **1986**, *86*, 857.

[7] I. Fleming in (Eds.: D. H. R. Barton and W. D. Ollis) *Comprehensive Organic Chemistry*, vol. 3., Pergamon Press, Oxford, **1979**, 541.

[8] H. Sakurai, *Pure Appl. Chem.* **1982**, *54*, 1.

[9] D. Schinzer, *Synthesis* **1988**, 263.

[10] H. Bürger, *Angew. Chem.* **1973**, *85*, 519. Angew. Chem. Int. Ed. Engl. **1973**, *12*, 474.

[11] H. Sakurai in *Selectivities in Lewis Acid-Promoted Reactions*, (Ed.: D. Schinzer), Reidel, Utrecht-Boston, **1988**

[12] M. Fujita, T. Hiyama, *J. Am. Chem. Soc.* **1984**, *106*, 4629.

[13] M. Chérest, H. Felkin, N. Prudent, *Tetrahedron Lett.* **1968**, 2199.

[14] D. J. Cram, D. R. Wilson, *J. Am. Chem. Soc.* **1963**, *85*, 1245.

[15] M. Kira, K. Sato, H. Sakurai, *J. Org. Chem.* **1987**, *52*, 949.

[16] M. Kira, K. Sato, H. Sakurai, Chem. Lett. **1987**, 2243.

[17] T. Hayashi, K. Kabeta, I. Hamachi, M. Kumada, *Tetrahedron Lett.* **1983**, 2865.

[18] M. T. Reetz, M. Sauerwald, *J. Org. Chem.* **1984**, *49*, 2293.

[19] C. H. Heathcock, *Science* **1981**, *214*, 395.

[20] M. Kira, M. Kobayashi, H. Sakurai, *Tetrahedron Lett.*, **1987**, 4081.

[21] M. Kira, K. Sato, H. Sakurai, *J. Am. Chem. Soc.* **1988**, *110*, 4599.

Oxidative Cleavage of Silicon–Carbon Bonds

Dieter Schinzer

The masking of functional groups as so-called "synthons" is a major strategy in modern organic synthesis for handling complex multi-functionality in long step sequences. [11]

The maneuver to provide the functionality needed from a synthon should be simple and efficient. In addition, the masked group should be stable during standard chemical operations so that it can be tolerated over several steps in a synthesis; given the complexity of modern stereoselective synthesis this is no easy goal. The technique of synthons should not replace the classical protecting group but should be added for planning a rational and convergent synthesis.

The following chapter deals with the masking of hydroxyl groups, which are important functional groups in almost any target molecule, such as macrolides or the natural product class of taxanes. [2] The analysis of potential groups that might be used for this purpose will almost automatically come up with a silyl group, because of its unique features: the electronic and steric behavior can be changed easily and "tuned" for the requirements needed.

The trimethylsilyl group is relatively stable towards most oxidizing reagents, but if one methyl group is replaced by a fluorine atom the silyl group can be transformed into a hydroxyl group under mild conditions (sila-Baeyer–Villiger rearrangement); [3] the same process can be achieved with higher coordi-nated silicon compounds. [4] The silyl group acts as the synthetic equivalent of a hydroxyl group (Scheme 1).

A related reaction has already been observed with acylsilanes in which an acid is

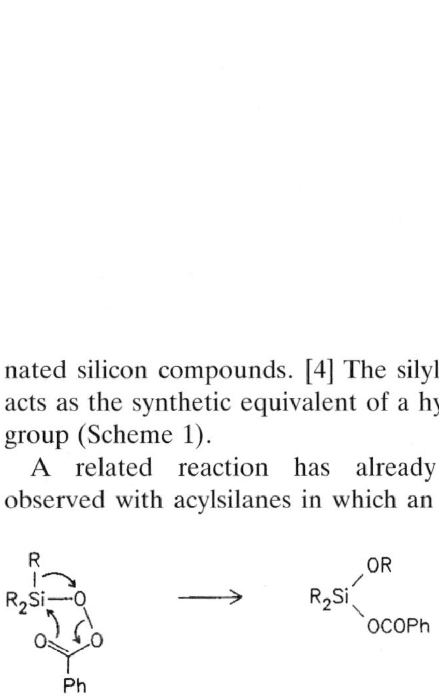

Scheme 1

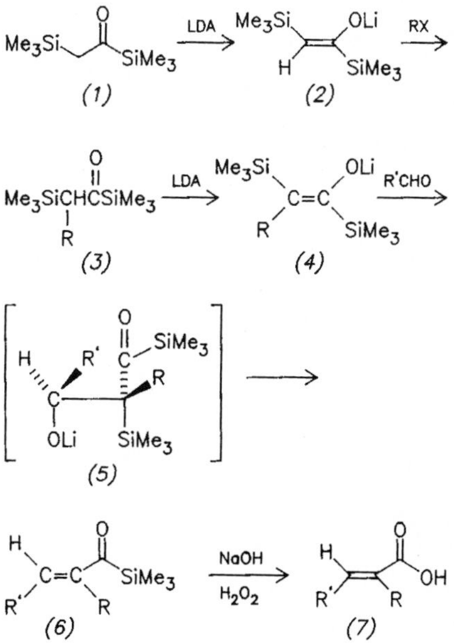

Scheme 2

formed regioselectively and under mild conditions in the presence of a peracid. [5, 6] Acylsilanes of type *(1)* could be transformed stereoselectively after a short sequence (deprotonation with LDA, alkylation, followed by Peterson olefination) [7] into trisubstituted α, β-unsaturated acids of type *(7)*.

Acylsilanes of type *(8)* can be transformed diastereoselectively into the *syn*-aldol products of type *(9)*. Acylsilane *(9)* was transformed directly in a one-pot procedure into β-hydroxy acid *(10)* with three defined stereogenic centers (Scheme 3). [8]

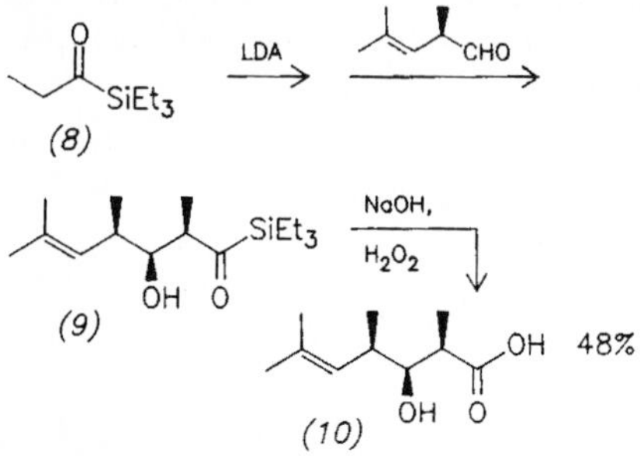

Scheme 3

In the same category the sila-Pummerer reaction can be discussed. An α-silylsulfide is deprotonated, alkylated, oxidized, rearranged into the silyl ether, and finally hydrolyzed into the aldehyde *(15)* (Scheme 4). [9]

An analogous reaction with selenium has been reported by Reich et al. (Scheme 5). [10]

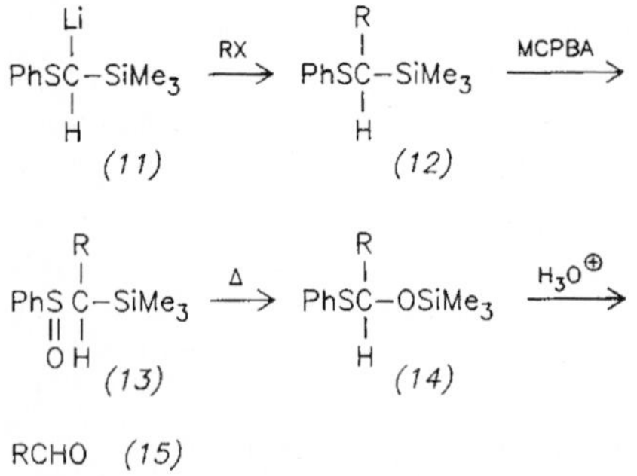

RCHO   *(15)*

R = H, Me, Et, Pr, Bu, s-Bu, PhCH$_2$

overall yield = 68 – 85%

Scheme 4

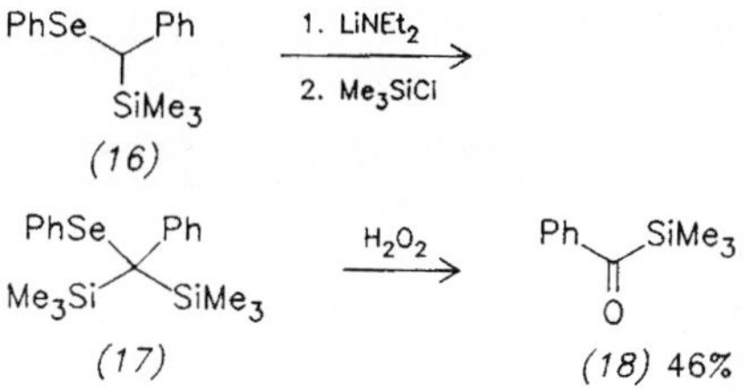

Scheme 5

The reaction of vinylsilanes with nitrile oxides leads to trimethylsilyl oxazolines. A cycloreversion is used to synthesize silylenol ethers *(21)* with retention of the olefin configuration (Scheme 6). [11]

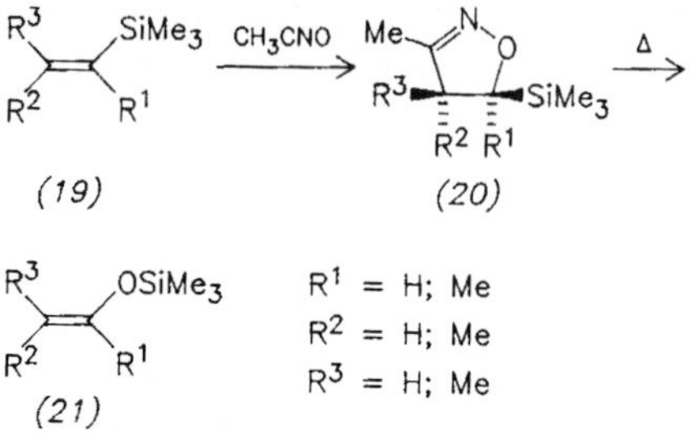

R^1 = H; Me
R^2 = H; Me
R^3 = H; Me

Scheme 6

In addition, the transformation of vinylsilanes to carbonyl compounds via α,β-epoxysilanes is a useful operation in synthesis: the reaction is directed by the regioselective opening of the epoxide (β-effect of the silyl group). Depending on the position of the silyl group, either aldehydes or ketones are obtained. Therefore, all reactions leading to vinylsilanes can, in principle, be used for the synthesis of carbonyl compounds (Scheme 7). [12, 13]

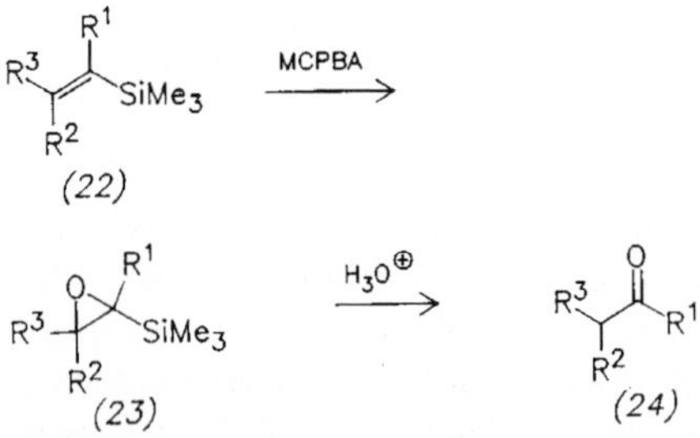

MCPBA = *m*-chloroperbenzoic acid

Scheme 7

Finally, the oxidation of allylsilanes to allylicalcohols via allylselenides should be mentioned. [14] The addition of phenylselenyl bromide depends only on steric interactions, as shown in the equation *(34)→(35)*. In the presence of H_2O_2 compounds of this type undergo an oxidative 2,3-sigmatropic rearrangement to the desired allylic alcohols such as *(36)* (Scheme 8).

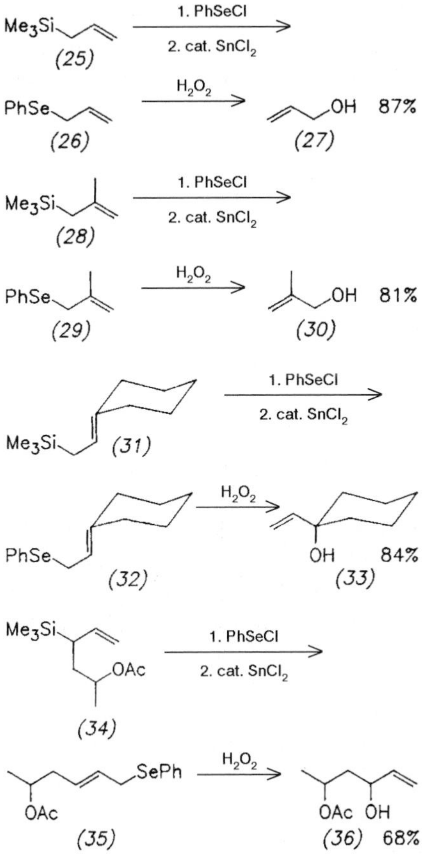

Scheme 8

The most important area using oxidative cleavage reactions of silyl groups is the treatment of "unactivated" alkyl- or aryl groups. Scope and limitations of this type of transformations will be presented in the following paragraph. Penta-coordinated silicates can be used as well as fluorosilanes. [4] In addition, a

further variation of an *anti*-Markownikoff hydration, via hydrosilylation of olefins (an alternative to hydroboration or hydroalumination), will be possible (Scheme 9). [15]

$$RCH=CH_2 + HSiCl_3 \xrightarrow{\text{catalyst}}$$

$$RCH_2CH_2SiCl_3 \xrightarrow[\text{2. MCPBA}]{\text{1. CuF}_2} RCH_2CH_2OH$$

Scheme 9

The stereochemical outcome of the sila-Baeyer-Villiger rearrangement was first studied with the norbornyl systems *(37)* and *(39)*. [16] Both, the *exo*- and *endo*-derivatives react stereospecifically with retention of configuration (Scheme 10).

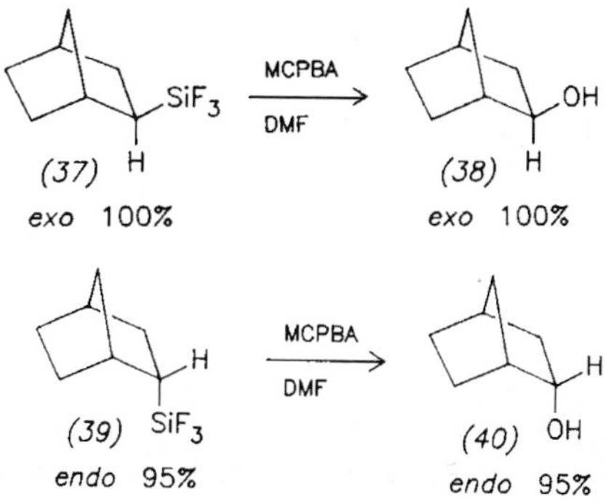

Scheme 10

Optically active alcohols can be obtained by an asymmetric hydrosilylation of norbornene *(41)* using a chiral catalyst and subsequent oxidation (Scheme 11).

A very useful application of this rearrangement chemistry has been used by Fleming et al. [17, 18] The addition of silyl cuprates to Michael systems and subsequent alkylation of the in situ formed enolate with methyl iodide provides the *anti*-compound *(46)* diastereoselectively, probably via a chelated transition state *(47)* (Scheme 12).

The phenyldimethylsilyl group used has the advantage of smooth protodesilylation [19] to form the fluoro compound, which can be transformed stereospecifically into the alcohol *(48)* (Scheme 13).

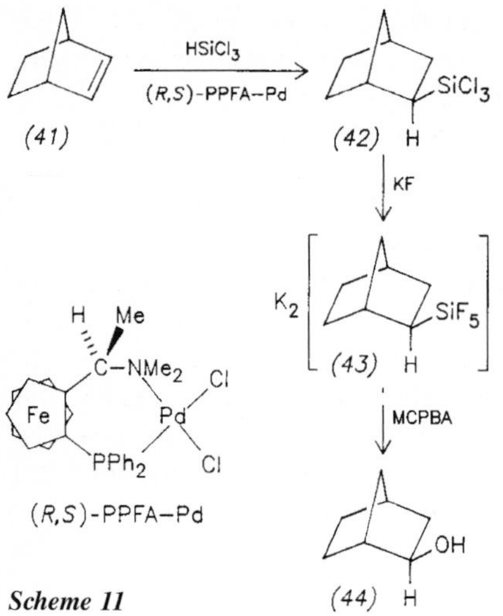

Scheme 11

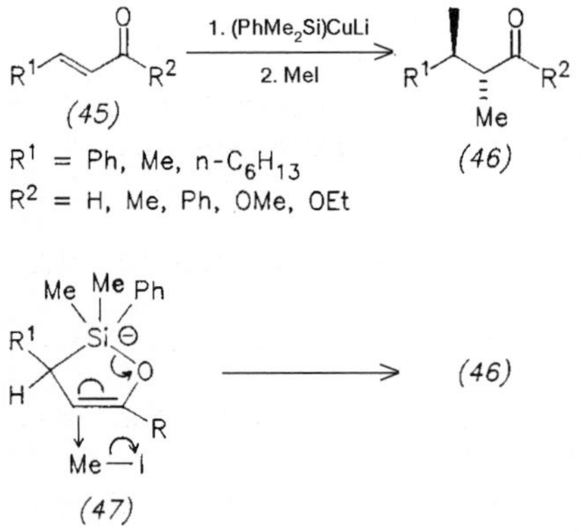

(41) → (42) HSiCl₃ / (R,S)-PPFA–Pd

(42) SiCl₃ H

KF

(R,S)-PPFA-Pd

(43) K₂ [SiF₅ H]

MCPBA

(44) OH H

Scheme 11

(45) R¹—CH=CH—C(O)—R² → 1. (PhMe₂Si)CuLi / 2. MeI → (46) R¹—C(O)—CH(Me)—R²

R¹ = Ph, Me, n-C₆H₁₃
R² = H, Me, Ph, OMe, OEt

(47) → (46)

Scheme 12

(46) PhMe₂Si ... → 1. HBF₄ / 2. MCPBA → (48) HO ...

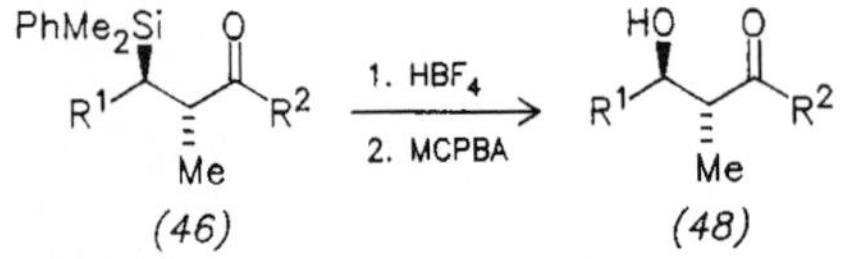

Scheme 13

Besides acyclic compounds, cyclic compounds like *(49)* or *(51)* can also be used. Again, synthetically useful intermediates *(50)* or *(51)* are obtained stereoselectively by simple operations (Scheme 14).

Furthermore, β-silyl enolate *(53)* yields aldol products diastereoselectively in reac-

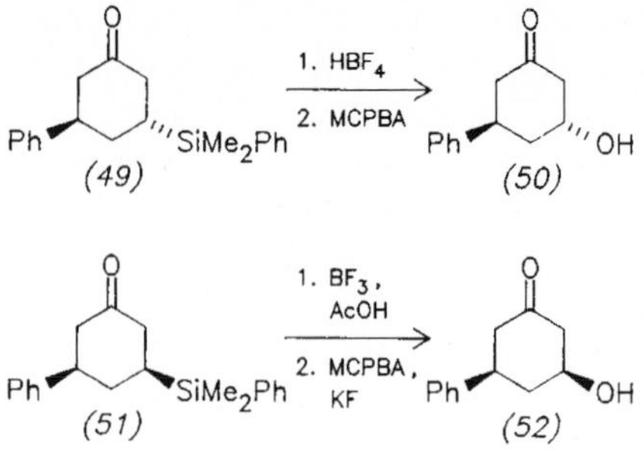

(49) Ph ... SiMe₂Ph → 1. HBF₄ / 2. MCPBA → (50) Ph ... OH

(51) Ph ... SiMe₂Ph → 1. BF₃, AcOH / 2. MCPBA, KF → (52) Ph ... OH

Scheme 14

(53) PhMe₂Si, OLi, R, OMe → R¹CHO → (54) PhMe₂Si, R, CO₂Me, R¹, OH

1. Ac₂O
2. BF₃, AcOH
3. MCPBA

→ (55) OAc, R, CO₂Me, R¹, OH

R = Ph, Me
R¹ = Ph, Me

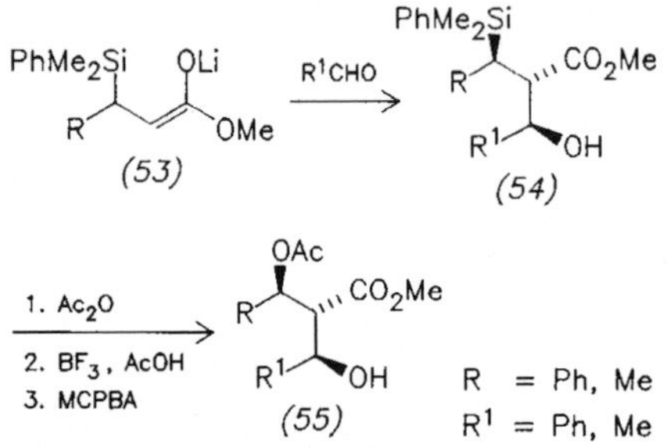

Scheme 15

tions with aldehydes. Therefore, a third chiral center can be controlled and was transformed into the important 1,3-diols *(55)* by a sila-Baeyer–Villiger reaction (Scheme 15). [20]

The phenyldimethylsilyl group is used for three purposes: 1) It stabilizes the silyl cuprate; 2) Smooth protodesilylation to form the fluoro compound; 3) Stereospecific rearrangement into the alcohol. Therefore, the properties of masked alcohols are completely different from those of alcohol protecting groups (silyl groups are more electropositive, have poor conjugating properties, and have no lone pairs available). Retro-Michael reactions are not possible with β-silyl-substituted ketones, in contrast to β-hydroxy ketones.

Instead of a phenyl group a furan ring has been used very successfully in a total synthesis of reserpine *(56)* reported by Stork et al. *(21)* A central problem in this synthesis is the construction of the highly functionalized E-ring *(63)* with its five asymmetric centers. Com-

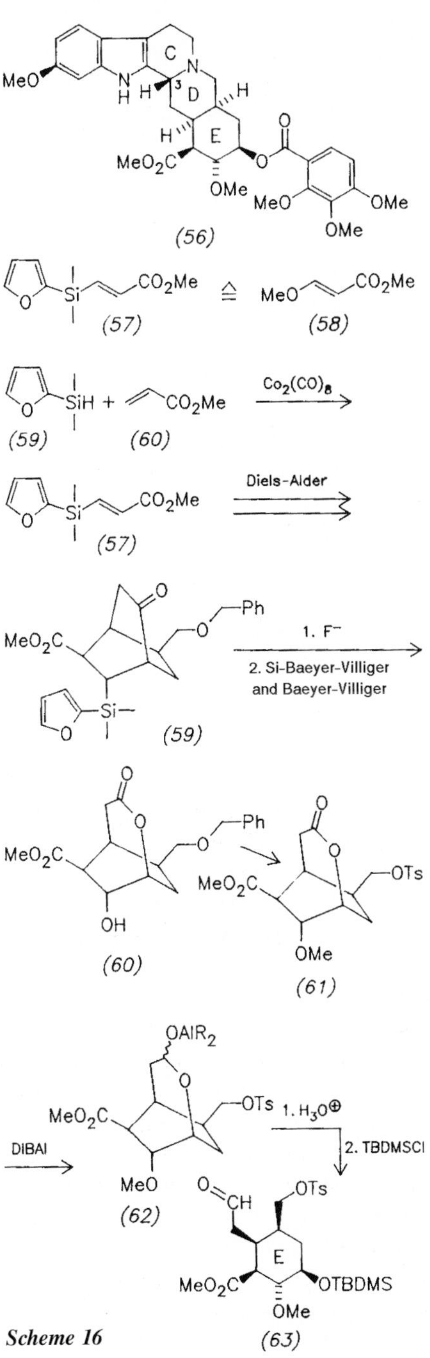

MeO

(56)

(57) ⇔ MeO⌒CO₂Me
              (58)

(59) SiH + CO₂Me →[Co₂(CO)₈]
       (60)

(57) →[Diels-Alder]

MeO₂C. Ph →[1. F⁻][2. Si-Baeyer-Villiger and Baeyer-Villiger]

(59)

MeO₂C. (60) → MeO₂C. OTs
OH   OMe
                (61)

DIBAl →

MeO₂C. (62) →[1. H₃O⊕][2. TBDMSCl]

MeO₂C. OTBDMS
OMe
(63)

Scheme 16

pound *(57)*, a synthetic equivalent of *(58)*, is first used in a Diels–Alder reaction, trans-

formed into the fluoride, and finally rearranged, both in sila-Baeyer–Villiger- and "classical" Baeyer–Villiger-fashion, to obtain *(60)*. The lactone *(61)* is cleaved with DIBAL into the monocyclic structure *(63)*, which contains five asymmetric centers. A master piece in strategy (Scheme 16)!

In summary, the oxidative cleavage of silicon–carbon bonds is a very useful tool in organic synthesis. It is the perfect technique to mask hydroxyl groups. An application in the total synthesis of a complex natural product has also been given that demonstrates the importance of this reaction sequence.

References

[1] D. Seebach, *Angew. Chem.* **1979**, *91*, 259. Angew. Chem. Int. Ed. Engl. **1979**, *18*, 239.
[2] D. Schinzer, *Nachr. Chem. Tech. Lab.* **1989**, 37, 172.
[3] E. Buncel, A. G. Davies, *J. Chem. Soc.* **1985**, 1550.
[4] D. Schinzer, *Nachr. Chem. Tech. Lab.* **1989**, 37, 28.
[5] G. Zweifel, S. J. Backlund, *J. Am. Chem. Soc.* **1977**, *99*, 3184.
[6] J. A. Miller, G. Zweifel, *J. Am. Chem. Soc.* **1981**, *103*, 6217.
[7] D. J. Peterson, *J. Org. Chem.* **1968**, *33*, 780.
[8] D. Schinzer, *Synthesis*, **1989**, 179.
[9] D. J. Ager, R. C. Cookson, *Tetrahedron Lett*, **1980**, *21*, 167.
[10] H. J. Reich, S. K. Shah, *J. Org. Chem.* **1977**, *42*, 1773.
[11] R. F. Cunico, *J. Organomet. Chem.* **1981**, *212*, C 51.
[12] G. Stork, E. W. Colvin, *J. Am. Chem. Soc.* **1971**, *93*, 2080.
[13] B.-G. Gröbel, D. Seebach, *Angew. Chem.* **1974**, *86*, 102. Angew. Chem. Int. Ed. Engl. **1974**, *13*, 83.
[14] H. Nishiyama, K. Itagaki, K. Sakuta und K. Itoh, *Tetrahedron Lett.*, **1981**, *22*, 5285.
[15] J. L. Speier, *Adv. Organomet. Chem.*, **1979**, *17*, 407.

[16] T. Hayashi, K. Tamao, Y. Katsuro, L. Nakae, M. Kumada, *Tetrahedron Lett.* **1980**, *21*, 1871.

[17] W. Bernhard, I. Fleming, D. Waterson, *J. Chem. Soc. Chem. Commun.* **1984**, 28.

[18] I. Fleming, R. Henning, H. Plaut, *J. Chem. Soc. Chem. Commun.* **1984**, 29.

[19] D. Häbich, F. Effenberger, *Synthesis* **1979**, 841.

[20] I. Fleming, J. D. Kilburn, *J. Chem. Soc. Chem. Commun*, **1986**, 305.

[21] G. Stork, "Merck-Schuchardt-Lectureship **1988**", Universität Hannover; Pure and Appl. Chem. **1989**, *61*, 439.

Temporary Silicon Connections

Martin Maier

Many chemical reactions can in principle lead to the formation of isomers. Therefore the efficiency of a certain transformation is determined not just by the yield but to what extent the selectivity can be controlled. The problem of selectivity can embrace many forms: *chemoselectivity* (differentiation between different functional groups of a molecule, *regioselectivity* (relative orientation of reacting components), *diastereoselectivity* (control of relative stereochemistry), and *enantioselectivity* (control of the absolute stereochemistry). [1] Usually, selectivity is determined by both reaction partners. Depending on which of the components exerts the dominant effect one refers to either of reagent- or substrate control. With a view to enantioselective syntheses, the development and application of chiral reagents that exert selectivity through reagent control is very important. On the other hand, in the synthesis of complex natural products, substrate control has been and still is the prominent strategy to induce, for example, regio- and diastereoselectivity. In this area, a high level of sophistication has been reached. [2] Nevertheless, there is still room for further progress since not all problems in the area of substrate control have been solved. [3] For example, the diastereoselective functionalization of a prochiral center across several bonds (remote stereocontrol) is only possible in special cases with high selectivity. Reactions in which the inherent substrate control leads to

the wrong isomer are also problematic in this regard.

An elegant possibility even to enforce regio- and diastereoselectivity involves using temporary connections, which through cyclic transition states or templates guarantee an unequivocal and predictable course of a reaction and which can be removed easily after the end of the reaction. Accordingly, carbon–heteroatom- and heteroatom–heteroatom bonds are be taken into consideration. Each of these bonds can appear as part of different functional groups. In and of itself, this strategy has been known for a long time, although recently, particularly through the use of silicon, there have been some interesting developments, which shall be described in this account.

The same regularities that make conventional intramolecular reactions so powerful also apply, of course, to the use of temporary connections, but in addition impose some restrictions. Because bifunctional substrates are used in ring formation, the formation of oligomers or polymers as byproducts is possible. By considering entropic effects and ring strain, some general rules for ring closure can be derived. [4] Cyclization reactions to give 5-7-membered rings proceed faster than the corresponding intermolecular reactions. Smaller rings (3- and 4-membered) form less readily because of ring strain in the transition state. The most difficult rings to close are medium-sized (8-11-membered) rings, because in these cases torsional and transannular interactions destabilize the transition state. Although larger rings (12- and higher membered) are almost free of strain, the unfavorable entropy prevents the reaction rates from being adequately high. Therefore, in the planning of temporary connections, the dependence of the activation barrier on the ring size must be taken into consideration. Depending on the number of atoms that are part of the temporary connection, one can easily reach a zone that is unfavorable for cyclization. There are other factors than ring size that influence the tendency of ring formation. Of great significance are steric effects on the chain connecting the reaction centers, such as backbone effects, [5] the *gem*-dialkyl effect, [6] and the oxygen effect. [4a] Stereoelectronic effects at the reaction centers that take part in the cyclization are primarily responsible for the selectivity, which makes temporary connections so attractive. Depending on the hybridization of the atom that is attacked in the course of the cyclization, we can distinguish between tetragonal, trigonal, and digonal processes. Reactions at such centers have ideal transition state geometries, which logically cannot be achieved for all ring sizes in a similiar way. Deviations from these ideal, strain-free geometries will considerably destabilize the corresponding transition state. Therefore, ring closing reactions that can give rise to isomers usually have a high preference in favor of one or other isomer. The ring closures rules, formulated by J. E. Baldwin, [7] are valid for nucleophilic, electrophilic, and radical reactions.

Why is silicon now especially so interesting for use as a position of cleavage? On the one hand, bonds of elements to silicon are longer than the corresponding bonds to carbon (cf. C–O = 143 pm; Si–O = 164 pm). Therefore, in cyclization reactions leading to small or middle-sized rings strain effects are smaller and the cyclization proceeds faster. On the other hand, silicon-containing connections can be installed very easily. Thus, dialkyldichlorosilanes are more reactive than the corresponding carbon compounds and are therefore quite appropriate for the preparation of silicon connections. In particular, silicon-carbon bonds, which can also be incorporated advantageously in temporary connections, may subsequently be used in several ways. For example, it is possible to cleave and functionalize such a bond by protodesilylation, halogenation, or by silicon-Baeyer–Villiger rearrangement. [8] Finally, reactions can be performed with silicon compounds that are not possible in the carbon series. An example is hydrosilylation, which can be used, particularly in the intramolecular variant, for the regio- and stereoselective functionalization of olefins and carbonyl groups.

However, this does not mean that other types of connections, such as those that contain C–O bonds, are less important. Given that various connections are predestined for certain applications, they complement each other. For example, one can use C–O connections in Wittig and Wittig–Horner reactions, respectively, whereby the intramolecular execution of the reaction secures the configuration of the newly formed double bond. [9]

In the area of cycloaddition reactions ester groups are frequently used to make the two components react in a regio- and stereocontrolled manner. [10] At the same time these tethers allow the incorporation of functional groups. Further possibilities for the application of temporary C–O connections lie in the regio- and diastereoselective functionalization of chiral olefins, such as homoallyl alcohols. [11]

In addition, temporary C–O connections are a valuable tool for stereocontrol in the area of radical reactions. [12] In one instance the electrophilic addition of allylic alcohols (often cyclic) to ethyl vinyl ethers is used to prepare mixed acetals *(1)* with a halogen atom in the α-position. The corresponding radicals

OR

Br

Bu₃SnH →

OR

(1)

(2)

OH

CHO

(3)

Si Br

Bu₃SnH →

O–Si

(4)

(5)

OH

X

X = H,
X = OH

(6)

Et

Br Si Cl

NEt₃ →

Et

OH

(7)

Br Si O

(8)

Bu₃SnH, AIBN ↓

Et

NaOH, H₂O₂ ←

Et

H

OH
OH

H

O
Si

(10)

(9)

which add with complete regio- and stereo-specificity to the double bond, can be generated from these halo acetals. [13] On the one hand, 5-*exo-trig*-cyclizations are favored compared with 6-*endo-trig*-cyclizations; on the other hand, a strain-free transition state is possible only in the case of a *cis*-annulation. A substantial extension of this methodology for the functionalization of allylic alcohols rests on the use of bromomethyldimethylsilyl ethers instead of the halogen acetals as radical precursors. [14] The corresponding radical cyclization also leads primarily to the 5-membered product *(5)*. Because this temporary connection can be dismantled to hydroxymethyl- or methyl groups, it represents a versatile procedure. It is precisely here that one recognizes how silicon connections expand the synthetic methodology.

Crimmins et al. used this method in the course of the synthesis of the spiroacetal talaromycin A *(10)*. The allylic alcohol *(7)* was converted to the ether *(8)* with bromomethyl-dimethylsilyl chloride. Radical cyclization led

to *(9)*, which furnished the hydroxymethyl group on oxidative cleavage of the carbon-silicon bond. [15] It is remarkable that the addition of the radical takes place to an enol ether, since carbon radicals exhibit rather nucleophilic character and react preferentially with electron-poor olefins. It speaks in favor of the addition that a radical stabilized with an oxygen atom is being formed. [16]

If the double bond to which the α-silyl radical adds is part of a ring, 5-*exo*-cyclizations are generally favored. The situation is less clear in case of exocyclic allyl alcohols, as shown by Koreeda. [17] With an α-silyl radical in an axial position and a *(Z)*-methyl group, 5-*exo*-cyclization (cf. *11* → *12*) is observed preferentially, except when this channel is obstructed on steric grounds. If the olefin has a *(E)*-configuration (cf. *13*), the two regioisomers *(12)* and *(14)* are formed in a 1:1 ratio, each of them stereoselectively. The resulting 5- and 6-membered siloxanes can be cleaved oxidatively to the corresponding diols. In the case of an equatorial α-silyl radical, the cy-

(11)

Bu3SnCl,
NaCNBH3,
AIBN
(90%)

(12)

(13)

ditto
(86%)

(14)

(15)

Cl–Si
Br
NEt3

(16)

(17)

(18)

H2O2,
KF (62%)

tBuOK
(62%)

(19)

(20)

NBS
(30%)

MeLi
(52%)

(21)

(22)

clization is less selective and proceeds in lower yield.

By adding a radical to a double bond, a new radical is formed, which for its part can add inter- or intramolecularly to another olefin. Such reaction cascades [18] can also be initiated from the appropriate bromomethyldimethylsilyl ethers and can be employed for the synthesis of polycyclic products. [19]

Another functionalized silicon reagent, 1-bromovinyldimethylsilyl chloride *(16)* can be used, after etherification with allylic alcohols, for the intramolecular addition of an α-silylvinyl radical to the double bond. Heating compound *(17)* with tributyltin hydride in the presence of catalytic amounts of azoisobutyronitrile (AIBN) furnishes mainly the 5-membered ring product *(18)* in addition to small amounts of the 6-membered ring and the reduced compound. The silicon connection may be removed or cleaved in various ways. Oxidation with hydrogen peroxide provides the acetyl derivative *(19)*, while treatment with base leads to the olefin *(20)*. Alternatively, conversions to the bromovinyl com-

pound *(21)* or the 1-trimethylsilyl compound *(22)* are possible. [20]

In order to control regio- and stereochemistry in the addition of an α-hydroxy radical to an allyl alcohol, Myers et al. made use of a connection via a dimethylsilyl acetal. Suitable cyclization substrates are available from aldehydes, for example *(23)*, phenylselenol, dimethyldichlorosilane and an allylic alcohol, for example *(24)*. Although, according to the Baldwin rules, 6-*exo*- and 7-*endo-trig*-cyclizations are stereoelectronically allowed, only formation of the 7-membered ring *(26)* is observed. Cleavage of the silyl acetal liberates the 1,4-diol structure. In connection with the synthesis of the antibiotic tunicaminyl uracil, this reaction served as a key reaction. The allylic alcohol *(27)* and the aldehyde *(28)* were first connected to compound *(29)*. After radical cyclization of *(29)* which, like in the model, gives the 7-membered ring, and subse-

acetonitrile with a ratio of 4:1. Removal of the protecting groups provided the natural product. [21]

In the above examples in which silicon ethers are used, the radical precursor is situated on the silicon side, that is α to the silicon atom. Conversely, silicon tethers can also be used to attach radical acceptors (olefins, acetylenes) to alcohols, which contain radical precursors. This strategy, pioneered by Stork, found use in the synthesis of prostaglandins and in the stereospecific synthesis of C-glycosides. [22] The appropriate substrates were prepared from phenylselenoglycosides, which bear a free hydroxyl group, and (phenylethinyl) dimethylsilyl chloride. The stereochemistry of the C-glycoside follows automatically from the stereochemistry (α or β) of the original OH group. Thus, radical cyclization of *(31)* followed by subsequent desilylation

quent hydrolytic removal of the temporary connection, the two diastereomers *(30a)* and *(30b)* were isolated. While the wrong diastereomer is formed primarily if toluene is used as a solvent, the desired isomer is preferred in with tetrabutylammonium fluoride gave the α-C-glycoside *(32)*. If the radical donor is in the β-position, the β-C-glycoside *(34)* is formed exclusively, although the substrate has to pass through a conformational change. The intramolecular *C*-glycoside formation proceeds satisfactorily even if in the course of the cyclization 6- or 7-membered rings must be formed. The additional expenditure for the preparation of the tethered substrates seems quite reasonable in the light of the good yield and the selectivity. It should be noted, however, that radicals at the anomeric center of

carbohydrates generally also react intermolecularly with olefins in a stereoselective manner. [23]

Temporary silicon connections may also be used advantageously in cycloaddition reactions. For acyclic dienes and dienopiles, respectively, which bear a stereocenter at the allylic position, the connection via a silicon acetal induces highly diastereoselective cycloaddition. Thermolysis of the substrate *(35)* leads primarily to the formation of the cycloadduct with *cis*-annulation of the two newly formed rings. To rationalize the observed selectivity, transition state *(38)* is suggested in which the methyl group points away from the reacting area, thereby evading destabilizing steric interactions. [24] If the stereocenter is in the dienophile part, the *cis*-product *(40)* can even be formed exclusively. In this work, it appeared, however, that the preparation of the silicon acetals from dialkyldichloro silanes was not without problems. In part the first connection step is not very selective, so that symmetrical acetals are formed to a significant extent as byproducts. Moreover, these acetals are not too stable towards silica gel chromatography. An essential improvement was proposed by Fortin et al. with the use of di-*tert*-butylsilylacetals. To guarantee a sequential attachment of alcohols to the silicon atom, di-*tert*-butylsilylchlorosilane monotriflate *(41)* was used as a reagent. Because of the graduated reactivity of the two leaving groups, selective functionalizations are possible. This was illustrated with the substrates *(44a)* and *(44b)*. First, the triflate was substituted with the alcohol *(42)* and, in a second step, the intermediate was treated with a dienolate to give the silicon acetal. The cycloaddition of *(44a)* and *(44b)* proceeds regiospecifically, although for electronic reasons the regioisomer would be favored. The observed stereoselectivity is explained in the *endo*-transition state *(47)*. [25] In the *exo*-transition state the residue R and one of the *tert*-butyl groups cross each others path.

(35)

toluene, 170 °C
(100%)

(36) 87   +   (37) 13

TS:

(38)

(39)

toluene, 170 °C
(100%)

(40)

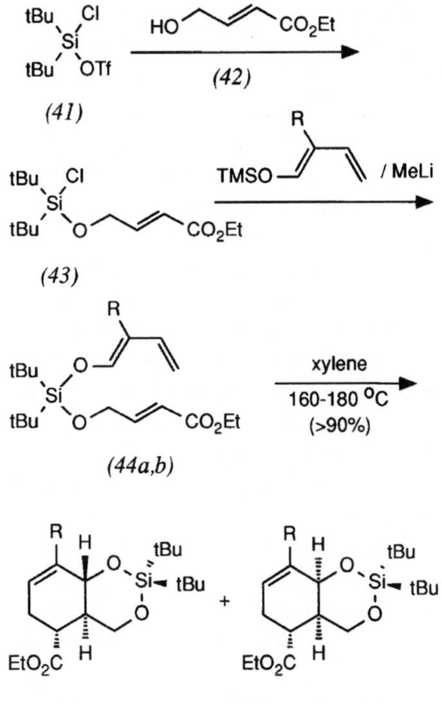

(41)   (42)

(43)

(44a,b)   xylene 160-180 °C (>90%)

(45a,b) + (46a,b)

	R	45/46
a	H	9 : 1
b	Me	99 : 1

TS: (47)

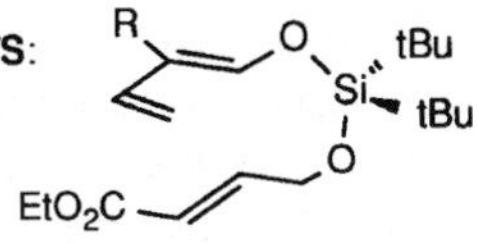

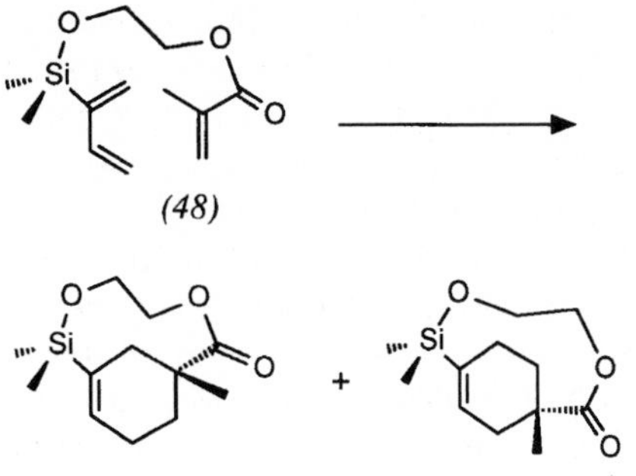

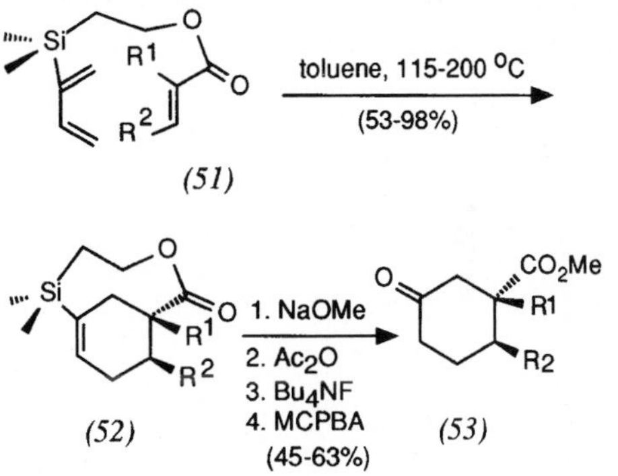

(48)

(49) 70   +   (50) 30

(51)   toluene, 115-200 °C (53-98%)

(52)   1. NaOMe 2. Ac₂O 3. Bu₄NF 4. MCPBA (45-63%)   (53)

R¹	R²
H	H
Me	H
H	Me
H	Ph
Me	Me
CN	Pr

Because both silicon-oxygen and silicon carbon bonds can be cleaved relatively easily, it is possible, if necessary, to vary the length of the connections. Thus, despite the fact that in the substrate *(48)* diene and dienophile are connected, one observes not only the *meta*-regioisomer *(49)* but also the *para*-isomer *(50)* which would be favored based on the polarization of the reactants. The complete pericyclic repolarization was successful only if the connection was shortened by one atom, such as in substrates of type *(51)*. Cyclohexanone deriv-

atives were obtained by oxidative cleavage of the vinyl silanes. [26]

In another variant of the temporary connection, the dienophile is connected via a carbon-silicon bond with the rest of the substrate. The corresponding substrates are easily accessible by etherification of dialkylvinylchlorosilanes with dienyl alcohols. Cycloaddition of *(54)* proceeds under mild conditions and delivers exclusively the *endo*-product *(55)*. The silicon connection in *(55)* can be removed by proto-desilylation or oxidative cleavage as occasion demands. These intramolecular cycloadditions are even possible without an activating group at the dienophile, although in these cases a somewhat higher temperature is necessary. Even so, it is possible to bring synthetic equivalents to a reaction that would normally not take place. Thus, by functionalization of the cycloadduct *(59)*, formal cycloaddition of ethylene or acetaldehyde-enol could be realized. [27]

The glycosylation of carbohydrates is also an area in which, due to the inherent selectivity of some glycosyl donors, not all glycosidic bonds are accessible with the same ease.

(54)

80 °C, 4-6 h
(90%)

CO₂Et

(55)

H₂O₂, F⁻
(80%)

Bu₄NF, DMF
(75%)

(56)

(57)

(58)

160 °C
(70%)

(59)

H₂O₂, F⁻
(85%)

Bu₄NF, DMF
(75%)

(60)

(61)

cis-1,2 / trans-1,2 = 70:30

These limitations hold true despite some very powerful leaving groups for the anomeric center of carbohydrates. [28] One of the problematic cases is the stereocontrolled construction of the β-D-mannopyranoside linkage. This represents a 1,2-*cis*-arrangement of the substituents that, for steric reasons and because of the anomeric effect, is disfavored compared with the 1,2-*trans*-stereochemistry. Accordingly, by intermolecular methods, this 1,2-*cis*-arrangement can only be reached with some expenditure. One possibility consists of forming a 1,2-*trans*-glycoside initially and subsequently inverting the configuration at C-2. [29] In a conceptually new procedure several groups now use temporary connections to transfer an aglycon intramolecularly to the anomeric center. [30] Because of the intramolecular reaction, the formation of 1,2-*cis*-glycosides is enforced. An example by Stork et al. might serve to illustrate the underlying principle: First, the 6-OH free compound *(62)* is converted with dichlorodimethylsilane to the chlorodimethylsilyl ether *(63)*. Subsequent reaction with the 2-OH free thioglycoside *(64)* yields the silicon acetal *(65)*. The essential glycosidation takes place via the sulfoxide *(66)*. Treatment with trifluorosulfonic acid anhydride in the presence of 2,6-di-*tert*-butylpyridine produce the β-D-mannopyranoside *(68)* exclusively. Presumably, the reaction proceeds through the intermediate *(67)*, which, after aqueous hydrolysis, leads directly to the glycoside *(68)*.

An important reaction from silicon chemistry is the transition-metal catalyzed hydrosilylation of olefins. [31] In combination with an oxidative cleavage of the silicon–carbon bond this procedure allows – similiar to the hydroboration – hydration of olefins. Particularly with regard to stereoselectivity, the intramolecular reaction is superior to the intermolecular variant. In this way higher substituted olefins can also react under mild conditions. On the other hand, this offers the opportunity to control the regioselectivity and additionally,

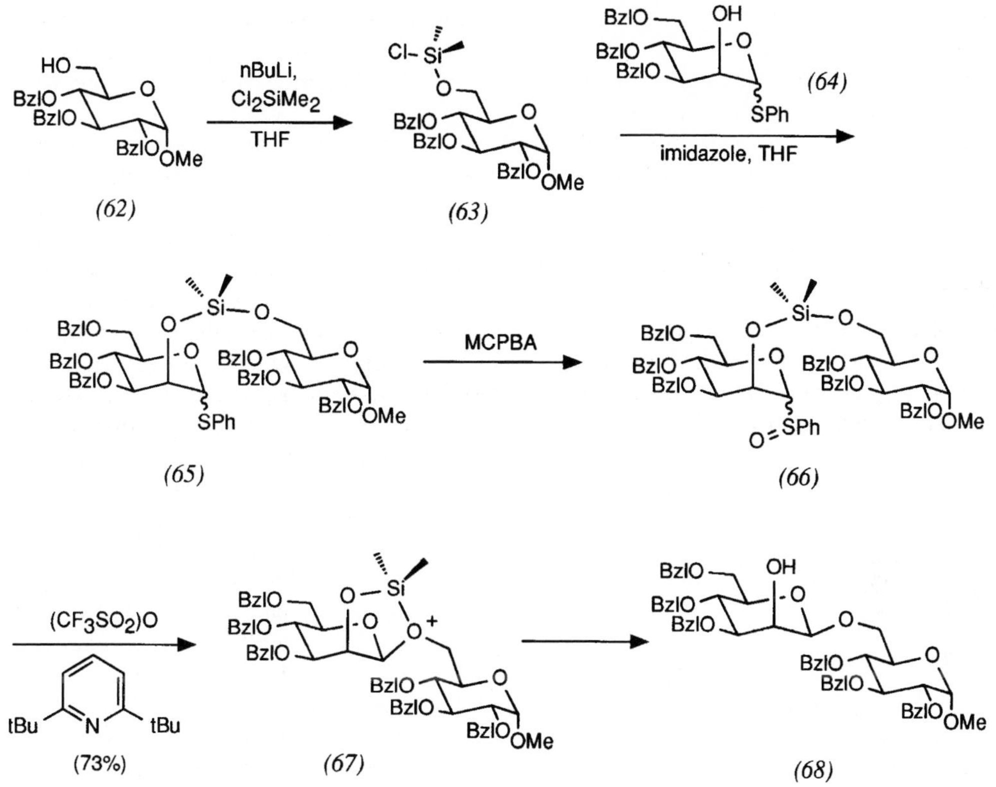

(62)　(63)　(64)

(65)　MCPBA　(66)

(CF₃SO₂)O (73%)　(67)　(68)

by using chiral olefins, to control diastereose-lectivity. As Tamao et al. could show in systematic studies, the intramolecular hydrosilylation of allyl- and homoallyl alcohols proceeds regioselectively, whereby 5- or 6-membered rings are formed preferentially. [32] In the case of chiral homoallylic alcohols that contain a Z-double bond or a trisubstituted double bond, one can achieve high stereoselectivity. [33] A typical substrate is the alcohol (69), which is first converted to the silyl ether (70). In the presence of a platinum catalyst, the 5-ring product (71) is formed, from which the diol (72) is obtained by subsequent oxidation. Mechanistically, the reaction proceeds through a hydrometalation, that is, first coordination of the olefin to the metal takes place, followed by insertion into the Si–H bond to give (73). Then the olefin inserts

into the metal-H bond with concomitant formation of the metallacycle (74), from which the catalyst is regenerated by reductive elimination. The stereochemical result is explained by assuming the transition state (73), in which the methyl group adopts a pseudoequatorial position.

Reasonably good selectivity can also be obtained in the hydrosilylation of allyl alcohols of type (75). In these cases the *syn*-isomers are formed preferentially. The method is therefore complementary to the intermolecular hydroboration, which furnishes the *anti*-isomers from such allylic alcohols. For the interpretation of the stereochemical result, the transition state (78) is invoked.

If chiral α-hydroxy enol ethers are used, this method opens the way to polyhydroxylated compounds. [34] For example, hydrosilylation

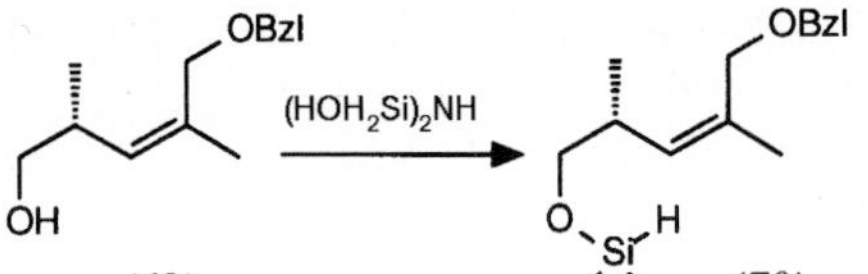

(69)

(70)

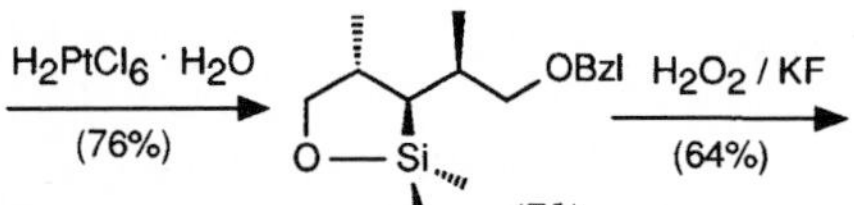

H₂PtCl₆ · H₂O
(76%)

(71)

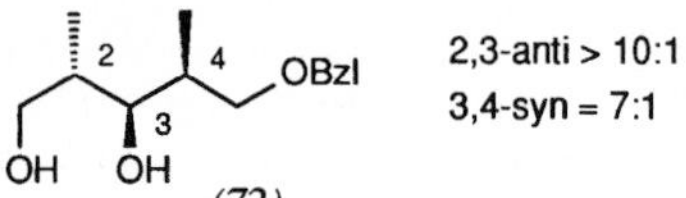

OBzl   H₂O₂ / KF
(64%)

(72)

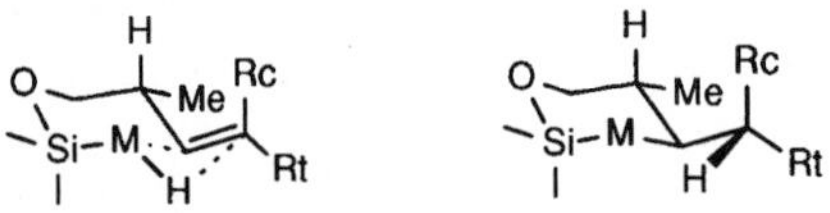

2,3-anti > 10:1
3,4-syn = 7:1

(73)

(74)

(75)

ditto

(76)

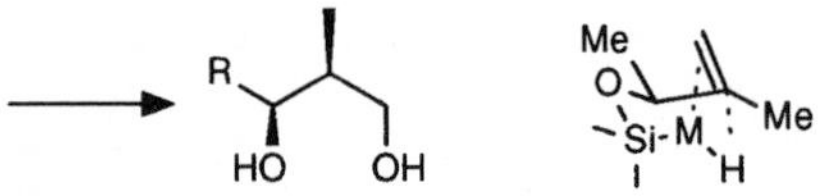

(77)

(78)

of (79) yields, after oxidative work-up, the triol derivative (81) stereoselectively. For these acid-labile enol ethers the use of a neutral platinum vinylsiloxane catalyst proved to be important. The stereochemistry can be rationalized by a transition state of type (78). The starting materials (cf. 79) are obtained by addition of metallated vinyl ethers to aldehydes. However, by this route they are not easily available in optically active form. In particular, the addition of metallated vinyl

ethers to chiral α-alkoxy aldehydes is unselective. Accordingly, this method has found almost no use in the synthesis of complex natural products. However, an alternative method for the preparation of α-hydroxy enol ethers, which is based on the Tebbe-olefination of chiral α-alkoxy esters, should offer advantages in this regard. Holmes et al. used this route in the synthesis of the enol ether (82) with an exocyclic double bond. Using a rhodium-catalyst, intramolecular hydrosilylation of (82) provided the stereoisomer (84) almost exclusively. Interestingly, the diastereoselectivity could be inverted by changing the catalyst and the reaction conditions. [35]

Instead of a hydroxyl group one can also use an amino group for the attachment of the silicon functionality. In contrast to the hydro-

(79)

1. (HMe₂Si)₂NH
2. Pt-catalyst

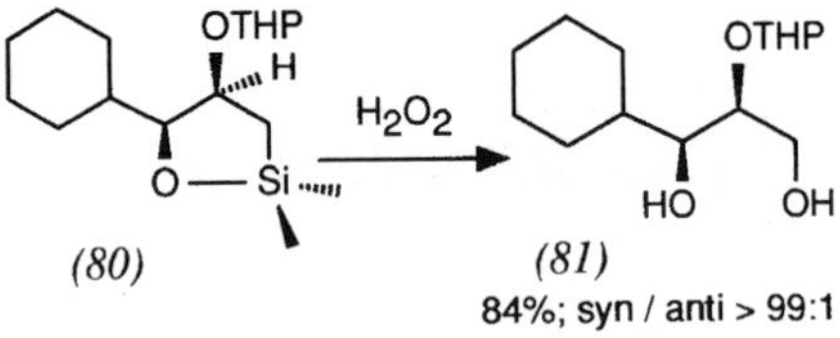

(80)

H₂O₂

(81)
84%; syn / anti > 99:1

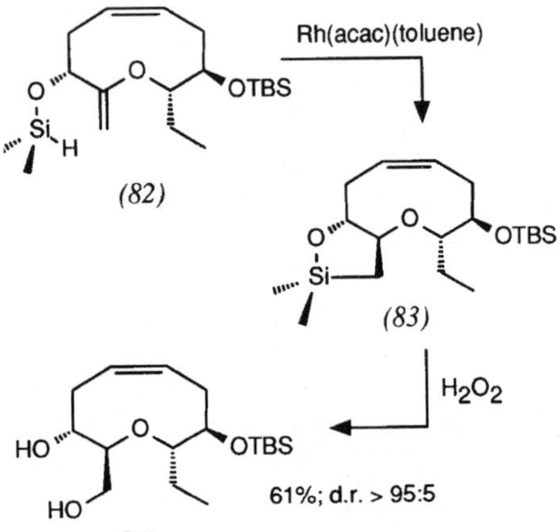

(82)

Rh(acac)(toluene)

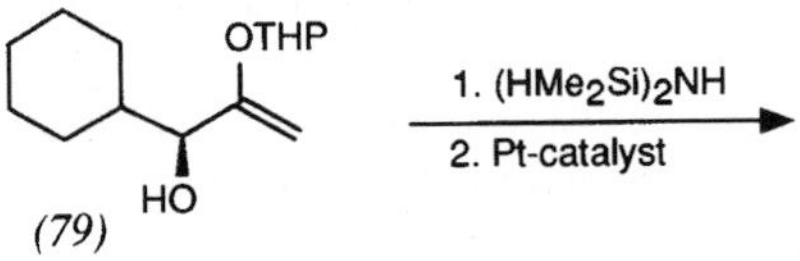

(83)

H₂O₂

(84)

61%; d.r. > 95:5

silylation of allylic alcohols, 1-aza-2-silacyclobutanes are obtained from allyl-amines. These heterocycles are converted to 1,2-amino alcohols on oxidation with hydrogen peroxide. [36]

The 1,3-arrangement of hydroxy groups is a common structural element, particularly in macrolides. Therefore, methods are in demand that allow the selective generation of either *syn-* or *anti*-diols. [37] The reduction of β-hydroxy-ketones *(85)* to *anti*-diols can be performed, among several methods, by intramolecular hydrosilylation. First an organosilane is attached to the hydroxy group and subsequently reduction of the carbonyl group is induced by Lewis-acid catalysis. The 1,3-diols *(88)* can be liberated from the *trans*-siladioxanes *(87)*, which are formed with high selectivity. For the interpretation of the intramolecular hydride transfer, the authors suggest a chair-like transition state. [38]

As this summary demonstrates, temporary silicon connections can be used in manifold ways for the solution of regio- and stereo-chemical problems. Certainly, this strategy is not restricted to the types of reactions and substrates presented herein, but may be broadened widely in its application. [39]

References

[1] B.M. Trost, *Science* **1991**, *254*, 1471–1477 and references therein.

[2] N. Anand, J.S. Bindra, S. Ranganathan, *Art in Organic Synthesis*, 2nd ed., Wiley, New York, **1988**; b) Synthesis of erythronolide B: J. Mulzer, H.M. Kirstein, J. Buschmann, C. Lehmann, P. Luger, *J. Am. Chem. Soc.* **1991**, *113*, 910–913.

[3] For a review on substrate-directable chemical reactions, see: A.H. Hoveyda, D.A. Evans, G.C. Fu, *Chem. Rev.* **1993**, *93*, 1307–1370.

[4] a) G. Illuminati, L. Mandolini, *Acc. Chem. Res.* **1981**, *14*, 95–102; b) M.A. Winnik, *Chem. Rev.* **1981**, *81*, 491–524; c) B. Capon, S.P. McManus, *Neighboring Group Participation*, Vol. 1, Plenum, New York, **1976**.

[5] D.F. DeTar, *J. Am. Chem. Soc.* **1974**, *96*, 1255–1256.

[6] a) M.E. Jung, J. Gervay, *J. Am. Chem. Soc.* **1991**, *113*, 224–232; b) see also: M.E. Jung, I.D. Trifunovich, N. Lensen, *Tetrahedron Lett.* **1992**, *33*, 6719–6722.

[7] J.E. Baldwin, *J. Chem. Soc., Chem. Commun.* **1976**, 734–736.

[8] Review: D. Schinzer, *Nachr. Chem. Tech. Lab.* **1989**, *37*, 263–266.

[9] D.A. Evans, E.M. Carreira, *Tetrahedron Lett.* **1990**, *31*, 4703–4706.

[10] [4+2]-Cycloadditions: D. Craig, *Chem. Soc. Rev.* **1987**, *16*, 187–238; [3+2]-Cycloadditions: B.M. Trost, T.A. Grese, *J. Am. Chem. Soc.* **1991**, *113*, 7363–7372; 1,3-Dipolar Cycloadditions: a) S. Takano, Y. Iwabuchi, K. Ogasawara, *J. Am. Chem. Soc.* **1987**, *109*, 5523–5524; b) M. Ihara, M. Takahashi, K. Fukumoto, T. Kametani, *J. Chem. Soc., Chem. Commun.* **1988**, 9–10; c) J.M. Sisko, S.M. Weinreb, *J. Org. Chem.* **1991**, *56*, 3210–3211.

[11] J.J.-W. Duan, P.A. Sprengler, A.B. Smith, III, *Tetrahedron Lett.* **1992**, *33*, 6439–6442 and references cited therein.

R¹	R²	anti/syn	yield (88)
iPr	iPr	120:1	67%
Bu	Bu	40:1	44%
iPr	Me	50:1	61%

[12] a) M. Regitz, B. Giese, *Methoden Org. Chem. (Houben-Weyl)* **1986**, Band E 19a and references cited therein; b) C. P. Japserse, D. P. Curran, T. L. Fevig, *Chem. Rev.* **1991**, *91*, 1237–1286; c) D. P. Curran in *Comprehensive Organic Synthesis* (Eds.: B. M. Trost, I. Fleming), Vol. 4 (Ed.: M. F. Semmelhack), Pergamon, Oxford, **1991**, 779–831.

[13] G. Stork, *Bull. Chem. Soc. Japan* **1988**, *61*, 149–154.

[14] a) H. Nishiyama, T. Kitajima, M. Matsumoto, K. Itoh, *J. Org. Chem.* **1984**, *49*, 2298–2300; b) G. Stork, M. J. Sofia, *J. Am. Chem. Soc.* **1986**, *108*, 6826–6828.

[15] M. T. Crimmins, R. O'Mahoney, *J. Org. Chem.* **1989**, *54*, 1157–1161.

[16] J. C. Lopez, B. Fraser-Reid, *J. Am. Chem. Soc.* **1989**, *111*, 3450–3452 and references cited therein.

[17] M. Koreeda, D. C. Visger, *Tetrahedron Lett.* **1992**, *33*, 6603–6607; see also: K. Matsumoto, K. Miura, K. Oshima, K. Utimoto, *Tetrahedron Lett.* **1992**, *33*, 7031–7034.

[18] For a review on sequential reactions in organic synthesis, see: L. F. Tietze, U. Beifuss, *Angew. Chem.* **1993**, *105*, 137–170; *Angew. Chem. Int. Ed. Engl.* **1993**, *32*, 131–163.

[19] M. Jounet, W. Smadja, M. Malacria, *Synlett* **1990**, 320–321.

[20] K. Tamao, K. Maeda, T. Yamaguchi, Y. Ito, *J. Am. Chem. Soc.* **1989**, *111*, 4984–4985.

[21] A. G. Myers, D. Y. Gin, K. L. Widdowson, *J. Am. Chem. Soc.* **1991**, *113*, 9661–9663.

[22] G. Stork, H. Suh Suh, G. Kim, *J. Am. Chem. Soc.* **1991**, *113*, 7054–7056.

[23] B. Giese, *Pure Appl. Chem.* **1988**, *60*, 1655–1658.

[24] a) D. Craig, J. C. Reader, *Tetrahedron Lett.* **1992**, *33*, 4073–4076; b) D. Craig, J. C. Reader, *Tetrahedron Lett.* **1992**, *33*, 6165–6168.

[25] J. W. Gillard, R. Fortin, E. L. Grimm, M. Maillard, M. Tjepkema, M. A. Bernstein, R. Glaser, *Tetrahedron Lett.* **1991**, *32*, 1145–1148.

[26] K. J. Shea, A. J. Staab, K. S. Zandi, *Tetrahedron Lett.* **1991**, *32*, 2715–2718.

[27] a) G. Stork, T. Y. Chan, G. A. Breault, *J. Am. Chem. Soc.* **1992**, *114*, 7578–7579; b) S. M. Sieburth, L. Fensterbank, *J. Org. Chem.* **1992**, *57*, 5279–5281.

[28] a) R. R. Schmidt in *Comprehensive Organic Synthesis* (Eds.: B. M. Trost, I. Fleming), Vol. 6 (Ed.: E. Winterfeldt), Pergamon, Oxford, **1991**, 33–64; b) H. Waldmann, *Nachr. Chem. Tech. Lab.* **1991**, *39*, 675–682.

[29] H. B. Boren, G. Ekborg, K. Eklind, P. J. Garegg, A. Pilotti, C.-G. Swahn, *Acta Chem. Scand.* **1973**, *27*, 2639; see also: H. Paulsen, O. Lockhoff, *Chem. Ber.* **1981**, *114*, 3102.

[30] a) G. Stork, G. Kim, *J. Am. Chem. Soc.* **1992**, *114*, 1087–1088; b) M. Bols, *J. Chem. Soc., Chem. Commun.* **1992**, 913–914; c) F. Barresi, O. Hindsgaul, *Synlett* **1992**, 759–760; d) Y. C. Xin, J.-M. Mallet, P. Sinay, *J. Chem. Soc., Chem. Commun.* **1993**, 864–865.

[31] T. Hiyama, T. Kusumoto in *Comprehensive Organic Synthesis* (Eds.: B. M. Trost, I. Fleming), Vol. 8 (Ed.: I. Fleming), Pergamon, Oxford, **1991**, 763–792.

[32] K. Tamao, T. Tanaka, T. Nakajima, R. Sumiya, H. Arai, Y. Yto, *Tetrahedron Lett.* **1986**, *27*, 3377–3380.

[33] K. Tamao, T. Nakajima, R. Sumiya, H. Arai, N. Higuchi, J. Ito, *J. Am. Chem. Soc.* **1986**, *108*, 6090–6093; see also: M. R. Hale, A. H. Hoveyda, *J. Org. Chem.* **1992**, *57*, 1643–1645.

[34] K. Tamao, Y. Nakagawa, H. Arai, N. Higuchi, Y. Yto, *J. Am. Chem. Soc.* **1988**, *110*, 3712–3714.

[35] N. R. Curtis, A. B. Holmes, *Tetrahedron Lett.* **1992**, *33*, 675–678.

[36] K. Tamao, Y. Nakagawa, Y. Ito, *J. Org. Chem.* **1990**, *55*, 3438–3439.

[37] D. A. Evans, J. Gauchet-Prunet, E. M. Carreira, A. B. Charette, *J. Org. Chem.* **1991**, *56*, 741–750.

[38] S. Anwar, A. P. Davis, *J. Chem. Soc., Chem. Commun.* **1986**, 831–832.

[39] a) Intramolecular Dötz reaction: B. L. Balzer, M. Cazanoue, G. Finn, *J. Am. Chem. Soc.* **1992**, *114*, 8735; b) Intramolecular amidoalkylation: H. Hioki, M. Okuda, W. Miyagi, S. Ito, *Tetrahedron Lett.* **1993**, *34*, 6131–6134.

D. Enzymes in Organic Synthesis

Enzymatic Carbon–Carbon Bond Formation

Herbert Waldmann

Methods for the stereoselective formation of C–C bonds lie at the heart of organic synthesis. For this purpose powerful chiral auxiliaries and, recently, chiral catalysts have been developed during the last decades. Moreover, these methods are being complemented by biocatalyzed C–C bond forming reactions [1] employing either isolated enzymes or whole cells (e.g. yeast) as biocatalysts.

Aldolases

Practically every organism has developed aldolases for the execution of stereoselective aldol reactions. More than twenty C–C bond forming enzymes belonging to this class have been characterized in detail, and some of them have already been used for synthetic purposes. The catalytic protein that has been studied most thoroughly and that has been applied most often on a preparative scale is fructose-1,6-diphosphate aldolase from rabbit muscle (FDP aldolase, RAMA). It catalyzes the stereoselective addition of dihydroxyacetone phosphate *(1)* (DHAP) to various aldehydes (Scheme 1). [1a, b, 2, 4] In addition, enzymes displaying similar substrate specificity and stereoselectivity, but showing a higher long-term stability were isolated from *E. coli* [3a] and from *Staphylococcus carnosus*. [3b] The enzyme from rabbits is fairly specific for

the ketophosphate, but it accepts aldehydes with broadly varying structure, so that, overall, more than 75 of these carbonyl compounds could be identified as substrates. These studies revealed that sterically unhindered aliphatic and α-heterosubstituted aldehydes are rapidly converted to the aldol adducts, whereas tertiary, α,β-unsaturated and, in particular, aromatic compounds are not accepted or are only poor substrates. In all aldol reactions carried out on a preparative scale so far, the enzyme showed virtually complete stereospecificity and always formed a D-*threo*-diol. Recently, however, this has been questioned (see below). FDP aldolase has been used in manifold ways for the construction of monosaccharides embodying four to nine carbon atoms in the chain. [1a, b, 2a, 4] Thus, the D-*threo*-hexulose phosphate *(3)* could be built up on a 1 molar scale from DHAP *(1)* and propionaldehyde *(2)* (Scheme 1). [2a] The subsequent removal of the phosphate by means of acid phosphatase delivered the deoxygenated monosaccharide *(4)*. A further interesting example consists of the synthesis of 3-deoxy-D-*arabino*-heptulosonic acid-7-phosphate *(7)*, [2b] a central intermediate in the biosynthesis of aromatic amino acids via the shikimic acid pathway. In the course of this chemoenzymatic route the racemic aldehyde *(5)* derived from aspartic acid was converted by the enzyme to the desired aldol adduct *(6)*, which

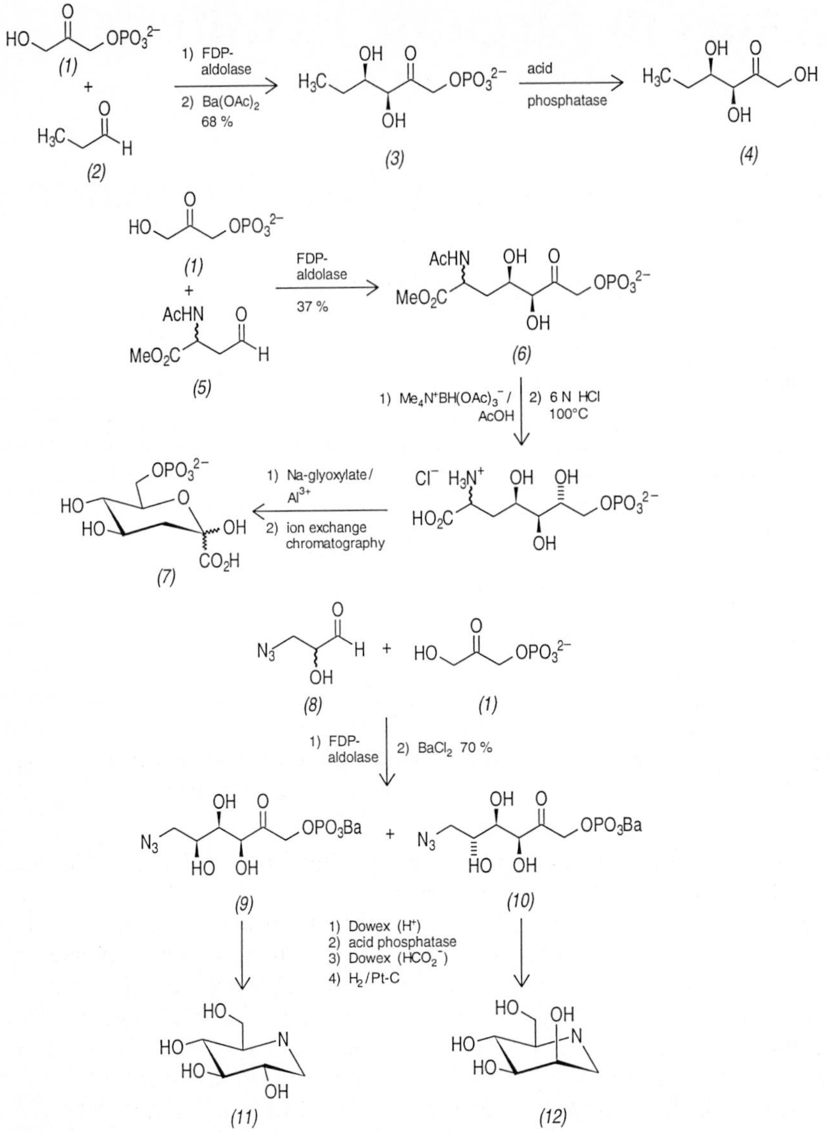

Scheme 1. Enzymatic aldol reactions employing FDP aldolase as biocatalyst.

was then transformed to the heptose *(7)*. A particularly valuable feature offered by the aldolase is that it allows rapid syntheses of polyfunctional carbohydrate analogues to be carried out, for example, deoxy-, [4a] fluoro-, [4b] alkoxy-, [4b] C-alkyl-, [4c, h] and aza-substituted [4a–g] saccharides. In particular,

the analogues mentioned last are of current interest, since the piperidines and pyrrolidines accessible from these compounds, for example, 1-deoxynojirimicin *(11)* and 1-deoxymannojirimicin *(12)* (Scheme 1), are potent glycosidase inhibitors that may be used in numerous different pharmacological studies.

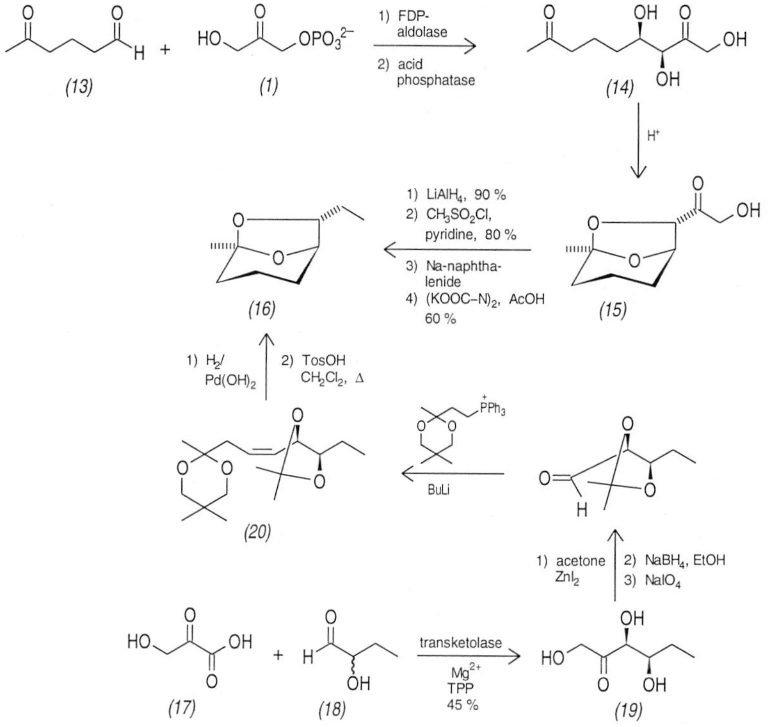

Scheme 2. Chemoenzymatic syntheses of (+)-*exo*-brevicomin employing RAMA or transketolase as biocatalysts.

To build up *(11)* and *(12)* [4d, e] the racemic aldehyde *(8)* and DHAP were condensed under the influence of the aldolase to give the diastereomeric 6-azidohexose phosphates *(9)* and *(10)* quantitatively and in a stereochemically defined way. These carbohydrate analogs were then separated chromatographically and the phosphoric acid esters hydrolyzed by means of acid phosphatase. In the course of the subsequent hydrogenolysis over platinum the azides were transformed into amines, which formed cyclic imines with the carbonyl group of the furanose, only to be hydrogenated further to the piperidines *(11)* and *(12)*.

The fact that aldolase can also be used advantageously for the construction of natural products that are structurally unrelated to carbohydrates was demonstrated by a chemoenzymatic synthesis of the bark beetle phero-

mone (+)-*exo*-brevicomin *(16)*. [5a] As the key step the enzyme-catalyzed conversion of 5-oxohexanal *(13)* was employed, which gave only the desired stereoisomer (Scheme 2). After removal of the phosphate and treatment with acid, the ketotriol *(14)* formed the bicyclic brevicomin precursor *(15)*. Subsequent reduction of the keto group and deoxygenation of the vicinal diol formed thereby delivered the enantiomerically pure pheromone. In a further application, also the naturally occurring spiroketal sphydrofuran [5b] and the C_{11}–C_{16} fragment of pentamycin [5c] were constructed in chemoenzymatic approaches employing FDP aldolase. Brevicomin *(16)* can also be constructed chemoenzymatically by means of the C–C bond forming enzyme transketolase. [6b] In the presence of thiamine pyrophosphate (TPP) and Mg^{2+}-ions this biocatalyst decarboxylates α-hydroxypyruvate *(17)* and transfers its α-hydroxyketone unit to the D-enantiomer of 2-hydroxyaldehydes to form D-*threo*-diols. [6] Taking advantage of this property the racemic 2-hydroxybutyraldehyde *(18)* was converted with simultaneous resolution of the racemate to the triol *(19)* in 45 % yield and with > 95 % diastereoselectivity. The geminal diol present in *(19)* was then protected and the α-hydroxyketone degraded to the aldehyde. After chain-extension via a Wittig reaction, reduction of the resulting olefin *(20)* and removal of the acetal protecting groups by treatment with acid, enantiomerically pure (+)-*exo*-brevicomin *(16)* was also obtained by this route.

Whereas the fructosediphosphate aldolase generates the D-*threo* stereochemistry (3S, 4R), the three remaining stereoisomers can be formed by employing the enzymes rhamnulose-1-phosphate aldolase (RhuA: L-*threo*; 3R, 4S), [7b, d] fuculose-1-phosphate aldolase (FucA: D-*erythro*; 3R, 4R) [7a, b, d] and tagatose-1-phosphate aldolase (TagA: L-

erythro; 3S, 4S) [7c] (Scheme 3). These biocatalysts are not commercially available, but must be isolated from microorganisms. In addition, for RhuA and FucA genetically engineered overexpression systems have been developed, [7a, b] so that these aldolases are available in larger amounts if desired. The aldol additions of DHAP to various aldehydes mediated by these enzymes proceed with complete stereoselectivity with respect to the stereogenic center at C_3; however, depending on the structure of the electrophile, varying amounts of the undesired stereoisomers may be induced (RhuA and FucA: up to 30 %, TagA: 10–90 %; Scheme 3). A closer inspection of the stereodifferentiation in the FDP-aldolase mediated transformations (see above) revealed that depending on the aldehyde used in these aldol-reactions – in contrast to earlier reports – up to 30 % of the "wrong" configuration at C_4 of the aldol adduct is formed. [7e] Although classical chemical aldol technology also often meets with varying success concerning its general applicability, these results clearly reduce the

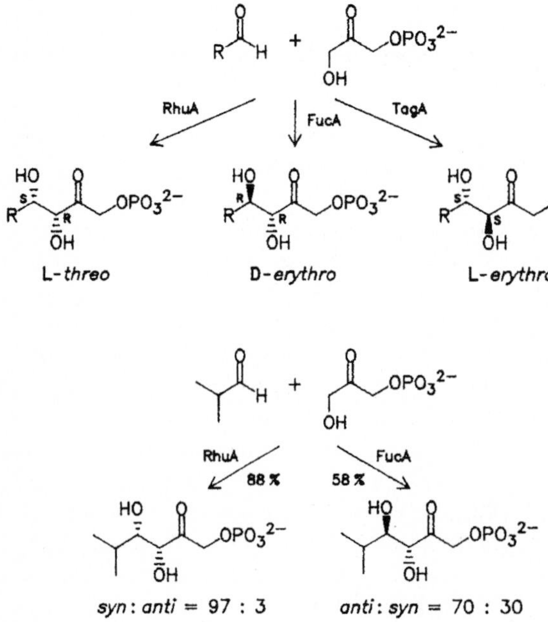

Scheme 3. Enzymatic aldol reactions mediated by FucA-, RhuA- and TagA aldolases.

willingness of synthetic organic chemists to establish enzymatic C–C bond formation as a standard methodology in organic synthesis, in particular if the biocatalyst must be isolated or cloned in a laborious enterprise.

In addition to the aldolases that use DHAP as nucleophile in aldol reactions, some enzymes that employ different CH-acidic compounds have also been applied for preparative purposes. Thus, 3-hexulosephosphate synthetase (HPS aldolase), [8a] 3-deoxy-D-*arabino*-heptulosonic acid phosphate synthetase (DAHP synthetase), [8b] 3-deoxy-D-*manno*-2-octulosonic acid-8-phosphate synthetase (KDO synthetase), [8c] and N-acetyl-neuraminic acid aldolase (NeuAc aldolase, NANA aldolase) [8d] have been used for the construction of complex carbohydrates (Scheme 4). Furthermore, enzyme-catalysed aldol reactions have been carried out with a cloned 2-deoxyribose-5-phosphate aldolase. [8h]

High expectations have been raised in particular concerning the application of NeuAc aldolase. This biocatalyst accelerates the reaction between pyruvate and N-acetyl-

mannosamine (ManNAc) *(22)* and accepts a variety of analogs of this monosaccharide as substrate. [8d] The product of the occurring aldol reaction is N-acetylneuraminic acid *(23)*, a complex carbohydrate that plays a decisive role in biological recognition phenomena. The availability of neuraminic acids with modified structure is of great interest for biological and immunological studies, and the preparation of these compounds by classical chemical techniques is very cumbersome. Continuous NeuAc aldolase catalyzed aldol reactions could be realized advantageously in an enzyme-membrane reactor, including the in situ generation of the expensive ManNAc from the inexpensive N-acetylglucosamine *(21)* by an epimerase in a coupled enzyme system. [8g] By means of this technology the enzymatic generation of N-acetylneuraminic acid on a large scale may become feasible.

(R)- and *(S)-*Oxynitrilase

(R)-Oxynitrilase from bitter almonds catalyzes the highly enantioselective addition of

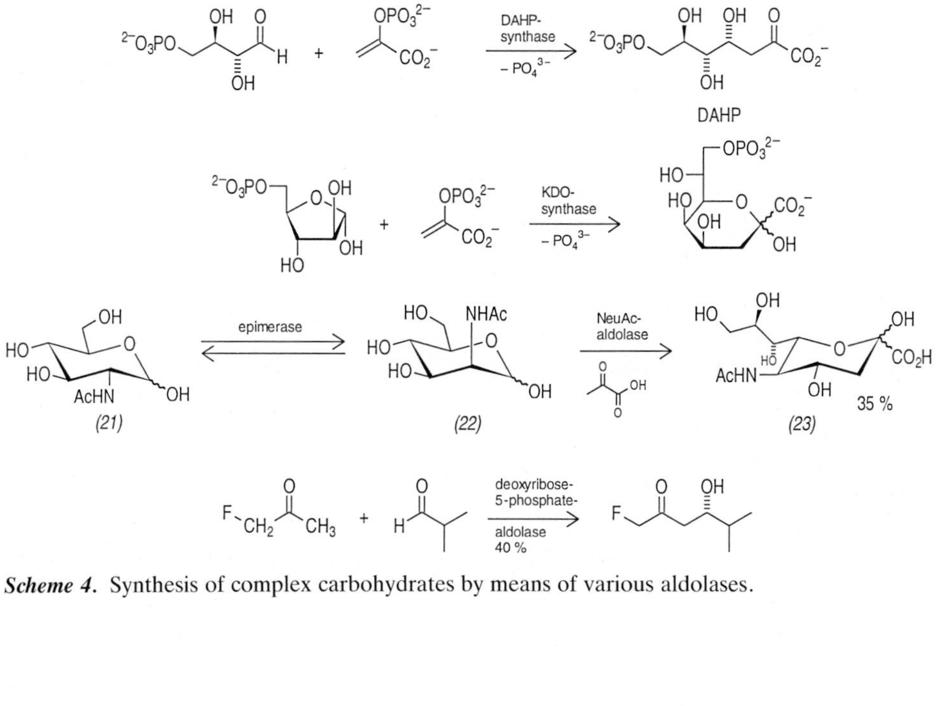

Scheme 4. Synthesis of complex carbohydrates by means of various aldolases.

(R)-oxynitrilase:
R = alkyl, aryl, 75 – 99 %, 14 – 99 % *ee*
(S)-oxynitrilase:
R = aryl, 61 – 97 %, 54 – 99 % *ee*

HCN to aromatic and aliphatic aldehydes to give the *(R)*-cyanohydrins *(24)*. [9a, b] Unfortunately, due to the competing nonenzymatic reaction the optical yield of the reaction often is low if aqueous reaction media are used. This drawback was successfully overcome by adsorbing the enzyme onto cellulose and using it in this immobilized form as biocatalyst in ethyl acetate, a solvent in which the nonenzymatic cyanohydrin formation proceeds only slowly. [9b] However, in the cases of sterically demanding carbonyl compounds (e.g. pivalaldehyde) and aldehydes carrying a basic atom (e.g., nicotinaldehyde) the non-biocatalyzed reaction cannot be suppressed completely even in organic solution, so that the enantiomeric excess remains unsatisfactory if such electrophiles are used. By employing an oxynitrilase from *Sorghum bicolor* [9c, d] or from *Hevea brasiliensis* [9f] the enantiomeric *(S)*-cyanohydrins *(25)* are also accessible with moderate to high *ee* values. The biocatalyst from *Sorghum bicolor* accepts only aromatic aldehydes as substrates. Once more ethyl acetate as solvent is recommended to achieve a high enantiomeric excess. [9c] However, in these cases the use of an aqueous solvent, the pH of which is lowered to 3–4, is more advantageous. [9d] *(R)*-Oxynitrilase is also capable of synthesizing ketone cyanohydrins with high stereoselectivity. [9e] For the enantioselective enzymatic synthesis of ali-

phatic *(S)*-cyanohydrins the enzyme from *Hevea brasiliensis* is currently the best biocatalyst available. [9f]

Baker's Yeast

The fermentation of suitable substrates with baker's yeast opens a route to various C–C bond formations. [1c, d] The yeast-mediated acyloin condensation of aromatic and aliphatic aldehydes with an acetaldehyde equivalent (Scheme 5) has been the subject of intensive studies. [10] The enzyme responsible for this transformation is pyruvate decarboxylase. Similar to transketolase (see above) in the presence of Mg^{2+} ions and TPP this biocatalyst transfers a C_2-unit, which formally resembles acetaldehyde, to the aldehyde substrate. Thereby, initially α-hydroxyketones like *(26)* are formed, which may either be isolated under suitable conditions or which may be further reduced by the yeast (Scheme 5). Although the yields are low, in this easily executable reaction the diols *(27)* are formed as virtually pure *anti*-isomers with excellent *ee* values. Using benzaldehyde as carbonyl component, this process was applied for the commercial manufacture of D-(–)-ephedrine *(28)*. This route belongs to the first industrial processes in which microbial and classical-chemical steps were combined advantageously. [11] The enzymatic acyloin condensation is not restricted to aromatic aldehydes and acetaldehyde: aromatic and particularly α,β-unsaturated carbonyl compounds can also be used. Instead of acetaldehyde from the corresponding keto acids with simultaneous decarboxylation a fragment can be transferred that is equivalent to propionaldehyde or butyraldehyde. [12] Diols like *(27)* have found numerous applications as chiral building blocks in natural product syntheses. [1c, d] For instance, by means of this technique from cinnamic aldehyde several L-carbohydrates and pheromones were constructed, and α-

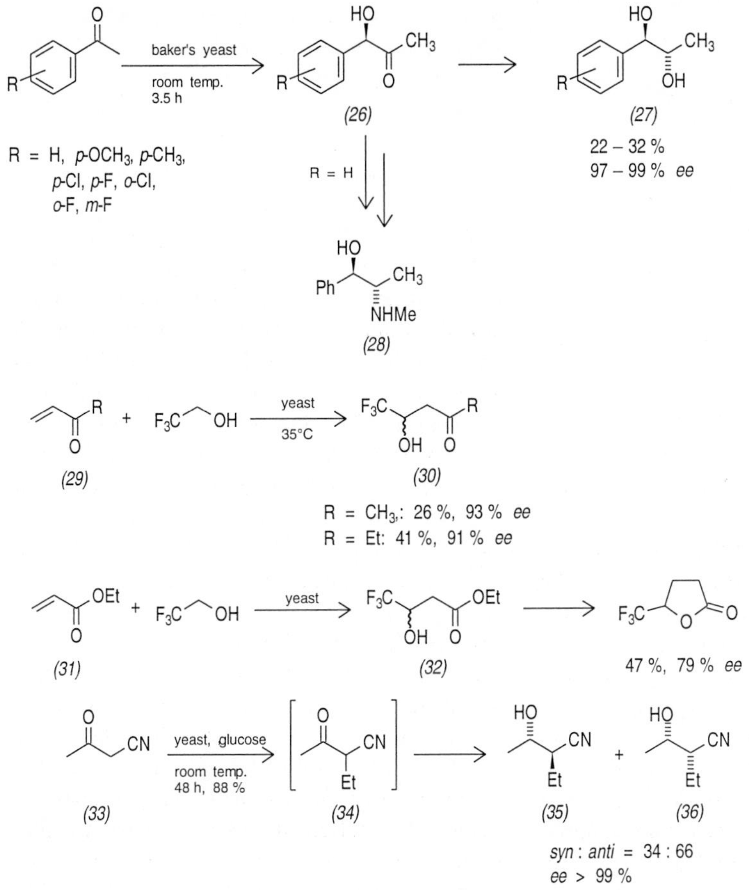

Scheme 5. Enzymatic C–C bond formation employing baker's yeast as biocatalyst.

methyl-β-2-furylacrolein could be converted to enantiomerically pure vitamin E. Moreover, processes that resemble a Michael addition or the alkylation of a CH-acidic compound can be carried out by yeast [13] (Scheme 5). If α,β-unsaturated esters and ketones like *(29)* and *(31)* are fermented in the presence of trifluoroethanol, the trifluoromethylcarbinols *(30)* and *(32)* are formed in yields of 26–47 % and with *ee* values of 79–93 %. [13a] Yeast also transforms α-cyanoacetone *(33)* to a mixture of the nitriles *(35)* and *(36)*. [13b] The ketone *(34)* is formed as an intermediate that is reduced with high

stereoselectivity. The enzymes responsible for these conversions are not known, but it is prudent to speculate that in the first case pyruvate decarboxylase initiates a transformation in the sense of a Stetter reaction followed by a reduction step. In the second case a yeast-mediated reduction of a C=C bond formed by aldol condensation between *(33)* and acetaldehyde may be assumed.

Furthermore, baker's yeast contains a sterolcyclase activity which can be advantageously employed to effect the cyclization of the squaleneoxides *(37)*. [14] After ultrasound treatment the steroid frameworks *(38)* are

(37)

baker's yeast
pH 7.4
ultrasound

R = CO₂CH₃: 28 h, 62 %
R = CH₃: 12 h, 84 %

R = CO_2CH_3: 28 h, 62 %
R = CH_3: 12 h, 84 %

(38)

constructed in the enzyme-catalyzed reactions with simultaneous resolution of the racemates. Finally, the steric steering of Diels–Alder reactions by using baker's yeast was reported. [15] Maleic acid *(39)* and cyclopentadiene in organic and aqueous media usually react to give predominantly the *endo*-cycloadduct *(40)* whereas in the presence of the microorganism the *exo*-isomer *(41)* is formed exclusively. However, it remains uncertain whether this effect is indeed caused

(39)

(40)   (41)

endo   *exo*

organic medium:	endo : exo = 80 : 20
aqueous medium without ylast:	endo : exo = 93 : 7
aqueous medium with ylast:	endo : exo = 0 : 100
	yield 74 %

by the influence of a specific biocatalyst or whether unspecific hydrophobic proteins are responsible (such observations have, for instance, been made for bovine serum albumine [16]).

In conclusion, the examples highlighted demonstrate that by means of appropriate biocatalysts a variety of synthetically important C–C bond formations can be carried out stereoselectively in a convincing way. However, for most of the reaction types discussed powerful classical chemical alternatives are also available. Both methodologies represent important tools available from the arsenal of organic synthesis. The choice as to which is to be preferred must be directed by the demands posed by the problems at hand and must be critically evaluated in each individual case.

References

[1] Reviews: a) E.J. Toone, E.S. Simon, M.D. Bednarski, G.M. Whitesides, *Tetrahedron* **1989**, *45*, 5365; b) D.G. Drueckhammer, W.J. Hennen, R.L. Pederson, C.F. Barbas III, C.M. Gautheron, T. Krach, C.-H. Wong, *Synthesis* **1991**, 499; c) S. Servi, *Synthesis* **1990**, 1; d) R. Csuk, B.I. Glänzer, *Chem. Rev.* **1991**, *91*, 49.

[2] a) A comprehensive discussion of the properties of FDP-aldolase relevant to organic synthesis is given in: M.D. Bednarski, E.S. Simon, N. Bischofberger, W.-D. Fessner, M.-J. Kim, W. Lees, T. Saito, H. Waldmann, G.M. Whitesides, *J. Am. Chem. Soc.* **1989**, *111*, 627 and references given therein; b) N.J. Turner, G.M. Whitesides, *J. Am. Chem. Soc.* **1989**, *111*, 624; c) W.J. Lees, G.M. Whitesides, *J. Org. Chem.* **1993**, *58*, 1887 and references therein.

[3] a) C.H. von der Osten, A.J. Sinskey, C.F. Barbas III, R.L. Pederson, Y.-F. Wang, C.-H. Wong, *J. Am. Chem. Soc.* **1989**, *111*, 3924 and references therein; b) H.P. Brockamp, M.R. Kula, *Tetrahedron Lett.* **1990**, *49*, 7123.

[4] a) C.-H. Wong, F.P. Mazenod, G.M. Whitesides, *J. Org. Chem.* **1983, 48**, 3493; b) J.R.

Durrwachter, D.G. Drueckhammer, K.No-
zaki, H.M. Sweers, C.-H.Wong, *J. Am.
Chem. Soc.* **1988,** *108*, 7812; c) J.R. Durrwachter, C.-H.Wong, *J. Org. Chem.* **1988,** *53*, 5175;
d) A.Straub, F.Effenberger, P.Fischer, *J. Org.
Chem.* **1990,** *55*, 3926; e) R.L. Pederson, M.-
J.Kim, C.-H.Wong, *Tetrahedron Lett.* **1988,**
37, 4645; f) T.Kajimoto, K.K.-C. Liu, R.L.
Pederson, Z.Zhong, Y.Ichikawa, J.A. Porko,
Jr, C.-H.Wong, *J. Am. Chem. Soc.* **1991,** *113*,
6187; g) R.H. Hung, J.A. Straub, G.M.
Whitesides, *J. Am. Chem. Soc.* **1991,** *56*, 3849;
h) W.-D. Fessner, C.Walter, *Angew. Chem.*
1992, *104*, 643; *Angew. Chem. Int. Ed. Engl.*
1992, *31*, 614.

[5] a) M.Schultz, H.Waldmann, H.Kunz,
W.Vogt, *Liebigs Ann. Chem.* **1990,** 1010; b)
B.P. Maliakel, W.Schmid, *Tetrahedron Lett.*
1992, *33*, 3297; c) M.Shimagaki, H.Mune-
shima, M.Kubota, T.Oishi, *Chem. Pharm.
Bull.* **1993,** *41*, 282.

[6] a) J.Bolte, C.Demuynck, H.Samaki, *Tetrahedron Lett.* **1987,** *28*, 5525; b) D.C. Miles,
P.J. Andrulis III, G.M. Whitesides, *Tetrahedron Lett.* **1991,** *32*, 4835; c) Y.Kobori, D.C.
Myles, G.M. Whitesides, *J. Org. Chem.* **1992,**
57, 5899; d) F.Effenberger, V.Null, T.Ziegler,
Tetrahedron Lett. **1992,** *33*, 5157.

[7] a) A.Ozaki, E.J. Toone, C.H. von der Osten,
A.J. Sinskey, G.M. Whitesides, *J. Am. Chem.
Soc.* **1990,** *112*, 4970; b) W.-D.Fessner, G.Sine-
rius, A.Schneider, M.Dreyer, G.E. Schulz,
J.Badia, J.Aguilar, *Angew. Chem.* **1991,** *103*,
Angew. Chem. Int. Ed. Engl. **1991,** *30*, 555;
596; Angew. Chem. Int. Ed. Engl. **1991,** *30*,
555; c) W.-D. Fessner, O.Eyrisch, *Angew.
Chem.* **1992,** *103*, 76; Angew. Chem. Int. Ed.
Engl. **1992,** *31*, 56; d) W.-D. Fessner, J.Badia,
O.Eyrisch, A.Schneider, G.Sinerius, *Tetra-
hedron Lett.* **1992,** *33*, 5231; e) W.-D. Fessner,
Fourth Chemical Congress of North America,
New York, **1991,** Abstracts of Papers BIOL 9;
W.-D. Fessner, personal communication.

[8] a) R.Beisswenger, G.Snatzke, J.Thiem, M.-
R. Kula, *Tetrahedron Lett.* **1991,** *32*, 3159; b)
L.M. Reimer, D.L. Conley, D.L. Pompliano,
D.L. Frost, *J. Am. Chem. Soc.* **1986,** *108*,
8010; c) M.D. Bednarski, D.C. Crans, R.Di-
Cosimo, E.S. Simon, P.D. Stein, G.M. White-

sides, *Tetrahedron Lett.* **1988,** *29*, 427; d)
C.Augé, S.David, C.Gautheron, A.Malleron,
B.Cavayé, *New. J. Chem.* **1988,** *12*, 733; e) M.-
J. Kim, W.J. Hennen, H.M. Sweers, C.-H.
Wong, *J. Am. Chem. Soc.* **1988,** *110*, 6481; f)
A.Schrell, G.M. Whitesides, *Liebigs Ann.*
Chem. **1990,** 1111; g) U.Kragl, D.Gygax,
O.Ghisalba, C.Wandrey, *Angew. Chem.* **1991,**
Angew. Chem. Int. Ed. Engl. **1991,** *30*, 827;
103, 854; h) C.F. Barbas, Y.-F.Wang, C.-
H.Wong, *J. Am. Chem. Soc.* **1990,** *112*, 2013; i)
C.-H.Lin, T.Sugai, R.L. Halcomb, Y.Ichi-
kawa, C.-H.Wong, *J. Am. Chem. Soc.* **1992,**
114, 10138.

[9] a) W.Becker, E.Freud, E.Pfeil, *Angew. Chem.*
1965, *77*, 1139; Angew. Chem. Int. Ed. Engl.
1965, *4*, 1079; b) F.Effenberger, T.Ziegler,
S.Förster, *Angew. Chem.* **1987,** *99*, 491;
Angew. Chem. Int. Ed. Engl. **1927,** *26*, 458; c)
F.Effenberger, B.Hörsch, S.Förster, T.Zieg-
ler, *Tetrahedron Lett.* **1990,** *31*, 1249; d) U.
Niedermeyer, M.-R.Kula, *Angew. Chem.*
1990, Angew. Chem. Int. Ed. Engl. **1987,** *26*,
458; *102*, 423; e) F.Effenberger, B.Hörsch,
F.Weingart, T.Ziegler, S.Kühner, *Tetrahedron
Lett.* **1991,** *32*, 2605; f) N. Klempier, H.
Griengl, M. Hayn, Tetrahedron Sctt. **1993,** *34*,
4769.

[10] a) C.Fuganti, P.Grasselli, F.Spreafico, C.J.
Zirotti, *J. Org. Chem.* **1984,** *49*, 543; b) C.Fu-
ganti, P.Grasselli, *Chem. Ind.* **1977,** 983; c)
H.Ohta, K.Ozaki, J.Konishi, G.Tsuchiashi,
Agric. Biol. Chem. **1986,** 1261.

[11] A.H. Rose *Industrial Microbiology,* Butter-
worths, Washington DC, 1961, p. 264.

[12] C.Fuganti, P.Grasselli, G.Poli, S.Servi,
A.Zorzella, *J. Chem. Soc., Chem. Commun.*
1988, 1619.

[13] a) T.Kitazume, N.Ishikawa, *Chem. Lett.* **1984,**
1815; b) T.Itoh, Y.Takagi, T.Fujisawa, *Tetrahedron Lett.* **1989,** *30*, 3811.

[14] a) J.Bujons, R.Guajardo, K.S. Kyler, *J. Am.
Chem. Soc.* **1988,** *110*, 604; b) A. Krief, P.Pa-
sau, L.Quere, *Bioorg. Med. Chem. Lett.* **1991,**
1, 365 and references therein.

[15] K.Rama Rao, T.N. Srinivasan, N.Bhanuma-
thi, *Tetrahedron Lett.* **1990,** *31*, 5959.

[16] S.Colonna, A.Manfredi, R.Annunziata, *Tetrahedron Lett.* **1988,** *29*, 3347.

Enzymatic Synthesis of *O*-Glycosides

Herbert Waldmann

Carbohydrates are involved in a multitude of important biological processes and, consequently, the chemistry of these multifunctional natural products has been intensively studied and is subject to a pronounced interest also from the view point of medicinal chemistry. To solve the central goal of carbohydrate chemistry, that is, the regio- and stereoselective formation of glycosidic bonds, a variety of efficient classical chemical methods is available. [1] Nevertheless, syntheses of oligosaccharides generally are laborious and time-consuming multistep sequences. According to a recent estimation [2] for the attachment of each saccharide unit seven weeks' work is required. However, the classical chemical techniques are increasingly being complemented by promising enzymatic methods. [3]

In principle, two classes of enzymes can be used for biocatalyzed glycosylations: glycosidases and glycosyl transferases.

Glycosidases

Glycosidases usually cleave glycosidic bonds (1)→(3). They display a high specificity for the terminal non-reducing carbohydrate and the type of the glycosidic bond, but the structure of the aglycon may vary within a wide range. By shifting the equilibrium of the cleavage reaction the hydrolysis can be reversed and

thereby the synthesis of saccharides can be achieved. Furthermore, glycosides (4) may be built up by trapping the activated intermediates (2) with nucleophiles. The process mentioned first, the so-called reverse hydrolysis, is thermodynamically controlled and is influenced by suitable measures, for example, increasing the concentration of the educts or removing the products from the equilibrium. In the second case, the so-called transglycosylation, the formation of the product is kinetically controlled. Activated substrates such as *p*-nitrophenyl glycosides, disaccharides and glycosyl fluorides are employed as glycosyl donors and the extent of glycoside formation (1)→(4) depends on the velocity of the competing hydrolysis (1)→(3) of the donor substrate. In general, the transglycosylation proceeds with higher yield than the reverse hydrolysis and, therefore, is usually preferred.

By means of this technique for instance Thiem et al. synthesized fucosyl- [4a] and N-acetylneuraminyl glycosides [4b] (Scheme 1). Thus, in the presence of the p-nitrophenyl α--L-fucoside (6) the β-methyl galactoside (5), which was used in 2.5 fold excess, was regiose-

lectively glycosylated by α-fucosidase at the 2-OH group (to give (7)) and the 6-OH group (to give (8)) in a combined yield of 16.5%. Similarly, (5) was coupled with the p-nitrophenyl glycoside (9) of N-acetylneuraminic acid in the presence of sialidase to give the α-(2-3)glycoside (10) (5% yield) and α-(2-6) glycoside (11) (15% yield). Glycosyl moieties can also be transferred to non-

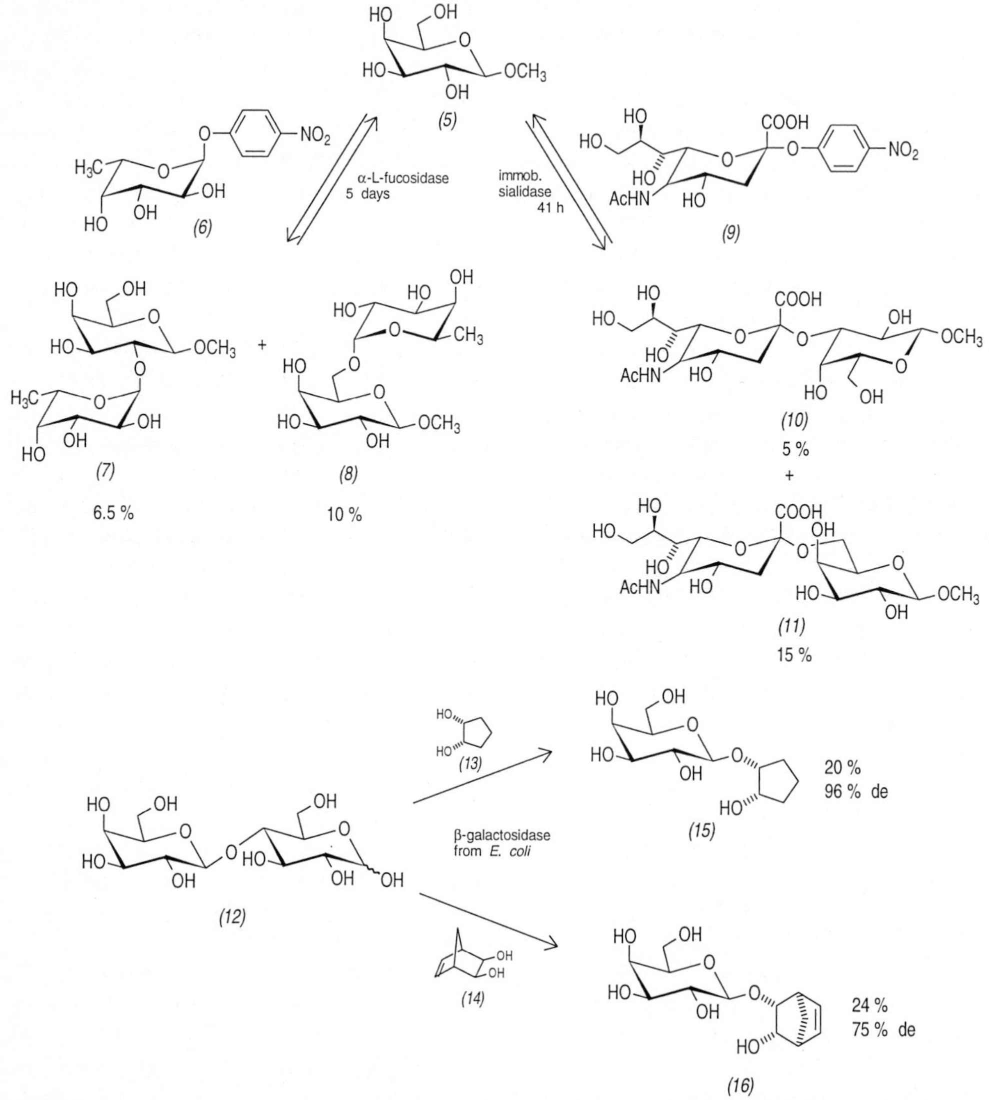

Scheme 1. Glycosidase mediated transglycosylations according to Thiem et al. [4a, b] and Gais et al. [4c].

carbohydrates. For instance, Gais et al. exploited the β-galactosidase from *E. coli* to transfer the galactose unit present in lactose *(12)* to *meso* substrates, for example, *(13)* and *(14)* (Scheme 1). In these processes the desired glycosides *(15)* and *(16)* were formed in 20 % and 24 % yield, respectively, and the transglycosylations proceeded with acceptable to high stereoselectivity. These and numerous further examples (for an extensive tabular survey see ref. 3a) demonstrate that glycosidases allow for regioselective one-step glycosylations to be carried out. However, they clearly show the limitations of the method, that is, the yields are generally low and mixtures of regioisomers are formed that have to be separated. Both disadvantages can be overcome by employing glycosyltransferases for the formation of the glycosidic bonds.

Glycosyltransferases

The overwhelming majority of the more than 100 glycosyltransferases characterized so far transfers a monosaccharide from a nucleoside-diphospho carbohydrate *(17)* to a glycosylacceptor *(18)* to form the desired glycoside *(19)* and a nucleoside diphosphate *(20)*. An exception is given by *N*-acetylneuraminic acid, which is activated as a nucleoside monophosphate [CMP-NeuAc *(17h)*]. To build up

the bewildering variety of the required oligosaccharides, the biosynthetic machinery of mammals needs eight activated building blocks: uridine-5'-diphospho glucose *(17a)* (UDP-Glc), -glucuronic acid *(17b)* (UDP-GlcUA), – *N*-acetylglucosamine *(17c)* (UDP-GlcNAc), -galactose *(17d)* (UDP-Gal), -N-acetylgalactosamine *(17e)* (UDP-GalNAc), guanosine-5'-diphospho mannose *(17f)* (GDP-Man) and -fucose *(17g)* (GDP-Fuc) as well as cytidine-5'-monophospho *N*-acetylneuraminic acid *(17h)* (CMP-NeuAc). Each glycosyl donor is transferred by different enzymes to various glycosyl acceptors. In general, these enzymes are specific for the activated carbohydrates, the type of the glycosidic bond, the glycosyl acceptor and a specific hydroxy group in the glycosyl acceptor. From these findings the notion was deduced that in each case one specific enzyme is responsible for the formation of a particular glycosidic bond (one enzyme – one linkage concept [5]). However, systematic evaluations of the substrate specificity of some glycosyltransferases indicate that this may not be the case (see below).

In order to employ glycosyltransferases on a preparative scale, the glycosyl donors *(17)* must be available in sufficient amounts. For this purpose, classical-chemical, enzymatic and chemoenzymatic processes were developed so that *(17a–17h)* now can be prepared in amounts of 1 g up to more than

100 g. [6] However, the large-scale synthesis and use of these frequently also very sensitive substances itself is laborious, time-consuming and expensive, and, in addition, many glyco-syltransferases are inhibited by the liberated nucleoside diphosphates *(20)* resulting in low yields of the glycosides. Therefore, it is highly desirable to employ the cofactors not in equi-molar amounts, but rather to regenerate them in situ by suitable enzymatic transformations from the products *(20)*. Both methodologies

have been realized successfully with different glycosyltransferases.

Galactosyltransferase

The enzyme most frequently used for glyco-side syntheses on a preparative scale is a galactosyltransferase isolated from milk. This biocatalyst transfers a β-galactose from UDP-Gal *(17d)* to the 4-OH group of *N*-

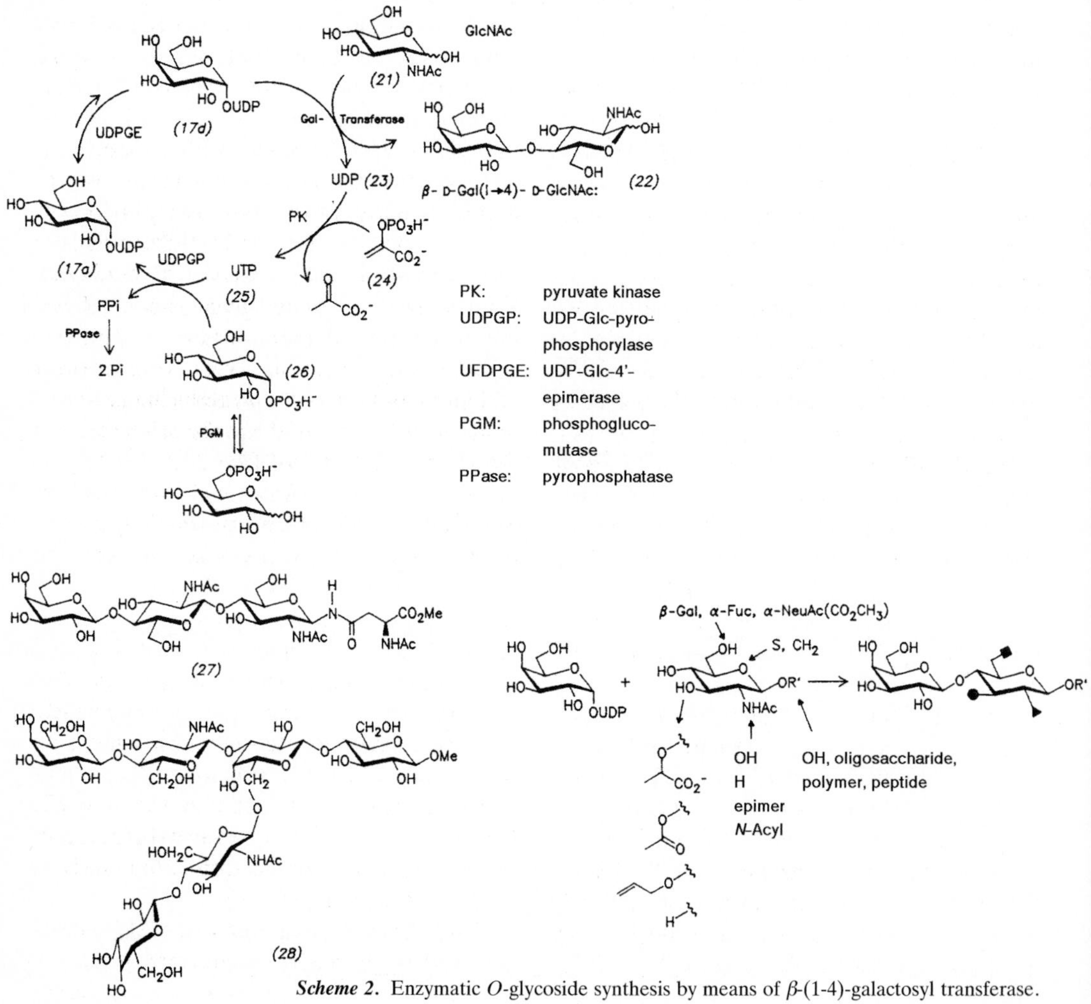

Scheme 2. Enzymatic *O*-glycoside synthesis by means of β-(1-4)-galactosyl transferase.

acetylglucosamine residues. [7, 8a–c, 8e–f] If these enzymatic transformations are carried out with stoichiometric amounts of the nucleotide cofactor often only unsatisfying yields are recorded. Higher yields are obtained if a regeneration system for UDP-Gal is employed, which was developed by Whitesides et al. [7a] In this reaction cascade the galactose incorporated in UDP-Gal *(17d)* is transferred to the glycosyl acceptor *(21)* to give the saccharide *(22)* and UDP *(23)* (Scheme 2). Subsequently, UDP is converted to UTP *(25)* by pyruvate kinase (PK) which employs phosphoenol pyruvate *(24)* as phosphorylating reagent. UTP then is further coupled enzymatically with glucose-1-phosphate *(26)* to give UDP-glucose *(17a)*. To shift the equilibrium in the last mentioned step to the desired side, the pyrophosphate formed is also cleaved by means of pyrophosphatase. Finally, an epimerase converts the activated glucose derivative *(17a)* to the analogous galactosyl compound *(17d)* to complete the cycle. This pioneering method by which, for instance, *N*-acetyllactosamine was built up on a 10 g scale, has proven its efficiency in a variety of applications. Thus, it allowed for the construction of galactosylated glycopeptides, for example, *(27)* [7d] and, by simultaneous galactosylation of two GlcNAc residues, the hexasaccharide *(28)* [7c] could be constructed (for a detailed review of the glycosylations carried out with glycosyl transferases see ref. 3a). In particular, this methodology opens up a route to modify or repair unwanted glycan chains of glycoproteins that may be isolated from natural sources or that may have been produced by genetic engineering. To this end, GlcNAc residues are unmasked on the surface of the macromolecules by enzymatic removal of undesired carbohydrates, and subsequently galactose is attached enzymatically. [7k] Since the galactosyltransferase displays a relatively broad substrate tolerance as far as the glycosyl acceptor is concerned, it may be widely applied for various purposes. For instance, it

tolerates the presence of different substituents at the OH-groups in position 1, 2, 3 and 6 of the acceptor and the ring oxygen may also be replaced by other heteroatoms (NH,S) or a CH_2 group [7d–i] (Scheme 2). Furthermore, in the presence of α-lactalbumine the enzyme is capable of glycosylating the 4-OH group of glucose. [7d]

Sialyltransferases

The non-enzymatic synthesis of sialylated oligosaccharides often proceeds only with low yield and unsatisfying stereoselectivity. Therefore, the use of biocatalysts for this purpose might serve to circumvent a real bottle-neck in carbohydrate chemistry. Sialyltransferase mediated glycosylations were developed independently and almost simultaneously by the groups of Paulson, [8a–c] Thiem [8d] as well as David and Augé. [7b, 8e–f] By means of this method in the presence of equimolar amounts of the activated glycosyl donor CMP-NeuAc *(17h)* various enzymes were applied to transfer sialic acid to the 3-OH group or the 6-OH group of terminal, non-reducing galactose units in order to build up the trisaccharides *(29)*, [7b, 8a, 8d] *(30)* [8a] and *(31)* [8e] in satisfying yields (Scheme 3). Particularly impressive are the chemoenzymatic syntheses of the complex oligosaccharides *(32)* and *(33)* as well as *(34)* and *(35)* by Paulson et al. [8c] and Jennings et al., [9a, b] respectively. In the construction of *(32)* and *(33)* in both cases a trisaccharide, which was built up using classical methodology, was first galactosylated twice and then sialylated twice enzymatically. In the case of the decasaccharide *(35)* an octasaccharide was synthesized non-enzymatically and the α-(2-3) linked *N*-acetylneuraminyl residues were then attached simultaneously by means of a sialyltransferase.

If stoichiometric amounts of CMP-NeuAc are used in the enzyme catalyzed glycosylations the CMP formed inhibits the glycosyl-

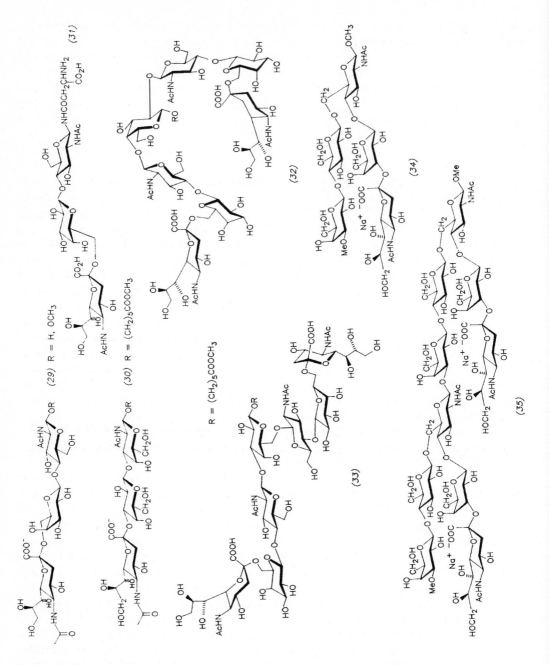

Scheme 3. Complex oligosaccharides which were built up by enzymatic glycosylation employing sialyl transferases.

galactosyl-
transferase,
pH 7.4
——————→
phosphatase

+ uridine + 2 P$_i$

α-2,6-sialyl-
transferase
pH 7.4
phosphatase

+ Cytidin   + P$_i$

(36): R = Aloc–Phe–Asn–Ser–Thr–Ile–OH
82.4 % overall yield

(37): R = H–Gly–Gly–Asn–Gly–Gly–OH
86 % overall yield

tioned step the equilibrium is shifted to the product side by enzymatic cleavage of the liberated pyrophosphate. By application of this multi enzyme system 2 g of the trisaccharide *(29)* was synthesized in 97 % yield. In addition, it was recently used in the chemical-enzymatic synthesis of the sialyl Lewis* tetrasaccharide and derivatives thereof. [10b] However, the enzymatic regeneration of cofactors offers further opportunities. Thus, Wong et al. [10c] developed an even more elaborate one-pot procedure in which the disaccharide *(22)*, which is the glycosyl acceptor in the sialyl transfer reaction, is built up in situ via enzymatic galactosylation (**C** in Scheme 4), as described above (see Scheme 2). In addition, *N*-acetylneuraminic acid *(41)* is generated in situ by an aldolase-catalyzed C–C bond formation from *N*-acetylmannosamine *(42)* (**B** in Scheme 4). In this complex system nine enzymes cooperate to give the trisaccharide *(29)* in 22 % overall yield starting from three monosaccharides.

transferase so that only low yields of the desired glycoside are obtained. This can be overcome by destroying CMP in situ by means of an alkaline phosphatase. Thus, Paulson et al. [8b] constructed the saccharide part of the *N*-glycopeptides *(36)* and *(37)* in a one-pot reaction via successive attachment of galactose and neuraminic acid. The efficiency of the enzymatic sialyl transfer is further enhanced if a cofactor regeneration system developed by Wong et al. [10a] is employed (**A** in Scheme 4). The CMP *(38)* liberated in the enzymatic glycosylation is converted to the diphosphate CDP *(39)* by means of the enzyme nucleoside monophosphate kinase (NMK) or myokinase (MK). CDP is then phosphorylated by pyruvate kinase to give CTP *(40)*. The activated monosaccharide CMP-NeuAc *(17h)*, which is required for a further acyl transfer step, is then formed by CMP-NeuAc synthetase, which was obtained by overexpression in *E. coli*. In this last men-

Glucuronic Acid Transferase and *N*-Acetylglucosaminyltransferases

Gygax et al. [11] developed the enzymatic transfer of glucuronic acid to alcohol acceptors (Scheme 5). UDP-GlcUA *(17b)* required for this purpose is generated in situ from UDP-Glc *(17a)* by a dehydrogenase, and the UDP liberated during the glycosylation is recycled to UDP-Glc by means of the enzymatic process highlighted above. This method provides an easy access to phenolic glycosides and steroid glycosides like *(43)* and *(44)* since all the enzymes required are present in a raw extract from pig or rabbit liver and therefore no isolation of the biocatalysts is necessary.

Hindsgaul et al. [2, 12] described the synthesis of the complex oligosaccharides *(46)* and *(47)* by enzymatic transfer of two *N*-acetylglucosamine units to the trisaccharide

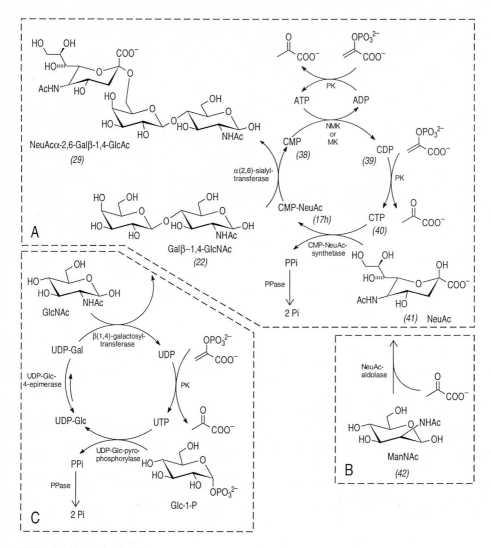

NMK: nucleosidemonophosphate kinase
MK: myokinase
PK: pyruvate kinase

Scheme 4. Enzymatic synthesis of sialylglycosides with in situ regeneration of CMP-NeuAc according to Wong et al. [10].

(45), which was built up non-enzymatically (Scheme 5). First *(45)* serves as substrate for *N*-acetylglucosamine transferase I, which selectively attaches a glucosamine to one of the two mannoses present. The tetrasaccha-

ride *(46)* formed thereby is then a good substrate for the second transferase, which generates the pentasaccharide *(47)*. In addition to UDP-GlcNAc, the enzyme mentioned first also accepts the respective activated 3'-, 4'- or

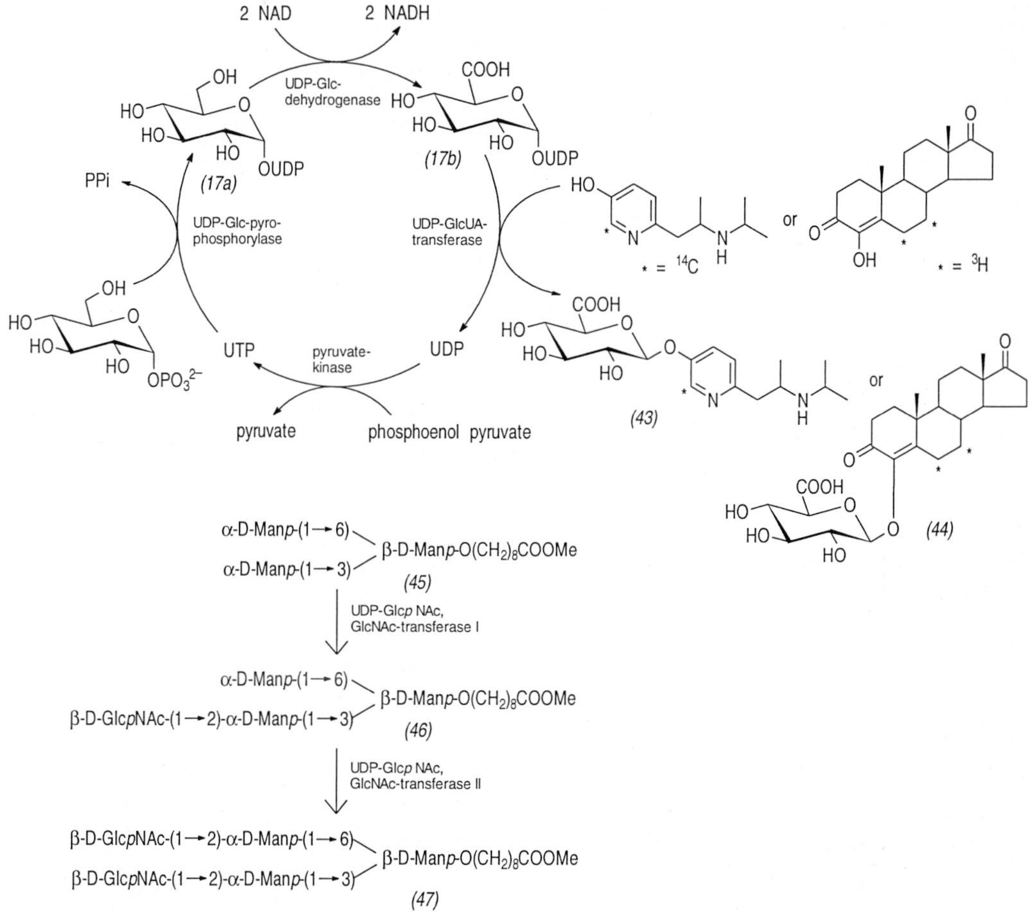

Scheme 5. Enzymatic synthesis of glucuronyl glycosides and *N*-acetylglucosaminyl glycosides according to Gygax et al. [11] and Hindsgaul et al. [12].

6'-deoxy hexosamine and thereby allows for the synthesis of selectively deoxygenated oligosaccharides.

In conclusion, enzymatic glycosylation methods provide interesting alternatives to classical chemical techniques. For the construction of complex oligosaccharides a combination of both techniques appears to be particularly advantageous. By means of non-enzymatic glycosylations first smaller saccha-rides may be generated. They then serve as substrates in enzymatic glycosylations during which glycosidic bonds are formed that are difficult to construct by the conventional methodology. At present, the wide application of this method is hindered by the limited availability of the glycosyl transferases required. However, by taking advantage of modern gene technology this problem should soon be overcome. [13]

References

[1] Review: K. Toshima, K. Tatsuta, *Chem. Rev.* **1993**, *93*, 1503.

[2] G. Srivastava, G. Alton, O. Hindsgaul, *Carbohydr. Res.* **1990**, *207*, 259.

[3] Reviews: a) E. J. Toone, E. S. Simon, M. D. Bednarski, G. M. Whitesides, *Tetrahedron* **1989**, *45*, 5365; b) D. G. Drueckhammer, W. J. Hennen, R. L. Pederson, C. F. Barbas III, C. M. Gautheron, T. Krach, C.-H. Wong, *Synthesis* **1991**, 499; c) U. Korf, J. Thiem, *Kontakte (Merck)* **1992**, (1), 3; d) K. G. I. Nilsson, *TIBTECH* **1988**, 256.

[4] a) S. C. T. Svensson, J. Thiem, *Carbohydr. Res.* **1990**, *200*, 391; b) J. Thiem, B. Sauerbrei, *Angew. Chem.* **1991**, *103*, 1521; *Angew. Chem. Int. Ed. Engl.* **1991**, *30*, 1503; c) H.-J. Gais, P. Maidonis, *Tetrahedron Lett.* **1988**, *29*, 5743.

[5] A. Hagopian, E. H. Eylar, *Arch. Biochem. Biophys.* **1968**, *128*, 422.

[6] Review: J. E. Heidlas, K. W. Williams, G. M. Whitesides, *Acc. Chem. Res.* **1992**, *25*, 307.

[7] a) C.-H. Wong, S. L. Haynie, G. M. Whitesides, *J. Org. Chem.* **1982**, *47*, 5416; b) S. David, C. Augé, *Pure Appl. Chem.* **1987**, *59*, 1501; c) C. Augé, C. Mathieu, C. Mérienne, *Carbohydr. Res.* **1986**, *151*, 147; d) J. Thiem, T. Wiemann, *Synthesis* **1992**, 141; e) J. Thiem, T. Wiemann, *Angew. Chem.* **1991**, *103*, 1184; *Angew. Chem. Int. Ed. Engl.* **1991**, *30*, 1163; f) M. Palcic, O. P. Srivastava, O. Hindsgaul, *Carbohydr. Res.* **1987**, *159*, 315; g) C.-H. Wong, T. Krach, C. Gautheron-Le Narvor, Y. Ichikawa, G. C. Look, F. Gaeta, D. Thompson, K. C. Nicolaou, *Tetrahedron Lett.* **1991**, *37*, 4867; h) C. Gautheron-Le Narvor, C.-H. Wong, *J. Chem. Soc., Chem. Commun.* **1991**, 1130; i) C.-H. Wong, Y. Ichikawa, T. Krach, C. Gautheron-Le Narvor, D. P. Dumas, G. C. Look, *J. Am. Chem. Soc.* **1991**, *113*, 8137; k) R. Schneider, M. Hammel, E. C. Berger, O. Ghisalba, J. Nuesch, D. Gygax, *Glycoconjugate J.* **1990**, *7*, 589.

[8] a) S. Sabesan, J. Paulson, *J. Am. Chem. Soc.* **1986**, *108*, 2068; b) C. Unverzagt, K. Kunz, J. C. Paulson, *J. Am. Chem. Soc.* **1990**, *112*, 9308; c) S. Sabesan, J. Duus, P. Domaille, S. Kelm, J. C. Paulson, *J. Am. Chem. Soc.* **1991**, *113*, 5865; d) J. Thiem, W. Treder, *Angew. Chem.* **1986**, *98*, 1100; *Angew. Chem. Int. Ed. Engl.* **1986**, *25*, 1096; e) C. Augé, C. Gautheron, H. Pora, *Carbohydr. Res.* **1989**, *193*, 288; f) C. Augé, R. Fernandez-Fernandez, C. Gautheron, *Carbohydr. Res.* **1990**, *200*, 257; g) see also: H. H. de Heij, M. Kloosterman, P. L. Koppen, J. H. van Boom, D. H. van den Eijnden, *J. Carbohydr. Chem.* **1988**, *7*, 209 and M. M. Palcic, A. P. Venot, R. M. Ratcliffe, O. Hindsgaul, *Carbohydr. Res.* **1989**, *190*, 1.

[9] a) V. Pozsgay, J.-R. Brisson, H. J. Jennings, *J. Org. Chem.* **1991**, *56*, 3377; b) V. Pozsgay, J. Gaudino, J. C. Paulson, H. J. Jennings, *Bioorg. Med. Chem. Lett.* **1991**, *1*, 391.

[10] a) Y. Ichikawa, G.-J. Chen, C.-H. Wong, *J. Am. Chem. Soc.* **1991**, *113*, 4698; b) Y. Ichikawa, Y.-C. Lin, D. P. Dumas, G.-J. Shen, E. Garcia-Junceda, M. A. Williams, R. Bayer, C. Ketcham, L. E. Walker, J. C. Paulson, C.-H. Wong, *J. Am. Chem. Soc.* **1992**, *114*, 9283; c) Y. Ichikawa, J. L. Chu, G. Shen, C.-H. Wong, *J. Am. Chem. Soc.* **1991**, *113*, 6300.

[11] D. Gygax, P. Spies, T. Winkler, U. Pfaar, *Tetrahedron* **1991**, *47*, 5119.

[12] K. J. Kaur, G. Alton, O. Hindsgaul, *Carbohydr. Res.* **1991**, *210*, 145.

[13] See for instance: a) J. C. Paulson, K. C. Colley, *J. Biol. Chem.* **1989**, *264*, 17615; b) L. K. Ernst, V. P. Rajan, R. D. Larsen, M. M. Ruff, J. B. Lowe, *J. Biol. Chem.* **1989**, *264*, 3436.

E. Cyclization Reactions

Electrophilic Cyclizations to Heterocycles: Iminium Systems

Dieter Schinzer

The stereoselective synthesis of heterocyclic ring systems – especially the constructions of medium-sized rings – is one of the major targets in synthesis. With this series of articles a systematic overview in the expanding area of cyclizations with iminium-, oxonium-, and sulfonium-ions will be presented. Parallels and significant differences in this family of reactions will be extracted.

In this initial account mostly stereoselective transformations of Mannich systems, i.e. iminium ions, will be discussed, with the major emphasis being placed on intramolecular processes. In these reactions regio- and stereoselectivity in particular can be controlled quite easily. In principle, two partners are involved in these reactions: the iminium system as acceptor and the nucleophilic π-system as donor.

Two of these reactions, the Pictet–Spengler and the Bischler–Napieralski reactions, proceed by a simple route in which the aromatic system terminates the reaction by proton elimination and rearomatization. [1, 2] On the other hand, the reaction path of olefins, like vinyl silanes is not so easy because the cation intermediate can be stabilized in various ways.

We shall first focus on a cyclization of Meyers et al. in connection with an asymmetric total synthesis of yohimbone. [3] Key reaction is the transformation (8) → (19), with optically active materials synthesized with optically active formamidines. The synthesis is based – except for the key step – on the previously published racemic synthesis of yohimbone from Winterfeldt et al. [4] If one starts with optically active (8), which induces asymmetry to the new chiral center in (10), an analysis can be made on the mechanistic path of this key reaction step (Scheme 1).

In this particular cyclization the question arises whether (10) is produced via a 3,3-sigmatropic rearrangement [5] with equilibration of the iminium ion, which would give rise racemic (10), or whether (8) cyclizes by a direct electrophilic reaction of the iminium ion yielding optically active (10) (Scheme 2).

Indeed, the reaction proceeds via equilibrium of the iminium ion, generating racemic (10). This particular equilibrium has been interrupted in Meyers' synthesis by reduction of the ketone (7) to the alcohol (11), which was obtained as a mixture of diastereomers. Assisted by the external nucleophile methanol the "donor power" to the iminium ion was increased, which resulted in a direct electrophilic cyclization to optically active (10), which was transformed in a short sequence to the natural product yohimbone (14) (Scheme 3).

A second, in this context, interesting study has been published by Overman et al. [6]

Cyclizations terminated by vinyl silanes are a powerful way in which to construct a large variety of systems. [7–11] The starting materials can be synthesized easily, by either carbo-

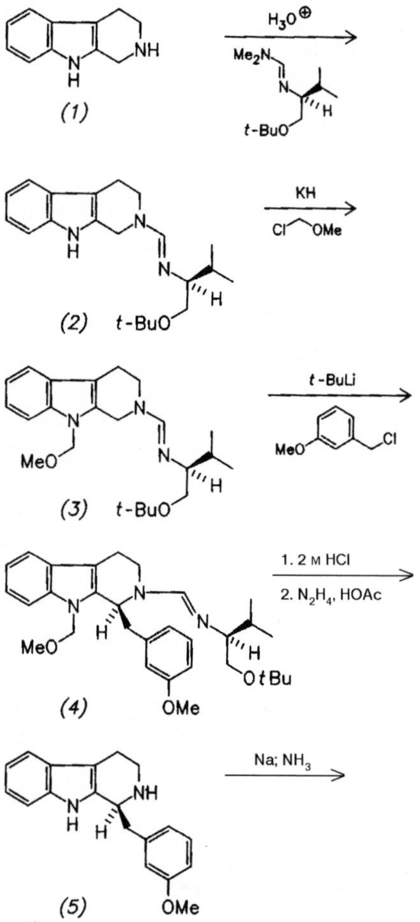

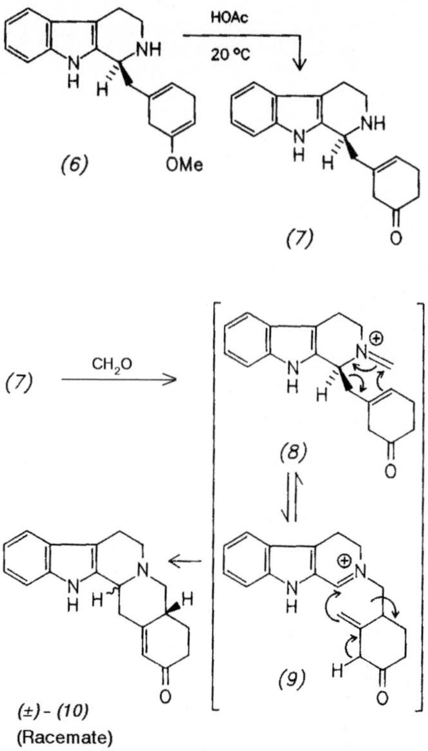

(1)

(2)   t-BuO

(3)   t-BuO

(4)   OMe

(5)   OMe

(6)   OMe

(7)

(7) → (8)

(8)

(9)

(±)-(10)
(Racemate)

Scheme 1

metallation [12], or hydroalumination [13] of triple bonds (Scheme 4).

The cyclization of (Z)-4-trimethylsilyl-3-butenylamine (18) with excess paraformaldehyde in the presence of camphor sulfonic acid proceeded regio- and stereoselectively to give

(8)   S(-)-(10)

Scheme 2

the desired 1,2,5,6-tetrahydropyridine (20). In general, only products built by the "endo-modus" [14] will be obtained, governed by the so called β-effect of silyl group, stabilizing the cation formed β to the silicon atom (Scheme 5).

The stereochemistry of the vinyl silane is not important in this type of reaction, as shown in the transformation of the (E)-isomer (21) to compound (22) (Scheme 6).

At the nitrogen atom unsubstituted tetrahydropyridine (25) can be obtained directly by proton catalyses via the protonated imine (24) (Scheme 7).

These cyclizations also have an interesting aspect. Again, two alternatives can be discussed: The simplest possibility would be the direct cyclization of the iminium ion (27) to the β-silyl cation (29) followed by desilylation

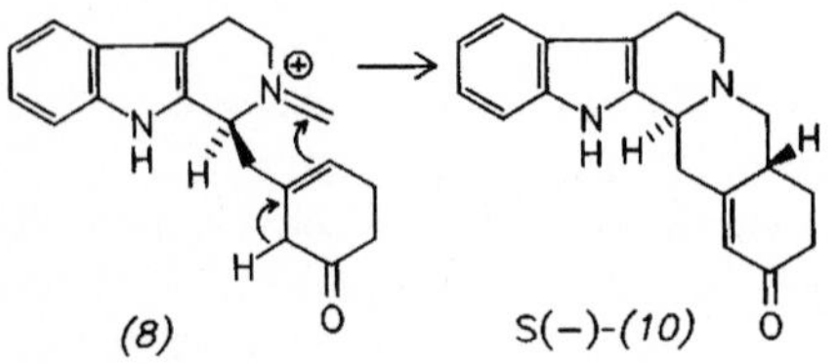

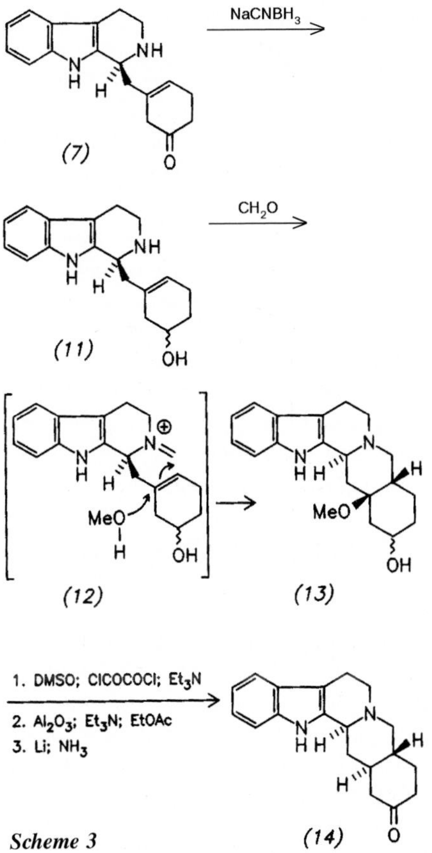

(7)

(18) → (19)

R_2CHO, $H^\oplus$

(11)

CH$_2$O

(20)

53-95% yield

R^1 = Alkyl, Aryl
R^2 = H, Alkyl
R^3 = H, Alkyl, SiMe$_3$
R^4 = H, Alkyl, Aryl

Scheme 5

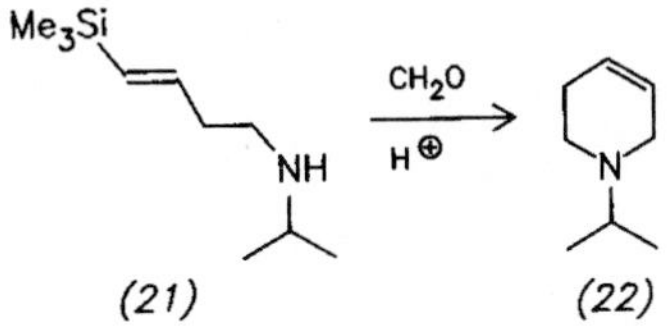

(12) → (13)

1. DMSO; ClCOCOCl; Et$_3$N
2. Al$_2$O$_3$; Et$_3$N; EtOAc
3. Li; NH$_3$

(14)

Scheme 3

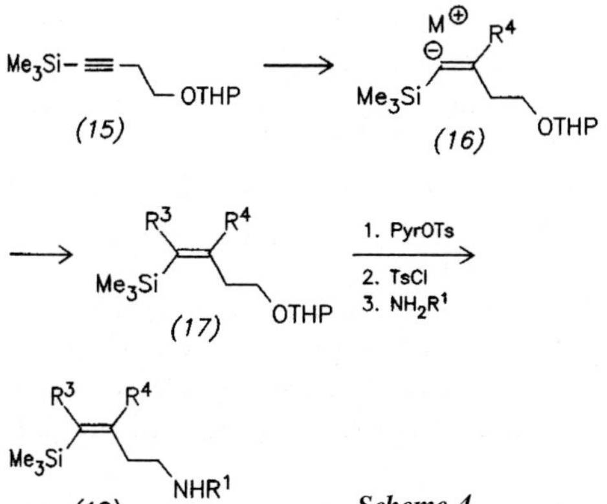

(21) → (22)

CH$_2$O, H$^\oplus$

Scheme 6

Me$_3$Si-≡ —OTHP

(15) → (16)

(23) → (24) → (25)

RCHO, H$^\oplus$

Scheme 7

(17)

1. PyrOTs
2. TsCl
3. NH$_2$R^1

(18)

Scheme 4

(26) → (27) → (29)

CH$_2$O, H$^\oplus$

(28) → (30)

to (30). Alternatively, compound (28) could cyclize via an aza-Cope rearrangement to (30) (Scheme 8).

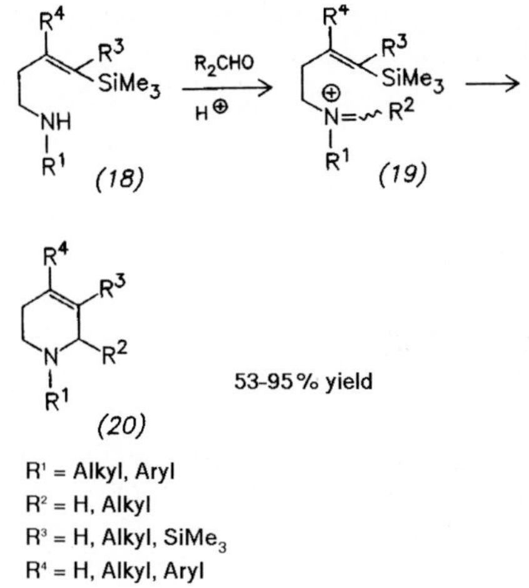

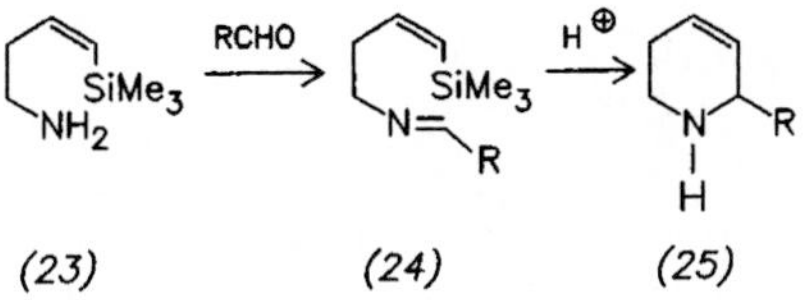

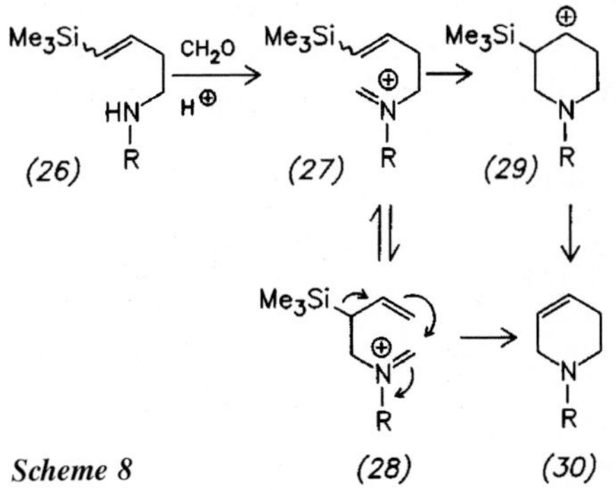

Scheme 8

Two experiments demonstrate elegantly that the isomerization of the allyl silane is faster than the direct cylization. The smooth cyclization (21) → (22), which proceeds under the same conditions as the cyclization of the Z-isomer, can only be explained by isomerization to the allylic silane (28) prior to cylization. The best argument for the allyl silane route is the following experiment (Scheme 9).

The unsaturated amino alcohol (32) is available by aminolysis of epoxide (31) as a 3:1 mixture with compound (33). The mixture was directly treated under acidic conditions with an excess of paraformaldehyde. The only product isolated was compound (36) (86%), which was converted into the more stable

alcohol (37). No trace of (38) could be detected. The experiment precludes direct cyclization of the vinyl silane because (36) was obtained from (35) via an intramolecular Mannich reaction. In addition, allyl silanes are more nucleophilic than vinyl silanes and therefore more reactive toward iminium ions.

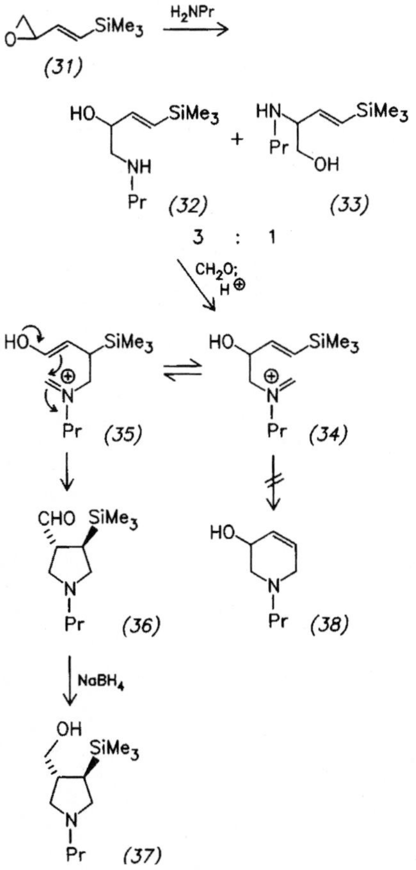

Another new aspect of these cyclizations is the assistance of external nucleophiles. Pioneering work in this area has been done by Winterfeldt et al., who published the first example in their carboline work. [15] Overman et al. have systematically investigated the reaction of simple alkynes, which are less reactive towards iminium ions in comparison with alkenes. [16] These results demonstrate that these reactions only work in the presence of external nucleophiles. Under these conditions alkynes are even more reactive than alkenes.

If hexynylamine (39) is treated with camphor sulfonic acid in the presence of paraformaldehyde no reaction occurs. On the other hand, if tetrabutylammonium bromide is added the exocyclic bromide (40) is obtained in 90% yield (Scheme 10).

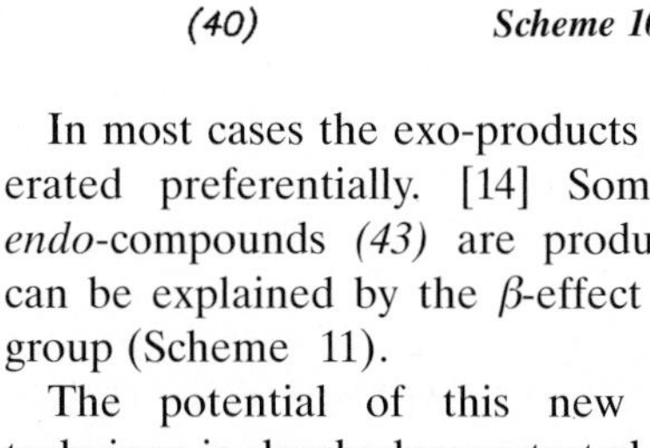

Scheme 9

Scheme 10

In most cases the exo-products will be generated preferentially. [14] Sometimes the endo-compounds (43) are produced, which can be explained by the β-effect of the silyl group (Scheme 11).

The potential of this new cyclization-technique is clearly demonstrated by the competitive cyclization of (44) (Scheme 12).

In the presence of water only (45) is produced, whereas in the presence of sodium

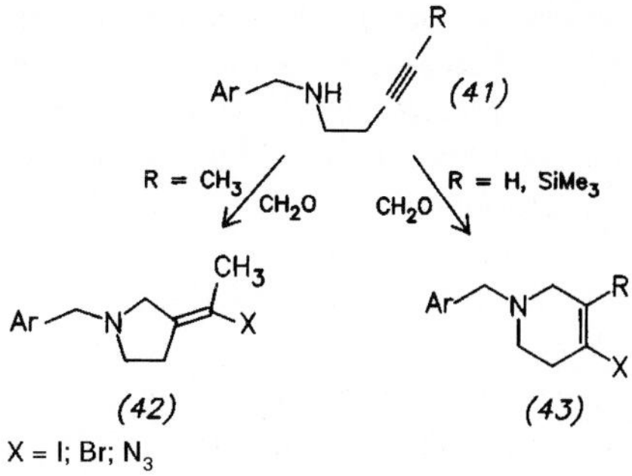

Ar—NH $\overset{R}{\underset{}{\Vert}}$ (41)

R = CH$_3$ / CH$_2$O   CH$_2$O \ R = H, SiMe$_3$

(42)   (43)

X = I; Br; N$_3$

Scheme 11

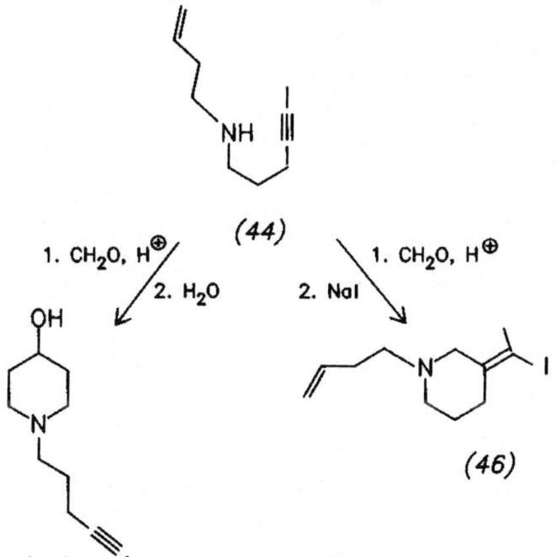

(44)

1. CH$_2$O, H$^\oplus$ / 2. H$_2$O   2. NaI \ 1. CH$_2$O, H$^\oplus$

(46)

(45)   *Scheme 12*

(47) $\xrightarrow[\text{2. MeOH}]{\text{1. BuLi}}$ (48)

$\xrightarrow{\text{LiMe}_2\text{Cu}}$ (49)

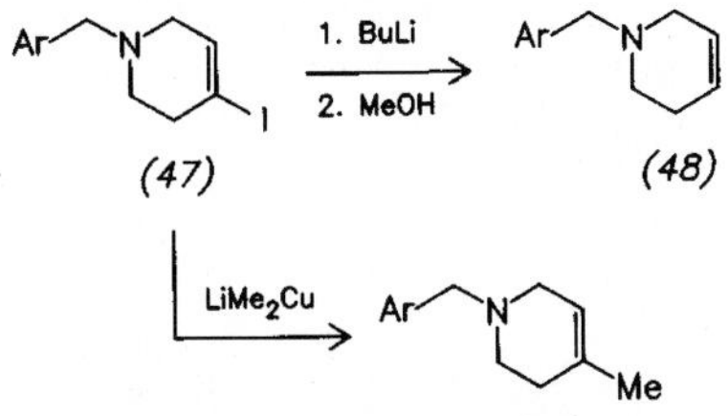

Scheme 13

iodide only *(46)* is generated. The halogen compounds obtained by this process can be transformed in a very flexible way as shown in the Scheme 13.

Finally, a new area in this context will be presented: the Diels–Alder reaction of iminium ions with dienes under Mannich conditions. Grieco et al. recently published two papers demonstrating the power of this method (Scheme 14). [17, 18]

Based on the high degree of stereospecificity these reactions appear to be more concerted than electrophilic cyclizations. [19] These reaction have already been applied toward syntheses of chinolizidine alkaloids, for example, the enantioselective synthesis of (–)-8a-epipulmiliotoxin C *(58)* (Scheme 15).

Starting material is the known chiral alcohol *(53)*, which was transformed in a number of steps to the aldehyde *(54)*. Under Mannich conditions the cycloadducts *(56)* and *(57)*

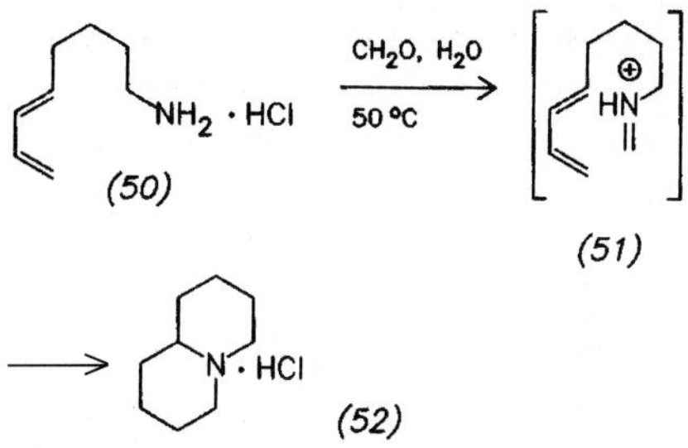

(50) $\xrightarrow[\text{50 °C}]{\text{CH}_2\text{O, H}_2\text{O}}$ (51)

$\longrightarrow$ (52)

Scheme 14

(53)   →   (54)

$\xrightarrow[\substack{\text{H}_2\text{O, EtOH} \\ 75\,°\text{C}}]{\text{NH}_4\text{Cl}}$ (55) $\longrightarrow$

(56)   +   (57)

$\xrightarrow{\text{H}_2, \text{Pd/C}}$ (58)

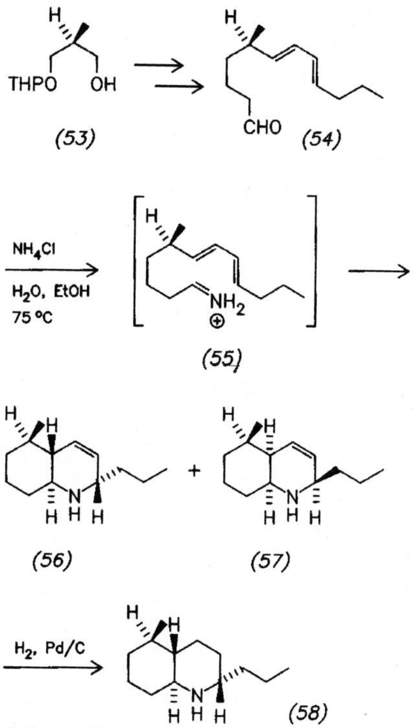

Scheme 15

were obtained (2.2:1). The stereoselectivity in this cyclization was only moderate and can be explained by a "chair-like" transition state *(59)*. The alternative *(60)* is sterically disfavored (Scheme 16).

Amino acids can also be used as chiral building clocks (via in situ formation of iminium ions). Waldmann et al. obtained several heterocycles in quite high diastereoselectivity (Scheme 17). [20]

The reactions presented in this article using iminium ions should be understood as additions to known cyclization reactions. Reactions with "hard electrophiles" like acyl imi-nium ions were not covered in this chapter and only reactions with "soft electrophiles" like iminium ions were presented. Their potential in synthesis can be extended by the use of external nucleophiles.

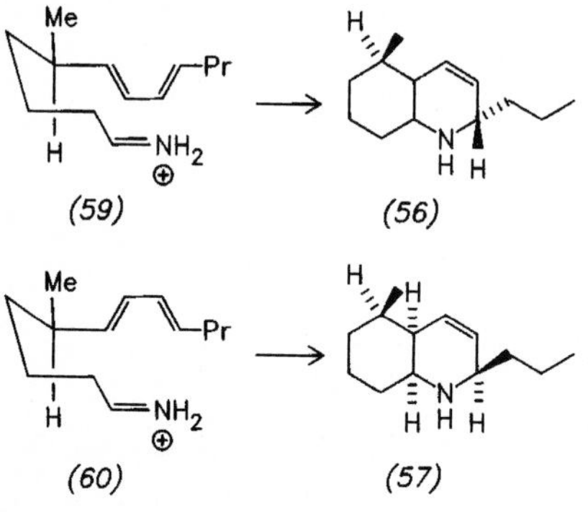

References

[1] W. M. Whaley, T. R. Govindachari, *Org. React.* **1951**, *6*, 74, 151.

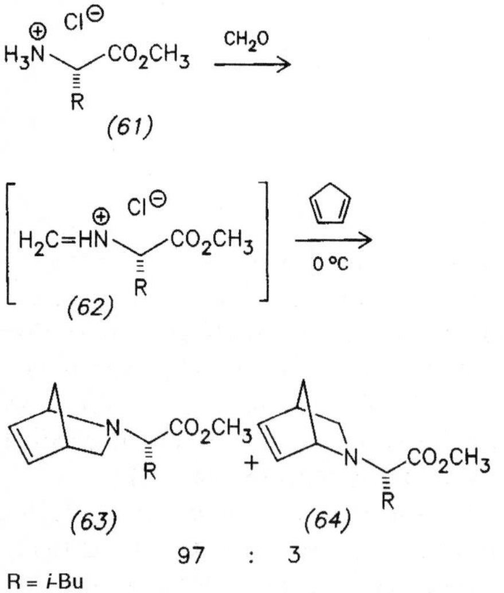

[2] A. Katritsky, C. W. Rees; *Comprehensive Heterocyclic Chemistry*, Vols. 1–6, Pergamon Press, Oxford, **1984**.

[3] A. I. Meyers. D. B. Miller, F. H. White, *J. Am Chem. Soc.* **1988**, *110*, 4778.

[4] W. Benson, E. Winterfeldt, *Chem. Ber.*, **1979**, *112*, 1913.

[5] L. E. Overman, M. Kakimoto, M. Okazaki, G. P. Meier, *J. Am. Chem. Soc.* **1983**, *105*, 6629.

[6] C. Flann, T. C. Malone, L. E. Overman, *J. Am. Chem. Soc.* **1987**, *109*, 6097.

[7] T. A. Blumenkopf, L. E. Overman, *Chem. Rev.* **1986**, *86*, 857.

[8] L. E. Overman, *Lec. Heterocycl. Chem.* **1985**, *8*, 59.

[9] L. E. Overman, T. C. Malone, G. P. Meier, *J. Am. Chem. Soc.* **1983**, *105*, 6993.

[10] L. E. Overman, R. M. Burk, *Tetrahedron Lett.* **1984**, *25*, 5739.

[11] L. E. Overman, N. H. Lin, *J. Org. Chem.* **1985**, *50*.

[12] J. E. Baldwin, *J. Chem Soc. Chem. Commun.* **1976**, 734.

[13] J. F. Normant, A. Alexakis, *Synthesis* **1981**, 841.

[14] G. Zweifel, J. A. Miller, *Org. React.* **1984**, *32*, 375.

[15] E. Winterfeldt, K. H. Feuerherd, V. U. Ahmad, *Chem. Ber.* **1977**, *110*, 3624.

[16] L. E. Overman, M. J. Sharp. *J. Am. Chem. Soc.* **1988**, *110*, 612.

[17] P. A. Grieco, D. T. Parker, *J. Org. Chem.* **1988**, *53*, 3325.

[18] P. A. Grieco, D. T. Parker, *J. Org. Chem.* **1988**, *53*, 3658.

[19] S. D. Larson, P. A. Grieco, *J. Am. Chem. Soc.* **1985**, *107*, 1768.

[20] H. Waldmann, *Angew. Chem.* **1988**, *100*, 307. Angew. Chem. Int. Ed. Engl. **1988**, *27*, 274.

(59)   *(56)*

(60)   *(57)*

Scheme 16

(61)

(62)

(63)   *(64)*

97 : 3

R = *i*-Bu

Scheme 17

Electrophilic Cyclizations to Heterocyles: Oxonium Systems

Dieter Schinzer

Oxacyclic natural products are isolated from a variety of sources in nature. Many different types of structures are known, combined with various types of rings, *exo-* and *endo-*cyclic functionality. The unusual halogen substituents, mostly located in natural products of marine origin are a challenge for synthesis. [1]

The synthesis of medium-sized rings is particularly highlighted in this chapter. Mostly new results will be presented in order to understand the different strategies used to sythesize this interesting class of natural products. Besides oxacycles, carbocycles can also be obtained, depending on the "planned frontier point" of the initiator. This chapter will only focus on cyclizations in which oxonium ions are involved.

The major reaction to generate oxonium ions from acetals is treatment with acids or Lewis acids. The oxonium ion formed is trapped in an intramolecular process by the double bond (Scheme 1). [2]

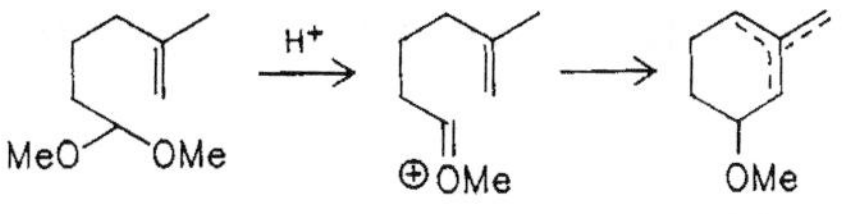

The regiochemistry of the double bond in such a carbocyclization cannot be controlled and leads to a mixture of products. This problem can be solved by the use of allylsilanes of type *(4)* (Scheme 2). [3]

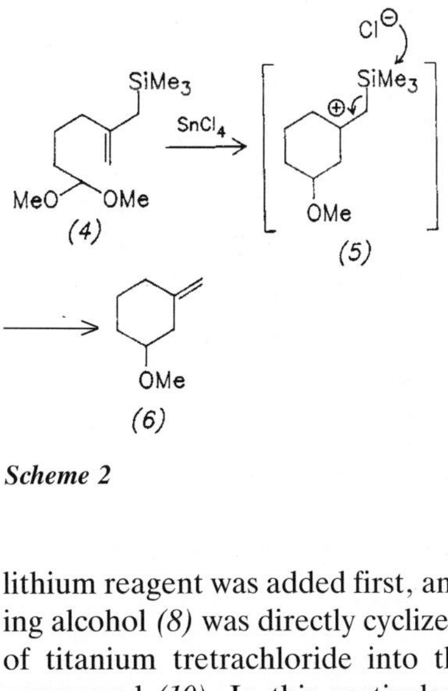

Fleming et al. showed that only *(6)* was obtained, in which the regiochemistry of the double bond was controlled by the β-effect of the silyl group. An interesting reaction leading to aromatic compounds has been described by Tius et al. [4] Starting with ketone *(7)* an allyl

Scheme 1

Scheme 2

lithium reagent was added first, and the resulting alcohol *(8)* was directly cyclized by the use of titanium tretrachloride into the aromatic compound *(10)*. In this particular cyclization the reaction was terminated by a vinyl silane, which has already been used extensively with iminium ion chemistry (see the preceeding article iminium systems (Scheme 3).)

A useful way to synthesize cyclic ethers has been reported by Itoh et al. Compound *(14)* can be obtained by the use of $SnCl_4$, promoting a regioselective cleavage of mono thioace-

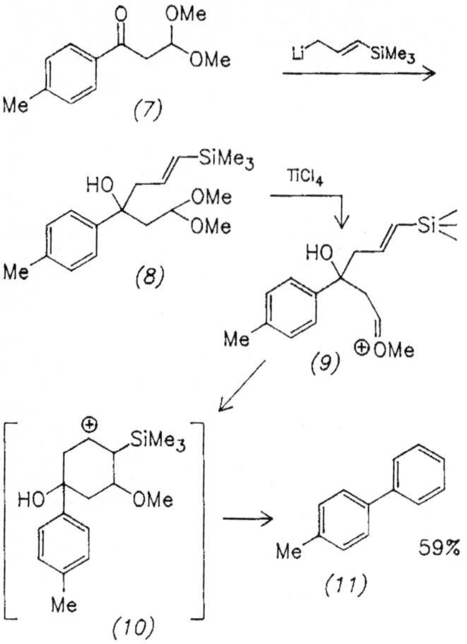

Me— (7)

(8)

(9) ⊕OMe

(10)

Me— (11) 59%

Scheme 3

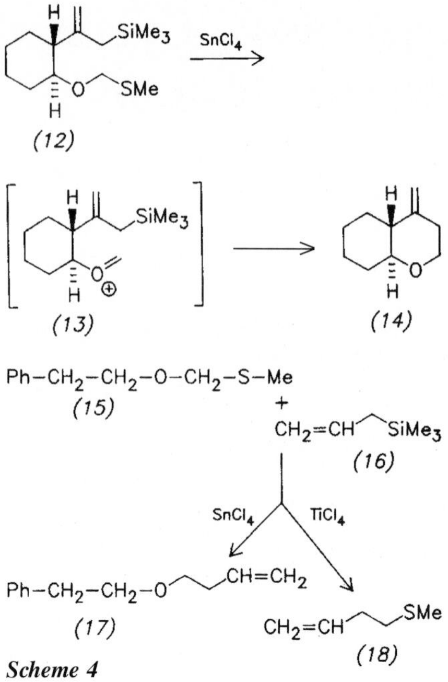

(12)

(13) (14)

Ph–CH₂–CH₂–O–CH₂–S–Me
(15)

CH₂=CH⌒SiMe₃
(16)

SnCl₄   TiCl₄

Ph–CH₂–CH₂–O⌒CH=CH₂
(17)

CH₂=CH⌒SMe
(18)

Scheme 4

n-C₆H₁₃⌒SiMe₃   TiCl₄
OMEM
(19)

n-C₆H₁₃
O
(20) 96%

Ph⌒SiMe₃   TiCl₄
OMEM
(21)

Ph
O
(22) 85%

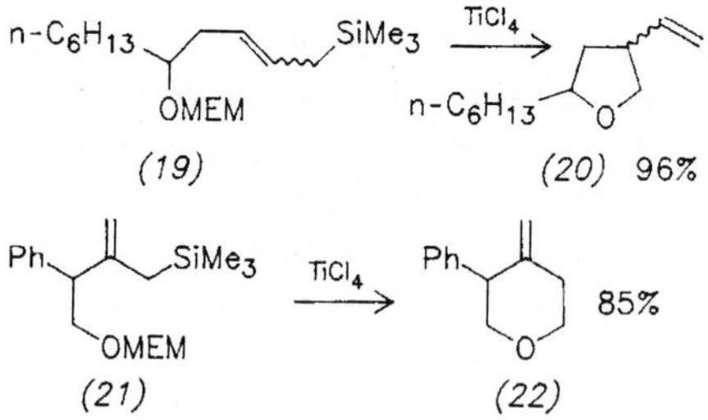

MEM = (2-Methoxyethoxy)methyl

Scheme 5

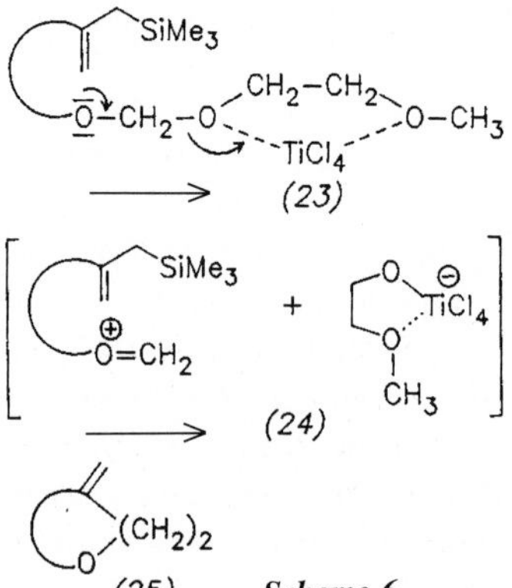

(23)

(24)

(25)   **Scheme 6**

tals. [5] The selective cleavage of the C–S bond can be explained by the high affinity of tin for sulfur. In contrast, the C–O bond can be cleaved smoothly with TiCl₄. These effects have been demonstrated by the transformation of *(15)* with SnCl₄ to give *(17)*, and with TiCl₄ to yield *(18)* (Scheme 4).

The non-symmetrical MEM-ethers can also be cleaved regioselectively. Again, TiCl₄ has been used successfully by Itoh et al. Allyl silanes have been used as electron-rich olefins (Scheme 5). [6]

The regioselective cleavage of the MEM-ether can be explained by the chelate *(23)*. The oxonium ion formed is trapped in situ by the allyl silane (Scheme 6).

Silyl enol ethers as terminators in intramolecular aldol reactions with acetals have been used successfully by Kocienski et al. [7, 8] In a short sequence *(27)* could be obtained in optically active form. The following Mukaiyama-type aldol reaction in the presence of TiCl₄ yields *(29)*, which can be used as an intermediate for the cytotoxic natural product pederin (Scheme 7).

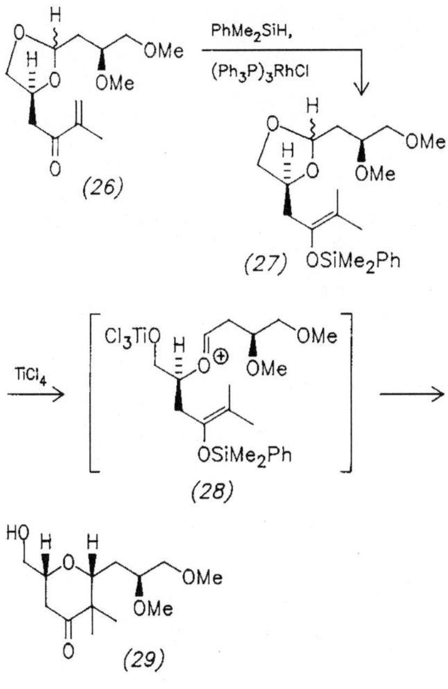

PhMe$_2$SiH,
(Ph$_3$P)$_3$RhCl

(26)

(27) OSiMe$_2$Ph

TiCl$_4$

(28)

(29)

Scheme 7

BF$_3 \cdot$ Et$_2$O;
HOCH$_2$CCl$_3$

(32)

(33)

(34)

I$_2$, HgO

Zn; HCO$_2$H

(35)

(36)

Scheme 9

Again, under regioselective cleavage the same group has synthesized seven- and eight-membered rings. The reaction proceeds only with TiCl$_4$ as the Lewis acid catalyst. The smooth formation of the medium-sized rings without high dilution can be explained by a template effect, because titanium can complex both the acetal oxygen and the oxygen of the silyl enol ether (Scheme 8):

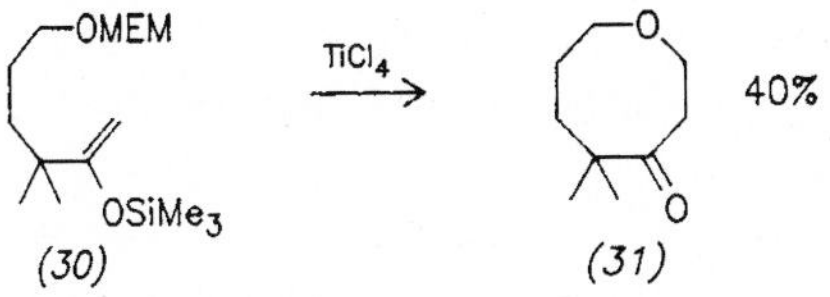

TiCl$_4$

40%

(30)

(31)

Scheme 8

The cleavage of tetrahydropyran ethers (obtained from homo allyl alcohols) under acid catalysis has been reported by Kay et al. [9] Products of type *(34)* are formed under stereoselective conditions. In a short sequence *(34)* could be transformed into the pheromone *(36)* of the olive gly *dacus oleae* (Scheme 9).

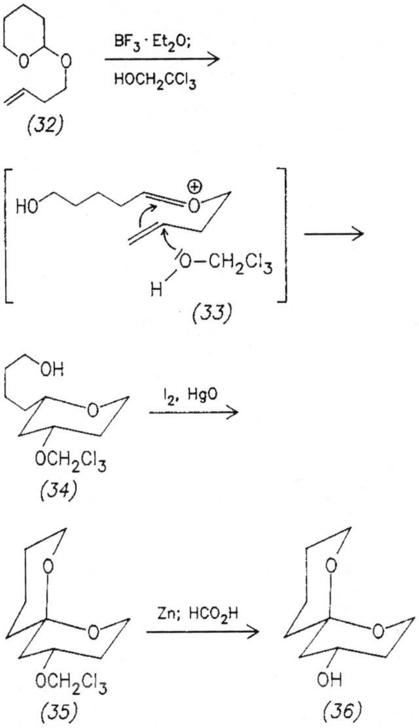

Fleming et al. were the first to study the influence of the stereochemistry of the vinyl silane in reactions with oxonium ions. [10] It has been demonstrated that *E*-vinyl silanes react much more slowly than the corresponding *Z*-vinyl silanes. This is a further result to

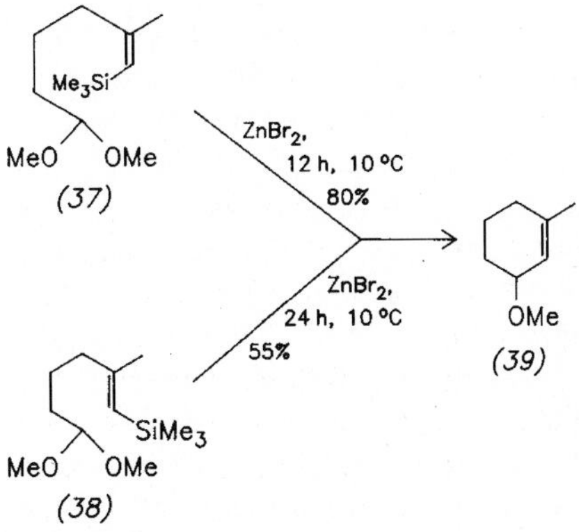

(37)

ZnBr$_2$,
12 h, 10 °C
80%

ZnBr$_2$,
24 h, 10 °C
55%

(39)

(38)

Scheme 10

show that vinyl silanes react in the presence of electrophiles with retention of configuration (Scheme 10). [11]

The first example in which an alkyne has been transformed in to a cyclic ether was reported by Bunnelle et al. [12] Similar to the reactions with iminium ions assistance of an external nucleophile – in this example the chloride ion of the Lewis acid – is required. The reaction proceeds again via a cation-π-cyclization, which was initiated by acetal cleavage (Scheme 11).

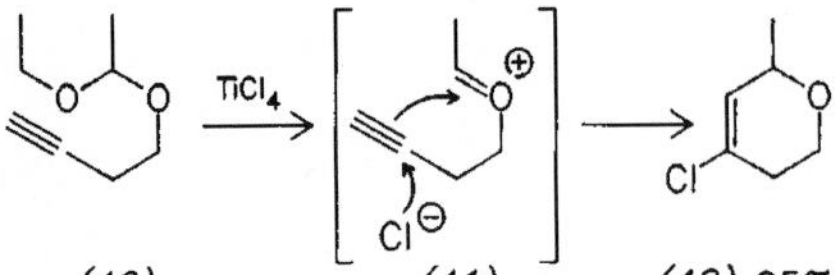

(40)   *(41)*   *(42) 95%*

Scheme 11

The most important contributions along these lines have been reported by Overman. [13] Besides development of synthetic methods many complex total syntheses have been reported from his laboratories. As terminating systems mostly vinyl silanes have been used, which were synthesized in a straightforward short sequence (Scheme 12).

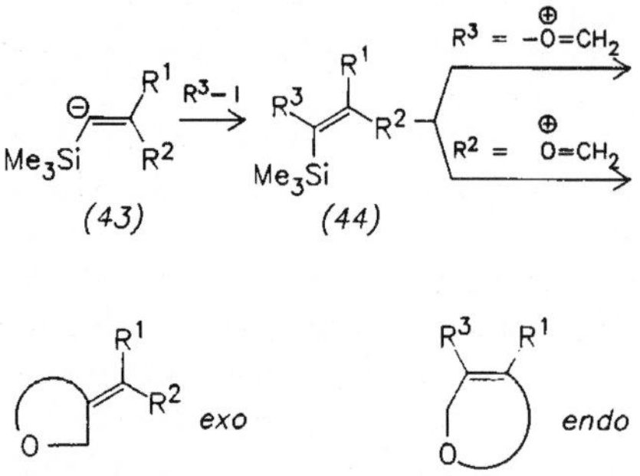

(43)   *(44)*

(45)   *(46)*

Scheme 12

The silylgroup has to perform three tasks: Stabilization of the carbanion in the α-position of the silyl group to hook on the side

chains that contain the initiator; Control of the regiochemistry during cyclization; and, finally, control of the configuration of the double bond in the cyclization product. Both *exo*- and *endo*-products are possible. [14] With this strategy in hand 5-, 6-, and 7-membered rings have been synthesized. To initiate the reactions the tried and trusted MEM-ethers were cleaved under Lewis acid conditions. The Lewis acid of choice in these cyclizations is $SnCl_4$ and the selectivity observed is always > 98 % (Scheme 13).

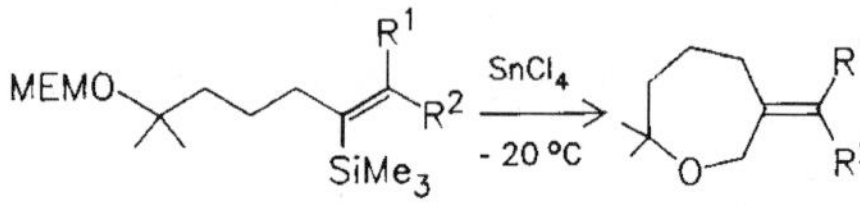

(47)   *(48)*

R^1 = H; R^2 = n-Bu   71%; > 98% Z

R^1 = n-Bu; R^2 = H   57%; > 98% E

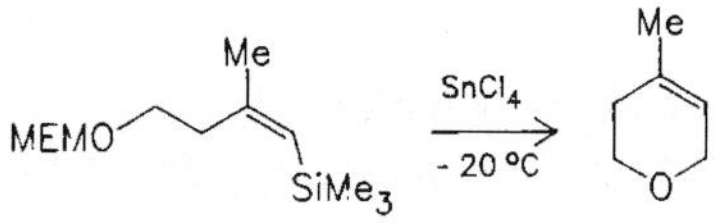

(49)   *(50)*

Scheme 13

E-Vinylsilane *(51a)* is transformed smoothly to the alkylidene tetrahydrofuran *(52)*, but *Z*-vinylsilane *(51b)* produced only tetrahydropyran *(54)*. This is the first example in which the stereochemistry of the double bond controls the ring size. The product is obtained via the six-centered transition state *(53)*, in which the butyl group is oriented equatorially. The destabilization of the α-cation is compensated by the attack of the chloride ion of the Lewis acid (Scheme 14).

Endo-cyclizations can be used to form eight-membered rings if suitable substitution patterns are present in the side chain and the initiator. [15] The mechanism of this particular cyclization is similar to that of the ene reaction, but the conditions are much milder (–20 °C). [16] The most important result in this

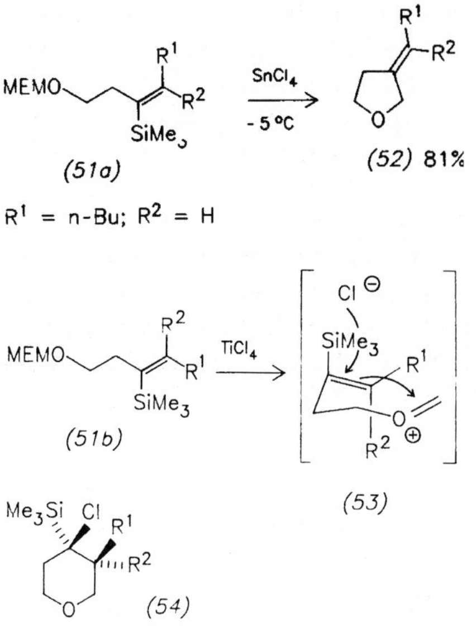

R^1 = n-Bu; R^2 = H

(51a) → (52) 81%

(51b) → (53) → (54)

cis : trans = 30 : 1

(55) → (56)

(57)

(58)

(59)

Scheme 14

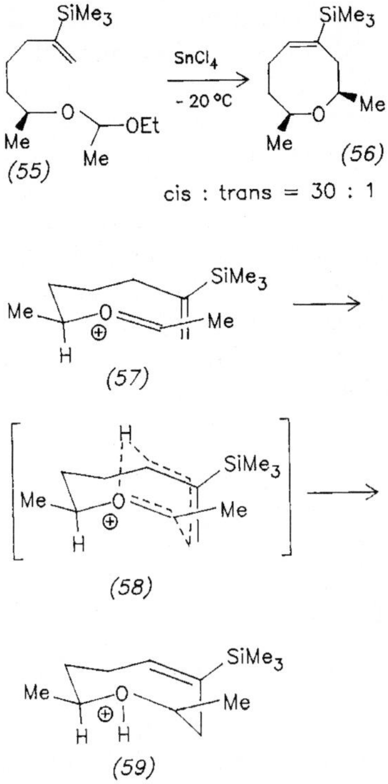

Scheme 15

study is the control over both configuration in (56). The result can be explained by the transition state (58), in which both substituents are pseudo-equatorial orientated, and the E-oxonium ion is attacked from the α-face (Scheme 15).

These results have opened the gate for the first and enantioselective total synthesis of the unusual marine natural product laurenyne (73). [17] In addition to the unusual eight-membered ether ring, the two sensitive unsaturated side chains and the chlorine substituent are noteworthy.

Various problems could be solved by Overman, except the low selectivity in the epoxide opening (64)→(65). The choice of the correct initiator initially caused some problems. Overman demonstrated that only (68) cyclized to the desired ring (69). All attempts to functionalize the acetal "in a better way" before

cyclization failed. In addition, the total synthesis of laurenyne led to a correction of the absolute configuration. The synthesis starts with the bromide (60), which was transformed in straightforward manner into (67). The acetal (68) cyclized in 37% yield to (69). The final steps of the synthis are the elaboration and functionalization of the side chains (Scheme 16).

The total synthesis presented demonstrates the potential of this new cyclization technique and allows an entry into a class of complex natural products with sensitive functionality. The cyclization can be mediated by a Lewis acid. The selection of the initiator was crucial for the success in the total synthesis.

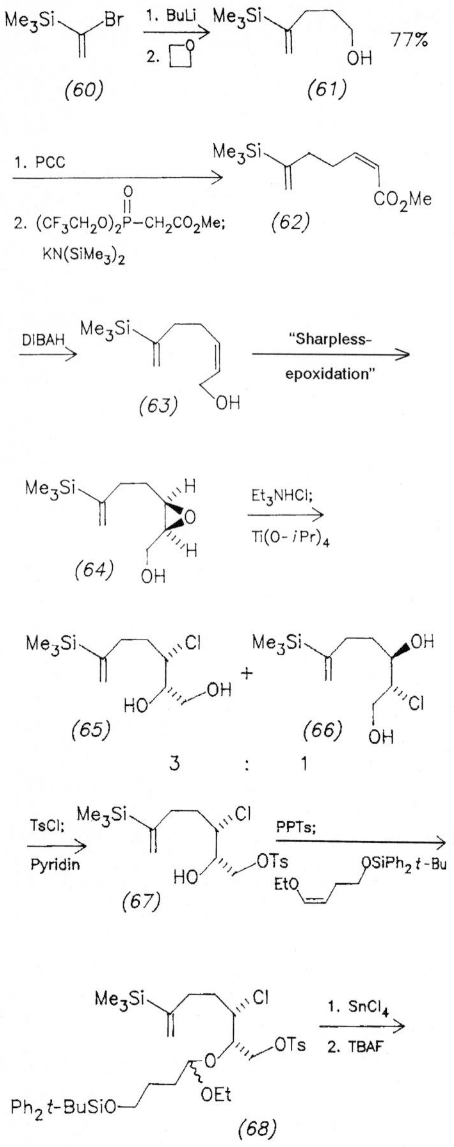

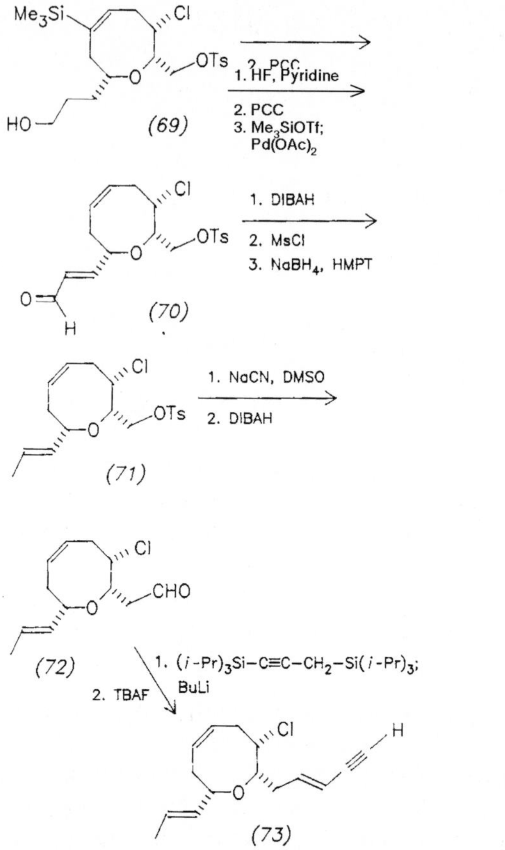

Me₃Si—⫝⫝⫝—Br 1. BuLi   Me₃Si—⫝⫝⫝ 77%
(60)   2. ▢O   OH
                                    (61)

1. PCC   Me₃Si
                                    CO₂Me
2. (CF₃CH₂O)₂P—CH₂CO₂Me;   (62)
   KN(SiMe₃)₂
       O

DIBAH   Me₃Si   "Sharpless-
                              epoxidation"
         (63)   OH

Me₃Si   H   Et₃NHCl;
                O
                      Ti(O-iPr)₄
    (64)   OH   H

Me₃Si   Cl   Me₃Si   OH
                  +
    HO   OH   Cl
   (65)   (66)   OH
       3   :   1

TsCl;   Me₃Si   Cl   PPTs;
Pyridin   OSiPh₂t-Bu
          HO   OTs   EtO
   (67)   Et

Me₃Si   Cl   1. SnCl₄
                OTs   2. TBAF
         O
Ph₂t-BuSiO   OEt
         (68)

Me₃Si   Cl
                       OTs   2. PCC
         O   1. HF, Pyridine
HO   (69)
                      2. PCC
                      3. Me₃SiOTf;
                         Pd(OAc)₂

Me₃Si   Cl
                       OTs   1. DIBAH
         O   2. MsCl
O   3. NaBH₄, HMPT
H   (70)

           Cl   1. NaCN, DMSO
                OTs   2. DIBAH
         O
       (71)

           Cl
                CHO
         O
       (72)   1. (i-Pr)₃Si—C≡C—CH₂—Si(i-Pr)₃;
                      BuLi
           2. TBAF

           Cl   H
                        ≡
         O
       (73)

Scheme 16

References

[1] D.J. Faulkner, *Nat. prod. Rep.* **1986**, 1.
[2] A. van der Gen, K. Wiedhaupt, J.J. Swoboda, H.C. Dunathan, W.S. Johnson, *J. Am. Chem. Soc.* **1973**, 95, 2656.

[3] I. Fleming, A. Pearce, R.L. Snowden, *J. Chem. Soc. Chem. Commun.* **1976**, 182.
[4] M.A. Tius, *Tetrahedron Lett.* **1981**, 22, 3335.
[5] H. Nishiyama, S. Marimatsu, K. Sakuta, K. Itoh, *J. Chem. Soc. Chem. Commun.* **1982**, 459.
[6] H. Nishiyama, K. Itoh, *J. Org. Chem.* **1982**, 47, 2496.
[7] K. Isaac, P. Kociensky, *J. Chem. Soc. Chem. Commun.*, **1982**, 460.
[8] G.S. Cockerill, P. Kocienski, *J. Chem. Soc. Chem. Commun.* **1983**, 705.
[9] I.T. Kay, E.G. Williams, *Tetrahedron Lett.* **1983**, 24, 5915.
[10] H.-F. Chow, I. Fleming, *J. Chem. Soc. Chem. Commun.* **1984**, 1815.

[11] T.H. Chan, P.W. K.Lau, W.Mychajlowskij, *Tetrahedron Lett.* **1977**, *17*, 3317.

[12] W.H. Bunnelle, D.W. Simon, D.L. Mohler, T.F. Ball, D.W. Thompson, *Tetrahedron Lett.* **1984**, *25*,. 2653.

[13] L.E. Overman, A.Castañeda, T.A. Blumenkopf, *J. Am. Chem. Soc.* **1986**, *108*, 1303.

[14] J.E. Baldwin, *J. Chem. Soc. Chem. Commun.* **1976**, 734.

[15] L.E. Overman, T..A. Blumenkopf, A.Castañeda, A.S. Thompson, *J. Am. Chem. Soc.* **1986**, *108*, 3516.

[16] H.M. R.Hoffmann, *Angew. Chem.* **1969**, *81*, 597. Angew. Chem. Int. Ed. Engl. **1969**, 8, 556.

[17] L.E. Overman, A.S. Thompson, *J. Am. Chem. Soc.* **1988**, *110*, 2248.

Electrophilic Cyclizations to Heterocycles: Sulfonium Systems

Dieter Schinzer

Thioacetals *(1)* are mostly known as protecting groups or in connection with *umpolung* of carbonyl groups. [1] These compounds can be deprotonated by a large number of bases, like butyl lithium, which makes them useful intermediates for a variety of transformations. Simple alkylation of *(1)* leads to starting materials for potential cyclization *(2)* in which both the initiator and the terminator are present (Scheme 1). [2]

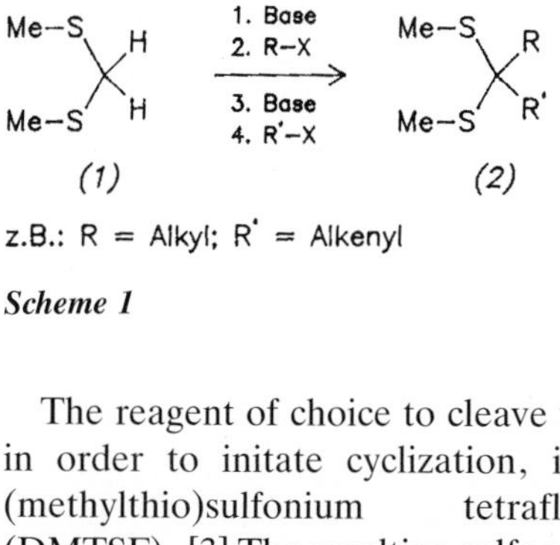

z.B.: R = Alkyl; R' = Alkenyl

Scheme 1

Scheme 2

Scheme 3

The reagent of choice to cleave thioacetals, in order to initate cyclization, is dimethyl (methylthio)sulfonium tetrafluoroborate (DMTSF). [3] The resulting sulfonium ion *(4)* can be trapped in intramolecularly by a variety of terminators (vinyl silanes, silylenol ethers, allyl stannanes) to form cyclic compounds (Scheme 2).

Sulfonium ions of type *(4)* have a long history and were reported first by Meerwein et al., [4] but it took almost 15 years to discover their potential in synthesis. Initially, sulfonium ions have been used to cleave thioacetals under mild conditions as an alternative to mercury-promoted hydrolysis (Scheme 3). [3]

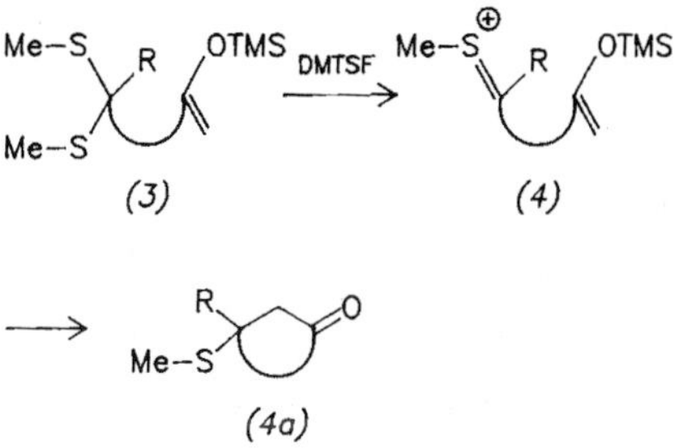

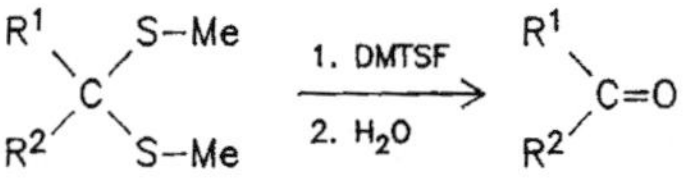

The research group of Trost first used intramolecular reactions with sulfonium ions. [2] The reactive intermediates can be trapped with nucleophilic olefins, e.g. vinyl silanes, *(7)*→*(9)*. The primary cyclization product *(8)* was not isolated, because it rearranged in situ to product *(9)* (Scheme 4). [2]

Compound *(10)* can be transformed chemoselectively in the presence of the vinyl silane to ketone *(11)*, which can be cleaved to the enone *(12)* with HgCl$_2$ and DBU. The overall process represents a selective aldol reaction (Scheme 5).

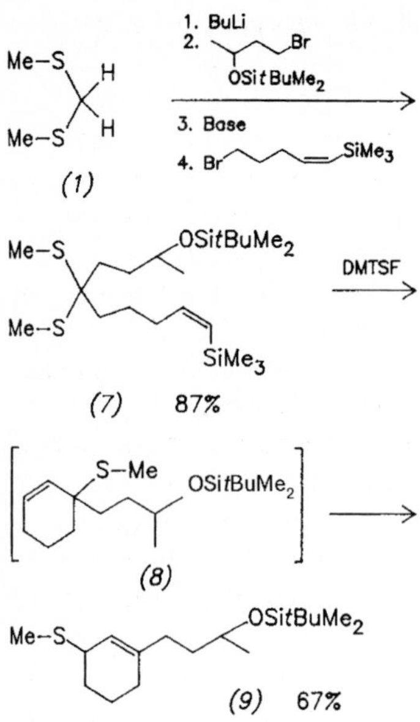

1. BuLi
2.

Me—S H
OSi*t*BuMe$_2$ Br

Me—S H

3. Base
4. BrSiMe$_3$

(1)

Me—S OSi*t*BuMe$_2$

DMTSF

Me—S

SiMe$_3$

(7)   87%

S—Me OSi*t*BuMe$_2$

(8)

Me—S OSi*t*BuMe$_2$

(9)   67%

Scheme 4

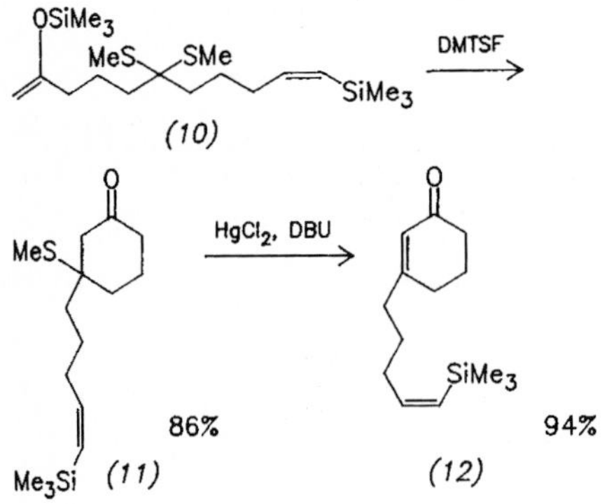

OSiMe$_3$ MeS SMe

DMTSF

SiMe$_3$

(10)

MeS

HgCl$_2$, DBU

SiMe$_3$

86%   94%

Me$_3$Si *(11)*   *(12)*

Scheme 5

Base

(13)   *(14)*

OSiMe$_3$

1.DMTSF

MeS SMe 2. HgCl$_2$, DBU

(15)   *(16) 90%*

Scheme 6

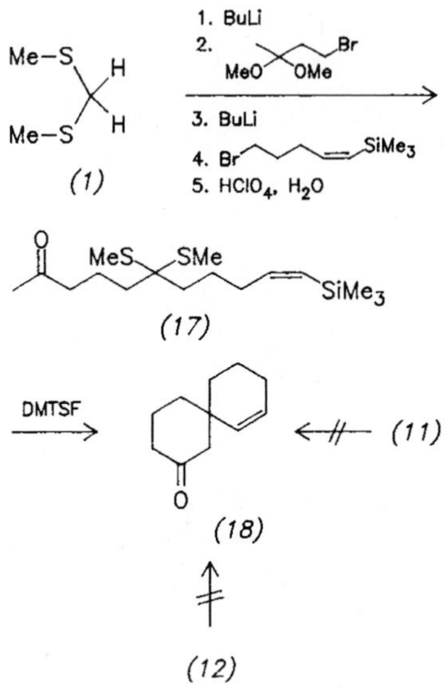

1. BuLi
2.

Me—S H
MeO OMe Br

Me—S H

3. BuLi
4. BrSiMe$_3$

(1) 5. HClO$_4$, H$_2$O

MeS SMe SiMe$_3$

(17)

DMTSF

(18)   ⟵⫫— *(11)*

⫫

(12)

Scheme 7

An intramolecular aldol reaction of *(13)* yielded exclusively 1-acetyl-2-methylcyclopentene *(14)*. [5] On the other hand, *(15)*, on addition of DMTSF, yielded only the seven-membered ring *(16)* (Scheme 6)!

Surprisingly, *(17)* cyclized to the spiro-system *(18)* in the presence of two terminators: a vinyl silane and the ketone. The primary cyclization occurred with the vinyl silane, because *(18)* cannot be obtained from *(11)* or *(12)* (Scheme 7).

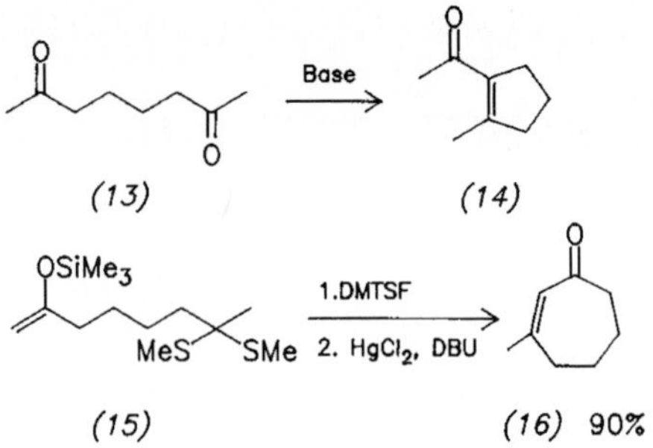

Another report of Trost's group provided an approach to the phenatrene skeleton. [4] In a short sequence starting from *(1)* *(20)* would be obtained, which was transformed with DMTSF into *(21)*. The latter could be transformed in three steps to *(24)* (Scheme 8).

Even in intermolecular reactions chemoselective transformations can be obtained with sulfonium ions. In addition to olefins, acetylenes, ester groups, and keto groups are tolerated. Very reactive terminators, like allyl stannanes, attack exclusively sulfonium ions (Scheme 9). [7]

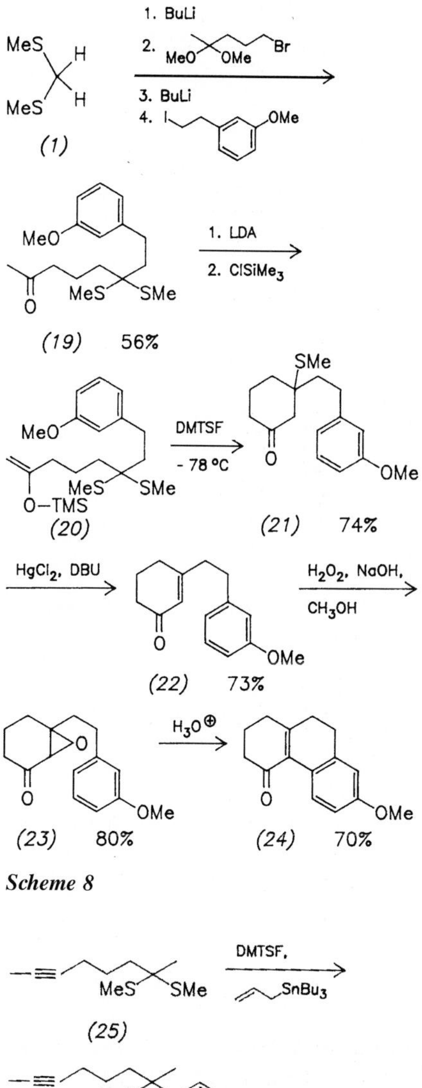

As a result of this outstanding chemoselectivity, sulfonium ions of type *(4)* can be regarded as "super-carbonyl-equivalents".

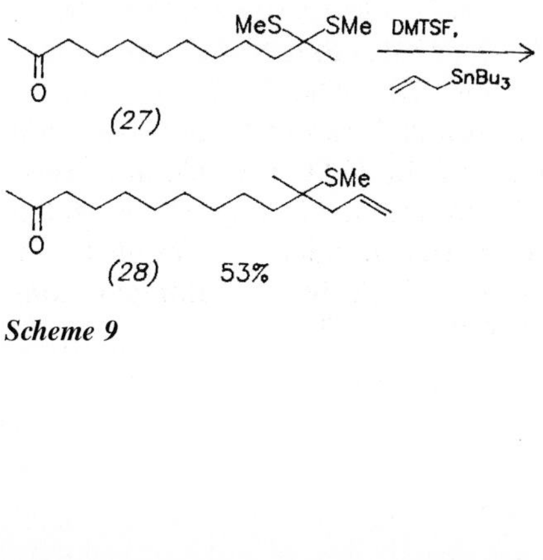

The synthesis of medium-sized rings is complicated and the synthesis of large rings is even more difficult. In particular, macrocyclizations that lead to macrolides are quite important. The transformation of *(31)*→*(32)* shows the potential of sulfonium ion initiated cyclizations. Compound *(32)* was obtained chemoselectively in 48 % yield (Scheme 10). [7]

The alternative reactivity of thioacetals is noteworthy: First, the thio group is used to stabilize a carbanion in the alkylation sequence. Second, a Lewis acid forms the sulfonium ion, which initiates a chemoselective reaction with nucleophiles.

Cationic intermediates are also known in the Pummerer reaction. [8] Compounds of type *(36)* react smoothly in intramolecular fashion with aromatic systems. Especially with anion stabilizing groups, like the cyano group,

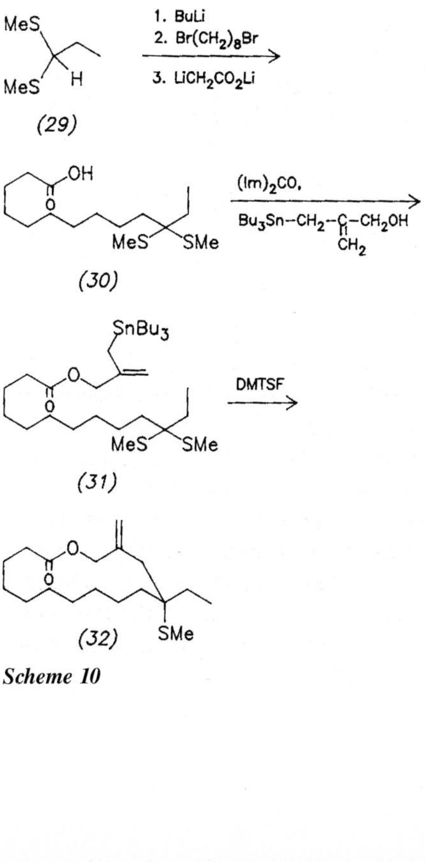

Scheme 8

Scheme 9

Scheme 10

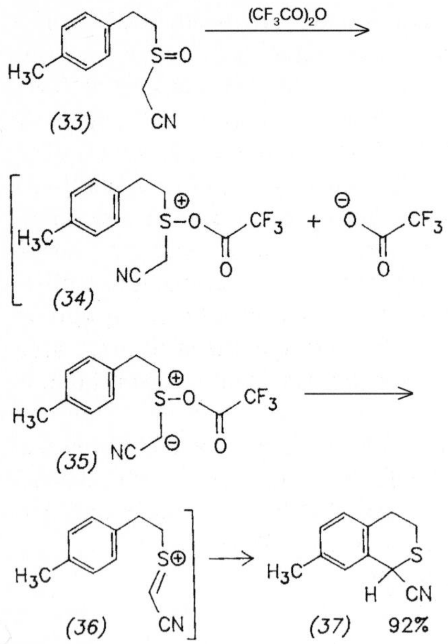

(33)

(34)

(35)

(36)   (37)   92%

Scheme 11

(38)   (39)

(40)

(41)   (42)

(43)   80%

Scheme 12

(44)   (45)   96%

Scheme 13

these reactions are quite powerful. The trifluoroacetate first attacks the sulfoxide *(33)* to form *(36)*, which spontaneously cyclizes to *(37)* (Scheme 11). [9]

The basic studies of Vedejs et al. have enormously extended the potential of this new cyclization technique. [10, 11] They have studied ring-expanding reactions of sulfur ylids to form medium and large rings. Even if these reactions can be regarded more as 2,3-sigmatropic rearrangements, they can be discussed here. The eight-membered ring *(43)* could be synthesized in 80% yield, from simple precursors (Scheme 12).

An almost quantitative yield is obtained in the formation of a nine-membered ring (Scheme 13)!

Finally, a complex seven-membered ring annulation in connection with an approach of the lathrane diterpenes from Fuchs will be discussed. [12, 13] Key reaction of the synthesis is a tetra cyclic intermediate, which is formed by a sulfonium ion-induced cyclization *(51)→(52)*.

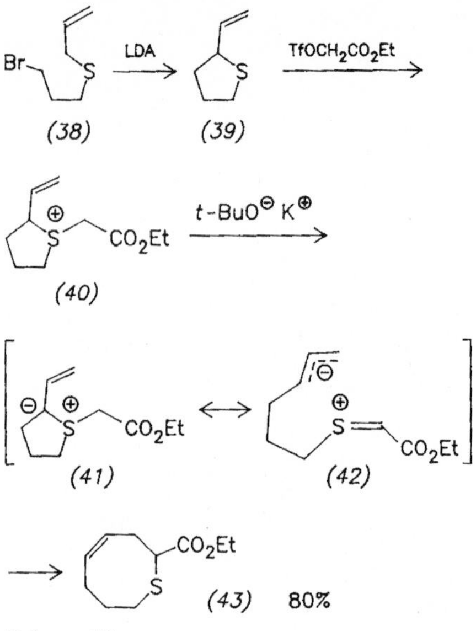

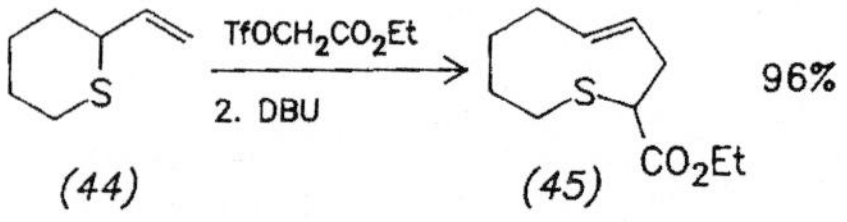

A model study was not necessary because the authors could make use of known results from Trost's group. [6] Starting from optically active *(46)*, cyclization precursor *(51)* could be obtained in a sequence of steps. The masked acetyl group is elegantly introduced as a lithiated vinyl ether, *(49)→(50)*. The enolate can be formed with potassium diisopropylamide, which was trapped with TMSCl to give the desired silyl enolether *(51)*. The latter was cyclized under standard conditions (DMTSF) to give *(52)* in 65% yield (Scheme 14).

This final sequence demonstrates that molecules with different functionality can also be selectively transformed into complex ring systems. This series of three chapters clearly shows the potential of cation-initiated cyclizations. The main emphasis in these articles was to point out the general principles and the slight differences in these cyclization tech-

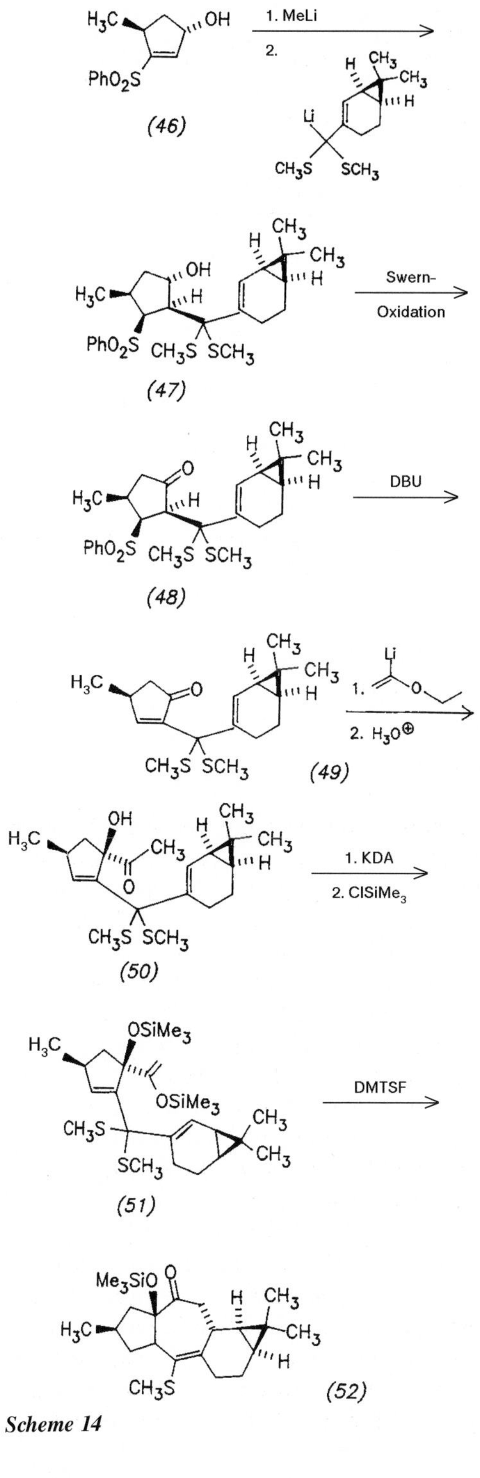

Scheme 14

niques. Iminium ions are mostly used to form five- and six-membered *N*-heterocycles, whereas oxonium ions can be used to form seven- or eight-membered ethers, as elegantly demonstrated by the total synthesis of laurenyne. Of course, five- and six-membered ethers can also be obtained! The situation with sulfonium ions is quite different. Carbocyclizations are mainly obtained in which the thio group occurs as a substituent that can be cleaved in a later stage of the synthesis. Large rings in particular were synthesised using this technique.

In conclusion, each type of cyclization has its own application in synthesis, but all reactions make use of the same terminators, like silylenol ethers, vinyl silanes, allyl silanes, and allyl stannanes.

References

[1] E. J. Corey, D. Seebach, R. Freedman, *J. Am. Chem. Soc.* **1967**, *89*, 434.

[2] B. M. Trost, E. Murayama, *J. Am. Chem. Soc.* **1981**, *103*, 6529.

[3] J. K. Kim, J. K. Paw, M. C. Caserio, *J. Org. Chem.* **1979**, *44*, 1544.

[4] H. Meerwein, K. F. Zenner, R. Gipp, *Liebigs Ann. Chem.* **1965**, *688*, 67.

[5] T. R. Marshall, W. H. Perkin, *J. Chem. Soc.* **1890**, *57*, 241.

[6] B. M. Trost, E. Murayama, *Tetrahedron Lett.* **1982**, *23*, 1047.

[7] B. M. Trost, T. Sato, *J. Am. Chem. Soc.* **1985**, *107*, 719.

[8] P. Welzel, *Nachr. Chem. Tech. Lab.* **1983**, *31*, 892.

[9] M. Hori, T. Kataoka, H. Shimitzu, M. Kataoka, A. Tomoto, M. Kishida, *Tetrahedron Lett.* **1983**, *24*, 3733.

[10] E. Vedejs, *Acc. Chem. Res.* **1984**, *17*, 358.

[11] E. Vedejs, D. Powell, *J. Am. Chem. Soc.* **1982**, *104*, 2046.

[12] T. F. Braish, J. C. Saddler, P. L. Fuchs, *J. Org. Chem.* **1988**, *53*, 3647.

[13] D. Uemura, K. Nobuhara, Y. Nakayama, Y. Shizuri, Y. Hirata, *Tetrahedron Lett.* **1986**, *27*, 4593.

Polycyclization as a Strategy in the Synthesis of Complex Alkaloids

Dieter Schinzer

In no other class of natural products are so many different types of structures found as in the field of alkaloids. Even hexa- or heptacyclic compounds with complex functionality are quite common. Since most of the biologically active compounds are alkaloids, many investigations concerning their synthesis have been performed. Many elegant syntheses are oriented on biosynthesis (so-called biomimetic synthesis), because quite simple types of reactions can be used combined with a high degree of "refinement" in the final products.

A great expert in biomimetic syntheses is Clayton Heathcock, who has used this technique in many total syntheses of complex structures. With a minimum of transformations (mostly so-called "low-tech" reactions are used) a maximum of success is reached in these strategies.

The first class of compounds that will be presented in this article are the lycopodium alkaloids. [1] This alkaloid family consists of about 100 compounds, of which lycopo-

dium itself is the most common member (Scheme 1).

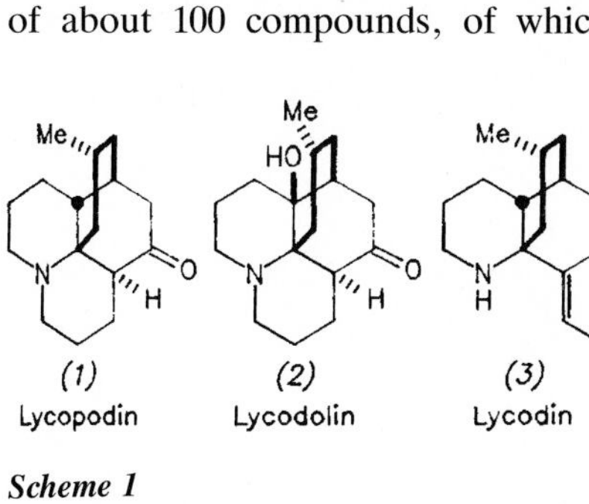

Lycopodium has been synthesized already several times. [2–4] The basic structure contains a tetracyclic framework combined with five asymmetric centers. The synthesis pub-

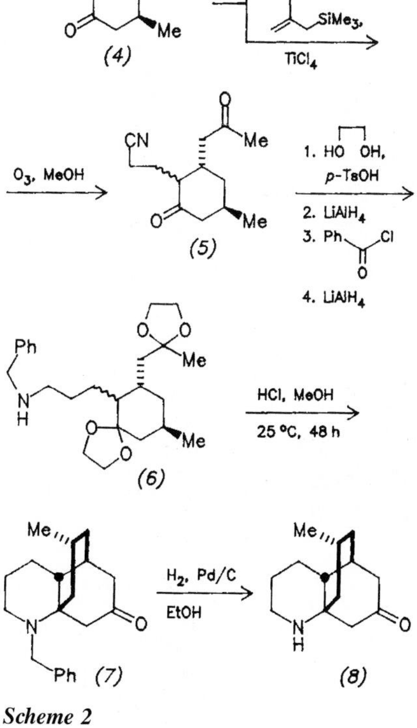

(1) Lycopodin *(2)* Lycodolin *(3)* Lycodin

Scheme 1

Scheme 2

lished by Heathcock is the most efficient; only 8 steps are required to obtain the natural product in 13 % overall yield. [5–8] A functionalized enone *(4)*, which is easily accessable from commercial material, was used as starting material. Compound *(4)* can be treated alternatively with a cuprate reagent or an allylsilane, whereas the Sakurai reaction provides the better chemical and stereochemical yield. [9] The sequence ozonolysis, ketalization and reduction of the nitrile to the amine *(6)* produces the key intermediate of the synthesis (Scheme 2).

After benzoylation of the amine and reduction of the benzamide the crucial step of the synthesis is performed: under thermodynamic conditions (MeOH, HCl) a Mannich reaction is used to stereoselectively produce the tri-

cyclic ketone *(7)*. The configuration of the methyl group was already established with the Sakurai reaction. Everything else is controlled by thermodynamics! The configuration of the side chain is of no interest, because only the axial epimer will cyclize into the iminium ion. The more stable equatorial epimer cannot reach the iminium ion for geometrical reasons. Since *(11)* and *(12)* are in equilibrium via *(9)* and *(10)*, only the "correct compound" *(11)* cyclizes to the tricyclic *(7)* (Scheme 3).

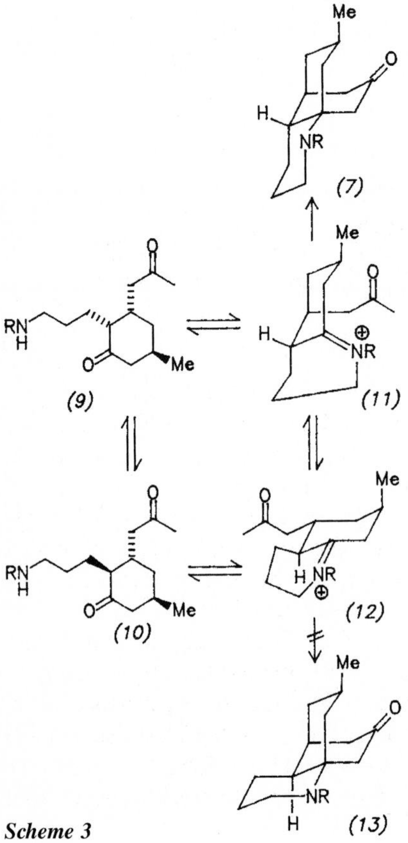

Scheme 3

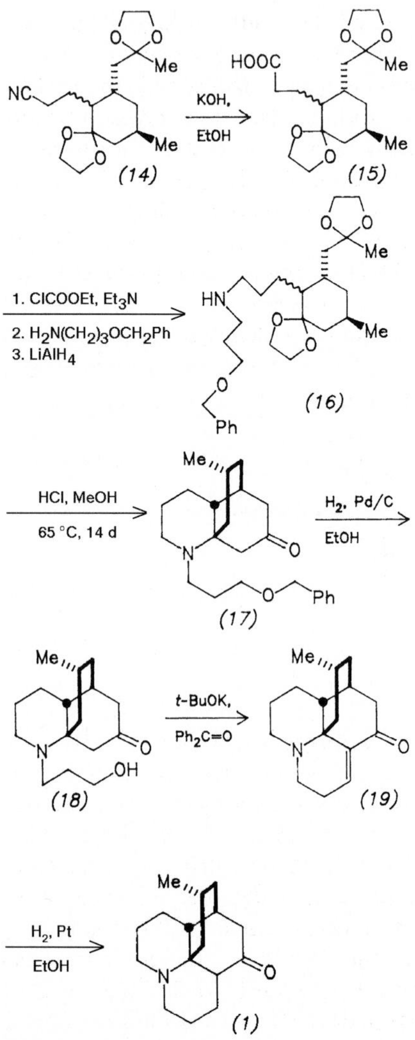

Scheme 4

The natural product can be synthesized by a modification of the amine side chain. But these details are important during the total synthesis, as shown in the transformation *(16)→(17)*. This particular transformation proceeded under more drastic conditions and took 14 d reaction time to yield *(17)* (64 % !). After hydrogenolysis of the benzyl ether, Oppenauer oxidation (benzophenone as oxidizing agent), [10] intramolecular aldol condensation to *(19)*, a final hydrogenation yielded the natural product *(1)* (Scheme 4).

Analysis of the synthesis shows that only simple C–C bond forming reactions have been used (only textbook reactions) to construct a complex molecule. The great accomplishment of this work is the strategy of the synthesis.

The next example will focus on the synthesis of 2,3,3-trialkylindoline alkaloids, which also belong to a complex family of compounds. [11] One has to remember the unique synthesis of strychnine by R. B. Woodward. [12] Again, Heathcock has synthesized vallesamidine *(20)* in a very efficient way (Scheme 5). [13]

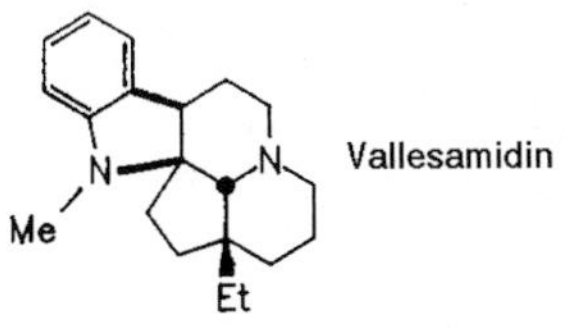

Vallesamidin

(20)

Scheme 5

Starting with the simple ketone *(21)* compound *(22)* can be obtained in a Michael addition. Key reaction is then a Michael addition of the tautomeric enamine *(24)*, which is lactamized to *(25)*. At the end of the sequence the indoline part is obtained by an intramolecular alkylation *(26)→(27)*. (Scheme 6).

Neither the Woodward nor the Stork strategy allows the synthesis of this type of alkaloids. [14] In only seven steps and 19 % overall yield this synthesis is also very efficient.

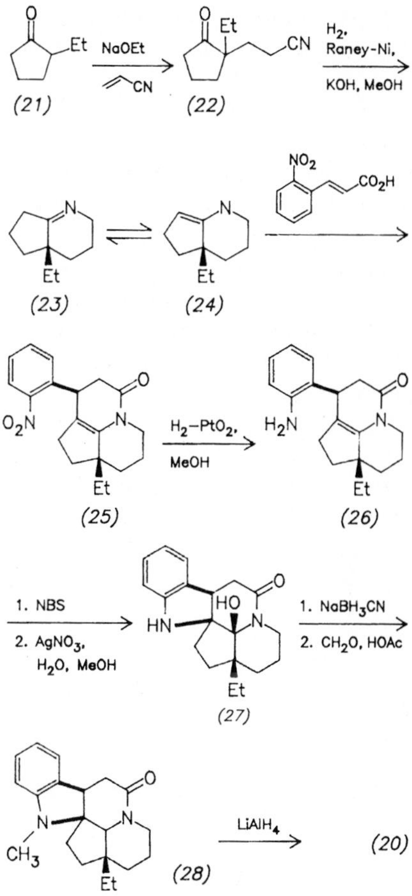

Scheme 6

Even more complex are the class of daphniphyllume alkaloids which got isolated from unusual Japanese trees. [15] The first total syntheses of these compounds were also reported by Heathcock et al. [16–18] The strategy is again based on very simple C–C bond forming reactions that proceed in one-pot operations in high chemical yield.

The daphniphyllium alkaloids are penta- and hexa-cyclic compounds with several asymmetric centers in which the construction of the frame-work looks quite hopeless (Scheme 7). The problem could be solved elegantly by Heathcock in two total syntheses (Scheme 8). [16–18]

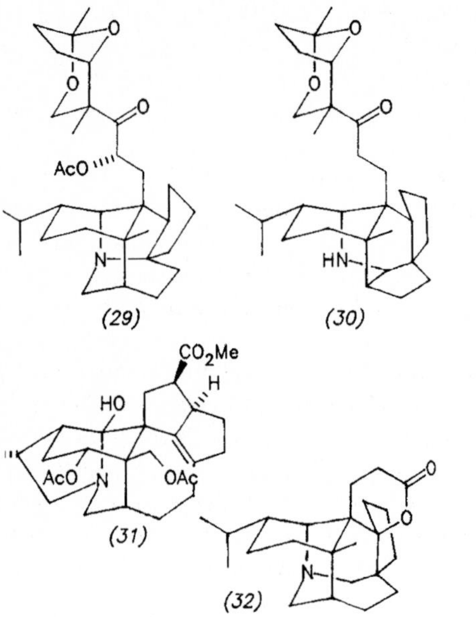

(29)   (30)

(31)

(32)

Scheme 7

The starting material of a 10 step synthesis of methylhomoseco daphniphllate *(40)* with an overall yield of 42 % is the amide *(33)*. A tandem reaction – Michael addition of the enolate of *(33)* with a cyclic unsaturated ester,

(33)

1. LDA

2. [cyclopentene-CO₂Me]

3. [chain structure]

(34)   (35)

followed by in situ alkylation with homogeranyl iodide – yielded *(34)*. DIBAL reduction and cleavage of the amide generated lactone *(35)*, which was reduced to the diol *(36)* with lithium aluminumhydride. The following Swern oxidation yields an unstable dialdehyde *(37)*, that is treated in situ first with ammonia

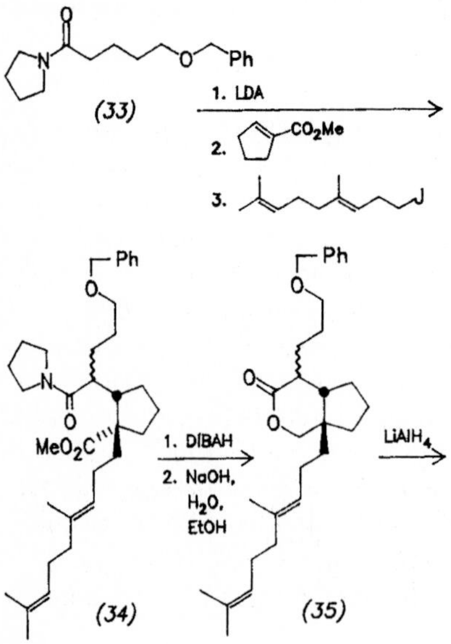

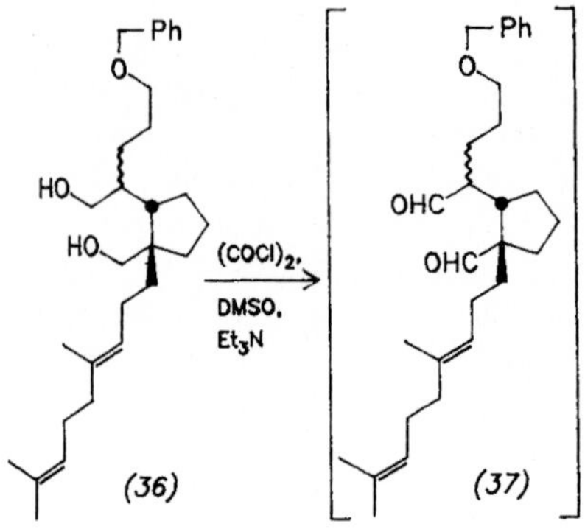

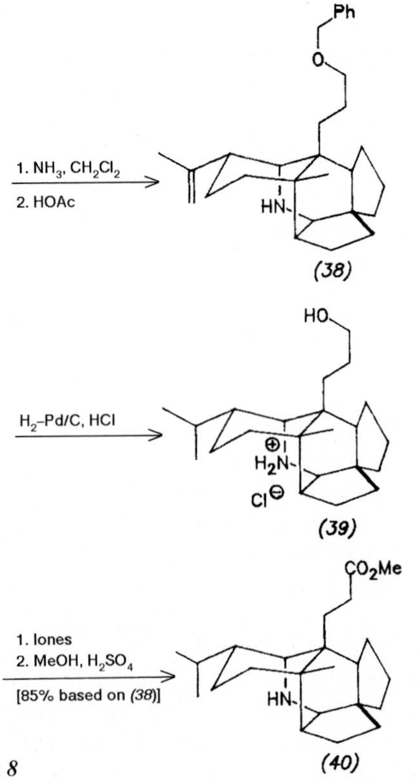

(36)   (37)

$(COCl)_2$, DMSO, Et_3N

(38)

1. NH_3, CH_2Cl_2
2. HOAc

(39)

H_2–Pd/C, HCl

(40)

1. Iones
2. MeOH, H_2SO_4

[85% based on *(38)*]

Scheme 8

(37)

Scheme 9

and then with acetic acid to give the stereo-chemically pure pentacyclic compound *(38)*. This compound can be transformed into the natural product *(40)* by simple operations (Scheme 8).

In the sequence *(37)→(38)* four rings are constructed in a one-pot procedure. In the presence of ammonia intermediate *(41)* is generated, which is trapped by an intramolecular aza-Diels–Alder reaction to form *(42)* (Scheme 9).

The iminium ion *(42)* closes the final ring by way of a π-cyclization to give *(38)*. The fascinating part in this synthesis is the interplay of electrophilic and nucleophilic properties of the reactive template, which generates a very complex skeleton in a one-pot procedure. The hexacyclic daphnilactone A *(32)* is available by simple modifications of this strategy.

The synthesis starts with ester *(43)*, which is deprotonated with potassium disilazide and alkylated with homogeranyl iodide. Reduc-

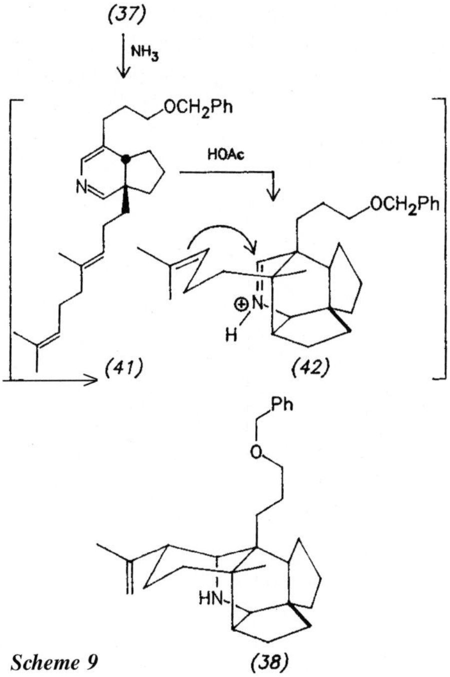

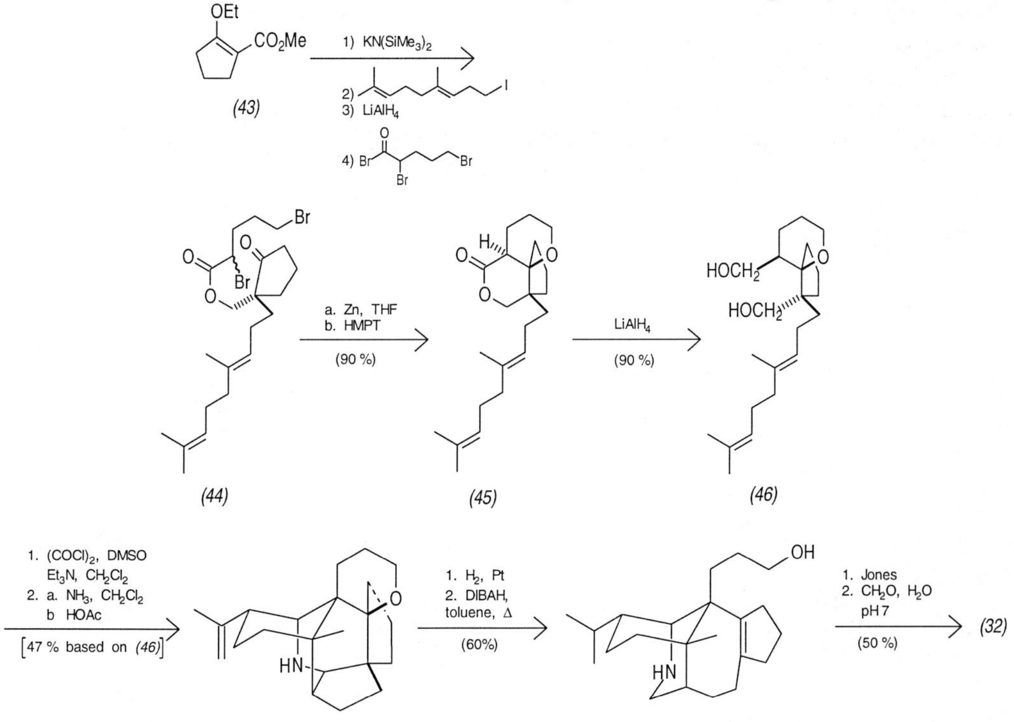

Scheme 10

tion of the ester group, and reaction with acetylbromide yields the highly functionalized ester *(44)*. The subsequent intramolecular Reformatsky reaction *(44)→(45)* generates the tricyclic lactone *(45)*.

The diol *(46)* is transformed into the hexacyclic compound *(47)* by the procedure described already for *(40)*. The aminoalcohol *(48)* is obtained by an interesting fragmentation induced by DIBAL *(47)→(48)*, followed by Jones oxidation, and treatment with formaldeyhde under biometic conditions, which gave the natural product *(32)* (Scheme 10).

The presentation of the last two syntheses has demonstrated how powerful biomimetic strategies can be applied in synthesis mostly using so-called "low-tech reactions" with simple and cheap reagents. By this technique very complex molecules could be synthesized in a very short and efficient way.

References

[1] K. Wiesner, *Fortschr. Chem. Org. Naturst.* **1962**, *20*, 271.
[2] G. Stork, R. A. Kretschmer, R. H. Schlesinger, *J. Am. Chem. Soc.* **1968**, *90*, 1647.
[3] W. A. Ayer, W. R. Bowman, T. C. Joseph, P. Smith, *J. Am. Chem. Soc.* **1968**, *90*, 1648.

[4] D. Schumann, H.-J. Müller, A. Naumann, *Liebigs Ann. Chem.* **1982**, 1700.
[5] C. H. Heathcock, E. F. Kleinman, E. S. Binkley, *J. Am. Chem. Soc.* **1978**, *100*, 8036.
[6] E. F. Kleinman, C. H. Heathcock, *Tetrahedron Lett.* **1979**, 4125.
[7] C. H. Heathcock, E. F. Kleinman, *J. Am. Chem. Soc.* **1981**, *103*, 222.
[8] C. H. Heathcock, E. F. Kleinman, E. S. Binkley, *J. Am. Chem. Soc.* **1982**, *104*, 1054.
[9] T. A. Blumenkopf, C. H. Heathcock, *J. Am. Chem. Soc.* **1983**, *105*, 2354.
[10] R. B. Woodward, H. L. Wendler, F. J. Brutschy, *J. Am. Chem. Soc.* **1945**, *67*, 1425.
[11] G. A. Cordell, J. E. Saxton, *Alkaloids* (N. Y.) **1981**, *20*, 1.
[12] R. B. Woodward, M. P. Cava, W. D. Ollis, A. Hunger, H. V. Daniker, K. J. Schenker, *J. Am. Chem. Soc.* **1954**, *76*, 4749.
[13] D. A. Dieckman, C. H. Heathcock, *J. Am. Chem. Soc.* **1989**, *111*, 1528.
[14] G. Stork, J. E. Dolfini, *J. Am. Chem. Soc.* **1963**, *85*, 2872.
[15] S. Yamamura, *The Alkaloids* (Ed.: A. Brossi) **1986**, *26*, 265.
[16] C. H. Heathcock, S. K. Davidson, S. Mills, M. A. Sanner, *J. Am. Chem. Soc.* **1986**, *108*, 5650.
[17] R. B. Ruggeri, M. M. Hansen, C. H. Heathcock, *J. Am. Chem. Soc.* **1988**, *110*, 8734.
[18] R. B. Ruggeri, K. F. McClure und C. H. Heathcock, *J. Am. Chem. Soc.* **1989**, *111*, 1530.

F. General Methods and Reagents for Organic Synthesis

Domino Reactions

Herbert Waldmann

For the total synthesis of complex organic compounds nowadays a rich arsenal of powerful synthetic methods is available. However, in comparison with the biosyntheses taking place in biological systems many total syn-

EDDA, CH₃CN
20 °C, 4h

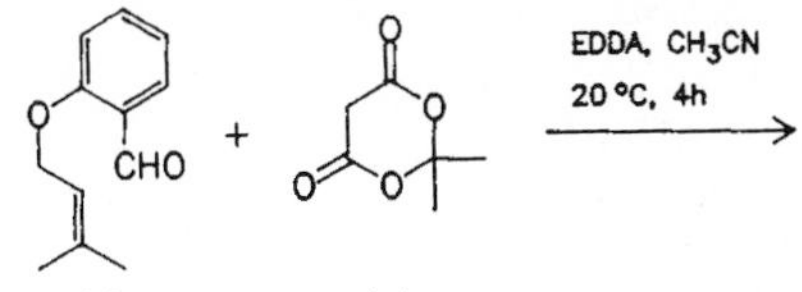

(1)   (2)

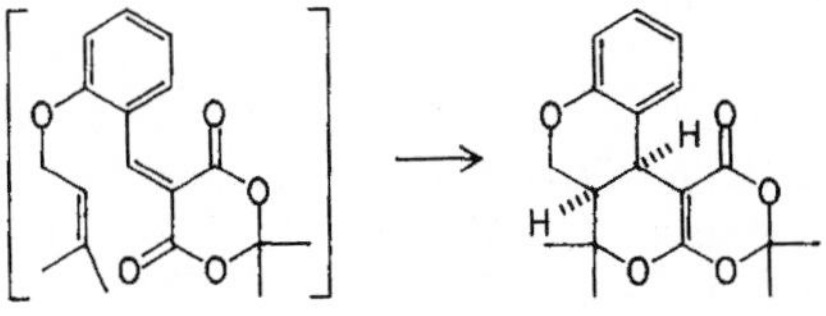

(3)   (4)   81 %

cis : trans = > 99 : 1

EDDA: H₃N⁺–CH₂–CH₂–⁺NH₃ (OAc⁻)₂

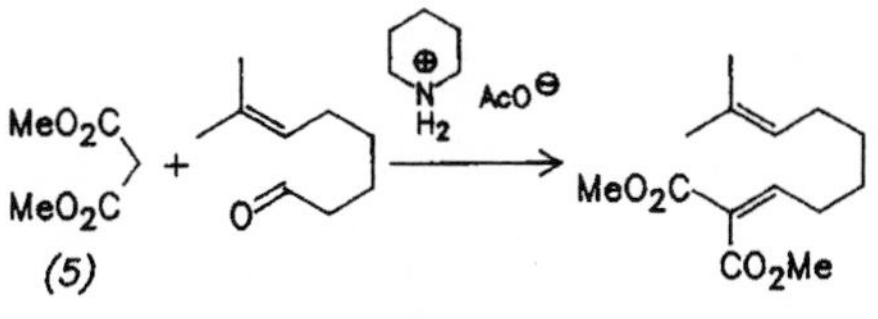

MeO₂C
         +
MeO₂C

(5)

MeO₂C
           CO₂Me

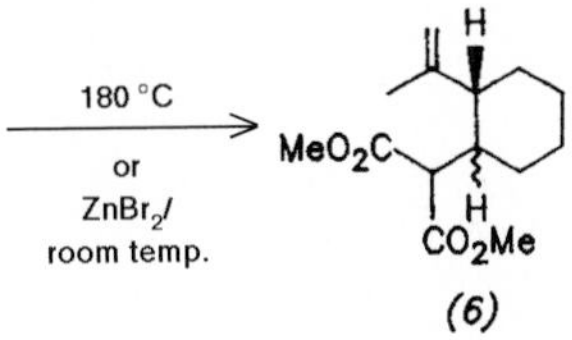

180 °C
or
ZnBr₂/
room temp.

MeO₂C
         CO₂Me

(6)

73-88 %

trans : cis = > 99 : 1

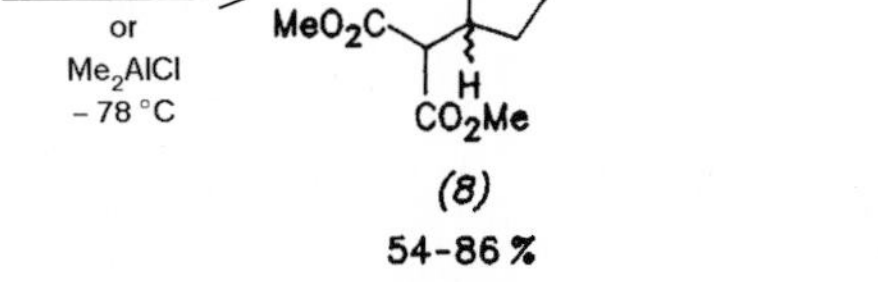

thetic endeavors aimed at the construction of natural products appear to be uneffective and circumstantial. In nature, biocatalysts often carry out sequential transformations, that is, a series of bond formations or scissions without the accumulation of intermediates. Although the multienzyme systems involved in these biosyntheses can advantageously be employed for preparative purposes, [1] the development of analogous nonenzymatic synthetic sequences opens up new and promising opportunities for organic chemistry. In particular, the so-called "domino reactions" (frequently also

SiMe₃

MeO₂C
         +
MeO₂C

SiMe₃

MeO₂C
         CO₂Me

(7)

ZnBr₂/
20 °C
or
Me₂AlCl
– 78 °C

MeO₂C
         CO₂Me

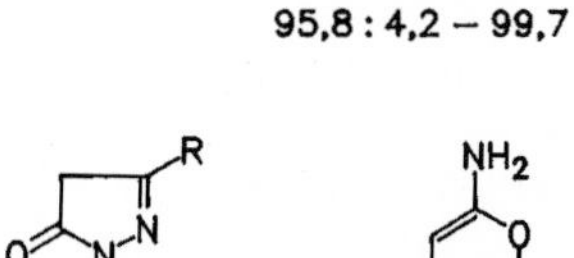

(8)

54-86 %

trans : cis =
95,8 : 4,2 – 99,7 : 0,3

(9)   (10)

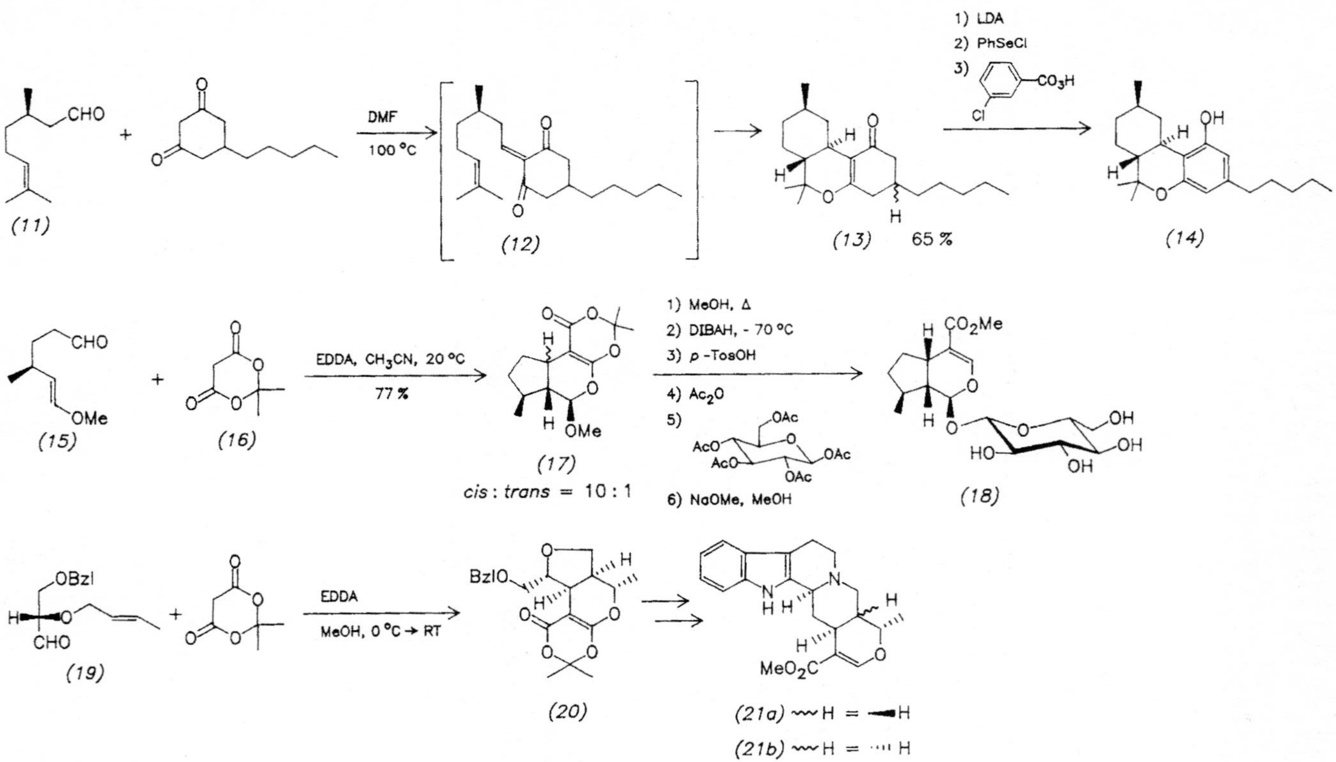

Scheme 1. Application of tandem Knoevenagel-hetero Diels-Alder reactions in the construction of natural products according to Tietze et al. [7, 8] and S. Takano et al. [9].

described as "tandem reactions" or "cascade reactions") prove to be very elegant and effective processes. The term "domino reactions" describes two or more subsequently occurring bond forming or bond breaking transformations, during which subsequent reactions occur at the functionalities generated in the respective preceeding step. [2]

Tietze et al. have developed three very flexible domino reactions. In the first case, in a tandem Knoevenagel–hetero-Diels–Alder sequence, an aldehyde, for example, *(1)*, undergoes condensation with a 1,3-dicarbonyl compound, for example, *(2)*, to give a 1-oxa-1,3-butadiene *(3)*, which undergoes cycloaddition with a dienophile to give *(4)*. [2a, b, e, 3] The dienophile may be accessible intramolecularly, as in *(1)*, or it may be added as a third component to the reaction mixture. If a 1,3-dicarbonyl compound that cannot react further in a hetero Diels–Alder reaction, for example, dimethyl malonate *(5)*, is introduced into the initiating Knoevenagel condensation, an ene reaction terminates the sequence [e.g., *(5)*→*(6)*]. [4] If, in addition, aldehydes are used that embody an allyl silane moiety, for example, *(7)*, the cascade is terminated by an allyl silane cyclization [e.g., *(7)*→*(8)*]. [5] A broad variety of aldehydes and CH-acidic compounds can be employed in the tandem Knoevenagel–hetero-Diels–Alder reactions, including such nucleophiles in which the dicarbonyl structure is hidden in a heterocycle, for example, *(9)* and *(10)*. The simple diastereoselectivity is generally high and by using enantiomerically pure aldehydes, [6a, 7–9] 1,3-dicarbonyl compounds [6b] or chiral Lewis acids [6c] asymmetric syntheses proceeding with excellent stereoselectivity can be carried out. Furthermore, this domino sequence has proven its efficiency in the construction of various natural products and pharmacologically active compounds (Scheme 1). For instance, the psychotropic (–)-hexahydrocannabinol *(14)* was obtained in a few steps in enantiomerically pure form by following this

method. [7] To this end, *(R)*-citronellal *(11)* was converted to *(12)* and after subsequent cycloaddition the Diels-Alder adduct *(13)* was formed, which could be transformed to the desired *(14)*. On the other hand, the aldehyde *(15)* and Meldrums's acid *(16)* deliver the tricyclic intermediate *(17)*, from which deoxyloganine *(18)* was obtained in six steps. [8] This iridoid glycoside plays a central role in the biosynthesis of numerous alkaloids. Takano et al. used this method to construct the cycloadduct *(20)* from the protected glycerol aldehyde *(19)*. [9] Compound *(20)* served as congener to the indole alkaloids (–)-ajmalicine *(21a)* and (–)-tetrahydroalstonine *(21b)*.

Domino reactions offer a wealth of new opportunities for alkaloid chemistry. For instance, Winterfeldt et al. [10] took advantage of the subsequently occurring diastereoselective conjugate addition of the enantiomerically pure enaminone *(22)* to the α,β-unsaturated aldehyde *(23)* and a following Pictet–Spengler reaction to generate the tetra-

(22)   (23)

CF₃COOH | 20d
– 78 °C → 20 °C | 59%
diastereomeric ratio 88 : 12

(24)   (25)

R* =

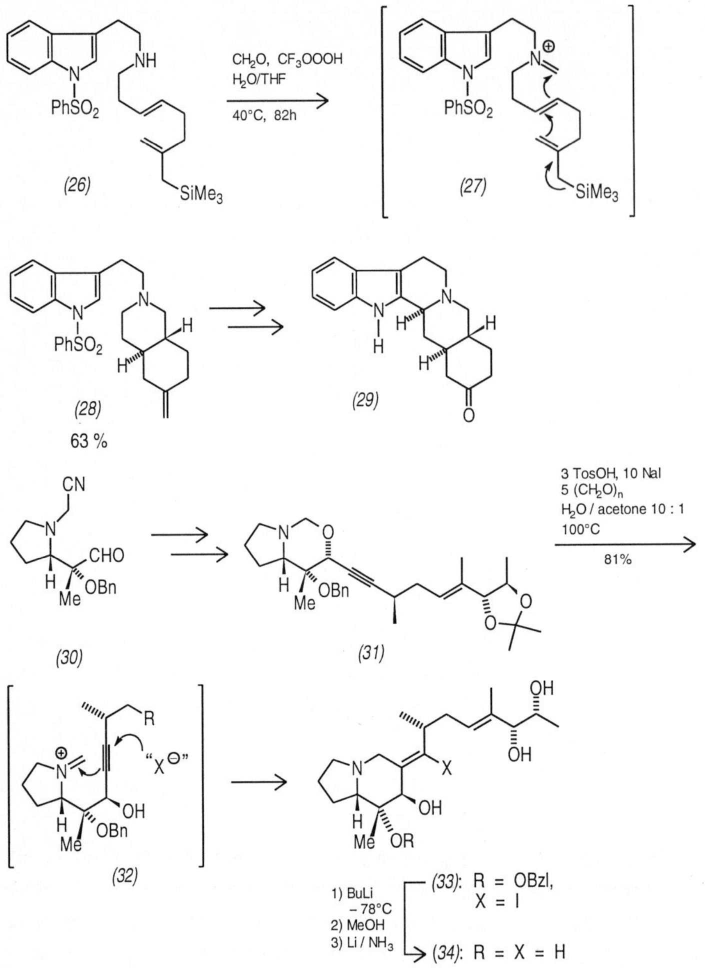

Scheme 2. Domino sequences which are initiated by an iminium salt formation according to Grieco et al. [11] and Overman et al. [12].

cyclic compound *(24)*, which is a central intermediate in the construction of the corynanthe alkaloids, for example, geissoschizine *(25)*. Grieco et al. [11] and Overman et al. [12] employed the Mannich reaction in very elegant ways to initiate cyclizations (Scheme 2). Thus, on treatment with formaldehyde in aqueous solution the secondary amine *(26)* forms an iminium intermediate *(27)* to undergo a polyene cyclization that is finally terminated by an allylsilane cyclization. [11]

The tertiary amine *(28)* formed thereby can be transformed to yohimbone *(29)*. A nucleophile-assisted iminium ion–alkine cyclization served as the key step in the synthesis of the alkaloid allopumiliotoxine 339A *(34)*. [12a] In the course of this sequence the oxazine *(31)*, which was built up from the proline derivative *(30)*, was converted to the iminium salt *(32)* by treatment with aqueous formaldehyde. Assisted by iodide as nucleophile the correctly placed acetylene in *(32)* then attacks

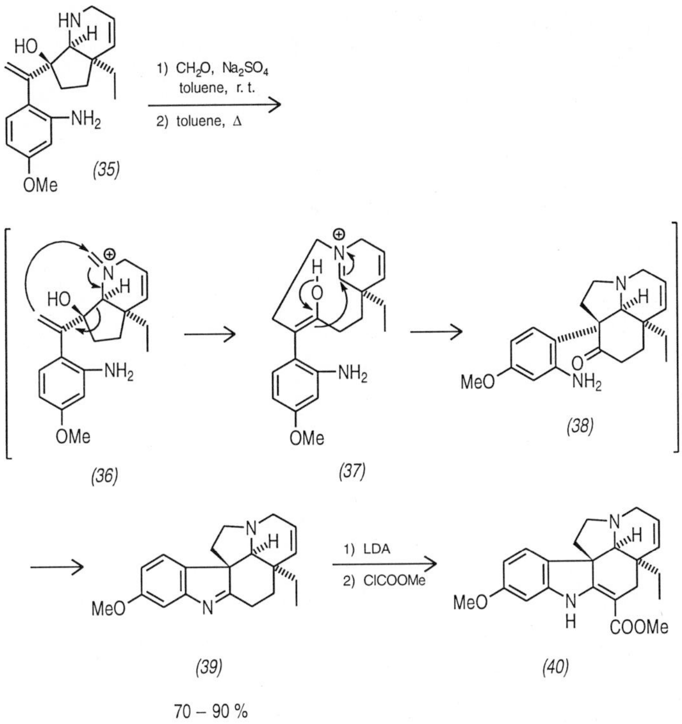

Scheme 2. continued

the heteroanalogous carbonyl compound and with simultaneous loss of the acetonide protecting group the intermediate collapses to give the heterocycle (33). In the course of the construction of the pentacyclic aspidosperma alkaloid 16-methoxytabersonine (40), [12b] Overman et al. combined the initiating iminium ion formation (35)→(36) even with an aza Cope rearrangement (36)→(37), a Mannich reaction (37)→(38) and the formation of a Schiff base (38) →(39). In addition, the authors recently used this impressive multistep reaction cascade as the key step in the first enantioselective total synthesis of strychnine. [12c]

A hetero Cope rearrangement also forms the core of the elegant and efficient domino syntheses developed by Blechert et al. (Scheme 3) [13] The sequence starts with in

situ generation of nitrons, which then react with acceptor-substituted allenes in 1,3-dipolar cycloadditions to give the adducts (41). These intermediates embody a 1-aza-1-oxa Cope system and by a spontaneous [3,3]-sigmatropic rearrangement followed by a retro Michael reaction and elimination of water they are converted to the 2-vinylindoles (42), which are viable congeners for the construction of complex alkaloids. If the moiety "R" in (42) incorporates an appropriately placed double bond, the reaction cascade can further be extended by an intramolecular Diels–Alder reaction and the perhydroellipticine derivative (43) is formed in a one-pot-reaction in 40 % overall yield. [13a] Furthermore, Blechert et al. developed new routes to iboga-, [13d] uleine- [13a] and tetrahydrocarbazole alkaloids. [13e]

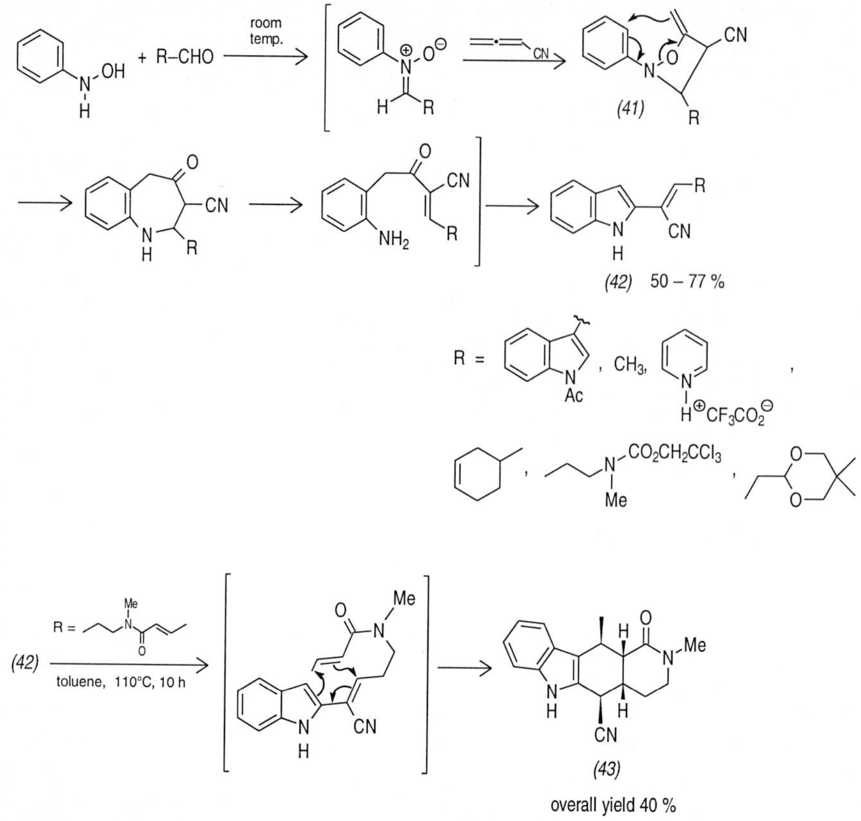

Scheme 3. Construction of complex frameworks of alkaloids by using domino reactions according to Blechert et al. [13].

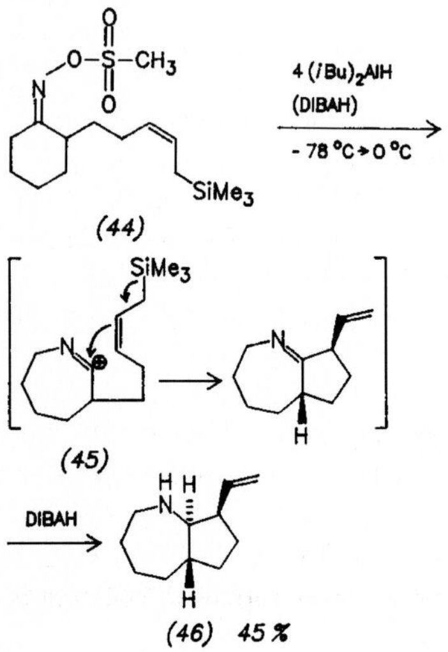

Schinzer et al. described the combination of a Beckmann rearrangement with an allylsilane cyclization. [14] On treatment of the oxime mesylate *(44)* with diisobutylaluminum hydride (DIBAH) at low temperature the Lewis acid first initiates a rearrangement leading to the intermediate *(45)*. The latter is then trapped by the allylsilane terminator to give an imine, which is directly reduced to the amine by DIBAH. Compounds such as *(46)* may prove to be interesting intermediates in the construction of azepine alkaloids.

Particularly interesting opportunities for the development of powerful domino reactions are opened up by the use of transition metal mediated transformations. An impressive example is provided by the cobalt-catalyzed directed cyclotrimerization of alkynes developed by Vollhardt et al. (Scheme 4). [15] For instance, bis(trimethylsi-lyl)acetylene *(47)* (BTMSA), which is unable to undergo self-trimerization, reacts with the 1,5-diacetylene *(48)* in the presence of the catalyst CpCo(CO)₂(Cp = cyclopentadienyl)

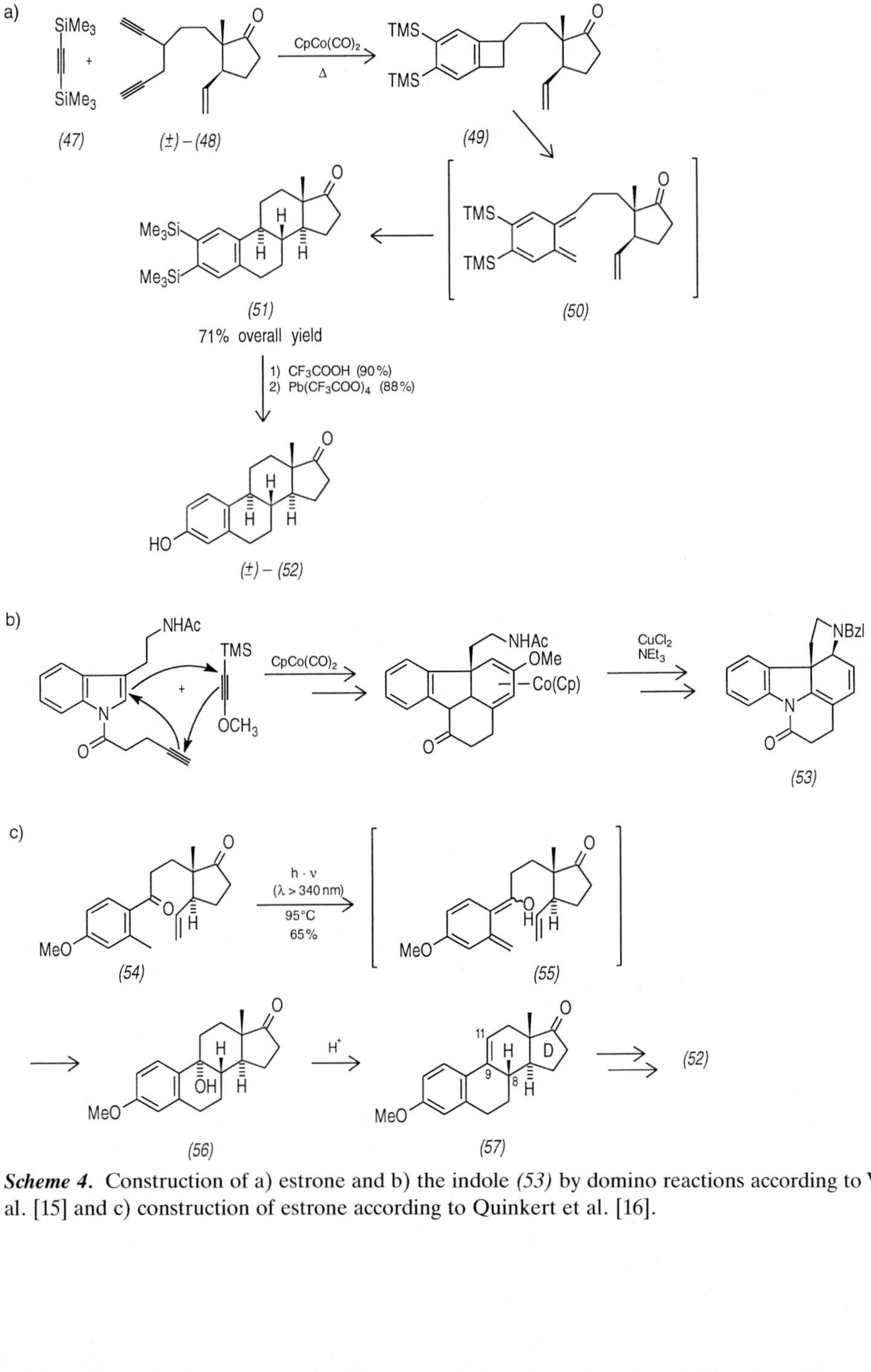

Scheme 4. Construction of a) estrone and b) the indole *(53)* by domino reactions according to Vollhardt et al. [15] and c) construction of estrone according to Quinkert et al. [16].

to give the isolable benzocyclobutane *(49)*. This strained compound is subject to a ring-opening reaction leading to the *ortho*-quinodimethane *(50)*, which under the reaction condition serves as diene in an intramolecular Diels–Alder reaction leading to *(51)*. The tetracyclic *(51)* was converted to *rac*-estrone *(52)* via regioselective protodesilylation at C-2 and oxidative cleavage of the C-3-Si bond. [15b] In the course of applying this methodology to the construction of various targets, the authors, for instance, built up the indole derivative *(53)*, which embodies the basic framework of the polycyclic indole alka-

loid strychnine (Scheme 4b). [15c] According to Quinkert et al. reactive *ortho*-quinodimethane intermediates can also be generated photochemically from alkyl-substituted benzophenones. [16] By irridation of *(54)* the tautomer *(55)* is formed, which is converted immediately to *(56)* by a stereoselective intramolecular Diels–Alder reaction. After dehydration *(57)* is formed, from which enantiomerically pure estrone can be synthesized (Scheme 4c).

By applying palladium-mediated transformations surprising polycyclization can be realized (Scheme 5). Thus, Oppolzer et al.

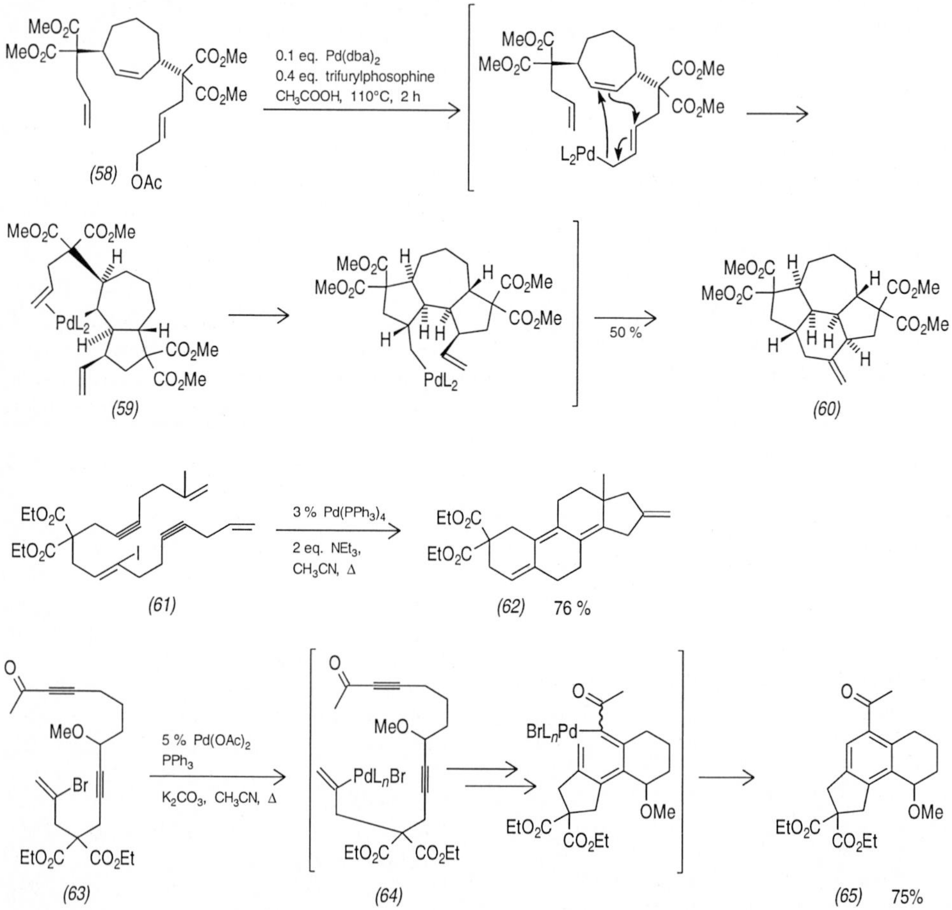

Scheme 5. Domino Heck reactions according to Oppolzer et al. [17], Negishi et al. [18] and de Meijere et al. [19].

described that the *trans*-1,4-cycloheptene *(58)* is converted in one step to the complex tetracyclic compound *(60)* by treatment with a palladium(0)-catalyst. [17] In the course of this reaction cascade a palladium ene reaction gives *(59)*, followed by two intramolecular Heck reactions. Finally, the sequence is terminated by reductive elimination of the palladium. Similarly spectacular are the domino Heck cyclizations of *(61)* to *(62)* and of *(63)* to *(65)*, developed by Negishi et al. [18] and de Meijere et al. [19] In both cases most probably first a vinylpalladium intermediate like *(64)* is formed, which then cyclizes with the neighbouring alkyne to form a new alkenyl metal species. In both cases the β-hydride elimination of palladium ends the tandem reaction.

The highlighted syntheses provide only a few selected examples out of a wealth of applications of domino reactions (for extensive reviews see in particular refs. [2a, g]). They demonstrate, however, that sequential transformations in many cases open up short, efficient, and at the same time, elegant routes to complex molecules. It is to be expected that this new synthetic strategy will establish itself as a powerful tool of organic synthesis.

References

[1] See, for instance.: a) G.M. Whitesides, C.-H. Wong, *Angew. Chem.* **1985**, *97*, 617; *Angew. Chem. Int. Ed. Engl.* **1985**, *24*, 617; b) H. Waldmann, *Nachr. Chem. Techn. Lab.* **1992**, *40*, 828.

[2] a) Reviews: L.F. Tietze, U. Beifuss, *Angew. Chem.* **1993**, *105*, 137; *Angew. Chem. Int. Ed. Engl.* **1993**, *32*, 131–163; b) L.F. Tietze, *J. Heterocycl. Chem.* **1990**, *27*, 47; c) F. Ziegler, *Chem. Rev.* **1988**, *88*, 1423; d) G.H. Posner, *Chem. Rev.* **1986**, *86*, 831; e) L.F. Tietze in *Selectivity – A Goal for Synthetic Efficiency* (eds.: W. Bartmann, B.M. Trost), VCH, Weinheim, **1984**, p. 299; f) H.M.R. Hoffmann, *Angew. Chem.* **1992**, *104*, 1361; *Angew. Chem. Int. Ed. Engl.* **1992**, *31*, 1332; g) T.-L. Ho, *Tandem Organic Reactions*, Wiley, New York, **1992**.

[3] L.F. Tietze und U. Beifuss in *Comprehensive Organic Synthesis*, Vol. 2 (ed.: B.M. Trost), Pergamon, **1991**, p. 341.

[4] a) L.F. Tietze, U. Beifuss, M. Ruther, *J. Org. Chem.* **1989**, *54*, 3120; b) L.F. Tietze, U. Beifuss, M. Ruther, A. Rühlmann, J. Antel, G.M. Sheldrick, *Angew. Chem.* **1988**, *100*, 1200; *Angew. Chem. Int. Ed. Engl.* **1988**, *27*, 1186; c) L.F. Tietze, U. Beifuss, *Angew. Chem.* **1985**, *97*, 1067; *Angew. Chem. Int. Ed. Engl.* **1985**, *24*, 1042; d) L.F. Tietze, U. Beifuss, *Liebigs Ann. Chem.* **1988**, 321.

[5] L.F. Tietze, M. Ruther, *Chem. Ber.* **1990**, *123*, 1387.

[6] a) L.F. Tietze, S. Brand, T. Brumby, J. Fennen, *Angew. Chem.*, **1990**, *102*, 675; *Angew. Chem. Int. Ed. Engl.* **1990**, *29*, 665; b) L.F. Tietze, S. Brand, T. Pfeiffer, J. Antel, K. Harms, G.M. Sheldrick, *J. Am. Chem. Soc.* **1987**, *109*, 921; c) L.F. Tietze and P. Saling, *Synlett* **1992**, 281.

[7] L.F. Tietze, G. v. Kiedrowski, B. Berger, *Angew. Chem.*, **1982**, *94*, 222; *Angew. Chem. Int. Ed. Engl.* **1982**, *21*, 221.

[8] L.F. Tietze, H. Denzer, X. Holdgrün, M. Neumann, *Angew. Chem.*, **1987**, *99*, 1309; *Angew. Chem. Int. Ed. Engl.* **1987**, *26*, 1295.

[9] a) S. Takano, *Pure Appl. Chem.* **1987**, *59*, 353; b) S. Takano, S. Satoh und K. Ogasaware, *J. Chem. Soc. Chem. Commun.* **1988**, 59.

[10] C. Bohlmann, R. Bohlmann, E.G. Rivera, C. Vogel, M.D. Manandhar, E. Winterfeldt, *Liebigs Ann. Chem.* **1985**, 1752.

[11] P.A. Grieco, W.F. Fobare, *J. Chem. Soc. Chem. Commun* **1987**, 185.

[12] a) L.A. Overman, L.A. Robinson, J. Zablocki, *J. Am. Chem. Soc.* **1992**, *114*, 368; b) L.E. Overman, M. Sworin, R.M. Burk, *J. Org. Chem.* **1983**, *48*, 2685; c) S.D. Knight, L.E. Overman, G. Pairaudeau, *J. Am. Chem. Soc.* **1993**, *115*, 9293.

[13] a) S. Blechert, *Synthesis* **1989**, 71; b) J. Wilkens, A. Kühling, S. Blechert, *Tetrahedron*, **1987**, *43*, 3237; c) S. Blechert, *Liebigs Ann. Chem.* **1985**, 673; d) S. Blechert, *Helv. Chim. Acta* **1985**, *68*, 1835; e) S. Blechert in *40 Jahre Fonds der Chemischen Industrie 1950–1990*, Frankfurt, **1990**, p. 41.

[14] D. Schinzer and Y. Bo, *Angew. Chem.* **1991**, *103*, 727; *Angew. Chem. Int. Ed. Engl.* **1991**, *30*, 687.

[15] a) K.P.C. Vollhardt, *Angew. Chem.* **1984**, *96*, 525; *Angew. Chem. Int. Ed. Engl.* **1984**, *23*, 539; b) R.L. Funk, K.P.C. Vollhardt, *J. Am. Chem. Soc.* **1980**, *102*, 5253; c) D.B. Grotjahn, K.P.C. Vollhardt, *J. Am. Chem. Soc.* **1986**, *108*, 2091 and **1990**, *112*, 5653.
[16] a) G. Quinkert and H. Stark, *Angew. Chem.* **1983**, *95*, 651; *Angew. Chem. Int. Ed. Engl.* **1983**, *22*, 637; b) G. Quinkert, U. Schwartz,

H. Stark, W.-D. Weber, F. Adam, H. Baier, G. Frank, G. Dürner, *Liebigs Ann. Chem.* **1982**, 1999.
[17] W. Oppolzer, R.J. de Vita, *J. Org. Chem.* **1991**, *56*, 6256.
[18] Y. Zhang, G. Wu, G. Agnel, E. Negishi, *J. Am. Chem. Soc.* **1990**, *112*, 8590.
[19] F.E. Meyer, A. de Meijere, *Synlett* **1991**, 777.

Group Selective Reactions

Martin Maier

There is no doubt that in the last decade asymmetric synthesis has reached such a high level that the stereocontrolled formation of any given arrangement of stereocenters in a molecule no longer poses an insurmountable challenge. [1] Thus, it is possible to functionalize prochiral centers in chiral starting materials by taking advantage of substrate control. In this area a large body of knowledge has been accumulated over the years. [2] Equally, for the transformation of achiral compounds into chiral, nonracemic products, one has a large array of chiral auxiliaries at one's command. Recently, these methods have become complemented by very efficient chiral catalysts. [3]

In most cases, the transformations that convert a prochiral center into a stereocenter take place at sp^2-hybridized atoms (e.g. olefins, dienes, carbonyl groups) or sp^2-hybridized intermediates (e.g. enolates). If one of the two faces of the molecule is attacked preferentially, one speaks of facial selectivity. In stereochemical terminology this means that in the course of such reactions, depending on the starting material, enantiotopic or diastereotopic faces are differentiated. However, stereocenters can also be generated by the discrimination of enantiotopic or diastereotopic *groups*, respectively. For example, the stereocenter in *(3)* may be generated either from the olefin *(1)* through a facially selective hydroboration or, alternatively, by selective reduction of one of the hydroxymethyl groups of the diol *(2)* (Scheme 1). In fact, such group selective reactions are common in nature. Some esterases are able to hydrolyze selectively one of two enantiotopic ester functions. Accordingly, enzymatic methods have found widespread use in the area of enantioselective synthesis. [4] It is the purpose of the following account to show that chemical methods are also very powerful for the differentiation of groups. In particular, these methods offer advantages that cannot be achieved with face-selective reactions.

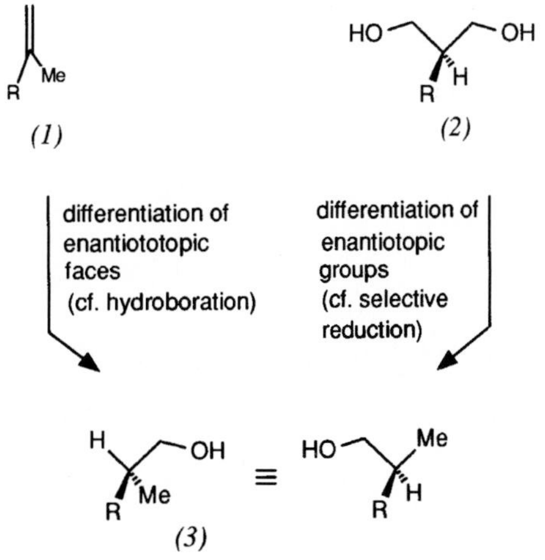

Scheme 1. Facial and group differentiating reactions for the generation of a stereo center. If R is chiral these are stereodifferentiating reactions.

The internal arrangement (topology) of constitutionally identical ligands (atoms and atomic groups) within a molecule may be classified according to symmetry criteria (Scheme 2). [5] If groups can be transformed into each other by internal C_n-symmetry operations, they are called *homotopic*. If a molecule possesses apart from a C_n-axis (n is usually 2, sometimes 3), no further symmetry elements, it is called dissymmetric, that is chiral [cf. compound *(4)*]. Logically, constitutionally equal groups that do not fulfill the criteria for homotopy are termed *heterotopic*. Heterotopic ligands may be further subdivided into enantiotopic and diastereotopic groups. If a molecule belongs to the point group C_s, that means the only symmetry element is a mirror plane, the groups to the left and right side of this plane are *enantiotopic*. [cf. molecules *(5)*, *(7)*, *(8)*]. Tetrahedral atoms that lie on the mirror plane are prochiral. However, the presence of a mirror plane in the molecule does not exclude elements of chirality in the molecule itself. Molecules that consist of equal numbers of chiral elements with opposite chirality, are called *meso*-compounds. These in turn can be classified into odd and even *meso*-compounds. In odd *meso*-compounds a central C-atom is termed a center of pseudoasymmetry. On the other hand, if constitutionally identical ligands are not superimposable on each other by symmetry operations, they are diastereotopic. It should be noted that diastereotopic groups can also exist in achiral molecules. Besides the above mentioned symmetry criteria for the classification of groups one can also verify the topology by mutually exchanging the ligands with test groups. [6]

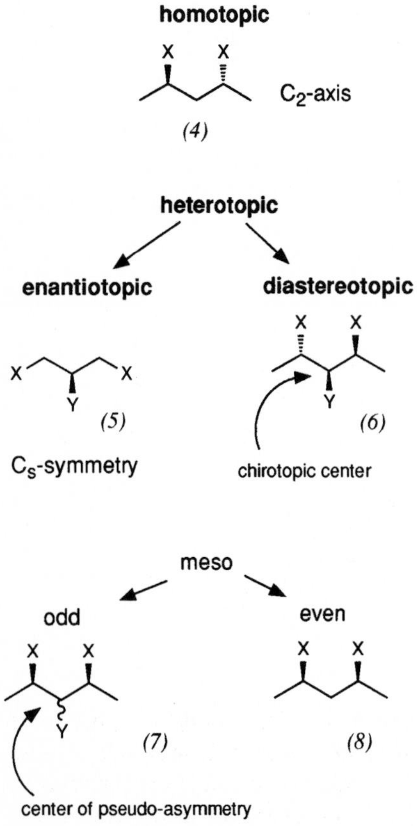

homotopic

(4)

heterotopic

enantiotopic   **diastereotopic**

C_s-symmetry   chirotopic center

meso

odd   even

center of pseudo-asymmetry

Scheme 2. Topology of groups.

Simultaneous Synthesis in two Directions

Particularly in the case of chain-like molecules the constitutionally symmetrical substrates for group selective reactions may be prepared very efficiently by applying a two-directional chain synthetic strategy (Scheme 3). [7] The major advantage of the simultaneous modification of a molecule in two directions is that the number of synthetic transformations is reduced compared with that of a one-dimensional synthetic strategy. Essentially, this means that more material can be processed. The application of a two-dimensional synthesis is recommended, according to Schreiber, [7] for three classes of symmetrical molecules: 1) For molecules with C_s-symmetry; 2) for molecules with C_2-symmetry; and 3) for molecules possessing pseudo-C_2-symmetry. In doing so one tries to keep the symmetry for as long as possible and to diffe-

C_S-molecules

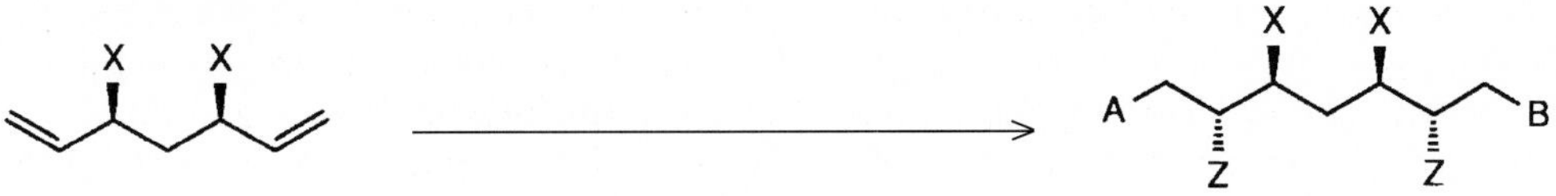

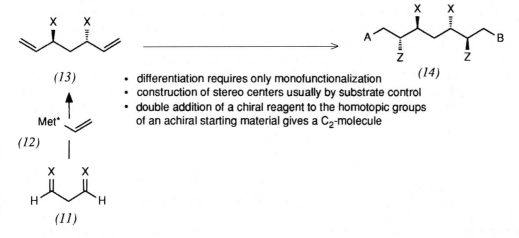

(9) (10)

- construction of the stereo centers by substrate control
- breaking of the C_s-symmetry requires reagent control;
- differentiation can deliver either enantiomer with high ee

C_2-molecules

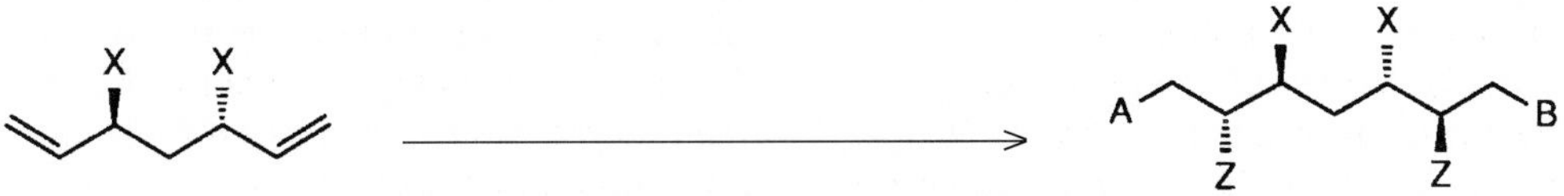

(13) (14)

- differentiation requires only monofunctionalization
- construction of stereo centers usually by substrate control
- double addition of a chiral reagent to the homotopic groups of an achiral starting material gives a C_2-molecule

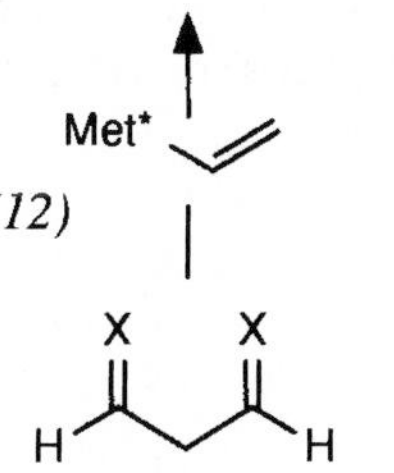

(12)

(11)

Pseudo-C_2-symmetrical molecules

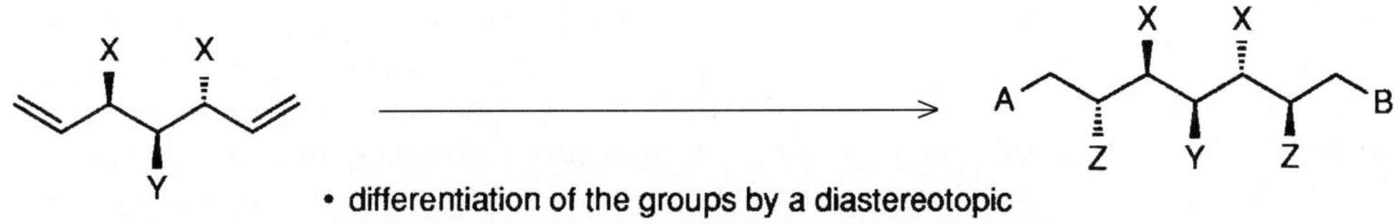

(15) (16)

- differentiation of the groups by a diastereotopic group selective reaction through substrate control
- in case of an achiral starting material (C_s-molecule) a double addition of a chiral reagent to the enantiotopic groups is necessary

Scheme 3. Two-dimensional synthesis.

rentiate the two ends at the latest stage possible in the synthetic sequence. The relative stereochemistry in the course of the construction of new stereocenters can be controlled as usual by internal asymmetric induction. For the preparation of an optically pure compound from a C_S-symmetric molecule a subsequent differentiation of the ends is necessary, whereby external chiral reagents (reagent control) must be used. Under certain circumstances, that is if both enantiotopic groups of the C_S-molecule may react principally with the chiral reagent, extremely high enantiomeric excesses are possible. If a two-dimensional synthesis of a C_2-symmetric molecule starts from an achiral starting material, an initial addition of a chiral reagent to both sides of the homotopic groups of the achiral substrate is necessary [cf. *(11)→(13)*]. The definite generation of new stereocenters in the two-dimensional elongation is also possible by substrate control. When the occasion arises, external chiral reagents must be used. In order to differentiate the homotopic ends of a C_2-symmetric molecule, monofunctionalization is sufficient. Finally, a two-dimensional synthesis can also be advantageous with pseudo-C_2-symmetric molecules. This description refers to a molecule that has a central C-atom that carries two different ligands as well as two identical chiral ligands. For example, reduction of a C_2-symmetrical ketone would lead to a pseudo-C_2-symmetrical molecule. Despite the fact that such molecules are chiral, the central atom is not an asymmetric center! That is, reduction can only lead to a single alcohol. Because this central atom is in a chiral environment, it is called *chirotopic*, according to Mislow. [8] Pseudo-C_2-symmetric molecules, such as *(15)*, can also be obtained from the chiral pool. Alternatively, they are available from odd *meso*-compounds by double addition of a chiral, non-racemic reagent to the ends of the molecule. Because the groups in a pseudo-C_2-symmetrical molecule are diastereotopic, internal differentiation of the two ends is possible. In the course of the differentiation process, the central atom becomes an asymmetric center.

Differentiation of Homotopic Groups

The groups in a C_2-symmetrical molecule are chemically identical, therefore differentiation

only requires monofunctionalization. Because of the statistical effect one can usually reach a differentiation with one equivalent of reagent. A selective monofunctionalization will also be successful if, for example, the first product is less reactive than the second. The best method, however, is to convert the difunctional compound into a derivative that can only react once. A classical example is the reductive ring opening of the acetal *(17)*, which is available from tartaric acid. [9] The product obtained after isopropylidenation can be functionalized in various ways. Strictly speaking, the groups (OH and ester) of compound *(17)* are not homotopic but rather diastereotopic because of the benzylidene protecting group. This does not, however, influence the course of the reaction. Finally, monofunctionalization is guaranteed if the intermediate from the first reaction step reacts subsequently with the other functional group. Thus, the Ni (0)-catalyzed cyclization of *(19)* in the presence of H-SiR$_3$ furnishes the cyclohexane derivative *(20)* in which the two olefins are differentiated. [10] The same principle is realized in the formation of the cyclopentane derivative *(22)* from the C_2-symmetrical α,β-unsaturated diester *(21)* by a double Michael-addition. [11] It is suggested that the reactive conformation is *(23)* in which the largest substituents, the OTBS groups occupy an antiperiplanar orientation. [12] In the depicted conformation the "outsides" of the olefinic faces are homotopic, so that it does not matter which of the olefins is attacked first. A similar conformation also explains the diastereoselective course in the hydroxylation of one of the double bonds of *(21)* [13] and in the cycloaddition with 1,3-dienes, [14] such as cyclopentadiene. It is obvious that compounds of type *(19)* and *(21)* are very easily accessible by a two-dimensional synthesis from the appropriate tartaric acid or from D-mannitol [15] [cf. *ent- (21)*]. [16] Using this concept of a two-dimensional synthesis of a C_2-molecule, an elegant synthesis of the antibiotic hikizimycin was developed (Scheme 5). [17] In this compound the nucleobase is connected *N*-glycosidic with an undecose. Hikiziymycin is attributed with significant antibiotic and anthelmintic properties. Starting with L-(+)-diisopropyl tartrate, in the first step, the hydroxyl groups are protected. Subsequently, the molecule is elongated in a one-pot reaction (reduction and Wittig–Horner reaction). A double hydroxylation, which is controlled by internal induction, furnishes *(25)* with eight carbon atoms and six stereocenters (Scheme 5). Treatment with DIBAH leads via a selective mono-reduction and thus a desymmetrization to *(26)*. The choice of the bulky TBS-protecting group is apparently responsible for the high selectivity. Through sequential elongation in two directions *(26)* is then transformed into *(27)*. First, the alcohol is oxidized to the aldehyde, which subsequently is converted with the Tebbe reagent into the vinyl group. The α, β-unsaturated ester functionality at the other end of the chain is manipulated by reduction, oxidation and a subsequent Wittig–Horner reaction. The elegance of the synthesis somehow suffers because the asymmetric induction in the dihydroxylation of *(27)* to *(28)* is unsatisfactory. In order to improve the diastereoselectivity in favor of the undecose *(28)*, the authors take advantage of an asymmetric dihydroxylation according to Sharpless. From *(28)* another 14 steps provide the target molecule *(29)*. It should be mentioned that optically active C_2-symmetric compounds generally are well suited, without differentiation of the groups for the preparation of interesting building blocks, if after a two-dimensional synthesis the molecule is cleaved in the middle. [18] For example, the preparation of the 2,3-isopropylidene aldehyde from D-mannitol falls into this category. [19]

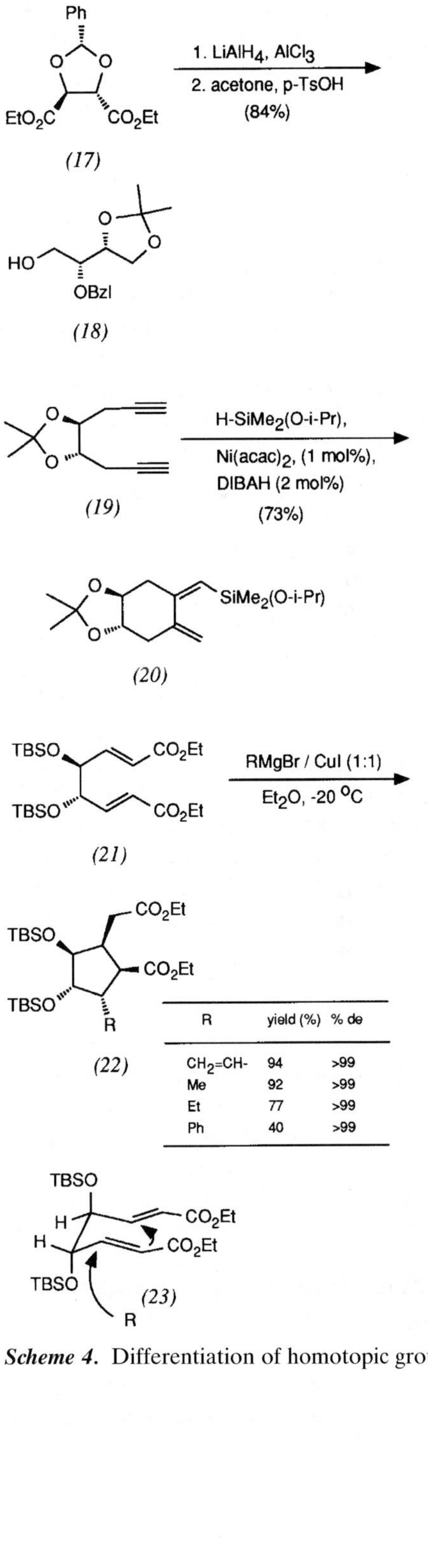

1. LiAlH$_4$, AlCl$_3$
2. acetone, p-TsOH
(84%)

(17)

(18)

H-SiMe$_2$(O-i-Pr),
Ni(acac)$_2$, (1 mol%),
DIBAH (2 mol%)
(73%)

(19)

(20)

RMgBr / CuI (1:1)
Et$_2$O, -20 °C

(21)

R	yield (%)	% de
CH$_2$=CH-	94	>99
Me	92	>99
Et	77	>99
Ph	40	>99

(22)

(23)

Scheme 4. Differentiation of homotopic groups.

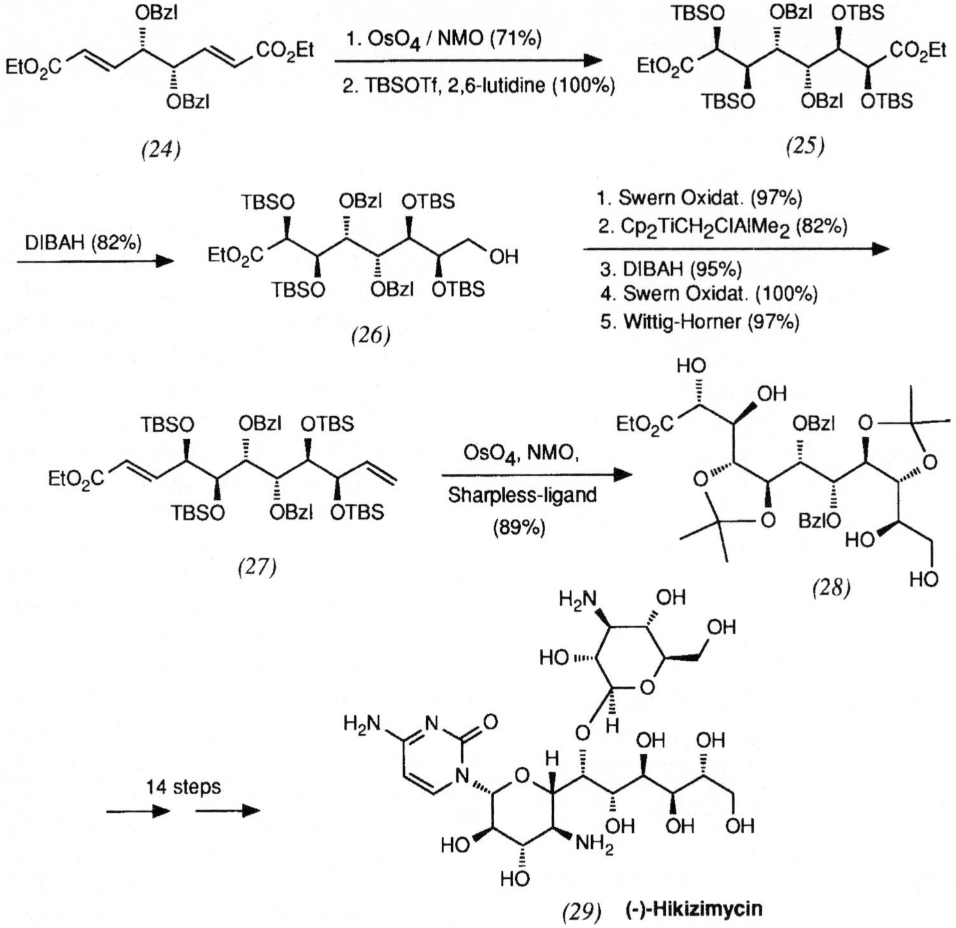

Scheme 5. Synthesis of (–)-hikizimycin with incorporation of a two-dimensional synthetic strategy.

Differentiation of Enantiotopic Groups

In molecules with a mirror plane as the only symmetry element, corresponding groups to both sides of this plane are enantiotopic. For the differentiation of such groups one has to refer to chiral auxiliary reagents or catalysts, which by formation of diastereomeric intermediates or transition states selectively render one of the enantiomers accessible. An illus-

trative example is the enantioselective lactonization of 4-hydroxypimelic acid (Scheme 6). [20] This reaction succeeds with high enantioselectivity by neutralization of the corresponding disodium salt *(30)* with 1-(S)-(+)-camphorsulfonic acid (CSA). The authors could show that the reaction probably takes place through an initial enantioselective protonation followed by lactonization and protonation. An opposite reaction is described by Heathcock et al., who open a cyclic compound, the C_s-symmetrical anhydride *(32)*,

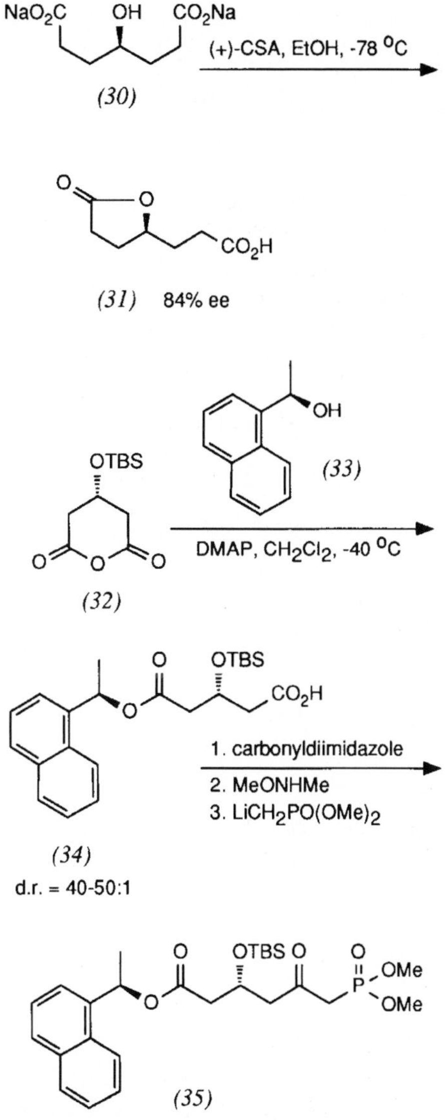

(30)

(31)   84% ee

(33)

(32)

(34)

d.r. = 40-50:1

(35)

Scheme 6. Differentiation of enantiotopic carboxylic groups.

with the chiral alcohol *(33)*. [21] Thereby the monoester *(34)* is obtained with high selectivity (40 to 50:1). By manipulation of the carboxylic group, *(34)* is converted to the ketophosphonate *(35)*, which was used, for example, for the synthesis of the lactone part of

HMG-CoA-reductase inhibitors. Interestingly, the optically active alcohol *(33)* was prepared by an enzyme catalyzed transesterification reaction from the racemate. These processes however, represent rather isolated examples. The differentiation of enantiotopic carboxyl or hydroxyl groups is clearly a domain of biochemical methods. It should be noted that there were earlier more or less successful attempts for the enantioselective acylation of glycerol derivatives. [22] Only recently, however, Ley et al. reported the enantioselective desymmetrization of glycerol using a C_2-symmetric disubstituted bis-dihydropyran (bis-enol ether). [23] In molecules with point group C_S with CH acidic groups, one can achieve a desymmetrization with chiral bases. Because these methods have been summarized recently, [24] only some representative examples are listed here. For example, the enantioselective deprotonation of enantiotopic methylene groups is possible in 4-alkylcyclohexanones and in several *meso*-ketones of various ring sizes with preparatively interesting selectivity. The lithium amide *(37)* turned out to be a useful base. With this base, enolization of ketone *(36)* proceeds with an enantioselectivity of 85 %. Usually the intermediate enolates are trapped as the corresponding silyl enol ethers, which can be functionalized in various ways. If one assumes that the chiral amine can be recycled, this procedure also appears to be interesting on a preparative scale. [25] Moreover, chiral lithium amides can be used for the enantioselective opening of *meso*-epoxides to give allylic alcohols. In the case of cyclic epoxides, the best results with up to 92 % *ee* are obtained with the amide *(40)* derived from L-proline (see Scheme 7). [26] The selectivity drops, however, with acyclic substrates. Possibly, this method may be improved by the use of other counter ions, for example aluminium, that from better complexes with the oxirane oxygen. Besides the base-induced opening of epoxides, there are practical methods at avail-

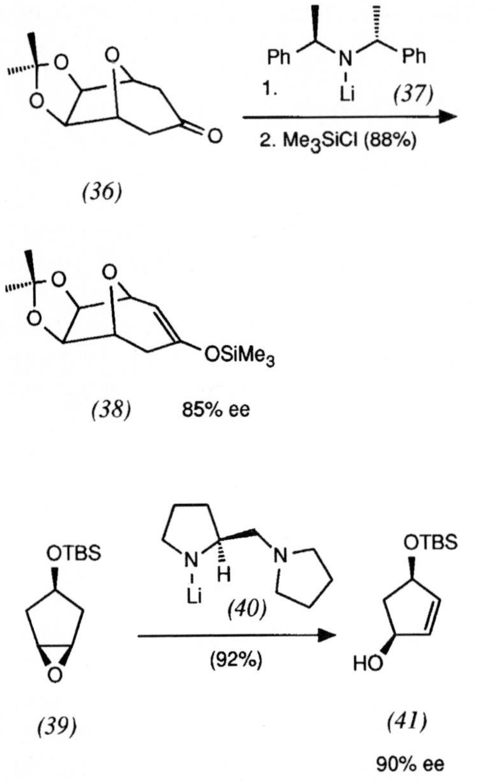

(36)

(38)   85% ee

Thus, Shibasaki described the palladium-catalyzed cyclization of the triflate *(43)* [available from the triketone *(42)*] to give the *cis*-annulated bicycle *(44) (Scheme 8)*. In this case the Noyori ligand (*S*)-BINAP, which induces an enantioselectivity of 80 %, functions as the source of the optical information. [29] Similiar to this is the cyclization of the biscarbamate *(45)* to the oxazolidinone *(47)*. Trost et al. found with optically active diphosphines such as *(46)* easily obtainable ligands, which, with palladium as the central metal, provide for an effective discrimination during the cyclization of *(45)*. [30] The highly functionalized cyclo-

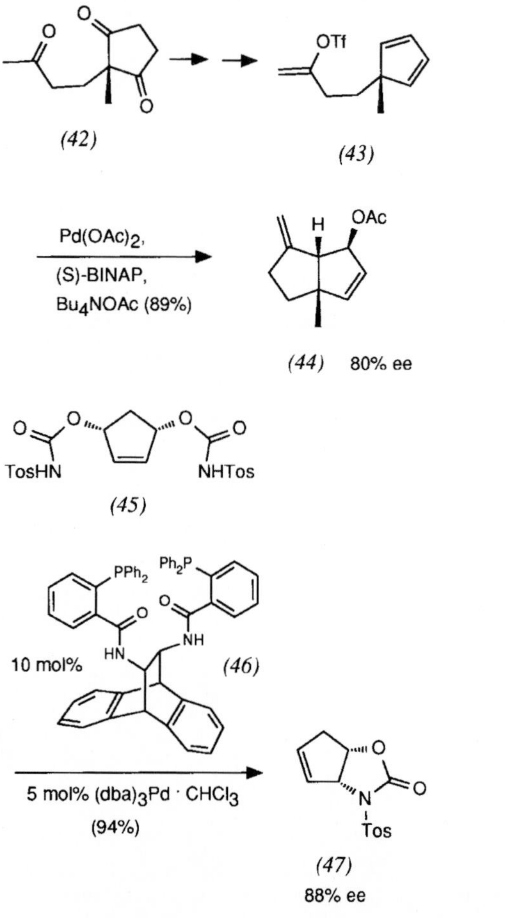

(39)   (92%)   (41)   90% ee

Scheme 7. Differentiation of enantiotopic methylene groups.

able that allow enantioselective opening of epoxides with nucleophiles. [27]

It is remarkable that enantiotopic protons of alcohols, the OH-function of which has been protected as the carbamate, can be deprotonated with high selectivity. As a base, the complex between *sec*-butyllithium and the chiral diamine (−)-sparteine finds application. Because the corresponding alkylation products generally are formed with very high selectivity, this method has enormous practical significance. [28]

If in the course of an enantiotopic group selective reaction ring closure has taken place with the addition to an sp^2-center, then one usually observes not only the enantioselectivity but also a complete diastereoselectivity.

(42)   (43)

(44)   80% ee

(45)

(46)

(47)   88% ee

Scheme 8. Differentiation of enantiotopic groups with concomitant diastereotopic facial selectivity.

pentene *(47)* obtained by this strategy served as a chiral starting material for the preparation of the mannosidase inhibitor mannostatine A. [31] Very high enantio- and diastereoselectivity are also observed in the intramolecular hydrosilation of di(2-propenyl)methanol in the presence of a Rh(I)-complex, which contains, for example, (R,R)-diop as a ligand. [32]

Although these examples are already quite impressive, the full power of the group differentiating reaction is only reached in a certain type of compound. These substrates are characterized by the fact that they can in principle react twice with the chiral, nonracemic reagent. Of course, this double reaction is not desired, although it can serve to increase the enantioselectivity with progressing reaction time. The textbook example is the asymmetric epoxidation of symmetrical divinyl carbinol. In the starting material there are four olefinic faces available for epoxidation (two pairs of enantiotopic faces; the two sides of an olefin are diastereotopic). If we assume, for the sake of simplicity, a highly diastereoselective epoxidation to the *erythro*-compound *(49)* (Scheme 9), then in the first step the forma-

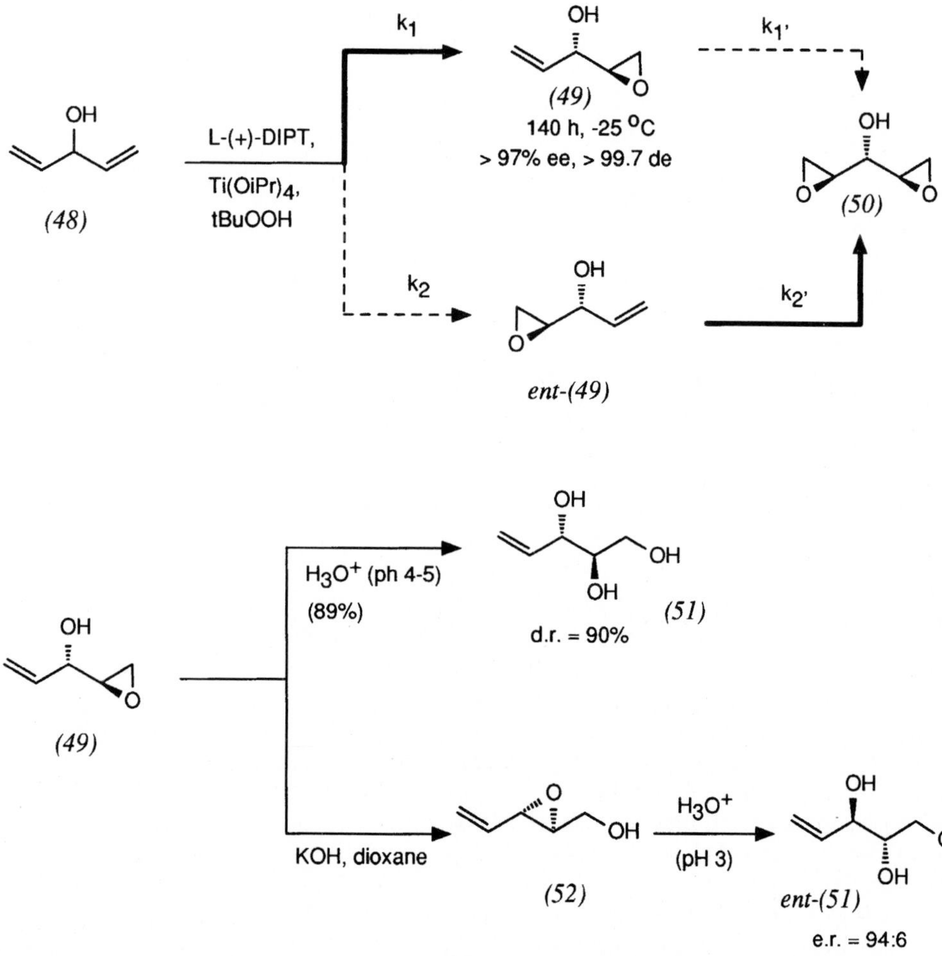

Scheme 9. Epoxidation of divinyl carbinols with enantiotopic group- and diastereotopic face selectivity.

tion of two enantiomeric epoxides is possible, of which one should be favored ($k_1 \gg k_2$). In the wrong isomer *ent-(49)* however, the double bond, which actually should have reacted, is available for a further fast epoxidation. Indeed, the wrong isomer is destroyed with increasing reaction time ($k_1' > k_2'$), resulting in the product *(49)* with very high enantiomeric excess. The first step converts the achiral substrate into a chiral, nonracemic product (e.g. 90:10 ratio), whereas the second reaction increases the enantiomeric excess in a kinetic resolution. This reaction of C_S-divinyl glycols was probably independently discovered by two groups. [33] A mathematical model that describes the kinetics of such coupled reaction is due to Schreiber. [34] Since both isomers of tartaric acid are available, entry to both enantiomeric series is possible. In addition to the choice of the tartrate for the asymmetric epoxidation there exists a further possibility of controlling the conversion of the epoxides to further products, as shown by Jäger et al. [35] Hydrolysis of *(49)* under acidic conditions yields the triol *(51)*. Alternatively, *(49)* can first be rearranged under basic conditions to the internal epoxide *(52)* with inversion at C-2. Subsequent acid hydrolysis leads via inversion at C-3 to the triol *ent- (51)*, the enantiomer of *(51)*. Jäger distinctively speaks of a dual system of stereocontrol.

Epoxides of type *(49)* indeed represent almost universal building blocks for natural product syntheses. Under carefully controlled conditions it is possible to protect the OH-group of *(49)* without observing a Payne rearrangement. Subsequently, the epoxide function ca be opened with various kinds of nucleophiles. For example, starting from *(49)* an enantioselective synthesis of the cyclohexyl part of FK506 was developed using this route. [36] Although these examples clearly appear very elegant, some practical problems during the isolation of *(49)* should not be kept secret: The compound is very water-soluble, very vol-

atile, and can only be obtained pure with considerable effort. This is also expressed in the various rotations that are given in the literature for *(49)* [Schreiber: [34] $[\alpha]^{23}_D$ + 48.8° ($c = 0.73$, CHCl$_3$); Jäger: [35] $[\alpha]^{22}_D$ + 61.1° ($c = 2.31$, CHCl$_3$). These problems obviously can be avoided if the substrate 1,5-bis-trimethylsilyl-1,4-pentadien-3-ol is used for the epoxidation instead of *(48)*. [37]

The principle of enrichment of the desired isomer by coupled asymmetric reactions is of course applicable to other substrates and reactions. Thus, achiral 5-alkyl-cyclopentadienes *(53)* can be converted with high optical purity via hydroboration to *trans*-alkyl-cyclopentenols *(55)*. [38] Compounds of this type represent important intermediates for the syntheses of terpenes, carbohydrates, and particularly prostaglandins. In the ene-reaction of the diene *(56)* with methyl glyoxylate in presence of the optically active Lewis-acid *(57)* the product *(58)* is supplied in very high optical purity. [39] This result might also be traced back to the destruction (double ene-reaction) of the isomer of *(58)* (*Scheme 10*).

A further logical development of this concept is the application of a two-dimensional synthesis for the preparation of larger molecules with C_S-symmetry (mostly *meso* compounds), which are subsequently desymmetrized with appropriate reagents. This combined strategy was used by Schreiber for the synthesis of several polyol systems. [40] The advantage in this case is that the desymmetrization will generate a large number of stereo centers in a single step! How fast one gets to bigger symmetrical molecules with the two-dimensional synthesis is demonstrated in an elegant piece of work by Wang et al. [41] (Scheme 11). From the C_3-building block epibromohydrin *(59)* the alcohol *(60)* was prepared by sequential epoxide opening. The alcohol *(60)* already contains all C-atoms of the *meso*-dialdehyde *(63)*. Partial *cis*-hydrogenation of the triple bonds with nickel boride and stereocontrolled epoxidation pro-

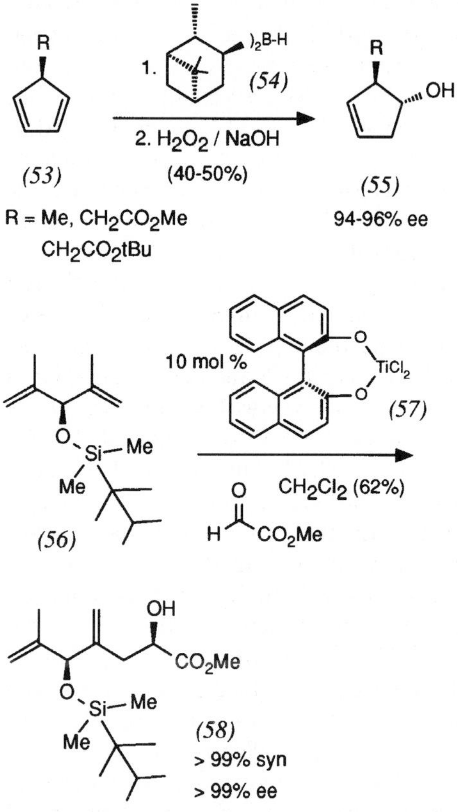

R = Me, CH$_2$CO$_2$Me
CH$_2$CO$_2$tBu

(55)
94-96% ee

Scheme 10. Differentiation of enantiotopic olefins by hydroboration and ene reaction.

vided the bis-epoxide *(61)*. In the subsequent Birch-reduction not only the aromatic rings are reduced but also the two epoxides are opened regiospecifically (benzylic opening). Ozonolysis liberated the β-ketoester groups. In a chelation-controlled reduction the diester with *syn*-arrangement of the hydroxyl groups was prepared from *(62)*. The C_S-symmetry is broken up at the stage of the aldehyde *(63)* by using one of the Brown reagents (+)- or (–)-diisopinylcampheyl allyl borane, which is added to both aldehyde groups. The choice of the reagent determines the configuration in the addition reaction. Simultaneously, the configuration of all the other stereocenters is established. In this double addition the selec-

tivities on each side multiply in the form of a polynomial, thereby delivering the major product with extremely high selectivity. [42] Of course with *(65)* an optically active compound (*dr* > 15:1, *ee* > 98 %) is on hand, yet it is still constitutionally symmetrical. That is, the chemical differentiation still remains to be done. An elegant solution comes from the odd number of hydroxyl groups. Cleavage of the silyl protecting groups liberates the corresponding heptaol, which is treated with acetone and a trace amount of camphorsulfonic acid. Inevitably, one of the hydroxyl groups must remain unprotected. Which one will result from the stability of the possible 1,3-acetonides. It is clear that the *syn*-hydroxy groups are protected preferentially, because *syn*-1,3-acetonides are thermodynamically more stable than *anti*-1,3-acetonides. The authors speak of a diastereotopic group selective acetonide formation. Compound *(66)* is formed with excellent selectivity (15:1). Subsequently, the two olefins can be differentiated very easily by a hydroxyl-directed epoxidation. Beforehand, however, the configuration of the free OH-group is inverted by a Mitsunobu reaction. Through a cuprate addition to the epoxide, followed by manipulation at the hydroxyl groups, the octamethoxy-1-alkene *(67)* is obtained, which represents the active principle of a toxin-producing algae.

Differentiation of Diastereotopic Groups

In principle, the differentiation of diastereotopic groups is a simple task, because their chemical environment is different. Consequently, one should expect different activation energies or different energies of the products in any functionalization reaction and therefore selectivity. Since, however, these differences usually do not allow unequivocal differentiation in practice, except perhaps by NMR-

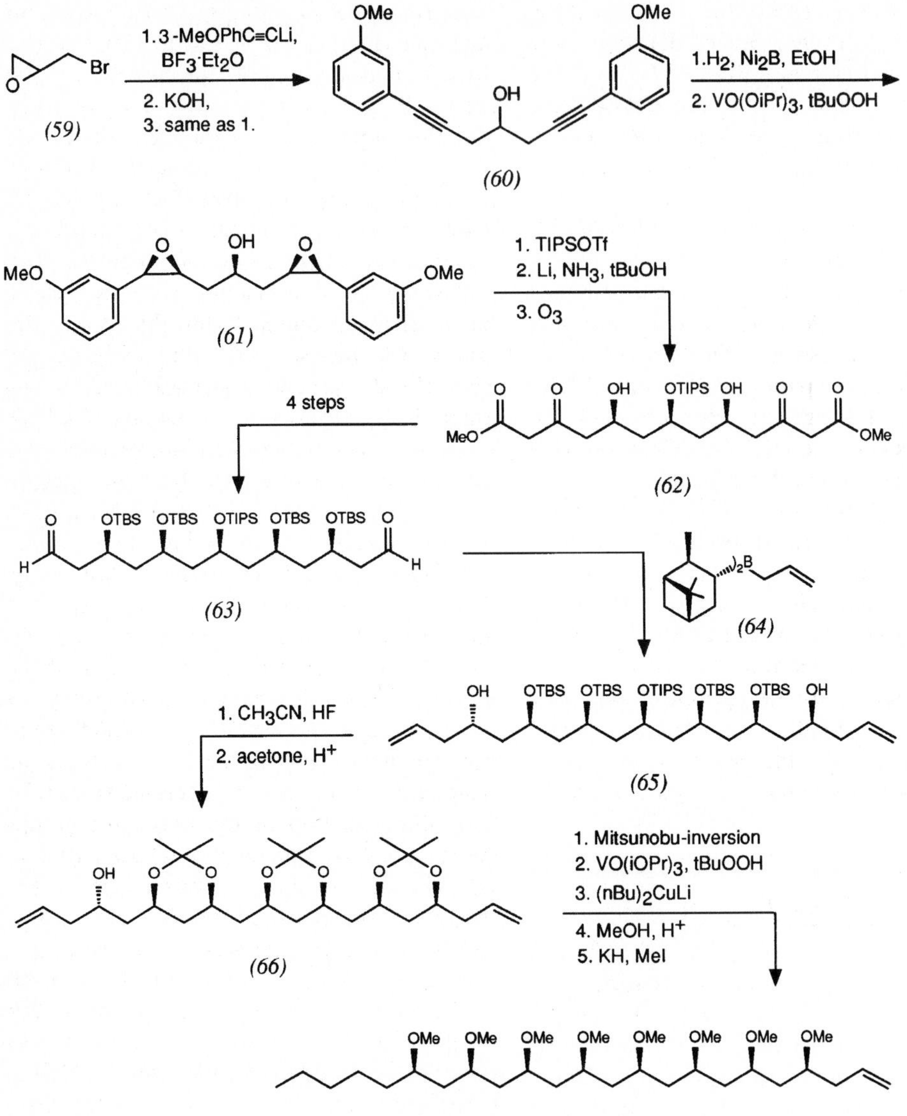

Scheme 11. Two-dimensional synthesis of an odd *meso*-polyol and diastereofacial and diastereotopic group-selective reactions.

spectroscopy, some structural presuppositions have to be present in the substrates. It is quite favorable if the molecule bears a further functional group in the middle that can react with one of the diastereotopic groups with concomitant ring formation. Thereby steric interac-

tions in the transition state or in the products are important. If there is no central group, one has to try to convert the groups into a derivative that only can react once in a subsequent reaction. Among the functional groups that can be differentiated quite easily are hydroxyl-, ester-, carboxylic-, and olefinic groups. The differentiation of diastereotopic groups can follow two purposes. On the one hand, it can proceed with a constitutional differentiation. On the other hand, one can use this method to ultimately differentiate enantiotopic groups.

An example with enormous practical significance is the differentiation of the two diastereotopic OH-groups in D-(–)-quinic acid *(68)* *(Scheme 12)*. In the reaction with carbonyl compounds exclusively the *cis*-acetal is formed with participation of the central OH-group. [43] Starting from compound *(69)* a large number of interesting cyclohexane derivatives are accessible. Danishefsky et al. prepared the cyclohexenone *(70)* [44], by reduction of the carboxylic function to the alcohol, subsequent periodate cleavage, mesylation and elimination. This optically pure compound is interesting as a dienophile for cycloaddition reactions and also as a Michael acceptor. From *(70)* the cyclohexenone *(71)* with the double bond on the other side can be obtained. Similarly, as in the case of the differentiation of diastereotopic OH-groups, in bis-

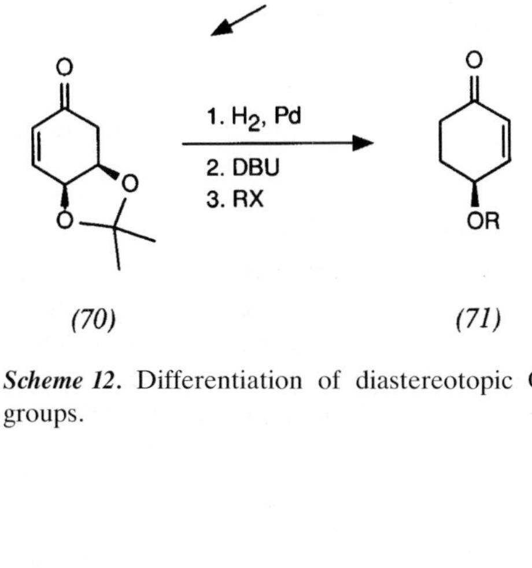

acids two carboxyl groups can be played off against each other if a central OH-group is present, because in a lactonization one of the carboxylic groups should react preferentially. [45] For applications in the area of natural product syntheses, the requisite bis-acids must, of course, be available in optically active form. We recall that pseudo-C_2-symmetrical molecules can be prepared efficiently by a two-dimensional synthesis. Schreiber et al. have demonstrated this in the context of the bis-ester *(78)* [36] (Scheme 13). The chiral (pseudo-C_2-symmetrical) L-(–)-arabitol *(72)* served as a starting material. This is converted with the Moffatt reagent *(73)* through the intermediate *(74)* to the epoxide *(75)*, which represents an ideal substrate for the two-dimensional elongation. By treatment of *(75)* with the lithium anion of ethoxyacetylene in the presence of boron trifluoro etherate and subsequent lactonization, the bis-lactone *(76)* was obtained. Double methylation of *(76)* proceeds with good selectivity (*dr* = 10:1) and leads to *trans*-stereochemistry in both lactones. By using standard reactions the bis-methyl ester *(78)* was generated from *(77)*. The differentiation of the two ester groups succeeds by an acid-catalyzed lactonization, whereby the lactone *(80)* was formed selectively. The preferential formation of this diastereomer may be rationalized by invoking a reactive conformation *(79)* with the smallest destabilizing interactions. In a chemoselective reaction, the lactone was reduced to the lactol, which was opened with dithiane. Simultaneously the lactone *(81)* was formed. Four more steps from *(81)* made the building block *(82)* available, which was used in the synthesis of FK506.

The procedures for the differentiation of hydroxyl groups rely exceptionally on the use of acetal templates. Thus, the hydrazone *(83)* was cyclized under acid catalysis to the spiro acetal *(84)* *(Scheme 14)*. [46] Because there is also a hydroxyl group in the "right" part of the molecule, only one of the "left" hydroxyme-

Scheme 12. Differentiation of diastereotopic OH-groups.

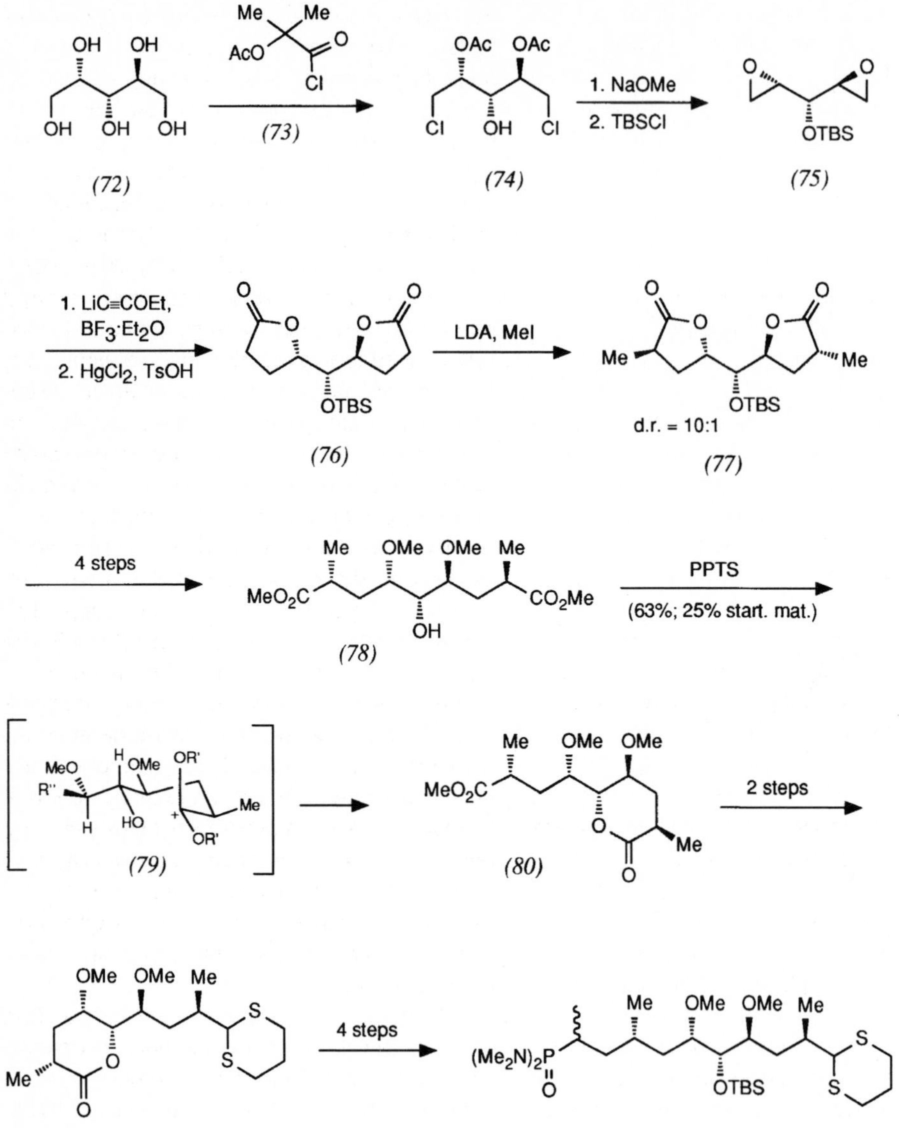

Scheme 13. Two-dimensional synthesis of a pseudo-C_2-symmetrical molecule and diastereotopic group differentiating reaction.

thyl groups can participate in acetalformation. The original stereocenter in the "right" part of (83) not only controls the differentiation of the hydroxymethyl groups, but also leads via enol ether intermediates to equilibration of the stereocenters in the α-position to the spiro acetal. Compound (84) is the most stable of all possible spiro acetals that can be formed from (83). In (84) there are two anomeric effects operating. In addition, the substituents of the

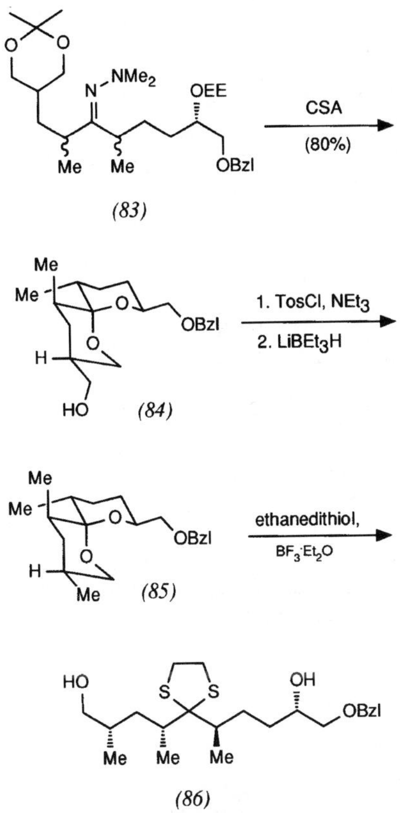

Scheme 14. Differentiation of diastereotopic hydroxymethylene groups by acetal formation.

rings occupy equatorial positions. Transformation of the free hydroxymethyl group into a methyl group and opening of the acetal with ethanedithiol/BF$_3$·$_2$O furnished *(86)*. In another version the hydroxyl groups to be differentiated are first incorporated in an acetal and subsequently one of the diastereotopic C–O bonds is opened selectively. [47]

Oku dexterously used the concept of the differentiation of diastereotopic C–O bonds in an acetal for the conversion of C_S-symmetrical diols into enantiomerically pure compounds (Scheme 15). The principle of this method consists of the acetalization of diols containing enantiotopic OH-groups with an optically active ketone. [48] (–)-Menthone *(87)* served as an appropriate ketone. Steric effects guar-

antee that the spiro scaffold [cf. *(89)*] exists in an unequivocal conformation. The chair conformation of the cyclohexane ring is fixed by the two alkyl groups. Hence, the chair conformation of the 1,3-dioxolane ring follows from this, in which the part of the menthone occupies the equatorial position that leads to the isopropyl group. The acetals of the various C_S-diols will form in such a way that allows their residues in the dioxolane ring to occupy equatorial positions. In some instances mixtures are formed, which must be separated, for example, if diols of type *(88a)* are used. The differentiations are based on the observation that of the two C–O bonds of the spiro acetal, the equatorial bond is cleaved in Lewis-acid-induced reactions with high selectivity throughout. A particularly intriguing example is the application of this method to 1,3-diols of type *(88c)*. Of the mixture of the *meso*- and *d,l*-diols *(88c)*, selectively only the *meso*-diol is acetalized, so that in this manner a simultaneous separation of *meso*- and *d,l*-diols can be achieved. Because of the two substituents of the *meso*-diol, only the diastereomer with residues in the equatorial position is formed. Ring opening in *(89)* takes place by reaction with the acetophenone trimethylsilyl ether in the presence of titanium tetrachloride. Subsequently, the free OH-group is protected and the chiral auxiliary is removed by β-elimination.

Alternatively, the acetal cleavage is possible with the reagent combination allyltrimethylsilane/titaniumtetrachloride whereby in this case the optically active products are set free under acidic conditions. This method even offers the possibility to acetalize *meso*-polyols enantioselectively, although the ratios are not too favorable.

On the contrary, optically active diols can be used to differentiate enantiotopic groups in C_2-symmetrical ketones. For example, such acetals may be converted in a group selective reaction to chiral enol ethers [49] by using appropriate Lewis acids.

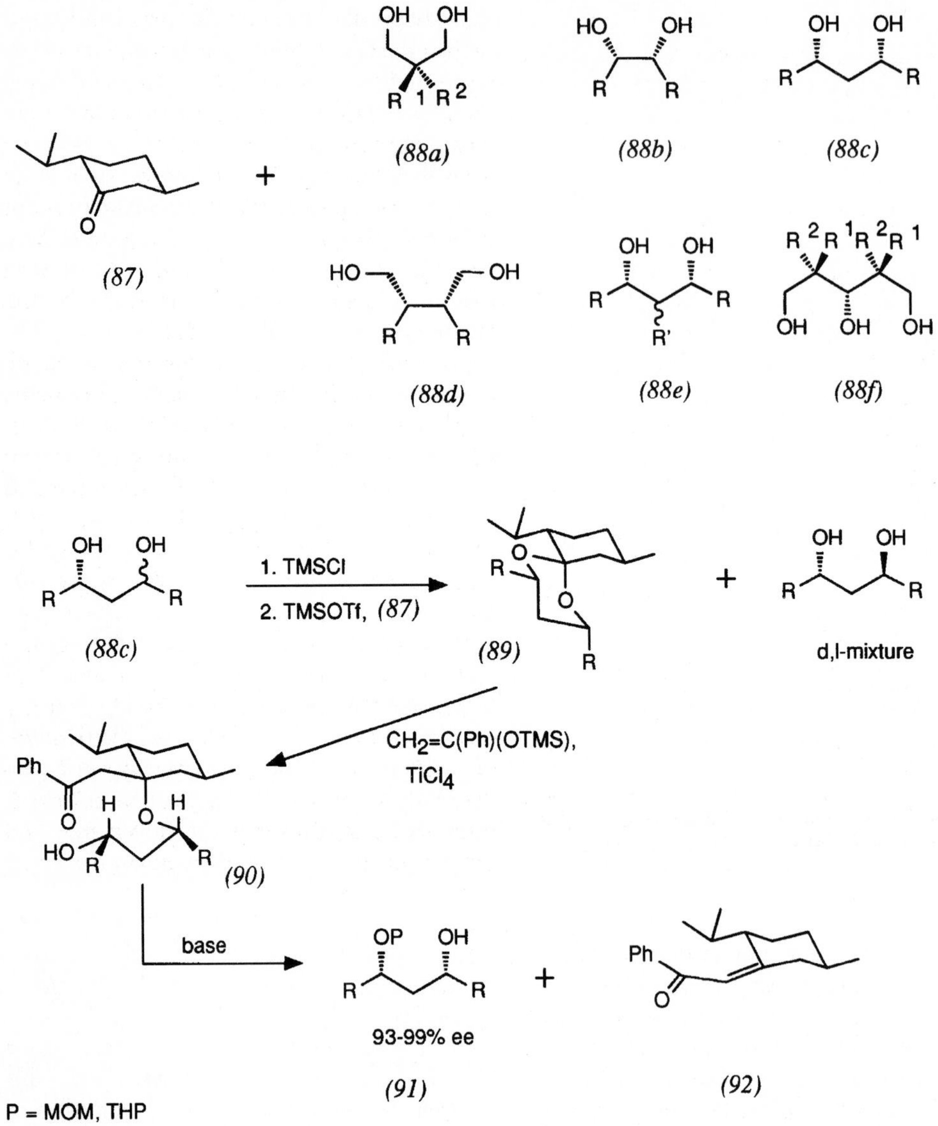

Scheme 15. Differentiation of diastereotopic hydroxy groups by acetalization with a chiral ketone.

Recently, examples have also become known of diastereoselective deprotonation. In this case chiral alcohols, the OH-function of which is protected as the carbamate, are deprotonated to the corresponding α-oxycarbanions. These in turn can be treated with various electrophiles. [50]

But what about the differentiation of diastereotopic olefins? In this case, group- and face-differentiating reactions can be combined (see Scheme 16). Let us consider as an example the pseudo-C_2-symmetrical heptadienoic acid *(93)*. Here, not only the two olefins are diastereotopic, but also the olefinic faces. That is, in

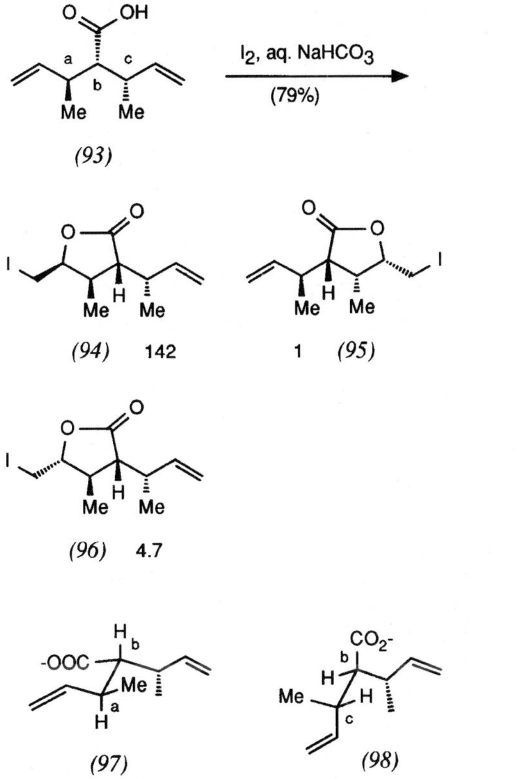

(93)

(94)   142   1   (95)

(96)   4.7

(97)   (98)

Scheme 16. Differentiation of diastereotopic olefins by simultaneous facial selectivity by halolactonization.

the lactonization of (93), formation of four products is, in principle, possible (neglecting regioisomers). The result of the lactonization of (93) is convincing. [51] The two olefins are differentiated with a group selectivity of 147:1 [(94) + (96):(95)]. Simultaneously, a high facial selectivity in the lactonization step [(94):(96) = 30:1] is observed. The preferential reaction of the "left" olefin is explained by conformational effects. In the favored conformation around the C_a–C_b bond, the carboxy group and the olefin are situated in close proximity, whereas in the energetically lowest rotamer around the C_c–C_b bond the corresponding groups are *anti*-periplanar and therefore unfavorable for the cyclization.

If carboxylic acids that contain enantiotopic olefins are converted with chiral amines to the corresponding amides, the two olefins and their faces also become diastereotopic. In the iodolactonization, there are four possible transition states and hence four possible products. However, since the chiral auxiliary is not present in the product two of the products are enantiomeric in nature. This means that with this method such carboxylic acids can be lactonized enantioselectively. [52]

Another class of bis-olefins can be generated very easily from aromatic precursors. Wipf et al. describe the oxidation of the tyrosine derivative (99) with a hypervalent iodine reagent to the dienone (100) in which the two olefins are diastereotopic (Scheme 17). [53] Methanolysis of the spiro lactone (100) under basic conditions provides out of four possible diastereomers exclusively the bicycle (101).

In summary, it can be noted that group selective reactions represent a valuable tool for asymmetric syntheses and that they ideally supplement enzymatic methods. Particularly in combination with a two-dimensional synthetic strategy high efficiency can be achieved, both with regard to material throughput and optical purity. It is therefore advisable, even

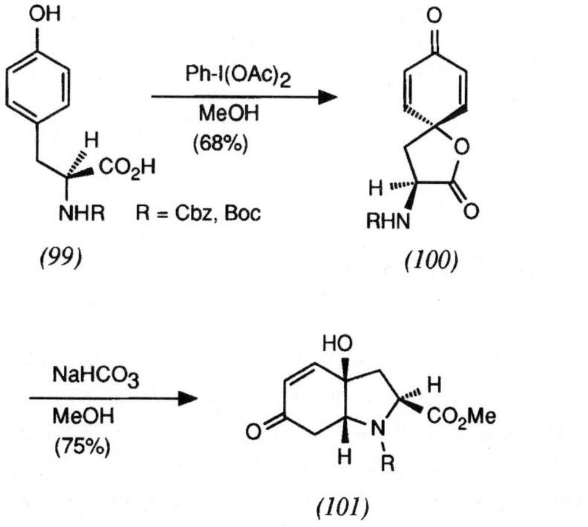

(99)

(100)

(101)

Scheme 17. Differentiation of diastereotopic olefins in dienones.

with bigger synthetic targets, to look for domains of constitutional symmetry. Sometimes, it can even be rewarding to invert, add, or remove stereocenters temporarily, to reach a certain symmetry that will allow the application of a group selective reaction. [54]

References

[1] See, for example: (+)-Calyculin A: D. A. Evans, J. R. Gage, J. L. Leighton, *J. Am. Chem. Soc.* **1992**, *114*, 9434–9453.

[2] See, for example: a) Stereocontrol by exploitation of 1,3-allylic strain: R. W. Hoffmann, *Chem. Rev.* **1989**, *89*, 1841–1860; b) Substrate-directed chemical reactions: A. H. Hoveyda, D. A. Evans, G. C. Fu, *Chem. Rev.* **1993**, *93*, 1307–1370.

[3] See, for example: *Chem. Rev.* **1992**, *92*, number 5; this issue is completely devoted to the topic of enantioselective synthesis.

[4] a) H.-J. Gais, H. Hemmerle, *Chem. Unserer Zeit*, **1990**, *24*, 239–248; b) J. Mulzer, H.-J. Altenbach, M. Braun, K. Krohn, H.-U. Reissig, *Organic Synthesis Highlights*, VCH, New York, **1991**; c) C. R. Johnson, A. Golebiowski, D. H. Steensma, *J. Am. Chem. Soc.* **1992**, *114*, 9414–9418 and references cited therein.

[5] Y. Izumi, A. Tai, *Stereo-Differentiating Reactions*, Kodansha, Tokyo/Academic Press, New York–San Franscisco–London, **1977.**

[6] E. Eliel, *J. Chem. Educ.* **1980**, *57*, 52–57.

[7] a) S. L. Schreiber, *Chemica Scripta* **1987**, *27*, 563–566; b) C. S. Poss, S. L. Schreiber, *Acc. Chem. Res.* **1994**, *27*, 9-17.

[8] K. Mislow, J. Siegel, *J. Am. Chem. Soc.* **1984**, *106*, 3319–3328.

[9] a) R. M. Wenger, *Helv. Chim. Acta* **1983**, *66*, 2308–2321; b) D. Seebach, E. Hungerbühler in *Modern Synthetic Methods* (Ed.: R. Scheffold), Salle und Sauerländer, Berlin, **1980**, 152; c) S. Valverde, B. Herradon, M. Martin-Lomas, *Tetrahedron Lett.* **1985**, *26*, 3731–3734; d) see also: P. Somfai, R. Olsson, *Tetrahedron* **1993**, *49*, 6645–6650.

[10] K. Tamao, K. Kobayashi, Y. Ito, *J. Am. Chem. Soc.* **1989**, *111*, 6478–6480.

[11] S. Saito, Y. Hirohara, O. Narahara, T. Moriwake, *J. Am. Chem. Soc.* **1989**, *111*, 4533–4535.

[12] S. Saito, O. Narahara, T. Ishikawa, M. Asahara, T. Moriwake, J. Gawronski, F. Kazmierczak, *J. Org. Chem.* **1993**, *58*, 6292–6302.

[13] S. Saito, Y. Morikawa, T. Moriwake, *Synlett* **1990**, 523–524.

[14] S. Saito, H. Hama, Y. Matsuura, K. Okada, T. Moriwake, *Synlett* **1991**, 819–20.

[15] S. Takano, A. Kurotaki, K. Ogasawara, *Synthesis* **1987**, 1075–1078.

[16] For a further example, see: H. Kotsuki, H. Nishikawa, Y. Mori, M. Ochi, *J. Org. Chem.* **1992**, *57*, 5036–5040.

[17] a) N. Ikemoto, S. L. Schreiber, *J. Am. Chem. Soc.* **1992**, *114*, 2524–2536; b) For another example of a two-dimensional synthesis of a C_2-symmetric chain molecule, see: C. S. Poss, S. D. Rychnovsky, S. L. Schreiber, *J. Am. Chem. Soc.* **1993**, *115*, 3360–3361.

[18] a) S. Saito, S.-I. Hamano, H. Moriyama, K. Okada, T. Moriwake, *Tetrahedron Lett.* **1988**, *29*, 1157–1160; b) T. Takahashi, T. Shimayama, M. Miyazawa, M. Nakazawa, H. Yamada, K. Takatori, M. Kajiwara, *Tetrahedron Lett.* **1992**, *33*, 5973–5976; c) M. Jayaraman, A. R. A. S. Deshmukh, B. M. Bhawal, *J. Am. Chem.* **1994**, *59*, 932–934; d) K. H. Ahn, D. J. Yoo, J. S. Kim, *Tetrahedron Lett.* **1992**, *33*, 6661-6664.

[19] a) C. R. Schmid, J. D. Bryant, M. Dowlatzedah, J. L. Phillips, D. E. Prather, R. D. Schantz, N. L. Sear, C. S. Vianco, *J. Org. Chem.* **1991**, *56*, 4056–4058; b) C. R. Schmid, J. D. Bryant, *Org. Synth.* **1993**, *72*, 6–12 and references therein.

[20] a) K. Fuji, M. Node, S. Terada, M. Murata, H. Nagasawa, T. Taga, K. Machida, *J. Am. Chem. Soc.* **1985**, *107*, 6404–6406; see also: Y. Yamamoto, A. Sakamoto, T. Nishioka, J. Oda, Y. Fukazawa, *J. Org. Chem.* **1991**, *56*, 1112–1119.

[21] a) P. D. Theisen, C. H. Heathcock, *J. Org. Chem.* **1988**, *53*, 2374–2378; b) for a similiar example, see: Y. Yamada, *Synlett* **1992**, 151–152.

[22] a) T. Mukaiyama, Y. Tanabe, M. Shimizu, *Chem. Lett.* **1984**, 401–404; b) J. Ichikawa, M. Asami, T. Mukaiyama, *Chem. Lett.* **1984**, 949–952.

[23] G.J. Boons, D.A. Entwistle, S.V. Ley, M. Woods, *Tetrahedron Lett.* **1993**, *34*, 5649–5652.

[24] H. Waldmann, *Nachr. Chem. Tech. Lab.* **1991**, *39*, 413–418.

[25] P.J. Cox, N.S. Simpkins, *Synlett* **1991**, 321–323; see also: M. Majewski, G.-Z. Zheng, *Synlett* **1991**, 173–175.

[26] M. Asami, *Tetrahedron Lett.* **1985**, *26*, 5803–5806.

[27] I. Paterson, D.J. Berrisford, *Angew. Chem.* **1992**, *104*, 1204–1205; *Angew. Chem. Int. Edn. Engl.* **1992**, *31*, 1179.

[28] a) M. Peatow, H. Ahrens, D. Hoppe, *Tetrahedron Lett.* **1992**, *33*, 5323–5326 and references therein; for a summary, see: P. Knochel, *Angew. Chem.* **1992**, *104*, 1486–1488; *Angew. Chem. Int. Ed. Engl.* **1992**, *31*, 1459.

[29] a) K. Kagechika, M. Shibasaki, *J. Org. Chem.* **1991**, *56*, 4093–4094; b) see also: K. Kondo, M. Sodeoka, M. Mori, M. Shibasaki, *Tetrahedron Lett.* **1993**, *34*, 4219–4222; c) for an early, classical example, see: Z.G. Hajos, D.R. Parrish, *Org. Synth.* **1985**, *63*, 26–36; d) K. Okrai, K. Kondo, M. Sodeoku, M. Shibasaki, *J. Am. Chem. Soc.* **1994**, *116*, 11737-11748.

[30] B.M. Trost, D.L. van Vranken, C. Bingel, *J. Am. Chem. Soc.* **1992**, *114*, 9327–9343.

[31] B.M. Trost, D.L. van Vranken, *J. Am. Chem. Soc.* **1991**, *113*, 6317–6318.

[32] K. Tamao, T. Tohma, N. Inui, O. Nakayama, Y. Ito, *Tetrahedron Lett.* **1990**, *31*, 7333–7336.

[33] a) S. Hatakeyama, K. Sakurai, S. Takano, *J. Chem. Soc., Chem. Commun.* **1985**, 1759–1761; b) B. Häfele, D. Schröter, V. Jäger, *Angew. Chem.* **1986**, *98*, 89–90; *Angew. Chem., Int. Ed. Engl.* **1986**, *25*, 87–88; c) see also: S. Takano, Y. Iwabuchi, K. Ogasawara, *J. Am. Chem. Soc.* **1991**, *113*, 2786–2787.

[34] a) S.L. Schreiber, T.S. Schreiber, D.B. Smith, *J. Am. Chem. Soc.* **1987**, *109*, 1525–1529; b) D.B. Smith, Z. Wang, S.L. Schreiber, *Tetrahedron* **1990**, *46*, 4793–4808.

[35] V. Jäger, D. Schröter, B. Koppenhöfer, *Tetrahedron* **1991**, *47*, 2195–2210.

[36] M. Nakatsuka, J.A. Ragan, T. Sammakia, D.B. Smith, D.E. Uehling, S.L. Schreiber, *J. Am. Chem. Soc.* **1990**, *112*, 5583–5601.

[37] F. Sato, Y. Kobayashi, *Synlett* **1992**, 849–857 and references therein.

[38] J.J. Partridge, N.K. Chadha, M.R. Uskokovic, *Org. Synth.* **1985**, *63*, 44–56.

[39] a) K. Mikami, S. Narsaw, M. Shimizu, M. Terada, *J. Am. Chem. Soc.* **1992**, *114*, 6566–6568; 9242. see also: b) J.K. Whitesell, D.E. Allen, *J. Org. Chem.* **1985**, *50*, 3025–3026.

[40] a) S.L. Schreiber, M.T. Goulet, G. Schulte, *J. Am. Chem. Soc.* **1987**, *109*, 4718–4720; b) S.L. Schreiber, M.T. Goulet, *J. Am. Chem. Soc.* **1987**, *109*, 8120–8122.

[41] Z. Wang, D. Deschenes, *J. Am. Chem. Soc.* **1992**, *114*, 1090–1091. See also: S.D. Burke, J.L. Buchanan, J.D. Rovin, *Tetrahedron Lett.* **1991**, *32*, 3961–3964.

[42] T.R. Hoye, J.C. Suhadolnik, *Tetrahedron* **1986**, *42*, 2855–2862.

[43] a) S. Hanessian, Y. Sakito, D. Dhanoa, L. Baptistella, *Tetrahedron* **1989**, *45*, 6623–6630; b) A.V. Rama Rao, T.K. Chakraborty, D. Sankaranayanan, A.V. Purandare, *Tetrahedron Lett.* **1991**, *32*, 547–550.

[44] a) J.E. Audia, L. Boisvert, A.D. Patten, A. Villalobos, S.J. Danishefsky, *J. Org. Chem.* **1989**, *54*, 3738–3740; b) L.O. Jeroncic, M.-C. Cabal, S.J. Danishefsky, G.M. Schulte, *J. Org. Chem.* **1991**, *56*, 387–395.

[45] T.R. Hoye, D.R. Peck, T.A. Swanson, *J. Am. Chem. Soc.* **1984**, *106*, 2738–2739.

[46] S.L. Schreiber, Z. Wang, *J. Am. Chem. Soc.* **1985**, *107*, 5303–5305.

[47] a) S.L. Schreiber, B. Hulin, *Tetrahedron Lett.* **1986**, *27*, 4561–4564; b) S.L. Schreiber, Z. Wang, G. Schulte, *Tetrahedron Lett.* **1988**, *29*, 4085–4088; c) C. Iwata, N. Maezaki, M. Murakami, M. Soejima, T. Tanaka, T. Imanishi, *J. Chem. Soc., Chem. Commun.* **1992**, 516–518.

[48] a) T. Harada, T. Hayashiya, I. Wada, N. Iwaake, A. Oku, *J. Am. Chem. Soc.* **1987**, *109*, 527–532; b) T. Harada, I. Wada, A. Oku, *Tetrahedron Lett.* **1987**, *28*, 4181–4184; c) T. Harada, K. Sakamoto, Y. Ikemura, A. Oku, *Tetrahedron Lett.* **1988**, *29*, 3097–3100; d) T. Harada, I. Wada, A. Oku, *J. Org. Chem.* **1989**, *54*, 2599–2605; e) T. Harada, Y. Ikemura, H. Nakajima, T. Ohnishi, A. Oku, *Chem. Lett.* **1990**, 1441–1444; f) T. Harada, I. Wada, J.-I. Uchimura, A. Inoue, S. Tanaka, A. Oku, *Tetrahedron Lett.* **1991**, *32*, 1219–1222; g) T. Harada, H. Kurokawa, Y. Kagamihara, S. Tanaka,

A. Inoue, A. Oku, *J. Org. Chem.* **1992**, *57*, 1412–1421; h) T. Harada, A. Inoue, I. Wada, J. Uchimura, S. Tanaka, A. Oku, *J. Am. Chem. Soc.* **1993**, *115*, 7665–7674; i) review: T. Harada, A. Oku, *Synlett* **1994**, 95–104.

[49] M. Kaino, Y. Naruse, K. Ishihara, H. Yamamoto, *J. Org. Chem.* **1990**, *55*, 5814–5815.

[50] J. Schwerdtfeger, D. Hoppe, *Angew. Chem.* **1992**, *104*, 1547–1549; *Angew. Chem. Int. Ed. Engl.* **1992**, *31*, 1505.

[51] M. J. Kurth, E. G. Brown, *J. Am. Chem. Soc.* **1987**, *109*, 6844–6845.

[52] a) K. Fuji, M. Node, Y. Naniwa, T. Kawabata, *Tetrahedron Lett.* **1990**, *31*, 3175–3178; b) For a similiar example: T. Yokomatsu, H. Iwasawa, S. Shibuya, *J. Chem. Soc., Chem. Commun.* **1992**, 728–729.

[53] P. Wipf, Y. Kim, *Tetrahedron Lett.* **1992**, *33*, 5477–5480.

[54] For further examples of group selective reactions, see: desymmetrization of a *meso*-anhydride by a chiral Grignard reagent: S. D. Real, D. R. Kronenthal, H. Y. Wu, *Tetrahedron Lett.* **1993**, *34*, 8063–8066; Differentiation of enantiotopic carbonyl groups by the Horner-Wadsworth-Emmons reaction: K. Tanaka, Y. Ohta, K. Fuji, T. Taga, *Tetrahedron Lett.* **1993**, *34*, 4071–4074; N. Kann, T. Rein, *J. Org. Chem.* **1993**, *58*, 3802–3804: Diastereoselective and enantioselective cyclization of symmetrical 3,4-disubstituted 4-pentanal using chiral rhodium(*I*) complexes: X.-M. Wu, K. Funakoshi, K. Sakai, *Tetrahedron Lett.* **1993**, *34*, 5927–5930; Oxazaborolidine catalyzed enantioselective reductions of cyclic *meso*-imides: R. Romagnoli, E. C. Roos, H. Hiemstra, M. J. Moolenaar, W. N. Speckamp, B. Kaptein, H. E. Shoemaker, *Tetrahedron Lett.* **1994**, *35*, 1087–1090; group selective ring expansion of prochiral ketones to the corresponding ring-expanded lactams: J. Aube, Y. Wang, M. Hammond, M. Tanol, F. Takusagawa, D. V. Velde, *J. Am. Chem. Soc.* **1990**, *112*, 4879–4891.

Hypervalent Iodine Reagents

Herbert Waldmann

In the course of multistep organic synthesis frequently difficult transformations can only be carried out by making use of heavy metal derivatives like Pd(0)-, Pd(II)-, Tl(III)-, Hg(II)-, Mn(IV)- and Cr(VI) reagents. However, the efficiency of these synthetic tools is unfavorably contrasted by their high toxicity and the fact that they are often not readily available. Therefore, alternative heavy metal-free reagents that display similar reactivity, or that allow equivalent transformations to be carried out, are of considerable interest to organic synthesis. A class of compounds that at least partly fulfills these demands is formed by the so called "hypervalent iodine reagents", for example, *(1)–(9)*. [1] The term "hypervalent" is attributed to molecules or ions of the 5th to the 8th main groups of the Periodic Table, the valence state of which exceeds the respective lowest stable valence of 3, 2, 1 and 0, respectively. [2] This definition was chosen to stress that the atoms involved use more binding electron pairs in the hypervalent compounds than are required for their stabilization according to the Lewis-Langmuir theory (formation of stable octets).

In the majority of cases the mechanisms by which the reactions of the hypervalent iodine reagents proceed can be explained by an initial attack of a nucleophile at the iodonium center, resulting in displacement of a substituent (e.g., **A → B**). Similarly, iodonium salts form tricoordinated iodine intermediates such as, **D**. In a subsequent step either iodobenzene is eliminated accompanied by a simultaneous radical or converted coupling of two ligands (**D → E**), or iodobenzene is released from the molecule in a process that may be regarded as an intramolecular nucleophilic substitution proceeding with retention of configuration (e.g., **B → C**). Various nucleophiles with widely varying structure can be used successfully in these reactions, for example, alcohols, amines, alkyl-, aryl-, alkenyl- and alkinyl anions, and enolates, silylenol ethers, and allyl silanes.

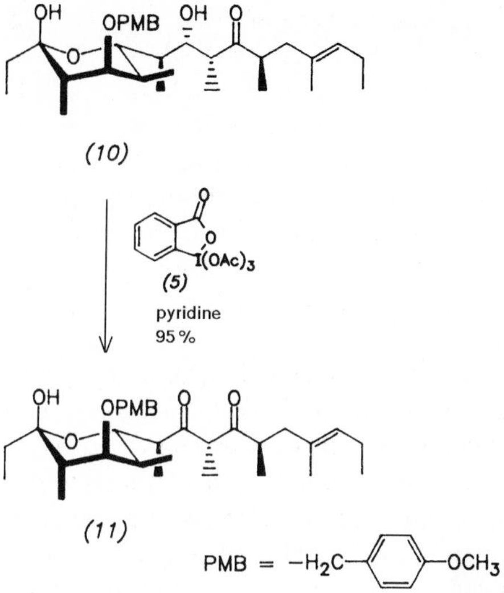

Oxidation

The phenyliodonium acetates *(2)* and *(3)* can be used advantageously to substitute lead tetraacetate in the oxidation of glycols to carbo-

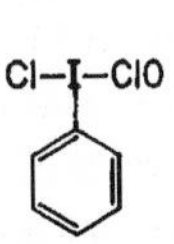

(10)

(5)

pyridine
95 %

(11)

PMB = —H$_2$C—⟨aryl⟩—OCH$_3$

(1) 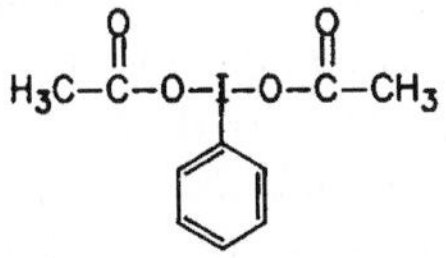   *(2)*

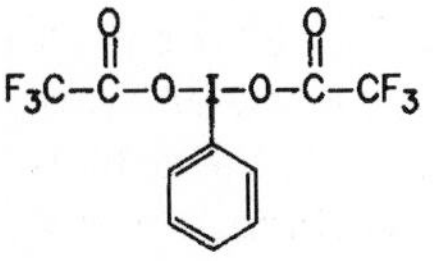

(3) 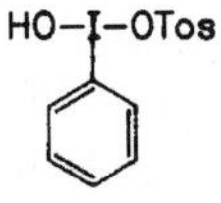   *(4)*

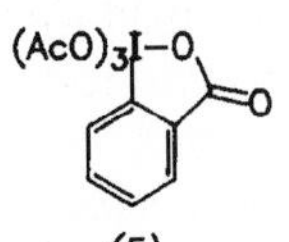

(5) 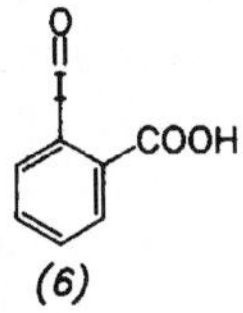   *(6)*

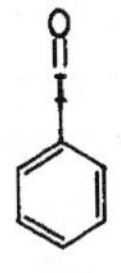

(7) 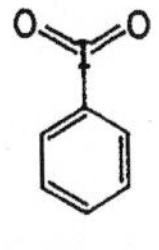   *(8)*

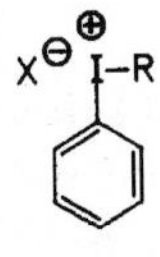

R = alkenyl, alkinyl, aryl,
perfluoroalkyl

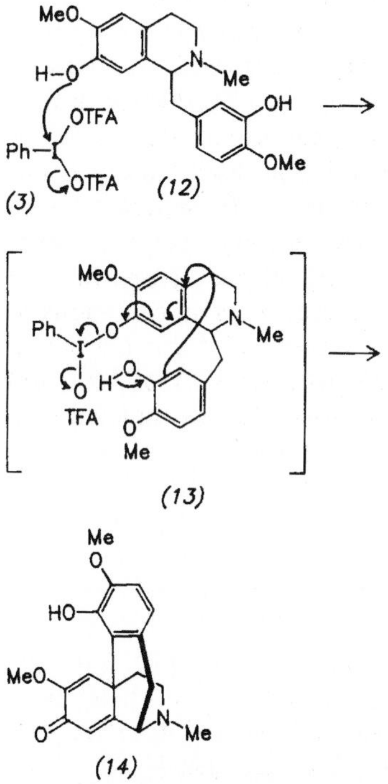

(9)

(12)

(13)

(14)

nyl compounds. [3] The oxidation of simple alcohols to aldehydes and ketones can also be carried out with these reagents. For this purpose the use of the Dess–Martin periodinane *(5)* in particular is recommended. [4] The respective reaction conditions are so mild that, for instance, in the synthesis of the polyketide natural products denticulatin A and B [5a] the β-hydroxyketone present in *(10)* could be converted to the 1,3-diketone *(11)* without racemisation of the 1,3-dicarbonyl compound formed. Furthermore, this reagent proved to be superior to the oxidants MnO_2 and pyridinium dichromate in transformations of sensitive enyne alcohols. [5b] Particularly interesting is the use of the bistrifluoroacetate *(3)* in the conversion of the 4-alkyl substituted phenol reticuline *(12)* to salutaridine *(14)*, [6a] a central intermediate in the biosynthesis of the morphine alkaloids. In the course of this transformation the phenol most probably attacks the iodonium center and the intermediate *(13)* formed thereby then cyclizes via formation of a C–C bond to give the spirodienone *(14)*. In this transformation the iodonium reagent clearly is superior to the oxidants thallium(III)-tris(trifluoroacetate) and lead tetraacetate. By analogy, spirodienones can be constructed that are congeners to the discorhabdin alkaloids. [6b]

Furthermore, oxidation of phosphites to phosphates under non-aqueous conditions in oligonucleotide synthesis [7a] and the opportunity to convert pyrazolones to alkinyl- and alkenylcarboxylic acid esters is of preparative interest. [7b]

Oxygenation

In addition to oxidation reactions, hypervalent iodine reagents can also be applied to carry out hydroxylations and to introduce acyloxy- and tosyloxy groups. Thus, simple alkenes like *cis*-2-pentene *(15)* react with the tosyloxyiodonium compound *(4)* to give vicinal ditosyloxyalkanes, [8] for example, the *erythro* compound *(17)*. The corresponding *trans* pentene delivers exclusively the *threo* isomer. If phenyl-substituted olefins are used

the stereospecificity is lost and competing rearrangements occur. To explain this unusual *cis* addition, which is formally equivalent to the *cis*-hydroxylation of an olefin with OsO_4, it is assumed that a non-ionic iodine(V) species *(16)* is formed as an intermediate, which transfers a tosyloxy group intramolecularly. After exchange of the remaining hydroxy ligand for a further tosyloxy group, the second functional group is also introduced with retention of configuration. If enols or, better, silyl-enol ethers or silylketene acetals are employed as nucleophiles, the transfer of sulfonyloxy groups to the α-position of esters, [9a] ketones [9b–d] and 1,3-dicarbonyl compounds [9b–d] also becomes feasible (Scheme 1). Whereas for simple unsymmetric ketones in these transformations the regioselectivity is low, the corresponding silyloxy compounds (e.g., *(18)*) react with complete selectivity and give higher yields. Furthermore, the use of silylenol ethers allows for the functionalization of acid-labile compounds, for example, the furane derivative *(19)*, without the side reactions that would otherwise occur. If the olefins or carbonyl compounds to be functionalized contain additional nucleophiles, cyclizations can be carried out via intramolecular attack on the iodonium intermediates, for example, *(20)*→*(21)* [10] and *(22)*→*(23)*. [1d]

α-Hydroxylations of esters and ketones can be achieved by employing iodosobenzene *(7)* or the iodonium-bis-acetate *(2)* in alkaline methanol. Under these conditions the bis-

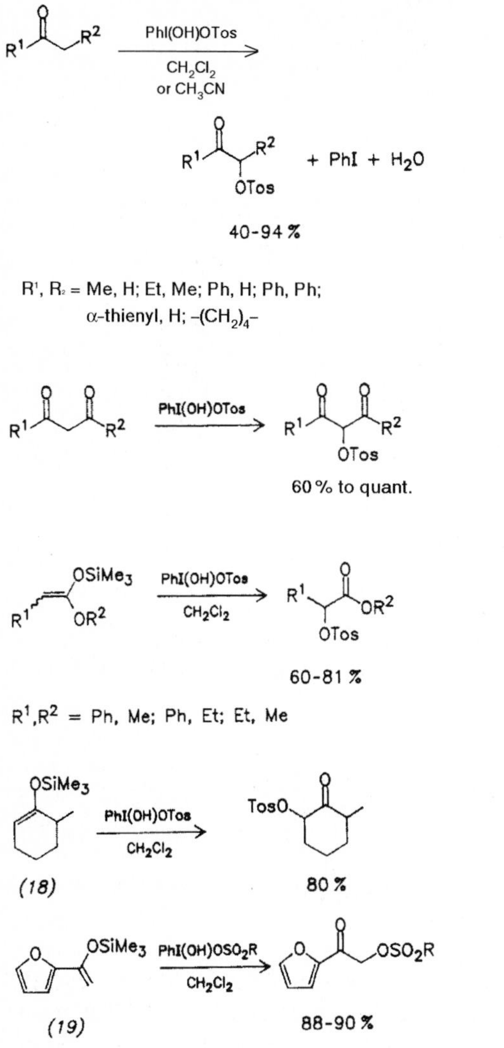

Scheme 1. α-Oxygenation of esters, ketones and 1,3-dicarbonyl compounds by means of hypervalent iodine reagents.

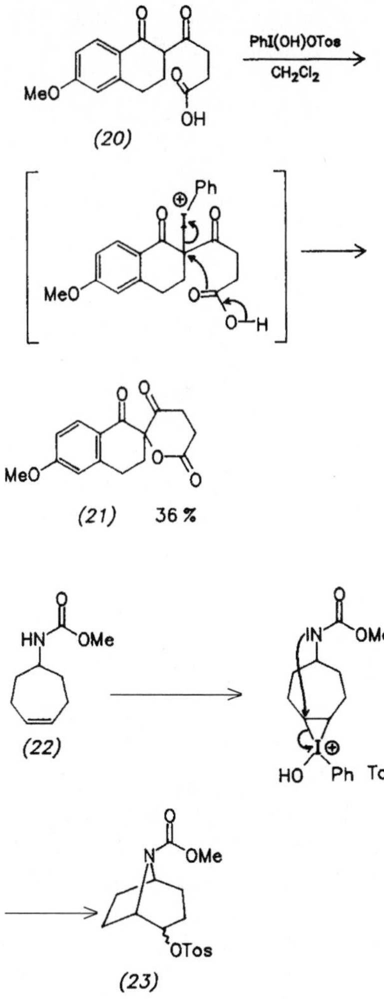

methoxy derivative $PhI(OCH_3)_2$ is formed. The latter is attacked at the iodine atom by the enol of the carbonyl compound and then transfers a methoxide to the carbonyl group. The tetrahedral intermediate (24) generated thereby eliminates iodobenzene to form a methoxy-substituted epoxide (25), which is opened by methanol to give a α-hydro-

xydimethylacetal (26). This transformation has proved itself in natural product chemistry [11] and has been, for instance, used to advantage for the conversion of pregnenes to glucocorticoids via transformation of the exocyclic acetyl groups to the dihydroxy acetone side chain of these steroids (27)→(28). In the course of this reaction double bonds and further secondary alcohols are not attacked. [11b] The reaction conditions are so mild that even oxidation-sensitive chromium carbonyl complexes like (29) remain unaffected. [11c] The steric course of the hydroxylation of (29),

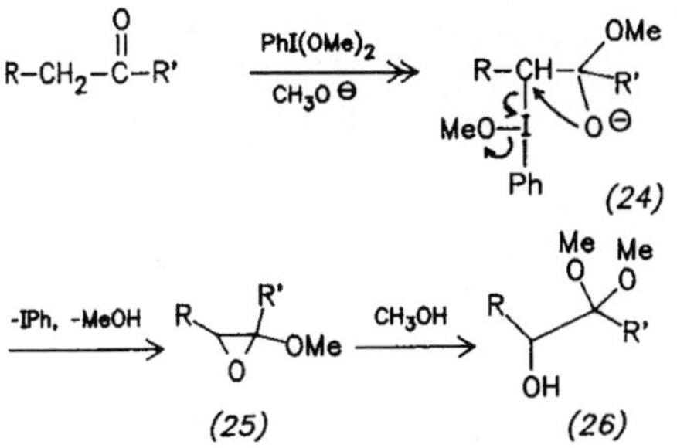

(24)

(25) (26)

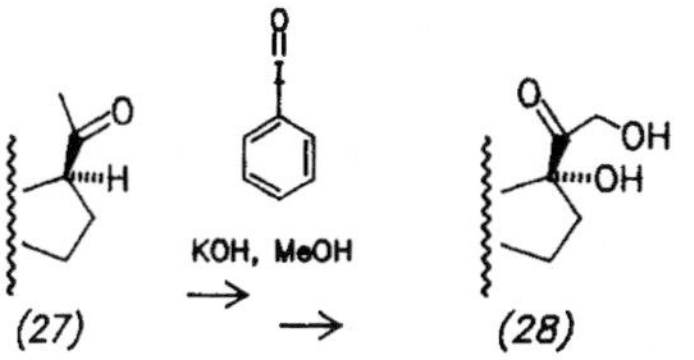

(27) KOH, MeOH → → (28)

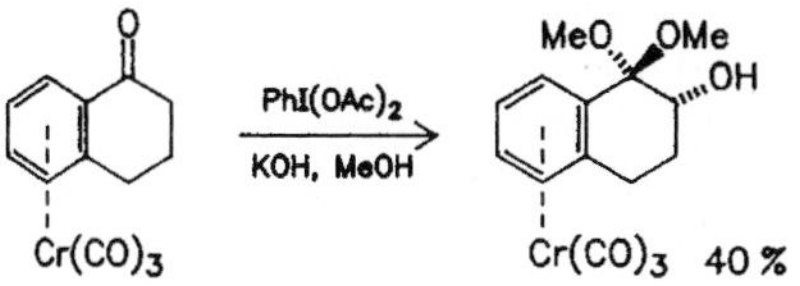

(29)

(30)

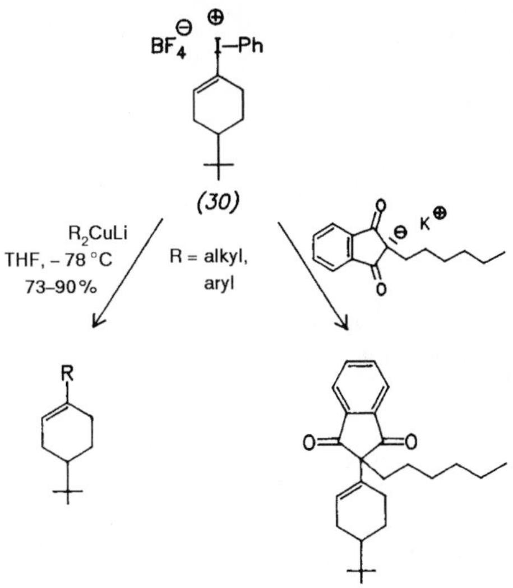

R₂CuLi
THF, − 78 °C
73–90%
R = alkyl, aryl

$$R-\!\!\equiv\!\!-I-Ph \quad \xrightarrow[\text{− 78 °C}]{\text{ether}}$$

(31) TosO⊖

−PhI

(32)
47–94 %

that is the OH group and the chromiumtricarbonyl fragment are on the same side, proves the mechanism proposed for this transformation. Accordingly, the iodonium compound first adds *trans* to the chromium and the epoxide corresponding to (25) subsequently must be formed by attack of the tetrahedral intermediate (24) from below.

Carbon–Carbon Bond Formation

The use of alkenyl-, alkinyl- and aryliodonium salts opens up interesting routes to the formation of C–C bonds. Vinyliodonium salts like (30) can be coupled with C-nucleophiles like organocuprates and anions of 1,3-dicarbonyl compounds to give substituted alkenes. [12] These transformations most probably follow the mechanistic scheme highlighted above,

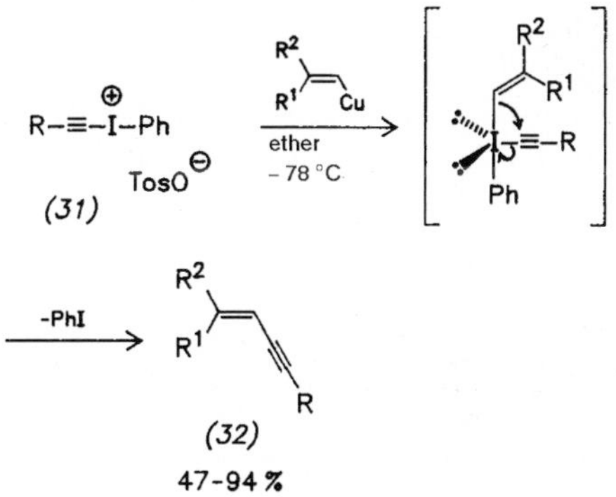

that is, the nucleophile initially attacks the iodonium center and then the C–C bond is formed via reductive elimination of iodobenzene. Similarly, alkinyliodonium salts (31) and alkenyl copper reagents can be transformed into enynes (32). [13] These reactions are formally equivalent to the Pd-mediated coupling of vinyl halides and organometallic reagents. Anions of 1,3-dicarbonyl compounds, on the other hand, add to the triple bond of alkinyliodonium reagents to generate unstable iodonium ylides like (33). These intermediates eliminate iodobenzene and thereby form highly reactive alkylidene carbenes, for example, (34), which give cyclopentenes by intra-

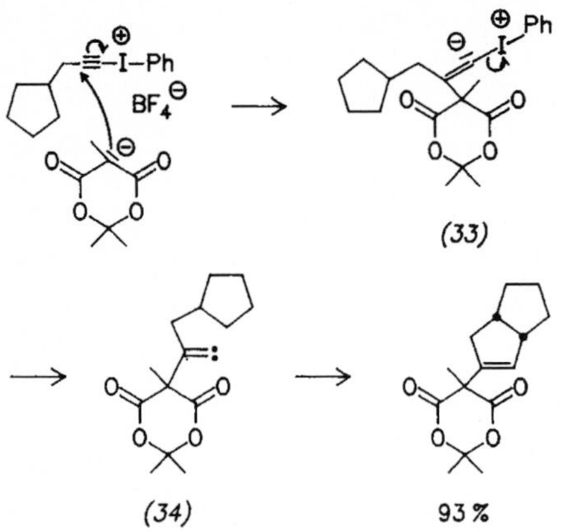

(33)

(34)   93 %

molecular C–H-insertion. [14] This tandem conjugate addition–carbene insertion sequence could also be used to synthesize highly substituted furans.

The transfer of aryl groups to a variety of C-nucleophiles can be realized by employing di-aryliodonium salts. [1b] Thus, products become accessible that can hardly be synthesized with other methods. Moreover, the transfer of aryl groups to O-, N-, S- and P-nucleophiles, like alkoxides, phenolates, carboxylates, amines, sulfides and phosphines can be carried out under mildest conditions by using these reagents (Scheme 2). [1b] These C–C bond forming reactions proceed in the sense of the formation of **D** and its conversion to **E**, which in many cases involves the formation of free radicals as intermediates. The anions of arylacetic acid esters, ketones, nitroalkenes, 1,3-dicarbonyl compounds, and enamines can be introduced as nucleophiles. Some representative examples are given in Scheme 2. If aryl Grignard and aryl lithium reagents are employed, triaryliodonium intermediates (35) are generated that react further via coupling to biaryls. [14] Particularly interesting is the use of the 3-indolyliodonium salt (36), which reacts with various nucleophiles to give 3-substituted indoles (37). Thereby an "umpolung" of the reactivity at the 3-position of the indole nucleus is achieved. [14c]

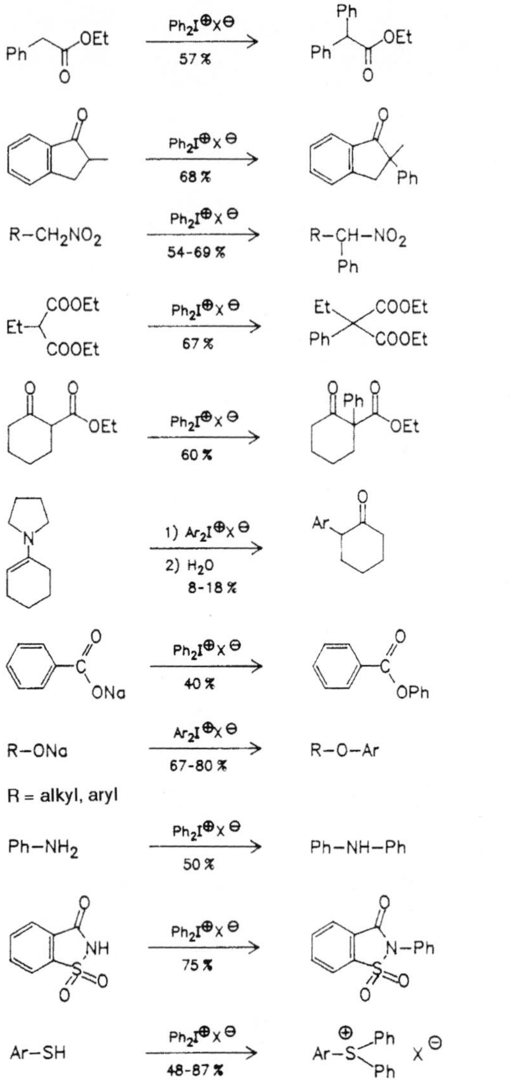

Scheme 2. Arylation of C-, O-, N- and S-nucleophiles by means of diaryliodonium salts.

Hofmann Rearrangement

By means of the bis-trifluoroacetate (3) [15] and the tosyloxyiodonium salt (4) [16] a transformation analogous to the Hofmann rearrangement can be initiated. In this process ali-

$$Ar-\overset{\oplus}{I}-Ar \quad X^{\ominus} \xrightarrow{Ar'MgX} \left[Ar \overset{Ar'}{\underset{Ar}{\overset{|}{I}}} \right] \longrightarrow Ar-Ar'$$

(35)

(36) $\xrightarrow[-78\ °C]{R-Li}$ (37)

R = Me, Bu, Ph, allyl, OMe
60–71 %

$$\underset{(38)}{R\overset{O}{\underset{}{\overset{||}{C}}}NH_2} + Ph-\overset{OH}{\underset{OTos}{\overset{|}{I}}} \xrightarrow{-H_2O}$$

$$\left[R\overset{O}{\underset{N}{\overset{||}{C}}} \overset{\oplus}{I}-Ph \right] TosO^{\ominus} \longrightarrow$$

(39)

R–N=C=O + PhI + TosOH

(40)

$\downarrow -CO_2 \quad | \quad H_2O$

$R-\overset{\oplus}{N}H_3 \ TosO^{\ominus}$

(41)

(42) 58 %

(43) 85 %

$$t\,BuO\overset{O}{\underset{}{\overset{||}{C}}}-NH \overset{Me}{\underset{Ph}{\cdots}} \overset{O}{\underset{H}{\overset{||}{N}}} \overset{}{\underset{O}{}} NH_2$$

(44)

$$\downarrow \begin{array}{c} CH_3CN/H_2O \\ 1;1 \\ R.T. \end{array} \quad PhI(OCOCF_3)_2$$

(45) 93 %

phatic carboxamides *(38)* are converted to amines *(41)* via the intermediate formation of N-phenyliodonium amides *(39)*, which rearrange to isocyanates *(40)*. A noteworthy feature of the reaction is that it proceeds under mildly acidic conditions (pH 1–3), which also prevent attack of the amine formed *(41)* on the isocyanate. This transformation is of particular interest as a tool for the generation of amines at bridgeheads, for example, in the case of the cubylammonium salt *(42)* and the adamantylammonium salt *(43)*. [16c] Goodman et al. [15c, d] used this reaction, which was first described by Loudon et al. [15a, b] for the construction of "retro-inverso" peptides. These compounds are isosteric peptide analogues in which a peptide bond is inverted [– HN – CR(R) – **C(O) – NH** – CH(R) – C(O)→– HN – CH(R) – **NH – C(O)** – CH(R) – C(O)–], and in which the absolute configuration of the former amino acid moieties is reversed. To build up retro-inverso peptides, amides like *(44)* are converted with complete retention of configuration at the migrating C-atom to the respective geminal aminoalkylamides *(45)* by treatment with the bistrifluoroacetate *(3)*. These acylated aminals can be cleaved by hydrolysis, but their stability is sufficient to allow for acylation with a further amino acid unit.

References

[1] Reviews: a) A. Varvoglis, *Chem. Soc. Rev.* **1981**, *10*, 377; b) A. Varvoglis, *Synthesis* **1984**, 709; c) R. M. Moriarty, O. Prakash, *Acc. Chem. Res.* **1986**, *19*, 244; d) R. M. Moriarty, R. K. Vaid, *Synthesis* **1990**, 431; e) R. M. Moriarty, R. K. Vaid, G. F. Koser, *Synlett* **1990**, 365; d) P. J. Stang, *Angew. Chem.* **1992**, *104*, 281; *Angew. Chem. Int. Ed. Engl* **1992**, *31*, 274; e) A. Varvoglis, *The Organic Chemistry of Polycoordinated Iodine*, VCH, Weinheim, **1992**. Int. Ed. Engl. **1969**, *8*, 54.

[2] J. I. Musher, **Angew. Chem. 1969,** *81*, 68. Int. Ed. Engl. **1969**, *8*, 54.

[3] a) D. F. Banks, *Chem. Rev.* **1966,** *66*, 243; b) R. Criegée *Oxidation in Organic Chemistry*, Academic Press, New York, 1965, p. 365; c) S. J. Angyal, R. J. Young, *J. Am. Chem. Soc.* **1959**, *81*, 5251.

[4] D. B. Dess, J. C. Martin, *J. Org. Chem.* **1983**, *48*, 4155.

[5] a) M. W. Andersen, B. Hildebrandt, R. W. Hofmann, *Angew. Chem.* **1991**, *103*, 90; Int. Ed. Engl. **1991**, *30*, 97; b) W. H. Okamura, M. L. Curtin, *Synlett* **1990**, 1.

[6] a) C. Szántay, G. Blaskó, M. Bárczai-Beke, P. Péchy, G. Dörnyei, Tetrahedron Lett. **1980**, *21*, 3509; b) Y. Kita, T. Yakura, H. Thoma, K. Kikuchi, Y. Tamura, *Tetrahedron Lett.* **1989**, *30*, 1119.

[7] a) J.-L. Fourrey, J. Varenne, *Tetrahedron Lett.* **1985**, *26*, 1217; b) R. M. Moriarty, R. K. Vaid, V. T. Ravikumar, T. E. Hopkins, P. Farid, *Tetrahedron* **1989**, *45*, 1605.

[8] a) G. F. Koser, L. Rebrovic, R. H. Wettach, *J. Org. Chem.* **1981**, *46*, 4324; b) L. Rebrovic, G. F. Koser, *J. Org. Chem.* **1984**, *49*, 2462.

[9] a) R. M. Moriarty, R. Penmasta, A. K. Awasthi, W. R. Epa, I. Prakash, *J. Org. Chem.* **1989**, *54*, 1101; b) G. F. Koser, A. G. Relenyi, A. N. Kalos, L. Rebrovic, R. H. Wettach, *J. Org. Chem.* **1982**, *47*, 2487; c) N. S. Zefirov, V. V. Zhdankin, Y. V. Dan'kov, A. S. Koz'min, O. S. Chizhov, *Zh. Org. Khim.* **1985**, *21*, 2461; d) J. S. Lodaya, G. F. Koser, *J. Org. Chem.* **1988**, *53*, 210.

[10] R. M. Moriarty, R. K. Vaid, T. E. Hopkins, B. K. Vaid, O. Prakash, *Tetrahedron Lett*, **1990**, *31*, 201.

[11] a) S. N. Suryawanshi, P. L. Fuchs, *Tetrahedron Lett.*, **1981**, *22*, 4201; b) R. M. Moriarty, L. S. John, P. C. Du, *J. Chem. Soc., Chem. Commun.* **1981**, 641; c) R. M. Moriarty, S. G. Engerer, O. Prakash, I. Prakash, U. S. Gill, W. A. Freeman, *J. Chem. Soc., Chem. Commun.* **1985**, 1715.

[12] a) M. Ochiai, K. Sumi, Y. Nagao, E. Fujita, *Tetrahedron Lett.* **1985**, *26*, 2351; b) M. Ochiai, K. Sumi, Y. Takaoka, M. Kunishima, Y. Nagao, M. Shiro und E. Fujita, *Tetrahedron* **1988**, *44*, 4095.

[13] P. J. Stang und T. J. Kitamura, *J. Am. Chem. Soc.* **1987**, *109*, 7561.

[14] a) F. M. Beringer, A. Brierley, M. Drexler, E. M. Gindler, C. C. Lumpkin, *J. Am. Chem. Soc.* **1953**, *75*, 2708; b) F. M. Beringer, J. W. Dehn, M. J. Winicov, *J. Am. Chem. Soc.* **1960**, *82*, 2948; c) R. M. Moriarty, Y. Y. Ju, M. Sultana, A. Tuncay, *Tetrahedron Lett.* **1987**, *28*, 3071.

[15] a) G. M. Loudon, M. E. Parham, *Tetrahedron Lett.* **1978**, *5*, 437; b) G. M. Loudon, A. S. Radharkrishna, M. R. Almond, J. K. Blodgett, R. H. Boutin, *J. Org. Chem.* **1984**, *49*, 4272; c) P. Pallai, M. Goodman, *J. Chem. Soc., Chem. Commun.* **1982**, 280; d) P. V. Pallai, S. Richman, R. S. Struthers, M. Goodman, *Int. J. Pept. Prot. Res.* **1983**, *21*, 84.

[16] a) I. M. Lazbin, G. F. Koser, *J. Org. Chem.* **1986**, *51*, 2669; b) A. Vasudevan, G. F. Koser, *J. Org. Chem.* **1988**, *53*, 5158; c) R. M. Moriarty, J. S. Khosrowshahi, A. K. Awasthi und R. Penmasta, *Synth. Commun.* **1988**, *18*, 1179.

Furan as a Building Block in Synthesis

Martin Maier

Expressions such as synthetic building blocks, synthetic design, architecturally interesting molecules, and so on, emphasize that the synthesis of natural products bears similarities to the construction of complex buildings. Yet, the synthesis of molecules is probably more complicated, because the starting point, that is the starting material, is not immediately evident and in the path to the target many possible variants and many obstacles exist. Of utmost importance for the success and the efficiency of a synthesis is not only an appropriate global synthetic strategy [1] (e.g. transform-based strategy) but also the proper choice of synthetic building blocks. Optimal building blocks are characterized by the fact that they lead in the course of a reaction directly to the structural prerequisites for another central synthetic operation so that laborious touch-ups, such as protecting group manipulations, the changing of oxidation states of functional groups or changes on the molecular backbone, are superfluous. These requirements are fulfilled, for example, in the so-called tandem reactions and cascade cyclizations. [2] Synthetic building blocks may be classified according to their number of carbon atoms (e.g. C_2-building blocks) and the number and nature of functional groups (e.g. 1,4-bis-electrophiles). In combination with heteroatoms and the principle of Umpolung, [3] it is possible to choose suitable building blocks for almost any situation. The impor-

tance of a certain building block depends to some extent on how it can function as a starting point for several fundamentally different synthetic transformations. One of these central building blocks is furan *(1)*. This molecule contains four C-atoms, a 1,4-difunctionalized diene, and an enol ether substructure. Thus, furan can undergo cycloaddition either as C_4-(diene) or C_2-component (dienophile, dipolarophile). [4] In addition, reactions are possible that are typical of enol ethers, such as the addition of electrophiles, carbenes, or metalation. [4] Although one might describe the use of furan as classical, [4] its synthetic potential is not by any means exhausted, as underscored by recent application that are the subject of this summary.

Cycloaddition Reactions

Furans can react with reactive dienophiles *(2)* by [4+2]cycloaddition to give 7-oxabicyclo[2.2.1]hept-5-ene *(3)*, whereby the furan takes the role of a 1,4-disubstituted electron-rich diene. The cyclic arrangement fixes the diene in the reactive conformation, thereby favoring the cycloaddition for entropic reasons. This reaction is, however, rather slow at room temperature due to the aromatic character and ring strain in the cycloadduct. Although a higher temperature favors the reverse reaction, it is possible to increase the

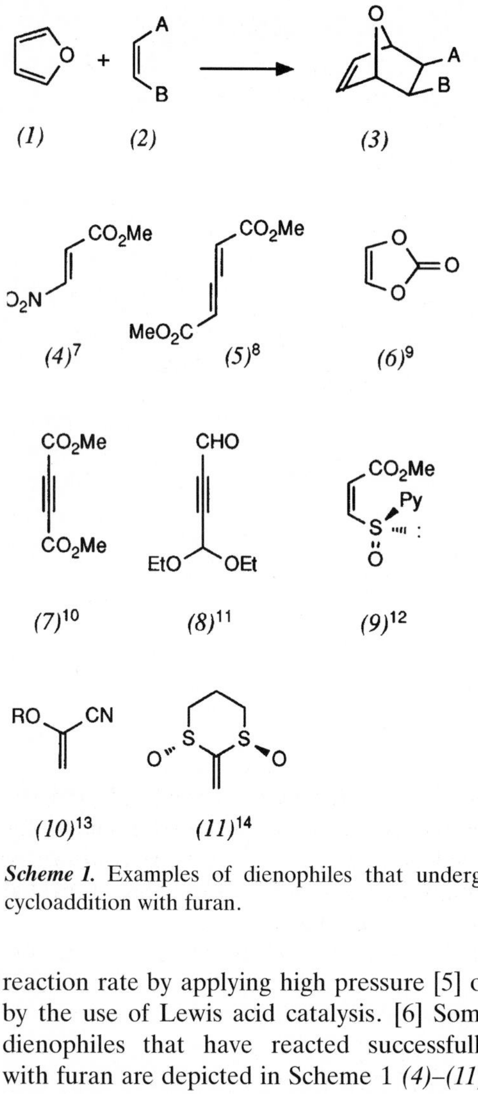

(1)   (2)   (3)

(4)[7]   (5)[8]   (6)[9]

(7)[10]   (8)[11]   (9)[12]

(10)[13]   (11)[14]

Scheme 1. Examples of dienophiles that undergo cycloaddition with furan.

reaction rate by applying high pressure [5] or by the use of Lewis acid catalysis. [6] Some dienophiles that have reacted successfully with furan are depicted in Scheme 1 *(4)–(11)*. [7–14] With regard to the preparation of optically active compounds, symmetrical dienophiles such as *(6)* or *(7)* proved to be particularly versatile, because they form *meso*-cycloadducts. From these in turn, the whole material can in principle be converted to optically active derivatives by group selective reactions. Of the unsymmetrical dienophiles, the α-acetoxyacrylonitrile *(10)*, R = Ac) [13] is of particular importance because it allows one to perform a formal ketene cycloaddition

that facilitates subsequent transformations. This reaction principle was further developed by Vogel et al., who developed two procedures for the synthesis of enantiomerically pure furan cycloadducts. One possibility is to initially hydrolyze the racemic *endo/exo*-mixture of the 5-acetoxy-5-cyano-7-oxabicyclo[2.2.1]hept-2-ene to the corresponding cyanohydrins. From these cyanohydrins one stereoisomer can be obtained by fractional crystallization with brucine. [15] In the other method, the chiral 1-cyanovinylcamphanate *(10,* R = *(S)*-camphanate) is used, whereby a diastereomeric mixture of four cycloadducts is obtained, from which the adduct *(12)* can be obtained in pure form by crystallization in 29 % yield. [16] An advantage of this method is that from the other three isomers the chiral dienophile can be recovered by thermolysis (retro-Diels–Alder reaction). Quite recently, Corey reported an enantioselective cycloaddition between furan and α-bromo-acrolein under the action of a chiral Lewis-acid, which represents an efficient route to numerous chiral 7-oxabicyclo[2.2.1]heptene derivatives. [17]

Because all these cycloadducts contain a dihydrofuran-subunit, they are predestined for conversion to carbohydrates and related natural products. Usually, the double bond is functionalized first, whereby the bicyclic ring system guarantees stereoselective attack of the reagent. Subsequent cleavage of the bicyclic ring system leads to highly functionalized tetrahydrofuran derivatives. This concept was exploited by Vogel et al. in connection with the synthesis of a large number of natural products. [18] A representative example is the preparation of D-allose *(18)* (Scheme 2). [19] The compound obtained from the cycloadduct *(12)* by dihydroxylation and isopropylidenation was hydrolyzed to the ketone *(13)* and subsequently converted to the silyl enol ether. By oxidation with a peracid, a stereospecific hydroxylation to *(15)* takes place, whereby the initially formed epoxide is opened by *m*-chloro perbenzoic acid, followed by migration

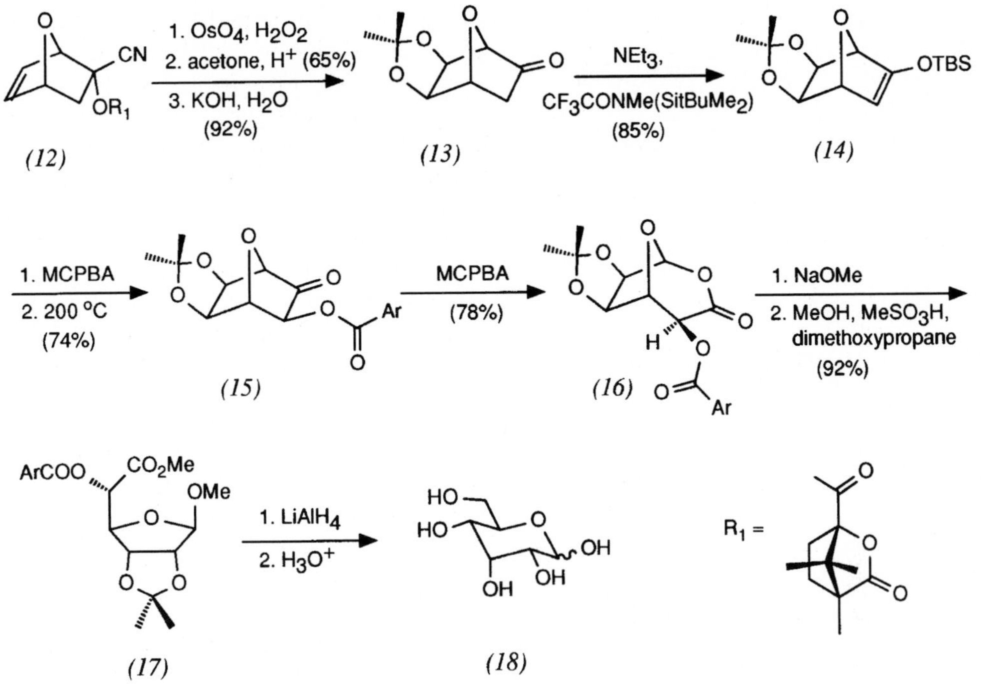

Scheme 2. Synthesis of D-allose from the cycloadduct *(12)*.

of the acyl group. Cleavage of the bicyclic ring was performed by a regiospecific Baeyer-Villiger oxidation, subsequent lactone hydrolysis, and acetalization to give compound *(17)*. Two further steps made D-allose *(18)* available.

Oxabicycles of type *(12)* not only can be opened to monocycles by cleavage of a C–C bond but also alternatively by cleavage of one of the C–O bonds. This opening is particularly facile if the original furan possesses a hetero substituent in the 2-position. Suzuki et al. used this strategy in an elegant synthesis of the antitumor compound gilvocarcin M (Scheme 3). [20] One of the key steps is the regiospecific cycloaddition of 2-methoxyfuran *(20)* to the arine, which was generated in situ by reductive elimination from the C-glycoside *(19)*. By opening the cycloadduct *(21)*, the naphtyl derivative *(22)* is formed in which the two hydroxyl groups are already differenti-

ated. After esterification to give *(23)*, the aryl residue is introduced by an intramolecular coupling. Cleavage of the benzyl protecting groups by catalytic hydrogenation provided the natural product *(24)*.

A conceptually different method for the cleavage of a C–O bond is to treat the oxabicycle with a nucleophile. This strategy recently was applied by Lautens et al. to oxa-bicyclo[2.2.1]-systems as well as the homologous oxa-bicyclo[3.2.1]-systems. [21] The latter are available very easily via a [4+3]-cycloaddition between a furan and a cationic 2-oxoallyl species. [22] Organo cuprates, organolithio compounds, and diisobutylaluminium hydride can be used for the opening of the bicyclic compounds. Thus, one obtains the cyclohexenes *(26)* in a regio- and stereospecific manner from *(25)* and alkyllithium compounds. [23] This method found application in a synthesis of the C_{17}–C_{23}-subunit of the anti-

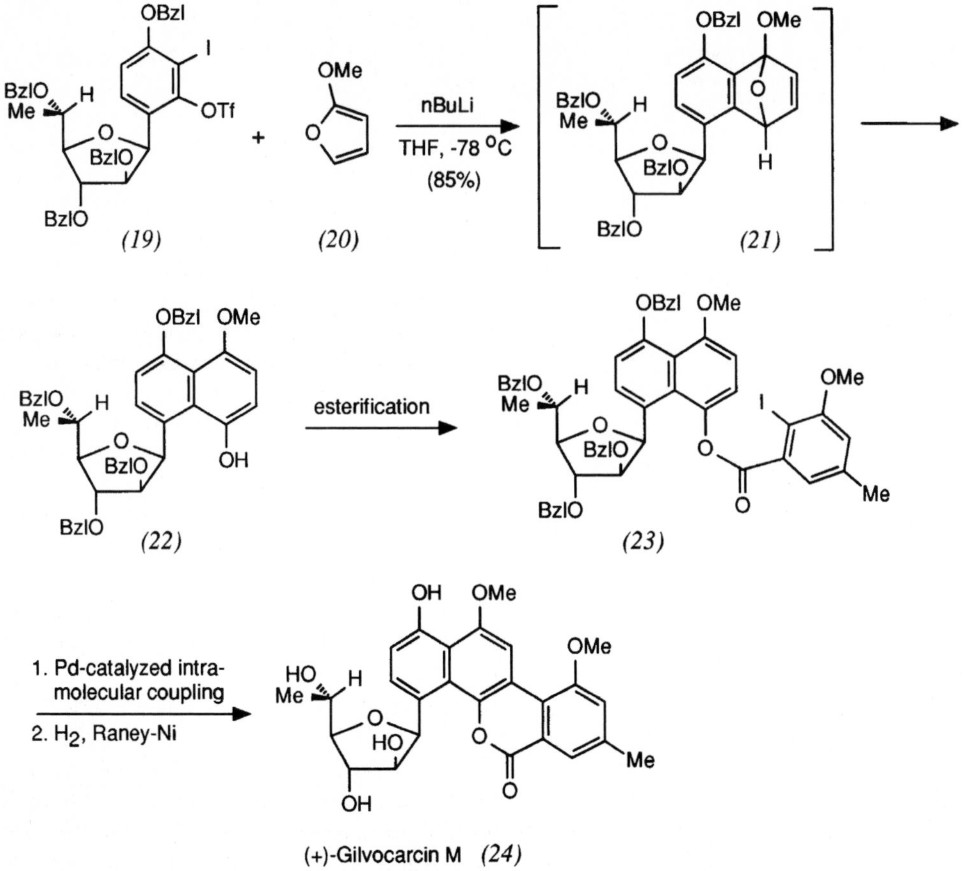

Scheme 3. Synthesis of (+)-gilvocarcin M through a furan–aryne cycloaddition.

biotic ionomycin. [24] Reductive opening of the oxabicyclo compound (27) furnished the cycloheptenol (28), which contains four stereocenters with defined configuration. After inversion of the stereochemistry at the C_{21}-center and protection of the OH-function, the ring was cleaved by ozonolysis. The corresponding open-chain compound could then be converted to the target molecule (29). Use of p-methoxybenzyl protecting group allowed differentiation of the primary OH-groups through an oxidative acetal formation. A disadvantage of this route is that optically active compounds are not available (yet) without a major effort.

Furans also proved quite versatile in intramolecular cycloaddition reactions [25] – particularly with a view to the synthesis of polycyclic target molecules. Because furans may be functionalized very easily, the syntheses of the substrates for the cycloaddition reactions are in general straightforward. This is expressed very well in a synthesis of bilobalide, which incorporates an intramolecular [2+2]-photocycloaddition as a key step (Scheme 4). [26] Starting with 3-furanaldehyde, the aldehyde (31) was constructed in four steps. This in turn was treated with the enolate (30) to give (32). From this addition product, the substrate (33) for the intramolecular cycloaddi-

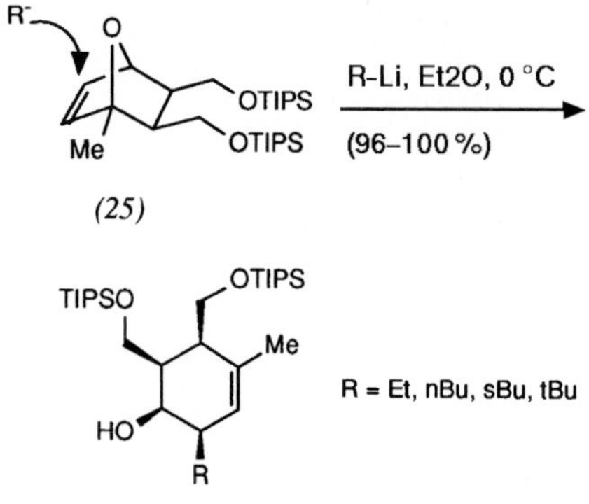

R-Li, Et2O, 0 °C

(96–100 %)

OTIPS
OTIPS
Me

(25)

TIPSO
OTIPS
Me
R = Et, nBu, sBu, tBu
HO
R

(26)

O
Me
Me OBzl

DIBAH

OBzl
Me
Me
21
OH

(27)   *(28)*

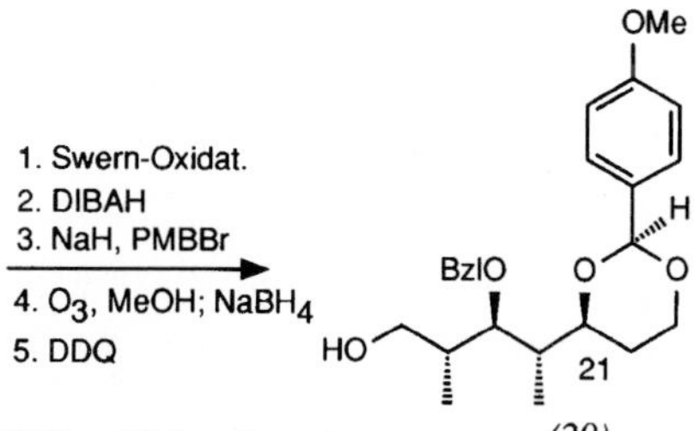

OMe

1. Swern-Oxidat.
2. DIBAH
3. NaH, PMBBr
4. O3, MeOH; NaBH4
5. DDQ

PMB = pMethoxybenzyl

H
BzlO   O   O
21
HO

(29)

tion could be obtained in two steps. Irradiation of *(33)* furnished compound *(34)* as the major product. Addition to the less substituted double bond was observed as a side reaction. Further key steps on the route to the target molecule are the oxidative cleavage of the cyclopentanone ring and a Baeyer–Villiger reaction at the four-membered ring. Altogether, this synthesis embraces only 17 steps, which is relatively few in light of such a highly complicated molecule. Bilobalide was isolated from gingko-trees, extracts of which have found use as a medicament.

Among the reactions of furans with heterodienes it is particularly the reaction with singlet-oxygen that is of great preparative significance. [27] For example, the preparation of 5-hydroxybutenolide *(39)* from 2-furancarboxylic acid or furfural is possible with this reaction. This oxidation proceeds via a [4+2]-cycloaddition with the formation of a bicyclic endoperoxide *(37)*, followed by fragmentation. A very elegant application of compound *(39)* is described by Feringa et al., who generated optically active butenolides by treatment of *(39)* with 1-menthol. Although both diastereomers *(40)* and *(41)* are formed in the course of the acetalization, the amount

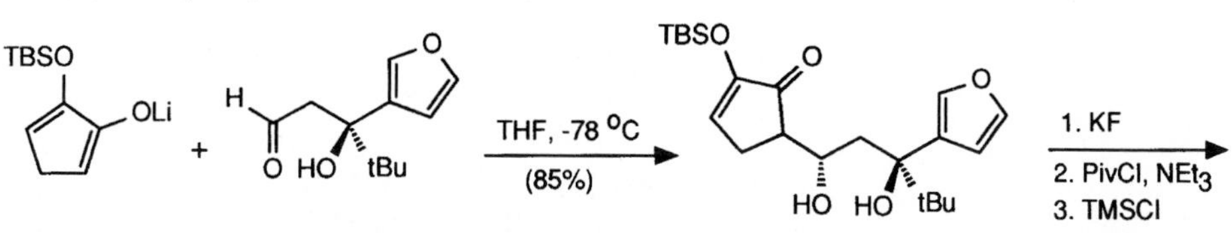

TBSO
OLi
+

H
O HO tBu

THF, -78 °C

(85%)

TBSO
O
O
HO HO tBu

1. KF
2. PivCl, NEt3
3. TMSCl

(50%)

(30)   *(31)*   *(32)*

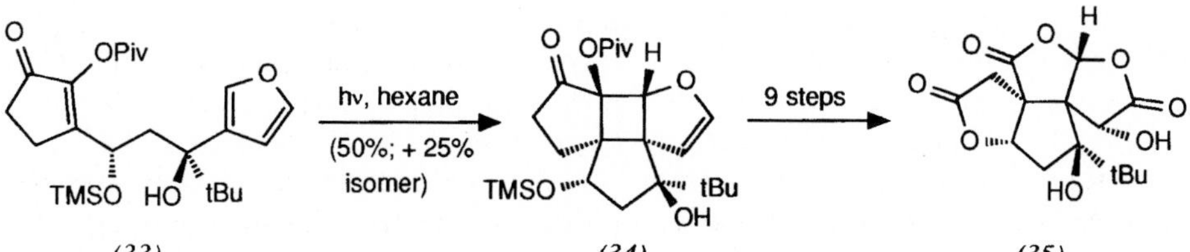

O
OPiv
TMSO HO tBu

hv, hexane

(50%; + 25% isomer)

O OPiv H
O
TMSO tBu
OH

9 steps

H
O O O
O
O
OH
HO tBu

(33)   *(34)*   *(35)*

Scheme 4. Synthesis of bilobalide by an intramolecular [2+2]-cycloaddition as key step.

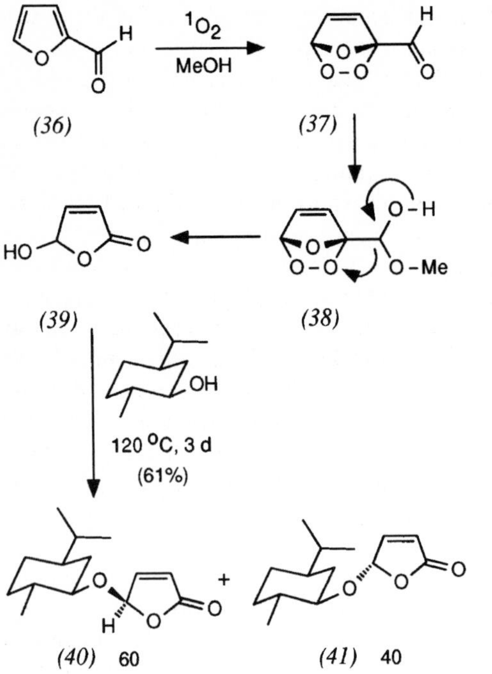

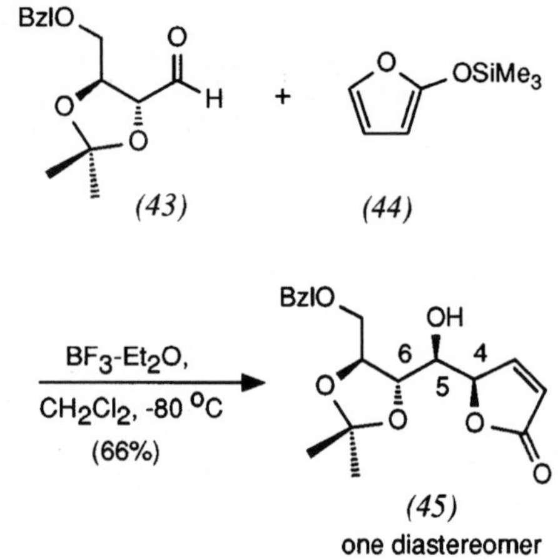

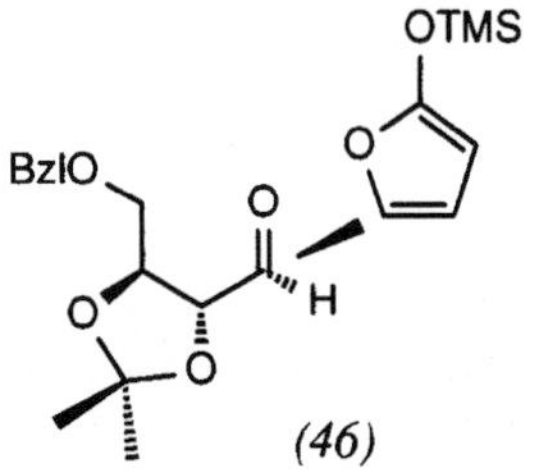

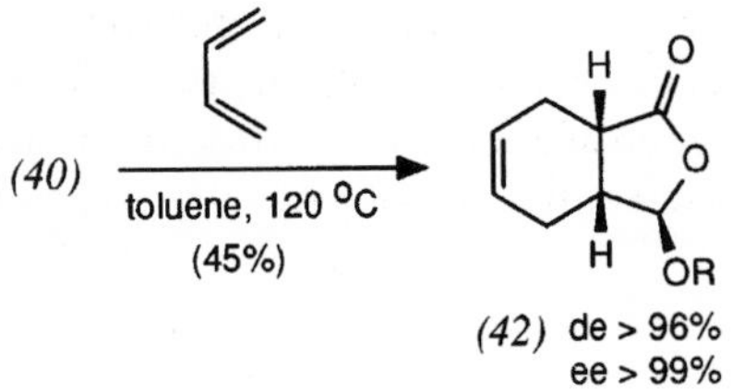

(36)

$^1O_2$ / MeOH

(37)

HO

(39)

(38)

O–H
O–Me

OH

120 °C, 3 d
(61%)

(40) 60

(41) 40

(40)

toluene, 120 °C
(45%)

(42) de > 96%
ee > 99%

of isomer (40) can be increased by crystallization. Compound (40) was used for asymmetric Diels–Alder reactions and Michael-additions. [28]

Reactions of Functionalized Furans

With functional groups at the periphery of the furan ring, more interesting synthetic building blocks are available that also open up additional reaction paths. For example, the easily

BzlO

(43)

+

OSiMe$_3$

(44)

BzlO   OH
6   4
5

BF$_3$-Et$_2$O,
CH$_2$Cl$_2$, -80 °C
(66%)

(45)
one diastereomer

OTMS

BzlO

H

(46)

accessible 2-(trimethylsilyloxy) furan (44) [29] can be used as a nucleophilic C$_4$-building block for the chain elongation of aldehydes and imines. These reactions proceed with high selectivity, whereby the two newly formed chiral centers possess syn(threo)-stereochemistry. [30] In the addition to α-alkoxy-aldehydes, one observes preferential Cram addition that is, the 4,5-syn, 5,6-anti-product is formed selectively. This result can be rationalized with a transition state such as (46), which corresponds to that of a Diels–Alder reaction of the furan and therefore is energetically favored. The so-formed butenolides may be further functionalized in various ways, for example by cis-hydroxylation. [31] Using this route, Casiraghi et al. prepared octopyranose derivatives by addition of (44) to the aldehyde (43).

Furans with a hydroxyalkyl group in the 2-position (furfuryl alcohols) (47) are also very interesting synthetic building blocks. If furfuryl alcohols are oxidized, whereby a large

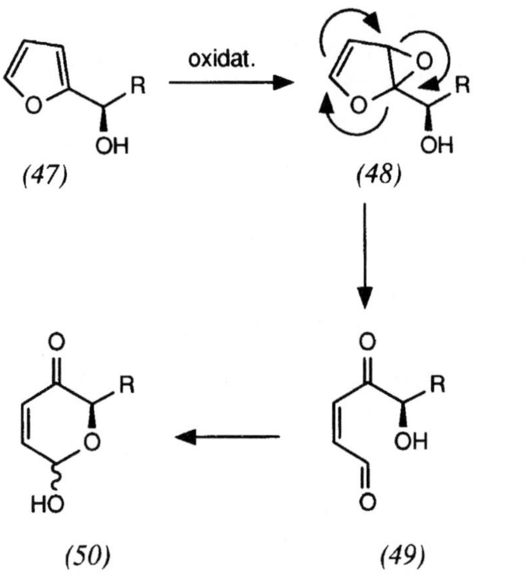

(47) (48)

(50) (49)

number of reagents such as MCPBA, Br₂, VO(acac)₂/tBuOOH, etc. can be used, rearrangement to ulose derivatives (hydroxy pyranones) *(50)* takes place. It is obvious that these pyranones are predestined for the synthesis of carbohydrates. [32] In addition, by application of this strategy, a large number of polyoxygenated natural products have been prepared. [33] In order to obtain furuyl alco-

hols of type *(47)* in optically active form, several methods are available. One is to add metallated furans to chiral aldehydes. However, the selectivity depends very much on the substrate and on the reaction conditions. Another important entry is addition of chiral *C*-nucleophiles to furfuryl aldehydes. Since there are now very powerful chiral reagents available, such as the aldol reagents of Evans *(34)* or allyl boranes, [35] the furfuryl alcohols can be obtained using this approach without any problem. Finally, it should be noted that optically active furfuryl alcohols can be obtained by resolution of racemates. [36] Two examples from the recent literature might serve to illustrate the application of furfuryl alcohols for the synthesis of natural products. In case of the synthesis of the C_{10}–C_{20} fragment of the immunosuppressive FK506 (Scheme 5), initially by adding a metalated furan (prepared from *(52)* by halogen/metal exchange) to the aldehyde *(51)*, the furfuryl alcohol *(53)* was prepared. Addition of zinc(II) bromide to the lithiated furan induces exclusive formation of the desired *syn*-isomer *(53)* by chelation control. Subsequent hydroxyl-directed epoxidation causes the rearrangement to the hydroxypyranone, the anomeric

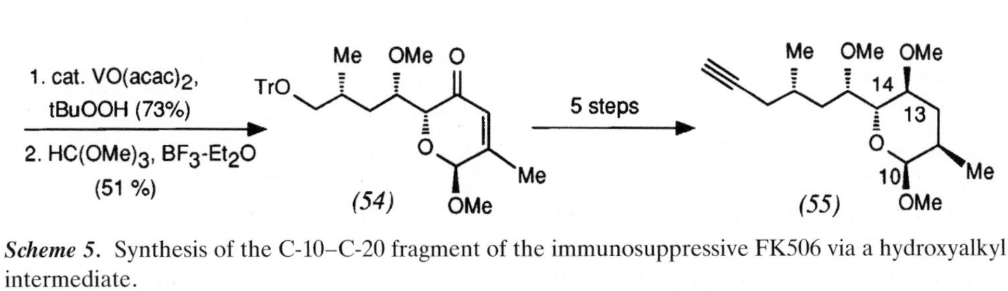

(51) (52) (53)

(54) (55)

Scheme 5. Synthesis of the C-10–C-20 fragment of the immunosuppressive FK506 via a hydroxyalkyl furan intermediate.

center of which was protected as the methyl glycoside *(54)*. The conformationally fixed pyran scaffold ensured a stereocontrolled introduction of the remaining two stereocenters. These were generated by reduction of the carbonyl group (NaBH₄) and hydrogenation of the double bond (H₂, Pd/Al₂O₃), whereby attack of the reagents took place from the α-face. This synthesis is the shortest known for such a building block. [37]

The other route, namely an asymmetric aldol reaction between the Evans reagent *(56)* and furfural *(36)* was followed by Martin et al. in the synthesis of the furfuryl alcohol *(57)* (Scheme 6). Oxidative rearrangement and protection of the anomeric hydroxy function with *tert*-butyldimethylsilyl triflate (TBSOTf) yielded the α-anomer *(58)* (α:β = 3:1) as the major product. The axial methyl group was introduced stereoselectively by conjugate addition of dimethyl cuprate to the enone *(58)*. The stereochemical course of the addi-

tion explains itself from the reactive conformation *(61)* with an axial OTBS and equatorial substituent R. For stereoelectronic reasons (chair-like transition state) the addition takes place *anti* to the OTBS-group. Subsequent reduction of the ketone *(59)* to the equatorial alcohol proved to be problematic. The reduction always favored the axial isomer, so that *(59)* was converted via a modified Mitsunobu reaction to the desired alcohol *(60)*. The latter served as an important intermediate in the synthesis of the ansamycin antibiotic (+)-macbecin I. [38] As these examples demonstrate, it is worthwhile to examine furfuryl alcohol based concepts if molecules must be synthesized that contain a diol structure (cf. centers 13 and 14 in *(55)*) within a complex stereochemical setting.

These hydroxy pyranones not only represent easily available templates for asymmetric synthesis, but they are furthermore precursors for carbonyl ylides. On this basis, Wender et

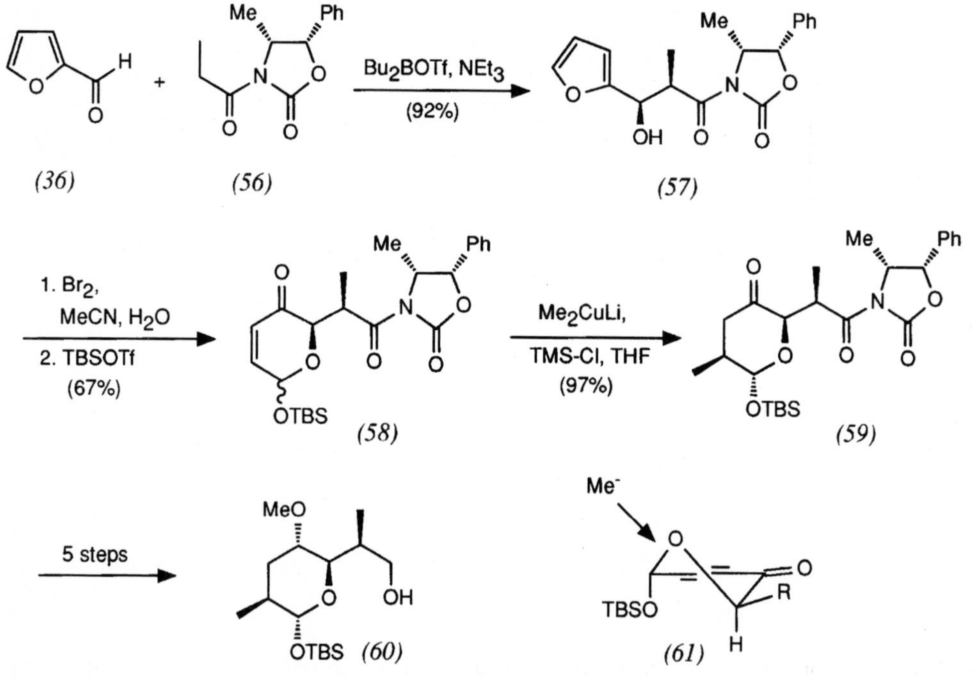

Scheme 6. Synthesis of an important intermediate for the antibiotic macbecin I through a furan route.

al. developed a synthesis of the tumor promoter phorbol (Scheme 7). [39] Beginning with furfuryl alcohol and proceeding via *(62)*, *(63)* was prepared as a mixture of diastereomers. Compound *(63)* was then oxidatively rearranged to the hydroxypyranone. On treatment of *(64)* with DBU, an oxidopyrilium intermediate is formed, which, as a carbonyl ylide, undergoes intramolecular cycloaddition to the double bond. Transition state *(65)* in which the side-chain assumes a chairlike conformation, explains the observed stereochemistry. In the further course of the synthesis, the cycloadduct is converted to the methylene ketone *(67)*. This, in turn, is subsequently

transformed to *(68)* via an intramolecular nitrile-oxide cycloaddition. From *(68)* another six steps made compound *(69)* available, which served as an entry to the synthesis of phorbol.

Furthermore, alkoxypyranones can also be rearranged to *trans*-4-alkoxy-5-hydroxy-cyclopentenones (e.g. *(70)*–*(73)*). This rearrangement is possible, for example, with the buffer system benzoic acid/potassium acetate although the yield is generally not too high. [40] This appears to be an attractive procedure for the preparation of functionalized cyclopentenones because of the ease of access to the corresponding educts. These cyclopente-

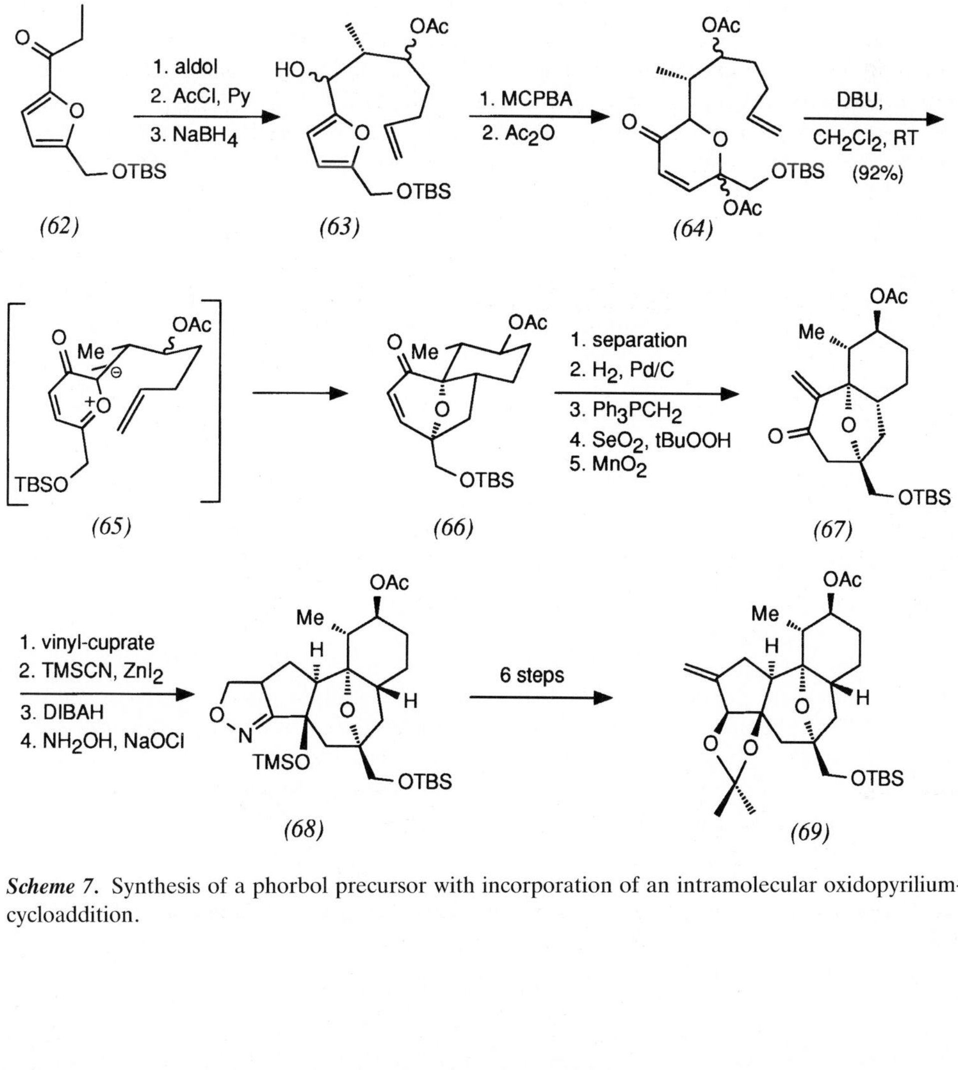

Scheme 7. Synthesis of a phorbol precursor with incorporation of an intramolecular oxidopyrilium-alkene cycloaddition.

(70)   *(73)*   *(71)*   *(72)*   *(74)*   *(75)*   *(76)*   *(77)*   *(78)*   *(79)*   *(80)*

nones, in turn, are interesting as building blocks in the context of prostaglandin syntheses. Mechanistically, this rearrangement probably proceeds through tautomerization to *(71)*, subsequent electrocyclic ring opening *(72)*, and, finally, an intramolecular vinylogous aldol reaction.

Interestingly, the rearrangement to cyclopentenones also succeeds directly with furfuryl alcohols, as shown by Dygos et al. [41] Thus, the reaction of *(74)* with zinc(II) chloride in water produces – probably via the intermediates *(75)* and *(76)* – the cyclopentenone *(77)*. This was used for the synthesis of a prostaglandin derivative. Of course, it would be better if one could perform these rearrangements enantioselectively.

Another class of functionalized furans are those that contain silicon residues as substituents, whereby the silyl group can serve, for example, as a protecting group. [42] In order to convert furans into butenolides, Pelter et al. used boron-substituted furans. First, the furans are metallated and the resulting anion is treated with chlorodimethoxyborane to give *(79)*. The furyldimethoxyboranes were then oxidized to 2-(3*H*)-butenolides by *m*-chloroperbenzoic acid. [43]

As this summary shows, furans are indeed versatile building blocks because they come into question as a platform for many different applications. From these furans, short syntheses result that are clearly not without elegance. The list of possible uses could be continued almost at will. [44] For example, furans can be used as C_1-building blocks [45] or ter-

R = Aryl, Alkyl

minators in cyclization reactions. [46] This review also shows that progress in synthesis can result from the refinement and a further development of known reactions and strategies.

References

[1] E. J. Corey, X.-M. Ming, *The Logic of Chemical Synthesis*, Wiley, New York, **1989**.

[2] L. F. Tietze, U. Beifuss, *Angew. Chem.* **1993**, *105*, 137–170; *Angew. Chem. Int. Ed. Engl.* **1993**, *32*, 131–163.

[3] D. Seebach, *Angew. Chem.* **1979**, *91*, 259–278; *Angew. Chem. Int. Ed. Engl.* **1979**, *18*, 239.

[4] For a review, see: B. Lipshutz, *Chem. Rev.* **1986**, *86*, 795–819.

[5] W. G. Dauben, H. O. Krabbenhoft, *J. Am. Chem. Soc.* **1976**, *98*, 1992.

[6] H. Takeyama, A. Iyobe, T. Koizumi, *J. Chem. Soc. Chem. Commun.* **1986**, 771 and references cited therein.

[7] G. Just, M. I. Lim, *Can. J. Chem.* **1977**, *55*, 2993.

[8] A. P. Kozikowski, A. Ames, *J. Am. Chem. Soc.* **1981**, *103*, 3923.

[9] R. R. Schmidt, A. Lieberknecht, *Angew. Chem.* **1978**, *90*, 821–822; *Angew. Chem. Int. Ed. Engl.* **1978**, *17*, 769.

[10] M. Ohno, Y. Ito, M. Arita, T. Shobata, K. Adachi, H. Sawai, *Tetrahedron* **1984**, *40*, 141–152.

[11] J. Gustafsson, O. Sterner, *I. Org. Chem.* **1994**, *59*, 3994-3997.

[12] T. Takahashi, A. Iyobe, Y. Arai, T. Koizumi, *Synthesis* **1989**, 189–191.

[13] a) R. R. Schmidt, C. Beitzke, A. K. Forrest, *J. Chem. Soc. Chem. Commun.* **1982**, 909–910; b) E. Vieira, P. Vogel, *Helv. Chim. Acta* **1982**, *65*, 1700–1706.

[14] V. K. Aggarwal, M. Lightowler, S. D. Lindell, *Synlett* **1992**, 730–732.

[15] K. A. Black, P. Vogel, *Helv. Chim. Acta* **1984**, *67*, 1612–1615.

[16] E. Vieira, P. Vogel, *Helv. Chim. Acta* **1983**, 66, 1865–1871.

[17] E. J. Corey, T.-P. Loh, *Tetrahedron Lett.* **1993**, *34*, 3979–3982.

[18] Review: P. Vogel, *Synlett* **1990**, 173–185.

[19] Y. Auberson, P. Vogel, *Helv. Chim. Acta* **1989**, *72*, 278–286.

[20] a) T. Matsumoto, T. Hosoya, K. Suzuki, *J. Am. Chem. Soc.* **1992**, *114*, 3568–3570; b) see also: T. Hosoya, E. Takashiro, T. Matsumoto, K. Suzuki, *J. Am. Chem. Soc.* **1994**, *116*, 1004–1015.

[21] M. Lautens, *Synlett* **1993**, 177–185.

[22] a) R. Noyori, *Acc. Chem. Res.* **1979**, *12*, 61–66; b) H. M. R. Hoffmann, *Angew. Chem.* **1984**, *96*, 29; *Angew. Chem. Int. Ed. Engl.* **1984**, *23*, 1. c) A. Hosomi, Y. Tominaga in *Comprehensive Organic Synthesis* (Eds.: B. M. Trost, I. Fleming), Vol. 5, Pergamon, Oxford **1991**, 593.

[23] M. Lautens, P. Chiu, *Tetrahedron Lett.* **1993**, *34*, 773–776.

[24] M. Lautens, P. Chiu, J. T. Colucci, *Angew. Chem.* **1993**, *105*, 267–269; Angew. Chem. Int. Ed. Engl. **1993**, *32*, 281.

[25] [4+2]-Cycloadditions: a) L. L. Klein, M. S. Shanklin, *J. Org. Chem.* **1988**, *53*, 5202–5209; b) E. Bovenschulte, P. Metz, G. Henkel, *Angew. Chem.* **1989**, *101*, 204–206; *Angew. Chem. Int. Ed. Engl.* **1989**, *28*, 202; c) H. Finch, L. M. Harwood, G. Robertson, R. Sewell, *Tetrahedron Lett.* **1989**, *30*, 2585–2588; d) B. L. Feringa, O. J. Gelling, L. Meesters, *Tetrahedron Lett.* **1990**, *31*, 7201–7204; e) L. M. Harwood, T. Ishikawa, H. Phillips, D. Watkin, *J. Chem. Soc. Chem. Commun.* **1991**, 527–530; f) D. P. Dolata, L. M. Harwood, *J. Am. Chem. Soc.* **1992**, *114*, 10738–10746; [4+2]-Cycloadditions of vinylfurans: a) J. A. Cooper, P. Cornwall, C. P. Dell, D. W. Knight, *Tetrahedron Lett.* **1988**, *29*, 2107–2110; b) K. Hayakawa, F. Nagatsugi, K. Kanematsu, *J. Org. Chem.* **1988**, *53*, 860–863; [4+3]-Cycloadditions: a) M. Harmata, S. Elahmad, *Tetrahedron Lett.* **1993**, *34*, 789–792, and referenced cited therein.

[26] M. T. Crimmins, D. K. Jung, J. L. Gray, *J. Am. Chem. Soc.* **1992**, *114*, 5445–5447.

[27] a) Review: B. L. Feringa, *Recl. Trav. Chim. Bas.* **1987**, *106*, 469–488; b) M. R. Kernan, D. J. Faulkner, *J. Org. Chem.* **1988**, *53*, 2773–2776; c) G. C. M. Lee, E. T. Syage, D. A. Harcourt, J. M. Holmes, M. E. Garst, *J. Org. Chem.* **1991**, *56*, 7007–7014.

[28] a) B.L. Feringa, J.C. de Jong, *J. Org. Chem.* **1988**, *53*, 1125–1127; b) B.L. Feringa, B. de Lange, J.C. de Jong, *J. Org. Chem.* **1989**, *54*, 2471–2475.

[29] M.A. Brimble, M.T. Brimble, J.J. Gibson, *J. Chem. Soc., Perkin Trans. 1*, **1989**, 179.

[30] a) D.W. Brown, M.M. Campbell, A.P. Taylor, X. Zhang, *Tetrahedron Lett.* **1987**, *28*, 985–988; b) C.W. Jefford, D. Jaggi, J. Boukouvalas, *Tetrahedron Lett.* **1987**, *28*, 4037–4040.

[31] a) G. Casiraghi, L. Colombo, G. Rassu, P. Spanu, G. Gasparri Fava, M. Ferrari Belicci, *Tetrahedron* **1990**, *46*, 5807–5824; b) G. Casiraghi, L. Colombo, G. Rassu, P. Spanu, *J. Org. Chem.* **1991**, *56*, 2135–2139; c) G. Rassu, P. Spanu, G. Casiraghi, L. Pinna, *Tetrahedron* **1991**, *47*, 8025–8030.

[32] a) A. Zamojski, G. Grynkiewicz in *Total Synthesis of Natural Products* (Ed.: J. ApSimon), Wiley, New York, **1984**, 141–235; b) P.G. Sammes, D. Thetford, *J. Chem. Soc., Perkin Trans. 1*, **1988**, 111–123 and references cited therein.

[33] a) K. Mori, H. Kisida, *Tetrahedron* **1986**, *42*, 5281–5290; b) S. Pikul, J. Raczko, K. Akner, J. Jurczak, *J. Am. Chem. Soc.* **1987**, *109*, 3981–3987; c) S.F. Martin, D.E. Guinn, *J. Org. Chem.* **1987**, *52*, 5588–5593; d) P. De-Shong, R.E. Waltermire, H.L. Ammon, *J. Am. Chem. Soc.* **1988**, *110*, 1901–1910; e) S.F. Martin, C. Gluchowski, C.L. Campbell, R.C. Chapman, *Tetrahedron* **1988**, *44*, 3171–3180; f) S.F. Martin, G.J. Pacofsky, R.P. Gist, W.-C. Lee, *J. Am. Chem. Soc.* **1989**, *111*, 7634–7636; g) K. Mori, H. Kikuchi, *Liebigs Ann. Chem.* **1989**, 963–967; h) M.F. Semmelhack, N. Jeong, *Tetrahedron Lett.* **1990**, *31*, 605–608; i) J. Raczko, A. Golebiowski, J. Krajewski, P. Gluzinski, J. Jurczak, *Tetrahedron Lett.* **1990**, *31*, 3797–3800; j) T. Honda, Y. Kobayashi, M. Tsubuki, *Tetrahedron Lett.* **1990**, *34*, 4891–4894; k) I. Paterson, M.A. Lister, G.R. Ryan, *Tetrahedron Lett.* **1991**, *32*, 1749–1752; l) S.F. Martin, P.W. Zinke, *J. Org. Chem.* **1991**, *56*, 6600–6606; m) S.J. Shimshock, R.E. Waltermire, P. DeShong, *J. Am. Chem. Soc.* **1991**, *113*, 8791–8796; n) S.F. Martin, H. Chen, C.-P. Yang, *J. Org. Chem.* **1993**, *58*, 2867–2873; o) T. Honda, K. Tomitsuka,

M. Tsubuki, *J. Org. Chem.* **1993**, *58*, 4274–4279.

[34] J.R. Gage, D.A. Evans, *Org. Synth.* **1989**, 68, 83–91, and references cited therein.

[35] U.S. Racherla, Y. Lao, H.C. Brown, *J. Org. Chem.* **1992**, *57*, 6614–6617, and references cited therein.

[36] a) Y. Kobayashi, M. Kusakabe, Y. Kitano, F. Sato, *J. Org. Chem.* **1988**, *53*, 1587–1590; b) H. Waldmann, *Tetrahedron Lett.* **1989**, *30*, 3057–3058; c) W.-S. Zhou, Z.-H. Lu, Z.-M. Wang, *Tetrahedron Lett.* **1991**, *32*, 1467–1470.

[37] M.E. Maier, B. Schöffling, *Tetrahedron Lett.* **1991**, *32*, 53–56.

[38] S.F. Martin, J.A. Dodge, L.E. Burgess, M. Hartmann, *J. Org. Chem.* **1992**, *57*, 1070–1072.

[39] a) P.A. Wender, H.Y. Lee, R.S. Wilhelm, P.D. Williams, *J. Am. Chem. Soc.* **1989**, *111*, 8954–8957; b) P.A. Wender, H. Kogen, H.Y. Lee, J.D. Munger, Jr., R.S. Wilhelm, P.D. Williams, *J. Am. Chem. Soc.* **1989**, *111*, 8957–8958.

[40] H.C. Kolb, H.M.R. Hoffmann, *Tetrahedron* **1990**, *56*, 5127–5144, and references cited therein.

[41] J.H. Dygos, J.P. Adamek, K.A. Babiak, J.R. Behling, J.R. Medich, J.S. Ng, J.J. Wieczorek, *J. Org. Chem.* **1991**, *56*, 2549–2552.

[42] a) E.J. Corey, Y.B. Xiang, *Tetrahedron Lett.* **1987**, *28*, 5403–5406; b) G. Beese, B.A. Keay, *Synlett* **1991**, 33–34.

[43] A. Pelter, M. Rowlands, *Tetrahedron Lett.* **1987**, *28*, 1203–1206.

[44] Synthesis of furan-3,4-diyl oligomers: Z.Z. Song, H.N.C. Wong, *J. Org. Chem.* **1994**, *59*, 33–41. Intramolecular Michael addition of an electron-rich furan to a cyclohexenone followed by an aldol condensation: M.E. Jung, C.S. Sieden, *J. Am. Chem. Soc.* **1993**, *115*, 3822–3823.

[45] S.J. Danishefsky, M.P. DeNinno, S. Chen, *J. Am. Chem. Soc.* **1988**, *110*, 3929–3940.

[46] a) S.P. Tanis, M.C. McMills, T.A. Scahill, D.A. Kloosterman, *Tetrahedron Lett.* **1990**, *31*, 1977–1980; b) S.R. Angle, M.S. Louie, *Tetrahedron Lett.* **1989**, *30*, 5741–5744.

Fluorine in Organic Synthesis

Rolf Bohlmann

Fluorine is the most reactive of all the elements. It has a higher natural abundance in the mantle of the earth than the other halogens and has pronounced properties in both elemental form and in its compounds. The extreme electronegativity, the high energy of the C–F bond, the low nucleophilicity of fluoride ions in protic solvents and the steric similarity with hydrogen are noteworthy. [1] The emphasis of this highlight is on the preparative aspects of monofluorinated compounds of biological importance. Despite the wide range of polyfluoro products and their often large scale of world-wide production they are not discussed here due to brevity. The number of known naturally occurring fluorinated products, such as the fluoro acetic from the South African "Gifblaar", is very small. Selective fluorination of biologically interesting compounds is of great interest in drug discovery, because of the close similarity of the van der Walls radii of fluorine and hydrogen (135 versus 120 pm). [2] This is the reason why fluoroanalogs of biomolecules are used as tools for the study of electronic effects in the absence of strong steric changes.

The preparation of fluorinated drugs requires safe methods for the directed monofluorination of the fluorine-free precursors. The search for safe, selective fluorine transferring reagents is a pursuit currently in progress. [3]

In protic solvents the fluoride anion forms the stable FHF anion, which is highly solvated and only mildly acidic and nucleophilic. The Olah reagent (70 % HF in pyridine) has been used for the introduction of fluoride into many acid stable compounds. [4] The solubility of inorganic fluorides in aprotic solvents is limited. Therefore, phase transfer catalysis or tetrabutyl ammonium fluoride (TBAF) is in common use for this purpose. TBAF is not stable, however, without water. [5] In contrast to the above mentioned FHF the strongly basic "naked" fluoride anion triggers elimination at the tetrabutylammonium cation with formation of amine *(1)* and butene. Tetramethylammoniumfluoride cannot undergo Hofmann elimination. Therefore it is possible to remove water from this reagent completely and it acts as a strongly basic nucleophile in acetonitrile [6] or dichloromethane. [7]

$$2 \ (nC_4H_9)_4N^+F^- \ \rightarrow \ (nC_4H_9)_4N^+FHF^- + (nC_4H_9)_3N + C_2H_5{-}CH{=}CH_2$$

$$nC_8H_{17}{-}OH + DAST \rightarrow nC_8H_{17}{-}F \ (90\,\%) - 50\,°C$$

The transformation of hydroxyl groups into fluorides by diethylaminosulfur trifluoride (DAST) has been widely used for several years now. [8]

Besides the reagents for the introduction of nucleophilic fluorine there is an urgent demand for reagents in which fluorine acts as an electrophile. Elemental fluorine itself is the simplest reagent of this type. Safe handling and selectivity of this element are much less simple, however. [9, 10] An explanation of the low selectivity often observed is the participation of radical reactions. Therefore stable and safe reagents, which act as sources of elemental fluorine, are of great value. A common feature of such reagents is an F–X bond. The X group in those reagents should be more electronegative than the carbon in the target molecule. Well known examples of this type are CF₃OF, FClO₃ or CF₃COOF. These gases are not without problems: CF₃OF is as dangerous and unselective as elemental fluorine, FClO₃ forms explosive reaction mixtures, and CF₃COOF is unstable, toxic, and reacts with low chemoselectivity.

Better selectivity has been observed by Rozen with acetyl hypofluorite (3). [11, 12] Many mild reactions of alkenes (2), (6), (7) and (14), aromatics (9) and (10) and ketones (12) with acetyl hypofluorite (3) are reported (Scheme 1). Not only electron-rich double bonds as in (14), but also normal alkenes as (2) and (6), and electron-deficient olefines like (7) undergo rapid addition of the hypo-

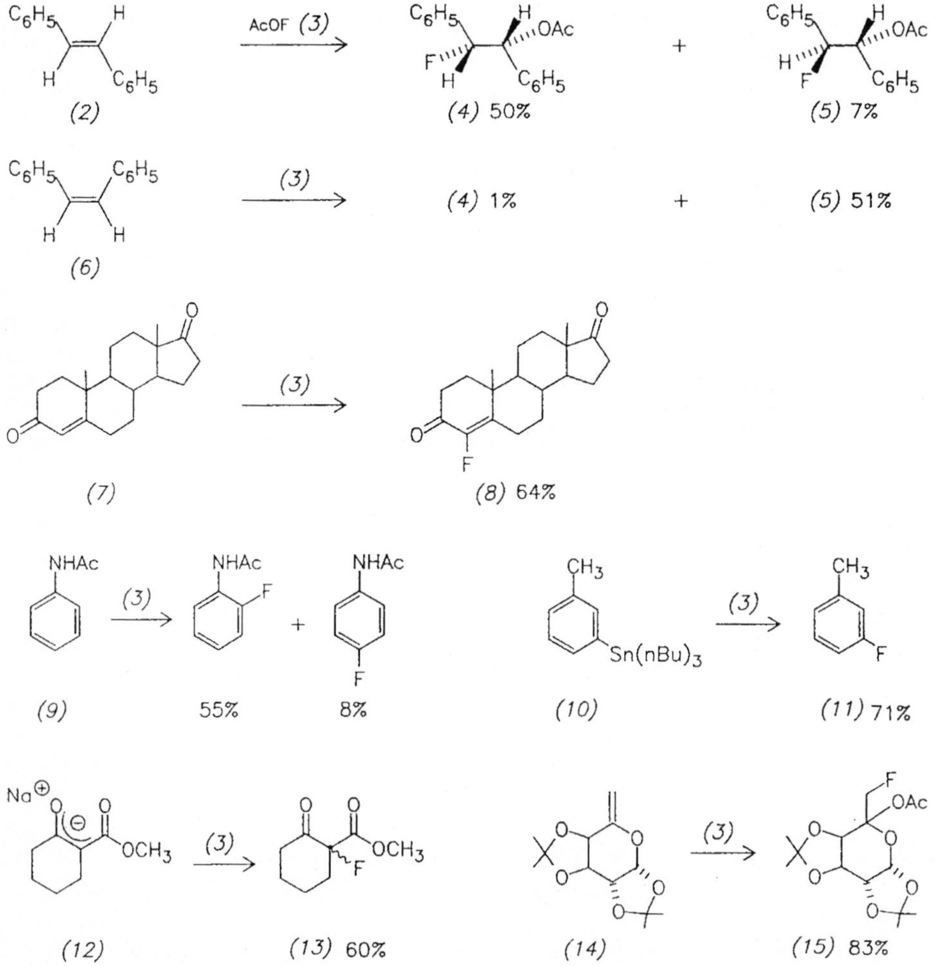

Scheme 1. Fluorination of different types of substrates with acetyl hypofluorite (3).

fluorite *(3)*. The reaction is regioselective, fluorination of the more nucleophilic center being preferred. Acetoxylation occurs accordingly at the site with higher stabilization for cations. The stereochemistry of the reaction is *syn*, as exemplified in the formation of *threo*-*(5)* and *erythro*-fluoroacetate *(4)* from the *cis*-*(6)* and *trans*-olefin *(2)* in the given ratios. Apart from addition to olefins other useful examples are the fluorination of *(12)* and *(10)* to fluoro compounds *(13)* and *(11)* by *(3)*. Direct fluorination of the aromatic ring *(9)* is also possible. Despite all these interesting results it must be mentioned that two authors report *unpredictable explosions* with acetyl hypofluorite *(3)*. [13, 14] The limited storage stability of *(3)* requires the reagent to be pro-

duced from elemental fluorine immediately before its use or in situ.

Safety hazards are a common problem with all known hypofluorites. The *N*-fluoro reagents *(17)*, *(19)*, *(21)* and *(24)* come much closer to the ideal profile. The *N*-fluorosulfonamides first described by Barton [15] and introduced as fluorinating agents by Barnette [16, 17] are now commercially available. Handling of these reagents is easy. It is even possible to purify them by chromatography on silica gel without decomposition. They are used without special precautions in organic solvents like THF, diethyl ether or toluene for the fluorination of reactive anions like *(16)*, *(18)*, *(20)*, *(22)* and *(23)*. Examples of the fluorination with *N*-fluorosulfonamides

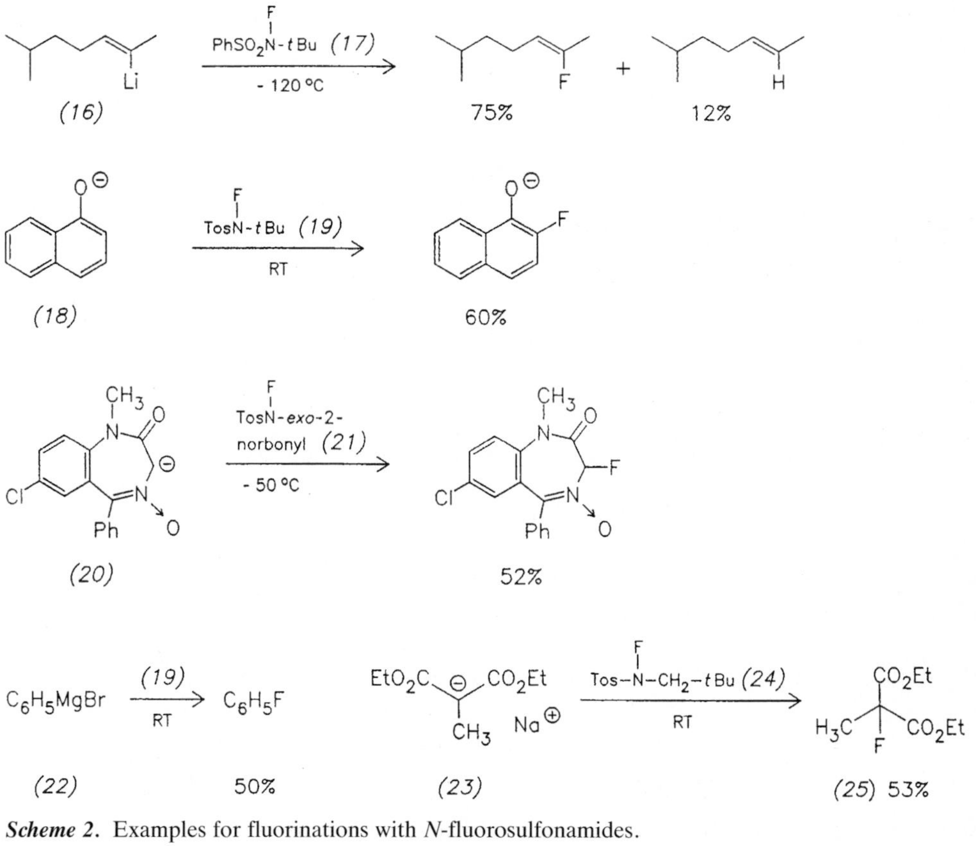

Scheme 2. Examples for fluorinations with *N*-fluorosulfonamides.

(17), *(19)*, *(21)* and *(24)* are shown in Scheme 2. The solubility of phenylsulfonamide *(17)* is better than that of the similar tosylamide *(19)*. For reactions at –120 °C this difference is very important. Anions with low reactivity and neutral enols, however, do not react with these mild reagents.

Reagents with higher reactivity are the *N*-fluorosulfonimides *(26)* first described by DesMarteau. [18] This reagent is activated by two strong acceptors at the central nitrogen atom and therefore shows increased electrophilicity of the attached fluorine. The anion

(23)

(25) 96%

50%

60%   40%

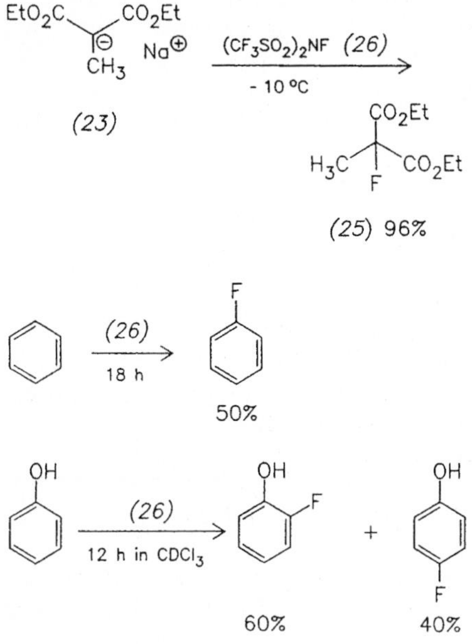

Scheme 3. Reactions with the highly reactive *N*-fluorosulfonamide *(26)*.

(23) reacts with sulfonimide *(26)* at lower temperature with higher yield to the fluoromethylmalonate *(25)* than with *N*-fluorosulfonamide *(24)*, as illustrated in Scheme 3.

Aromatic compounds are oxidatively fluorinated to fluoroaromatic compounds. The high reactivity of this reagent requires a careful choice of the solvent, which should not be fluorinated itself. The reagent is not commercially available. It is made by reaction with elemental fluorine.

The first reagent for the chiral electrophile fluorination of anions was reported by Lang and Differding. [19] The highest observed enantiomeric excess in the fluorinated product is 70 %. This is an important start, which leaves room for improvement (Scheme 4). The preparation of this reagent also involves the use of elemental fluorine.

The commercial availability of the *N*-fluoropyridinium salts, which were first described by Meinert [20] and introduced as fluorinating agents in organic synthesis by Umemoto, [21, 22] allows simple application of these reagents without access to elemental fluorine. The triflates *(30)*, *(32)* and *(34)* are more reactive than tetrafluoroborates, hexafluoroantimonates, or perchlorates also investigated. Donor substituents at the pyridine result in reduced reactivity, whereas acceptor substituents give increased reactivity. Therefore it is possible to find a reagent matching the reactivity of the substrate of interest. The scope of the reaction is wide despite the low

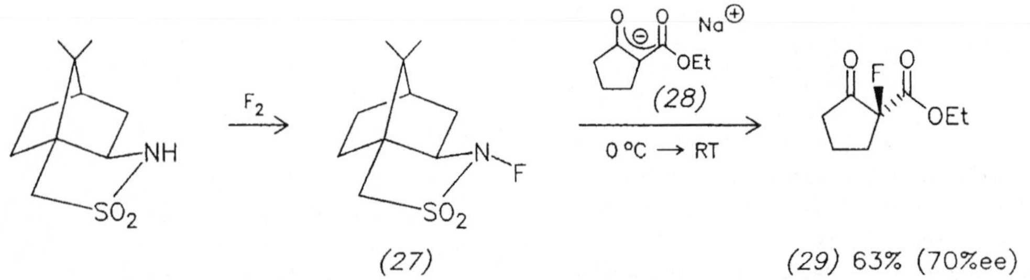

(27)   *(29) 63% (70%ee)*

Scheme 4. The camphor sultame *(27)* as chiral fluorinating agent.

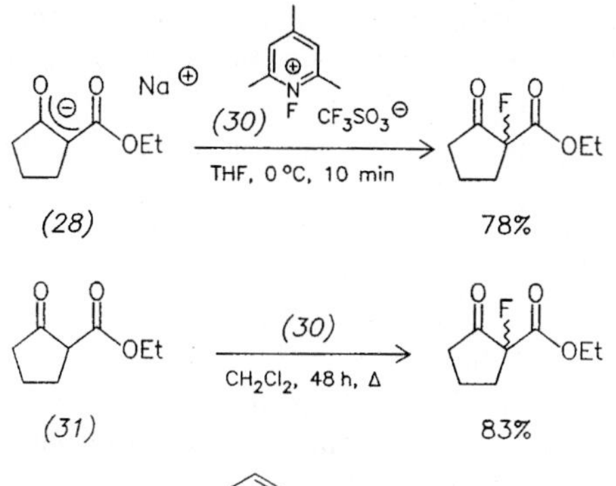

(28) → 78%

(31) → 83%

(33) → (34)

71%
($\alpha : \beta = 1 : 2$)

Scheme 5. Examples for fluorinations with several *N*-fluoropyridinium salts.

solubility of the pyridinium salts in organic solvents.

Some examples for fluorinations with *N*-fluoropyridinium salts (30), and (34) are given in Scheme 5. The reagents react not only with anions (28), but also with enols (31) and (33). Even benzene is oxidatively fluorinated to fluorobenzene by reagent (32). Amines and other oxidizable substrates are therefore not suitable for direct fluorination under these electrophilic conditions.

The "naked" fluorideion is not only nucleophilic but also a very strong base. Strong elec-

trophilic fluorinating agents are not only fluorinating but also strong oxidants. A single reagent cannot satisfy all requirements. It is exciting to see what new selective and safe fluorinating reagents and methods will offer in the future.

References

[1] H. Meinert *Fluorchemie*, Akademie Verlag, Berlin, **1979**.
[2] J.T. Welch, *Tetrahedron* **1987**, *43*, 3123.
[3] A. Haas, M. Lieb, *Chimia* **1985**, *39*, 134.
[4] G.A. Olah, J.T. Welch, Y.D. Vankar, M. Nojima, I. Kerekes, J.A. Olah, *J. Org. Chem.* **1979**, *44*, 3872.
[5] R.K. Sharama, J.L. Fry, *J. Org. Chem.* **1983**, *48*, 2112.
[6] T.J. Tewson, *J. Org. Chem.* **1983**, *48*, 3507.
[7] K.O. Christe, W.W. Wilson, *J. Fluorine Chem.* **1989**, *45*, 4.
[8] W.J. Middleton, *J. Org. Chem.* **1975**, *40*, 574.
[9] H. Vyplel, *Chimia* **1985**, *39*, 305.
[10] S.T. Purrington, B.S. Kagen, T.B. Patrick, *Chem. Rev.* **1986**, *86*, 997.
[11] S. Rozen, O. Lerman, M. Kol, *J. Chem. Soc., Chem. Commun.* **1981**, 443.
[12] S. Rozen, *Acc. Chem. Res.* **1988**, *21*, 307.
[13] M.J. Adam, *Chem. Eng. News* **1985**, *63*, 2.
[14] E.H. Appelman, M.H. Mendelsohn, H. Kim, *J. Am. Chem. Soc.* **1985**, *107*, 6515.
[15] D. H.R. Barton, R.H. Hesse, M.M. Pechet, H.T. Toh, *J. Chem. Soc., Perkin Trans. 1*, **1974**, 732.
[16] W.E. Barnette, *J. Am. Chem. Soc.* **1984**, *106*, 452.
[17] S.H. Lee, J. Schwartz, *J. Am. Chem. Soc.* **1986**, *108*, 2445.
[18] S. Singh, D.D. DesMarteau, S.S. Zuberi, M. Witz, H.N. Huang, *J. Am. Chem. Soc.* **1987**, *109*, 7194.
[19] E. Differding, R.W. Lang, *Tetrahedron Lett.* **1988**, *29*, 6087.
[20] H. Meinert, *Z. Chem.* **1965**, *5*, 64.
[21] T. Umemoto, K. Tomita, *Tetrahedron Lett.* **1986**, *27*, 3271.
[22] T. Umemoto, K. Kawada, K. Tomita, *Tetrahedron Lett.*, **1986**, *27*, 4465.

Part II. Applications in Total Synthesis

A. Synthetic Routes to Different Classes of Natural Products and Analogs Thereof

Synthesis of Hydroxyethylene Isosteric Dipeptides

Rolf Henning

Renin is a highly specific aspartic proteinase, cleaving the circulating serum glycoprotein angiotensinogen between the amino acids leucine-10 and valine-11. Subsequent removal of a C-terminal dipeptide by angiotensin converting enzyme (ACE) produces the octapeptide angiotensin II, which causes an elevation of blood pressure by, amongst others, direct vasoconstriction and stimulation of aldosterone release. Thus, inhibition of the enzyme renin represents a novel therapeutic approach for the treatment of hypertension. The search for renin inhibitors has greatly intensified during the past ten years. [1] The most promising chemical approach is based on the concept of analogy to the tetrahedral transition state intermediate involved in the enzyme-catalyzed hydrolysis of a peptide bond. [2] Structures that correspond to this concept are, among others, statine *(2)* and statine analogues, which already have been described here, [3] and the so-called hydroxyethylene isosteric dipeptides *(3)*. These compounds both represent non-cleavable mimics of the Leu-Val dipeptide *(1)* (Scheme 1).

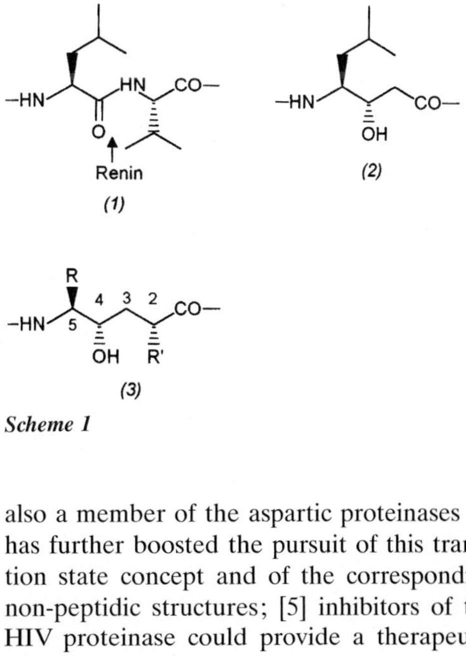

The fact that the virus-coded proteinase of the human immuno-deficiency virus (HIV) is

Scheme 1

also a member of the aspartic proteinases [4] has further boosted the pursuit of this transition state concept and of the corresponding non-peptidic structures; [5] inhibitors of the HIV proteinase could provide a therapeutic opportunity for the treatment of retrovirus-associated diseases such as AIDS.

The vast majority of synthetic approaches to *(3)* starts from chiral α-amino acid derivatives as N-terminal components, thus ensuring the stereochemical integrity and the type of the side chain at C-5. For the further assembly of the dipeptide framework, three strategies can be distinguished:

- Amino acid + C_1-electrophile + C_2-nucleophile (enolate)
- Amino acid + C_1-nucleophile + C_2-electrophile
- Amino acid + C_3-nucleophile (homoenolate)

In many cases, however, complete control of the configuration at C-4 (using aldol-type reactions or reduction) and at C-2 (by alkylation) could not be achieved, thus necessitating a double separation of diastereomers during synthesis.

Pioneering work on a whole series of dipeptide isosters was performed by Szelke and co-workers. The starting point of their synthesis of the hydroxyethylene isoster *(9)*, following the C_1-synthon/enolate strategy, [6, 7] is the diazo ketone *(4)*, itself easily obtainable from L-leucine. Reduction of the ketone *(5)* using sodium cyanoborohydride produces a 3:1-mixture *(2R/2S)* of the diastereomeric alcohols *(6)*. The isovaleryl unit (as substitute for valine) at C-1 is added in a multi-step sequence via a malonate alkylation to give *(8)*. Cleavage of the auxiliary ester function finally furnishes a mixture of the *(2S/2R)*-epimers *(9)* (Scheme 2).

Two shorter routes, while following the same strategy of enolate alkylation, employ different alkylation agents. Evans and co-workers use the aminoalkyl epoxide *(11)*, [8] which is available by the – non-diastereoselective – reaction of Boc-L-phenylalaninal *(10)* with dimethylsulfonium methylide [9] (1:1 epimeric mixture). Malonate alkylation using this epoxide smoothly yields the lactone-ester *(12)* after spontaneous cyclization; unfortunately, acetate enolates do not react with *(11)*. Incorporation of the second side chain in a separate alkylation step provides the key lactone *(13)* as a mixture of the C-2 diastereomers (Scheme 3).

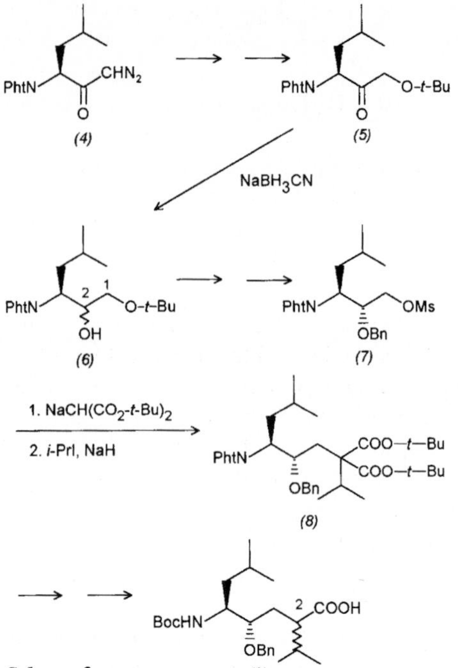

Scheme 2

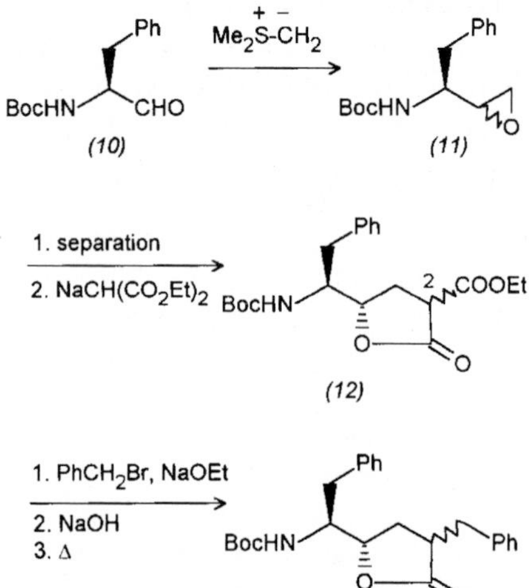

Scheme 3

In a synthesis developed by a Ciba-Geigy group, [10] a different side chain and a modified ylide help to improve the diastereoselectivity of the Corey–Chaykovsky reaction [9] to 5:1 in favor of the desired isomer *(15)*. In contrast to the method described above, *(15)* is not used directly as alkylation agent, but first transformed into the apparently more reactive iodide *(16)*. Reaction with the isovalerate enolate *(17)* in THF/HMPA then affords the dipeptide isoster *(18)* as a 1:1 mixture of the C-2 epimers (Scheme 4).

A significant improvement and simplification of the enolate method was recently achieved by chemists from Merck, Sharp & Dohme. [11] The *direct* coupling of the chiral amide enolate derived from *(20)* with epoxide *(11)* provides the product *(21)* with the desired *(2R)*-configuration. The reaction proceeds with both excellent chemical yield and diastereoselectivity (> 99:1), if the electrophilicity of epoxide *(11)* is enhanced by pre-lithiation of the carbamate moiety (using a second equivalent of butyllithium in a one-pot procedure). Epoxide *(11)* is alternatively accessible via peracid oxidation of the corresponding olefin *(19)*, [11, 12] again with improved diastereoselectivity (86:14) compared with the Corey–Chaykovsky reaction [9] (Scheme 5).

Key step of a synthetic route [13] related to the enolate method is the Wittig–Horner reac-

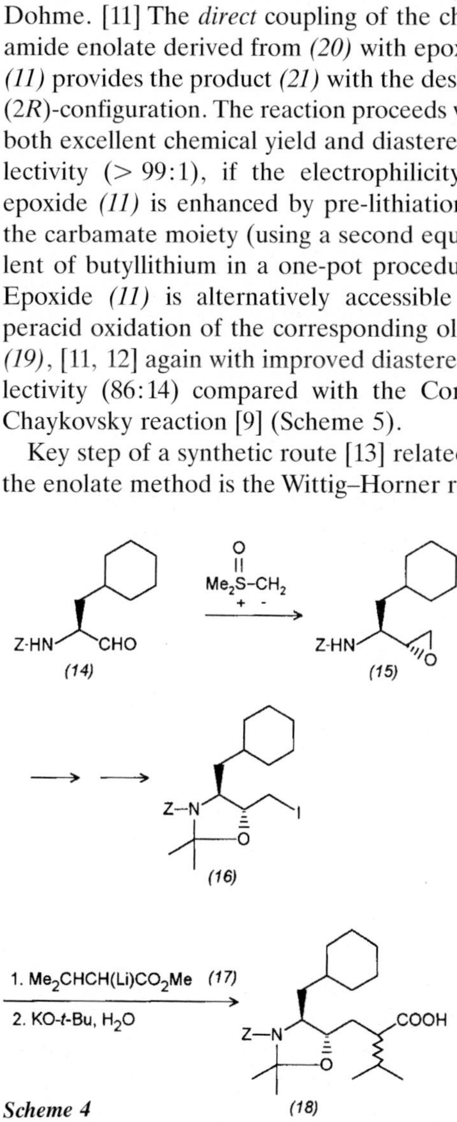

tion of aldehyde *(24)* with phosphonate *(25)*. Aldehyde *(24)* is available by ozonolysis of the vinyl adduct *(23)* and can be equilibrated (7:1) in favor of the desired *anti* isomer. However, catalytic hydrogenation of the double bond in *(26)* proceeds without significant diastereoselectivity with respect to C-2 (Scheme 6).

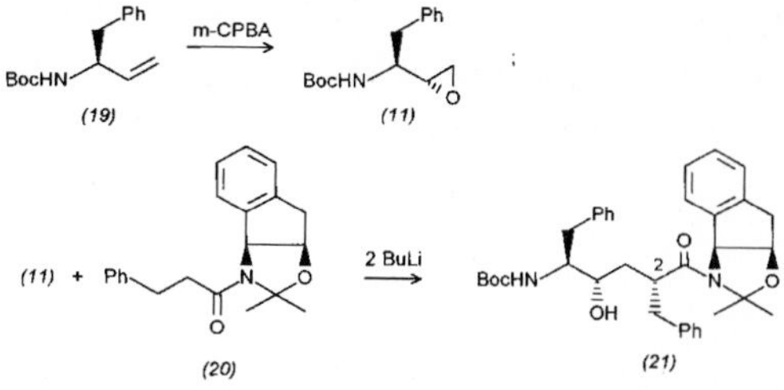

An alternative route [14] to aldehydes of type *(24)* involves the chelation-controlled, diastereoselective (> 10:1) addition of 2-furyllithium *(28)* to aldehyde *(27)* in the presence of zinc bromide. The desired compound *(31)* is then obtained by oxidative cleavage of the furan ring (Scheme 7).

In an original variation of the enolate strategy, [15] *(31)* is first converted into the *dihy-*

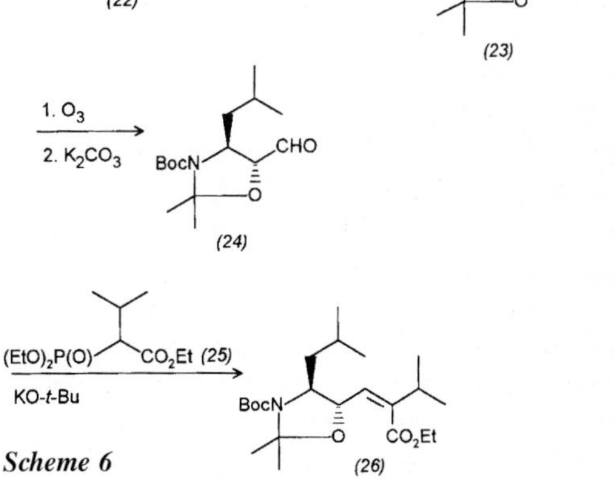

Scheme 5

Scheme 4

Scheme 6

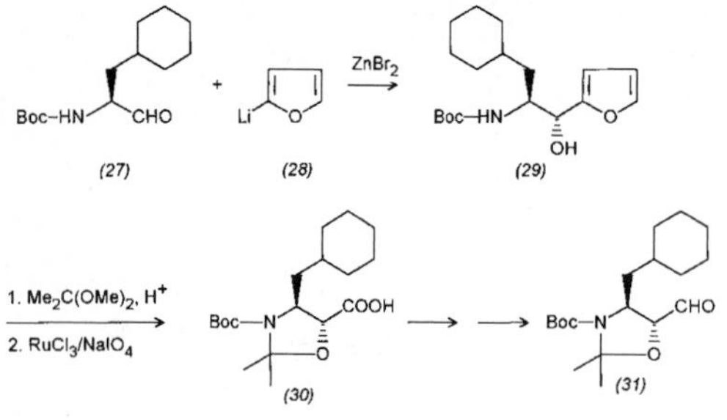

Scheme 7

Scheme 9

*droxy*ethylene isoster *(32)* [16] by a highly diastereoselective aldol reaction; Barton-deoxygenation [17] of the extra hydroxyl group then provides the hydroxyethylene isoster *(34)* with unambiguously defined configuration at C-2 (Scheme 8).

The "Umpolung" variation of the enolate method [18] makes use of the β-keto phosphonate *(36)* which is readily available from amino acid ester *(35)* by a Claisen condensation. Wittig-Horner reaction of *(36)* with α-keto ester *(37)* smoothly yields the enone *(38)*. Subsequent reduction of the keto group, cyclization and hydrogenation lead to the key lactone *(39)*, albeit with unfavorable diastereoselectivity at C-2 and C-4 (Scheme 9).

The use of homoenolate equivalents [19] enables a shortening of the reaction sequence. This strategy has increasingly gained attention during the last years. The first synthesis of this type was developed by Rich. [20] Addition of

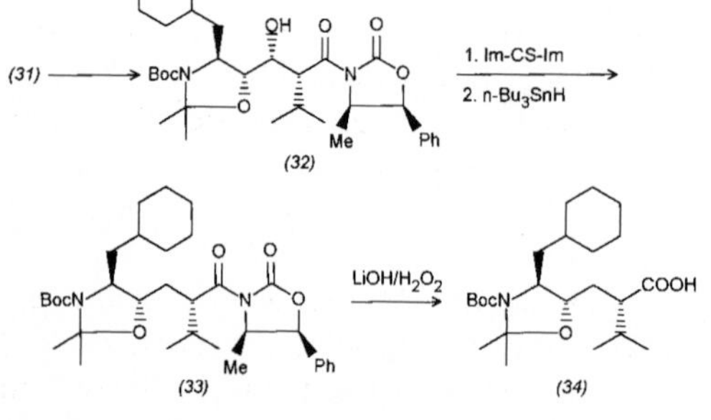

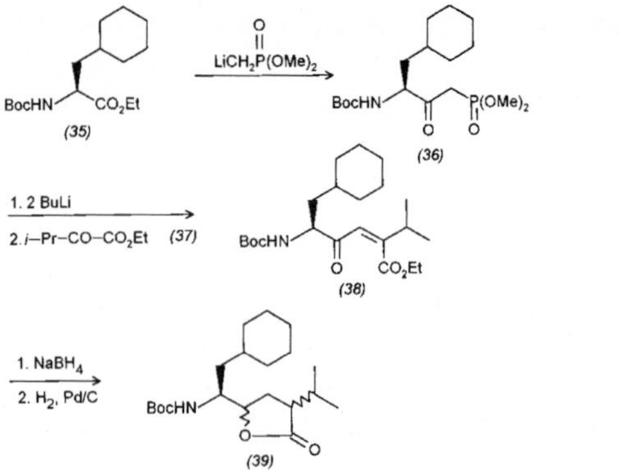

the chiral Grignard reagent derived from *(40)*, which is available in four steps applying the acyl oxazolidinone method of Evans, [21] and which is already endowed with the stereocenter C-2, to aldehyde *(22)* affords a 4:1 mixture *(4S/4R)* of the secondary alcohol *(41)*. Oxidation of the primary hydroxyl group then leads to *(42)* (Scheme 10).

Homoenolate equivalents related to *(40)* include propyl bromide *(43)* [22] and butenyl bromide *(44)*. [23] In these compounds, the phenyl substituent and terminal double bond, respectively, function as a masked carboxyl group on oxidative degradation (Scheme 11).

Kleinman [24] uses propiolate as a homoenolate synthon. Hydrogenation of the adduct *(45)* followed by cyclization furnishes lactone *(46)* as a 4.5:1 mixture in favor of the desired 4S isomer. [25] Reaction of the dianion of *(46)* with methallyl bromide produces the target

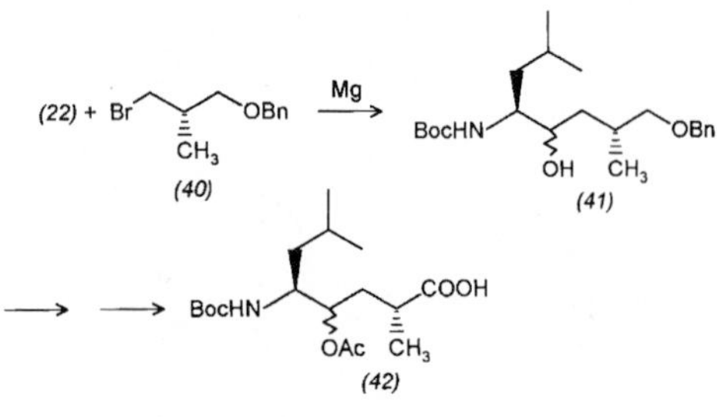

Scheme 8

Scheme 10

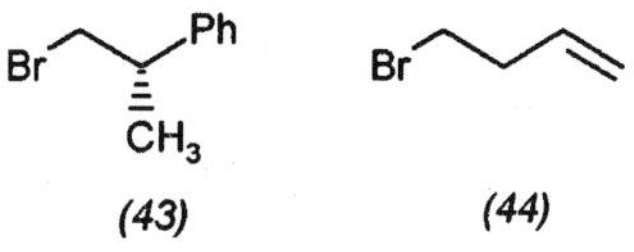

Br—CH(Ph)(CH₃)

(43)

Br—CH₂CH₂CH=CH₂

(44)

Scheme 11

CO–NLiMe

(10) + (48)   ClTi(O-i-Pr)₃ →

Li⁺

(48)

Ph

BocHN— ... —CONHMe   1. separation
OH   2. Δ

(49)

BocHN— ... (lactone with Ph)

(50)

—MgBr, CuCN →

BocHN— ... (lactone with Ph, isopropyl)

(51)

Scheme 13

compound *(47)* as the preferred (15:1) *trans* alkylation product (Scheme 12).

Kempf [26] employs the dianion *(48)* of *N*-methyl methacrylamide as homoenolate equivalent. Titanium(IV)-mediated addition to *(10)* gives a 1.6:1 mixture of hydroxy amide *(49)*. Incorporation of the second side chain can be accomplished by conjugate addition of alkyl cuprates to methylene lactone *(50)*, but diastereoselectivities are rather low (Scheme 13).

Allyltrimethylsilane *(52)* represents an additional homoenolate synthon. [27] Chelation-controlled reaction with aldehyde *(22)* in the presence of tin(IV) chloride preferably (20:1) yields the *threo* isomer *(53)*. Subsequent hydroboration and oxidation then reveal the masked carboxyl function (Scheme 14).

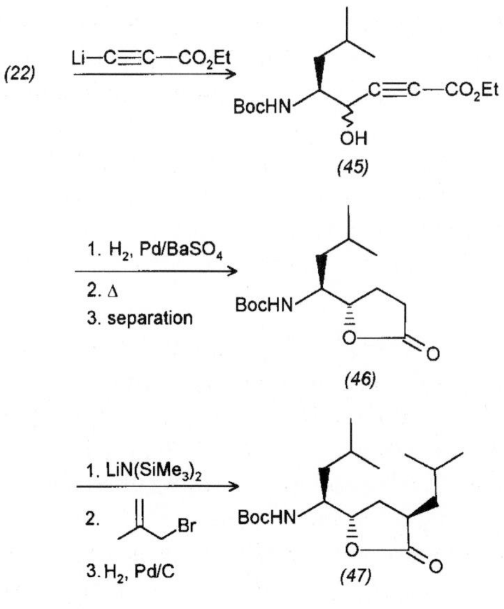

(22)   Li—C≡C—CO₂Et →

BocHN— ... —C≡C—CO₂Et
OH

(45)

1. H₂, Pd/BaSO₄
2. Δ
3. separation

BocHN— ... (lactone)

(46)

1. LiN(SiMe₃)₂
2. (allyl)Br
3. H₂, Pd/C

BocHN— ... (lactone)

(47)

Scheme 12

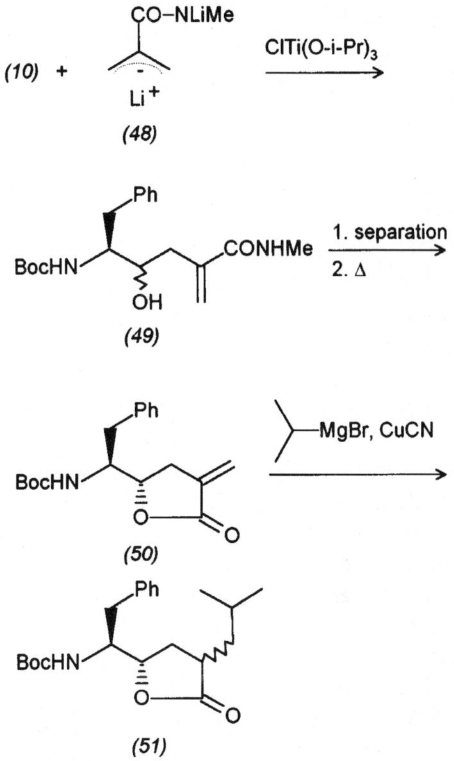

Explicit homoenolates, finally, provide a straightforward access to the central lactone intermediates. [28] An example [28a] is the chelation-controlled addition of titanium homoenolate *(55)* to aldehyde *(10)*, leading to the key compound *(56)* with a (4*S*/4*R*) diastereoselectivity of 16:1 (Scheme 15).

In this type of reaction, extent and direction of diastereoselectivity are strongly influenced by the nature of both amino protecting group and titanium ligands. [28, 29]

In order to unambiguously control the configuration at C-4 *and* C-5, three synthetic approaches use chiral precursors other than simple α-amino acid derivatives. In the first route, [30] statine *(2)* (obtained by hydrolysis of pepstatin) is "degraded" to aldehyde *(24)* [13] in an eight-step sequence. The second route [31] starts from commercially available

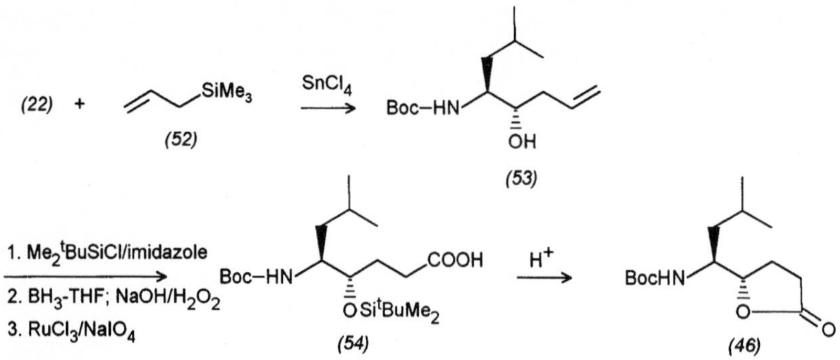

(22) + (52)   SiMe₃ — SnCl₄ → Boc–HN (53) OH

(54) Boc–HN COOH / OSi^tBuMe₂ — 1. Me₂^tBuSiCl/imidazole 2. BH₃-THF; NaOH/H₂O₂ 3. RuCl₃/NaIO₄ — H⁺ → BocHN (46)

Scheme 14

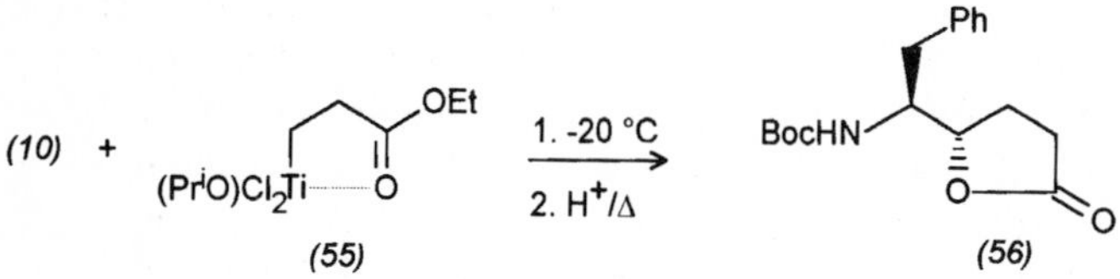

(10) + (55) (Pr^iO)Cl₂Ti····O OEt — 1. -20 °C 2. H⁺/Δ → BocHN (56) Ph

Scheme 15

(2) H₂N COOH OH — 8 steps → BocN CHO (24)

(57) OAc AcO OAc — 12 steps → t–BuMe₂SiO (58)

— 1. separation 2. HCl, MeOH → HO (59)

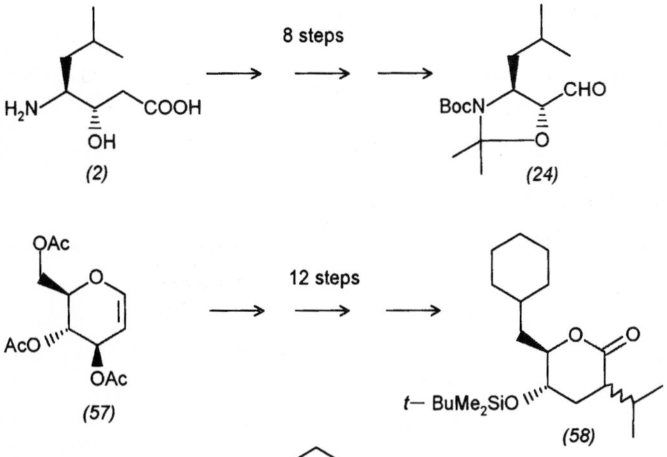

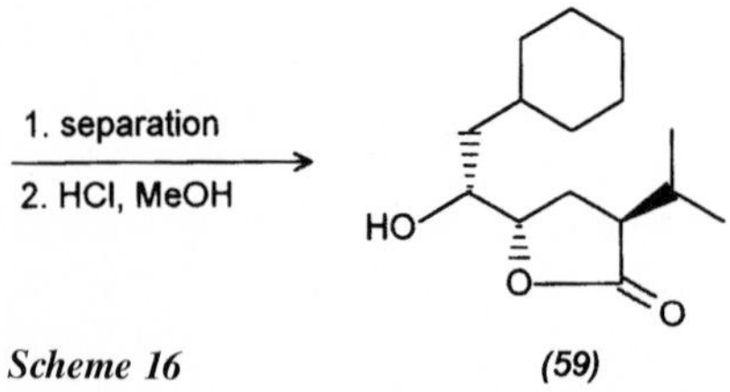

Scheme 16

D-glucal *(57)*, which is transformed in twelve steps to the valerolactone *(58)* and subsequently to the butyrolactone *(59)* (Scheme 16).

The third route [32] employs D-mannose *(60)* as chiral starting material. After a three-fold "deoxygenation" via an eight-step sequence, compound *(61)* is obtained. A regiospecific epoxide opening using phenyl magnesium bromide in the presence of copper(I) iodide then provides the hydroxy lactol *(62)* as the precursor to lactones of type *(13)* (Scheme 17).

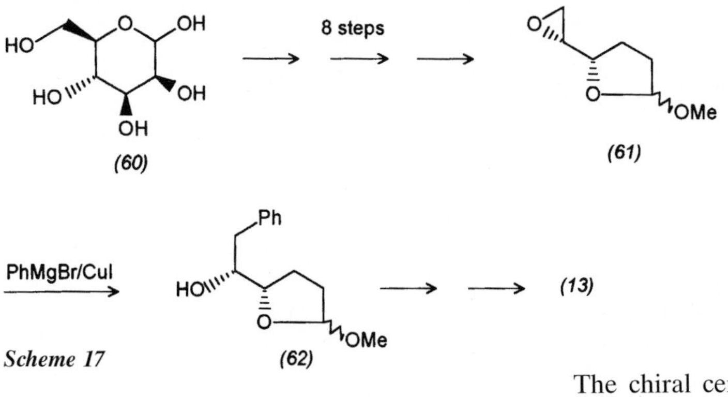

Scheme 17

An asymmetric synthesis of dipeptide isosters *(3)* that does *not* start from amino acid or carbohydrate derivatives, that controls the stereochemistry of all three chiral centers and that allows for a broad variation of the side chains, has been developed [33] by chemists from Ciba-Geigy. The key intermediate is the γ,δ-unsaturated acid *(64)*, readily available by a titanium(IV)-catalyzed Carroll rearrangement of the mixed malonate *(63)*. A classical, two-step resolution procedure provides the desired (*S*)-enantiomer with 90% *ee*. An alternative route is given by the diastereoselective (99:1) alkylation of acyl oxazolidinone *(66)* with allyl bromide *(65)*, following the work of Evans [21] (Scheme 18).

The chiral center in *(64)* now induces the configuration of the two centers that are not yet functionalized (C-4, C-5). The critical step is the diastereo- and regioselective (20:1 and 10:1, respectively) bromo lactonisation of dimethylamide *(67)*, leading to the thermodynamically more stable *trans* product *(68)*. S_N2 substitution at C-5 then gives the dipeptide isoster *(69)* with correct configuration at all

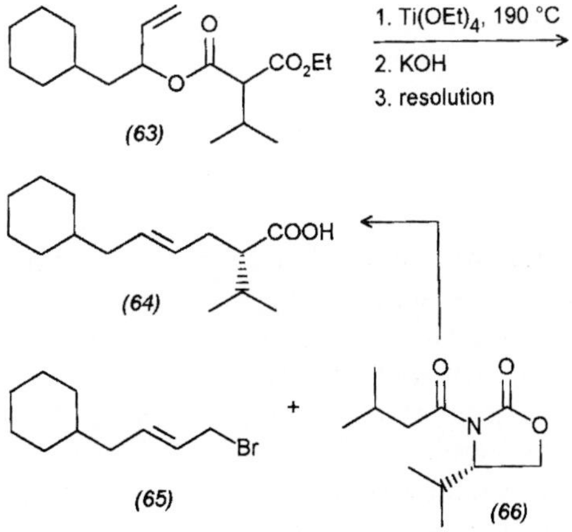

Scheme 18

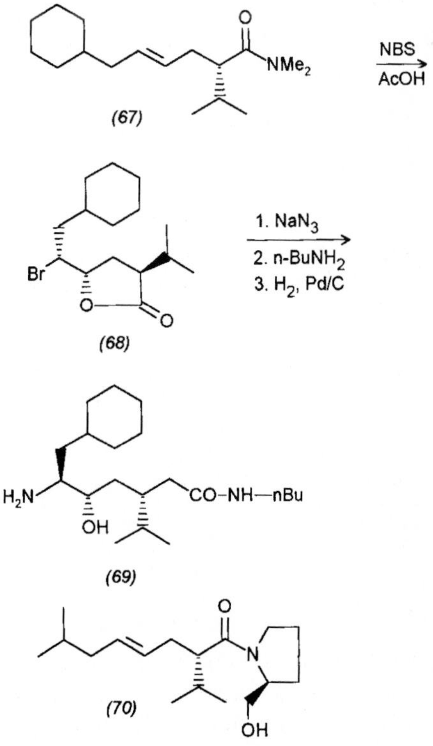

Scheme 19

three stereocenters. A related approach, using the prolinol derivative *(70)*, which also originates from Evans, [34] has been described in the patent literature [35] (Scheme 19).

The interest in selective enzyme inhibitors as potential new therapeutics [36] has steadily increased over the last few years as a consequence of a deeper understanding of biochemical and physiological correlations. This is resulting in new challenges to stereoselective synthesis that can be met with confidence on the grounds of the progressing refinement on synthetic repertoire, as this overview was to illustrate exemplarily.

References

[1] a) J. M. Wood, J. L. Stanton, K. G. Hofbauer, *J. Enzyme Inhibition* **1987**, *1*, 169; b) T. Kokubu, K. Hiwada, *Drugs of Today* **1987**, *23*, 101; c) J. Boger, *Trends Pharmacol. Sci.* **1987**, *8*, 370; d) W. J. Greenlee, *Pharm. Res.* **1987**, *4*, 364; e) W. J. Greenlee, *Med. Res. Rev.* **1990**, *10*, 173.

[2] a) R. Wolfenden, *Nature* **1969**, *223*, 704; b) R. Wolfenden, *Acc. Chem. Res.* **1972**, *5*, 10.

[3] H.-J. Altenbach, *Nachr. Chem. Tech. Lab.* **1988**, *36*, 756.

[4] L. H. Pearl, W. R. Taylor, *Nature* **1987**, *329*, 351.

[5] G. B. Dreyer, B. W. Metcalf, T. A. Tomaszek, T. J. Carr, A. C. Chandler, L. Hyland, S. A. Fakhoury, V. W. Magaard, M. L. Moore, J. E. Strickler, C. Debouck, T. D. Meek, *Proc. Natl. Acad. Sci. USA* **1989**, *86*, 9752.

[6] a) M. Szelke, D. M. Jones, B. Atrash, A. Hallet, B. Leckie, *Proc. 8th Am. Pept. Symp.* **1983**, 579; b) M. Szelke, D. M. Jones, A. Hallet, B. Atrash, EP 118.223, **1984**.

[7] S. L. Harbeson, D. H. Rich, *J. Med. Chem.* **1989**, *32*, 1378.

[8] B. E. Evans, K. E. Rittle, C. F. Homnick, J. P. Springer, J. Hirshfield, D. F. Veber, *J. Org. Chem.* **1985**, *50*, 4615.

[9] E. J. Corey, M. Chaykovsky, *J. Am. Chem. Soc.* **1965**, *87*, 1353.

[10] P. Bühlmayer, A. Caselli, W. Fuhrer, R. Göschke, V. Rasetti, H. Rüeger, J. L. Stanton, L. Criscione, J. M. Wood, *J. Med. Chem.* **1988**, *31*, 1839.

[11] D. Askin, M. A. Wallace, J. P. Vacca, R. A. Reamer, R. P. Volante, I. Shinkai, *J. Org. Chem.* **1992**, *57*, 2771.

[12] J. R. Luly, J. F. Dellaria, J. J. Plattner, J. L. Soderquist, N. Yi, *J. Org. Chem.* **1987**, *52*, 1487.

[13] P. G. M. Wuts, S. R. Putt, A. R. Ritter, *J. Org. Chem.* **1988**, *53*, 4503.

[14] M. A. Poss, J. A. Reid, *Tetrahedron Lett.* **1992**, *33*, 1411.

[15] S. A. Boyd, R. A. Mantei, C. Hsiao, W. R. Baker, *J. Org. Chem.* **1991**, *56*, 438.

[16] S. Thaisrivongs, D. T. Pals, L. T. Kroll, S. R. Turner, F. Han, *J. Med. Chem.* **1987**, *30*, 976.

[17] D. H. R. Barton, S. W. McCombie, *J. Chem. Soc., Perkin Trans. 1* **1975**, 1574.

[18] P. K. Chakravarty, S. E. de Laszlo, C. S. Sarnella, J. P. Springer, P. F. Schuda, *Tetrahedron Lett.* **1989**, *30*, 415.

[19] a) N. H. Werstiuk, *Tetrahedon* **1983**, *39*, 205; b) D. Hoppe, *Angew. Chem.* **1984**, *96*, 930; *Angew. Chem. Int. Ed. Engl.* **1984**, *23*, 932.

[20] a) M. W. Holladay, D. H. Rich, *Tetrahedron Lett.* **1983**, *24*, 4401; b) M. W. Holladay, F. G. Salituro, D. H. Rich, *J. Med. Chem.* **1987**, *30*, 374.

[21] D. A. Evans, M. D. Ennis, D. J. Mathre, *J. Am. Chem. Soc.* **1982**, *104*, 1737.

[22] D. M. Jones, B. Nilsson, M. Szelke, *J. Org. Chem.* **1993**, *58*, 2286.

[23] G. B. Dreyer, D. M. Lambert, T. D. Meek, T. J. Carr, T. A. Tomaszek, A. V. Fernandez, H. Bartus, E. Cacciavillani, A. M. Hassell, M. Minnich, S. R. Petteway, B. W. Metcalf, M. Lewis, *Biochemistry* **1992**, *31*, 6646.

[24] A. H. Fray, R. L. Kaye, E. F. Kleinman, *J. Org. Chem.* **1986**, *51*, 4828.

[25] T. Nishi, M. Kataoka, Y. Morisawa, *Chem. Lett.* **1989**, 1993.

[26] D. J. Kempf, *J. Org. Chem.* **1986**, *51*, 3921.

[27] J. V. N. Vara Prasad, D. H. Rich, *Tetrahedron Lett.* **1990**, *31*, 1803.

[28] a) A. E. DeCamp, A. T. Kawaguchi, R. P. Volante, I. Shinkai, *Tetrahedron Lett.* **1991**, *32*, 1867; b) S. Kano, T. Yokomatsu, S. Shibuya, *ibid.* **1991**, *32*, 233.

[29] a) M. T. Reetz, M. W. Drewes, A. Schmitz, *Angew. Chem.* **1987**, *99*, 1186; *Angew. Chem. Int. Ed. Engl.* **1987**, *26*, 1141; b) M. T. Reetz, M. W. Drewes, K. Lennick, A. Schmitz, *Tetrahedron Asymmetry* **1990**, *1*, 375.

[30] a) M. Nakano, S. Atsuumi, Y. Koike, S. Tanaka, H. Funabashi, J. Hashimoto, M. Ohkubo, H. Morishima, *Bull. Chem. Soc. Jpn.* **1990**, *63*, 2224; b) S. Atsuumi, M. Nakano, Y. Koike, S. Tanaka, K. Matsuyama, M. Nakano, H. Morishima, *Chem. Pharm. Bull.* **1992**, *40*, 364.

[31] a) M. Shiozaki, T. Hata, Y. Furukawa, *Tetrahedron Lett.* **1989**, *30*, 3669; b) M. Shiozaki, Y. Kobayashi, T. Hata, F. Furukawa, *Tetrahedron* **1991**, *47*, 2785.

[32] A. K. Ghosh, S. P. McKnee, W. J. Thompson, *J. Org. Chem.* **1991**, *56*, 6500.

[33] P. Herold, R. Duthaler, G. Rihs, C. Angst, *J. Org. Chem.* **1989**, *54*, 1178.

[34] D. A. Evans, J. M. Takacs, *Tetrahedron Lett.* **1980**, *21*, 4233.

[35] J. B. Hester, D. T. Pals, H. H. Saneii, T. K. Sawyer, H. J. Schostarez, R. E. Tenbrink, S. Thaisrivongs, EP 173.481, **1986.**

[36] M. Sandler, H. J. Smith (Eds.), *Design of Enzyme Inhibitors as Drugs*, Oxford University Press, New York, **1989**.

Synthesis of Natural Products for Plant Protection

Hans-Peter Fischer

Introduction

To provide acceptable solutions for crop production and to be successful as a supplier of pest management solutions, it is necessary to meet the steadily enhanced requirements that are requested by users of pesticides and consumers of agricultural goods. Besides the criteria for activities as herbicides, fungicides, insecticides and parasiticides, and for the exposure and risk assessments based on toxicological studies, much emphasis must be given to ecological considerations in research strategies directed to future agricultural products. [1]

New technologies and product profiles in plant protection markets aim at integrated pest management strategies (IPM). This implies the following:
- Low dose rates to reduce environmental impact. [2]
- Minimized degradation time to reduce accumulation in the ecosystem, including prevention of leaching into the ground water. [3]
- Optimal physico-chemical properties of applied chemicals with special consideration of soil migration and evaporation.
- A selective mode of action targeted solely at the pest organism and simultaneously exhibiting a high safety margin and tolerance in the environment for beneficial organisms, users and consumers.
- A target-directed placement of products by new, innovative, user-friendly formulation, packaging and application technology. Active ingredients must be effective when used at forecast and with threshold levels.
- A synergy with biological control methods, traditional breeding and gene technology [4] and with a fit into antiresistance management and appropriate agricultural techniques. [5]
- In addition, supporters of "soft" technologies suggest basing future pesticide production on renewable resources and safe processes with minimum liberation of persistent wastes. [6]

These challenges must be carefully balanced with economic criteria for the farmers, the distributors, the producers and the consumers.

Many of the desirable properties would clearly point to a renaissance of natural product chemistry and efficient fermentation processes, which are based on natural feedstocks. Indeed, many natural products catabolize and bioreintegrate into metabolic cycles rapidly, and accumulation of toxic residues can be avoided. A number of examples demonstrate that high pesticidal activities can be detected with microbial and plant metabolites. [7] Commercial products of the most important class of agrochemicals derived from natural products, the pyrethroid esters, are easily degraded and mineralized. [8] However, the toxicity of natural constituents may not be more favorable when compared with that of synthetic chemicals. A sophisticated statistical evaluation of these two groups with different origins was published recently. [9] Thus, similar criteria for the responsible selection of appropriate development candidates must be applied for natural and synthetic products. Nevertheless, environmental considerations might be the main driving force to increase efforts in metabolite research.

Many reviews give evidence of enormous efforts to change agricultural application practices to nature-based methods. [10] In order to arrive at these ambitious objectives, a stage plan must be developed and a critical mass for R & D must be reached. [11] Challenges of a natural product discovery programme include the following:

- To find new natural, active ingredients by systematic screening of broths and extracts of microbial cultures and of plant extracts. [11]
- To discover leads and templates, whose properties can be improved by chemical modifications. [11]

- To find new plant protection principles by studying chemical and ecological interactions between organisms and by understanding the underlying modes and mechanisms of action of signal compounds, natural regulators and metabolites that enhance resistance, induce defense and assist in the survival of the species. [12]
- To develop economic (enantioselective) processes with minimum waste precipitation using recycling or bioremediation and fermentation/biotransformation technology using biomass as media.
- To increase the market penetration with profitable products of natural origin from a level of $\approx$ \$ 1.9 billion ($\approx$ 8 % of the chemical plant protection market) to $\geq$ 12 % in the year 2000. Estimated market figures of natural products per se and compounds that are based on natural templates are given in Table 1.

Table 1. Estimated sales of natural products and mimics for agriculture

Use	Natural Products and Mimics	Estimated Sales 1991 $\$ \cdot 10^{-6}$
Herbicides	Phosphinothricin (1) Bialaphos[1] (2)	90
Fungicides / Bactericides	Diverse Antibiotics[1][13]	50
Insecticides / Acaricides	Abamectin[1] (45) Pyrethrins (Hygiene) Bac. thuringiensis preparations[1] Pyrethroids Juvenoids	160 100 1500 100
Animal Parasites	Avermectins[1] Milbemycins[1]	850
	Total market of natural product[2] (incl. mimics)	850

[1] Fermentation dependent products.

Production

Biotechnical processes, fermentations, and biotransformations compete with (enantioselective) synthesis for the production of natural products. In specific situations they may complement each other. Decisive factors for the selection of an industrial process are primarily based on economic considerations in production, available technical expertise of a producer, and patent coverage.

In the future chiral products might be developed more frequently. [14] Continuous progress in finding new economically feasible enantioselective synthetic methods has been observed. [15] In academia, the intellectual challenge to make a complex molecule with sophisticated sequences is still attractive. Examples are outlined in the following chapters and approaches of several investigators are compared. In the industrial environment, however, the use of cheap starting materials and reagents, a small number of steps and acceptable levels of investment are decisive. Because knowledge about biotransformations and how to recover metabolites is steadily increasing, there might be a shift toward better economic pre-conditions for fermentation processes in the agrochemical industry. [16] This trend is also likely because process safety, bioremediation and other appropriate ecologically acceptable waste disposal methods will absorb additional investment for new manufacturing plants based on chemical steps. [17]

As a result of the cost limitations of agricultural products, it is a difficult mission for industry to manufacture biologically active natural products competitively. Unit costs of only $ 10–20/acre/treatment are usual in the marketplace. These figures can be reached only, if natural products are found that can be applied at low dose rates of below 100 g/acre and production costs can be reduced, e.g. to less than $ 200/kg. [11]

Phosphinothricin

Phosphinothricin *(1)* is the hydrolysis product of the dipeptide bialaphos *(2)* and of phosalacine *(3)*. These metabolites were isolated from culture broths of *Streptomyces viridochromogenes* and *S. hygroscopicus* as well as of *Kitasatosporia phosalacinea sp.* nov. KA-338. [18] The racemic ammonium salt of *(1)* was introduced by Hoechst under the names glufosinate, HOE 39'866, Basta®, and the single isomer *(2)* under the names bialaphos, MW 801, MW 831, SF-1293 Na, Meiji Herbiace®, all as contact herbicides with a broad spectrum of weed control. [19]

Several syntheses of the racemate *(1)* were published before synthetic methods for the active *(S)*-isomer were investigated. [20] An efficient enantioselective synthesis by Zeiss [21] is based on Schöllkopf's amino acid synthesis via metallated bis-lactimethers. The lithium salt of the *(3R)*-bis-lactimether *(4)* is alkylated with isobutyl-2-chloroethyl-methylphosphinate *(5)* at –78 °C in THF with high enantioselectivity. After hydrolysis of *(6)* the natural *(S)*-(+)-phosphinothricin *(1)* is isolated with an optical purity of 93 % *ee* and a total yield of 50 %. The alternative strategy of Minova [22] is based on the asymmetric Michael addition of the potassium enolate *(10)*, made by metallation of the Schiff base *(9)*, to O-methyl-methyl-vinylphosphinate *(11)*. Compound *(9)* is easily obtained by reaction of glycine with the chiral auxiliary *(S)*-2-hydroxypinanone *(8)*. The amino acid *(1)* is obtained, after hydrolysis of *(12)* in 79 % optical purity. Recently *(1)* was also conveniently synthetized in six consecutive steps from L-aspartic acid with 94 % *ee* and 35 % overall yield. [23] *(S)*-Phosphinothricin *(1)* was also prepared in 70 % yield by biotransformation from racemic 4-(ethoxymethylphosphinyl)-2-acetamido-butanamide *(13)* by consecutive treatment with the commercial enzymes phosphodiesterase I, acylase I and glutaminase in 70 % yield. [24] In another bioconversion *(1)* was obtained by stereospecific transamination of 2-oxo-4-(hydroxy (methyl) phosphinyl) butyric acid with a phosphinothricin specific aminotransferase from *E. coli* K-12 and a number of amino-group donors. [25] Further progress in improving fermentation yields of bialaphos *(2)* can be expected based on the brilliant biosynthetic studies of Seto et al. [26] Since the reaction conditions and the details of the scale-up trials are trade secrets, it is difficult to give a judgement on the prefered industrial solutions (Scheme 1).

Pyrrolnitrin Analogs

The effectiveness of microbial antagonists as biocontrol organisms is based on a combination of several modes of interaction between a pathogen and the host. [5] In some well investigated examples it could be shown that the formation of certain antibiotics could be responsible for the dominance of a disease controlling biological over the pathogenic invader. [11] Antagonistic strains of *Pseudomonas fluorescens* with activity against *Rhi-*

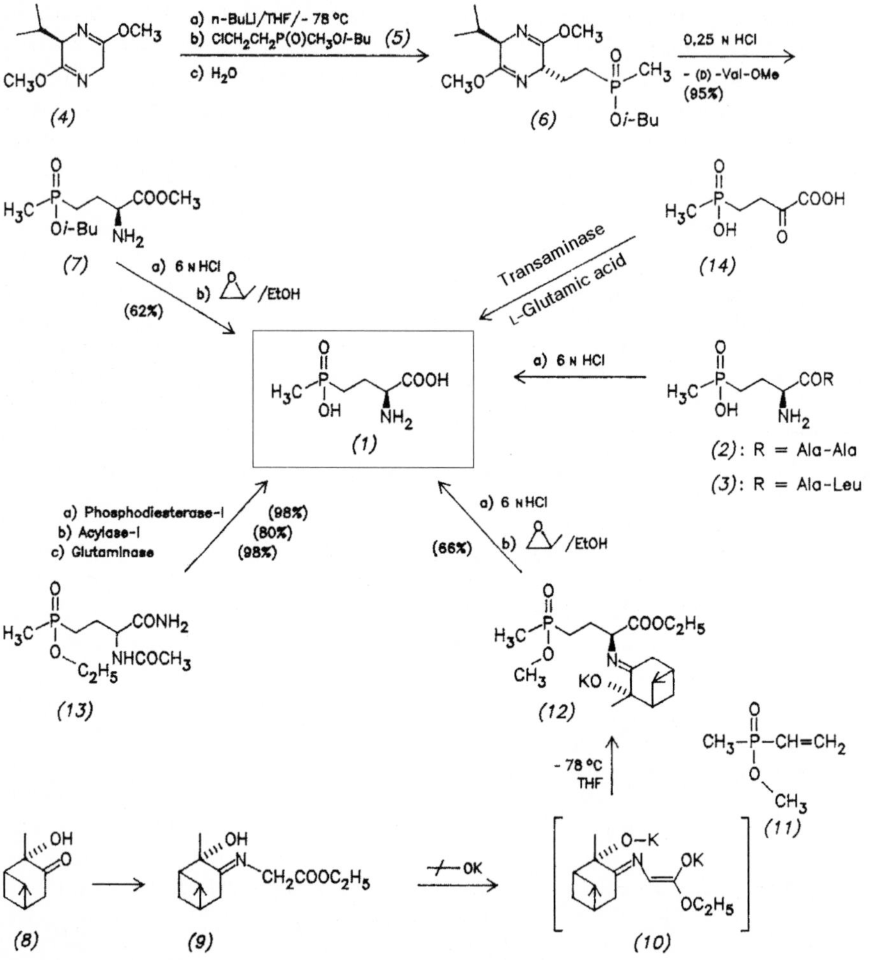

Scheme 1. Enantioselective syntheses of *(S)-(+)-phosphinothricin (1)* by Zeiss [21], Minowa [22], Natchev [24] and Schulz [25].

zoctonia solani are producing a number of antibiotics, among them pyrrolnitrin *(15)*. [27] By applying pyrrolnitrin *(15)* the control of a number of pathogens in the rhizosphere of cotton and an increase of the survival rate of seedlings is achieved.

The first synthesis of *(15)* involved six steps and gave an insufficient yield for making the antibiotic economically. [28] Later Nippon Soda described a two step synthesis to prepare the analog *(16)*. [29] In field trials, however, 3-chloro-4-phenyl-pyrrols like *(16)* proved to be too photolabile. [11, 30] A new synthetic approach by Van Leusen et al. opened an easy preparation method for photostable mimics of *(15)*, 3-cyano-4-phenyl-pyrroles, like *(18)*. [31] By using TosMIC *(17)* as a reagent preparation of analogeous α,α'-unsubstituted pyrrols and an optimization of the fungicidal activities became possible (Table 2). The [2+3]-cycloaddition opened a way to optimize the fungicidal activity of this class. Van Leusen's innovative synthetic step *(17)* → *(18)* led finally to a cereal seed dressing agent Beret® *(18)* [32] and Saphire®, a fungicide for soil and leaf-borne diseases. [11, 30]

Strobilurins

Strobilurin A *(19a)* was first isolated by Musilek et al. under the name of mucidin from the fungus *Oudemansiella mucida*. [33] Independently, Steglich and Anke purified strobilurin A *(19a)* and B *(19b)* from *Strobilurus tenacellus*. The antibiotic family was complemented by the isolation of related β-methoxyacrylates from other basidiomycetes and of the related oudemansins *(20)* (Table 3). [34] The strong fungicidal activities attracted a number of

Table 2. Comparison of syntheses of 3,4-disubstituted fungicidal pyrroles according to Gosteli [28], Ueda et al. [29] and Van Leusen [31].

Starting Materials; Critical Step	Steps (Yield)	Product	Light Stability (τ-Suntest lamp)
	6 (20%)	*(15)*	
	2 (20%)	*(16)*	0,5 h
[2+3]-cycloaddition	1 (80%)		48 h

Table 3. Structure of natural fungicidal β-methoxyacrylates [34].

Strobilurins			Oudemansins			Synthetic Mimic
(19)	R	R'	(20)			(21)
A	H	H	A			
B	MeO	Cl	B			
C	PrenylO	H				
X	H	MeO	X			

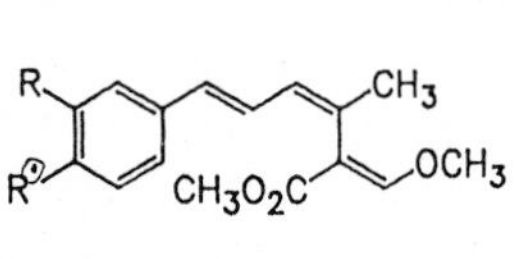

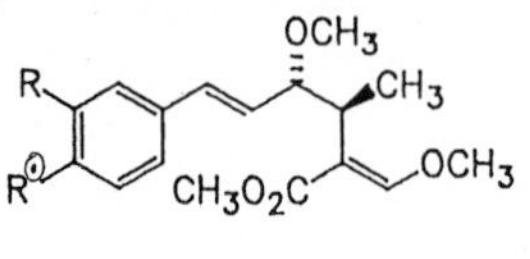

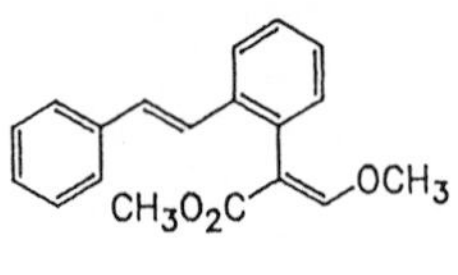

groups to carry out total syntheses and the synthesis of mimics.

The strategy of Sutter [35] for building the polyolefinic side chains of strobilurin B *(19b)* was based on the application of consecutive Wittig reactions (Scheme 2). Since both geometrical isomers of the C=C bonds are normally formed, the yield of the final product *(19b)* was very low, and the separation of the isomers by chromatography was laborious. However, all isomers were wanted for struc-

ture–activity considerations. The natural *(E,Z,E)*-isomer *(19b)* was most active in vivo against phytopathogens. Through the systematic modifications by Steglich it also became evident that the *(E)*-methoxyacrylate substructure was essential for high activity. Due to the photolability of the natural products only a short residual activity was found in field trials. Thus, mimics of the natural template like *(21)* were synthesized using Wittig technology (Scheme 3). A rich patent literature

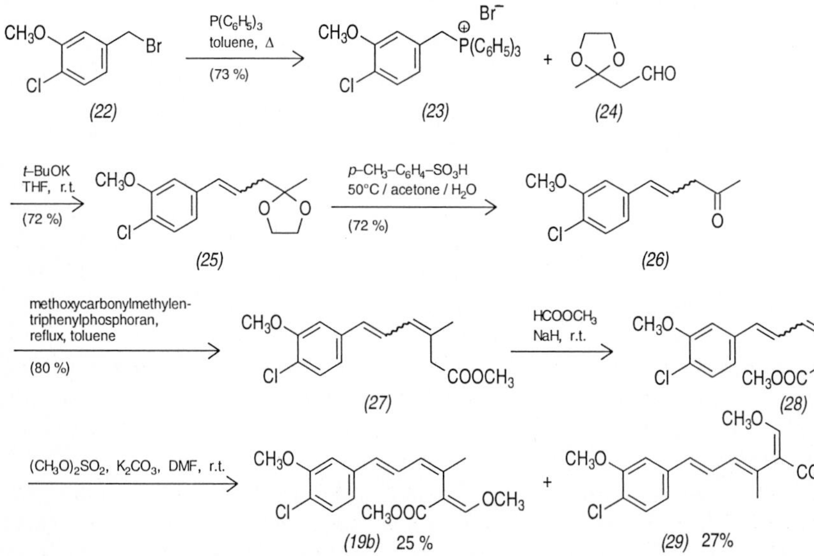

Scheme 2. Synthesis of strobilurin B *(19b)* [35].

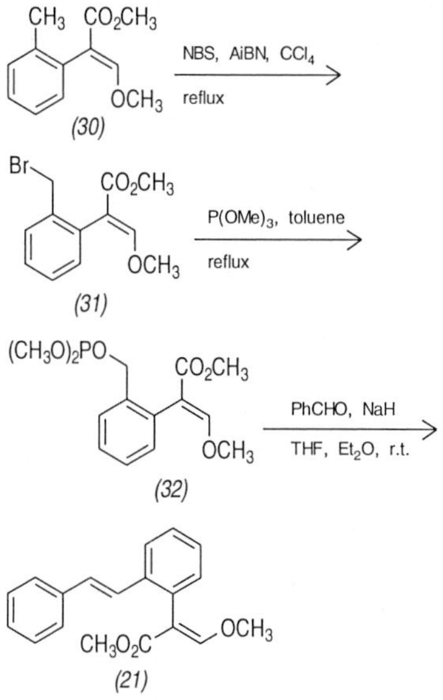

(30)

NBS, AIBN, CCl₄
reflux

(31)

P(OMe)₃, toluene
reflux

(32)

PhCHO, NaH
THF, Et₂O, r.t.

(21)

Scheme 3. Synthesis of strobilurin-mimic *(21)* by Steglich [34].

from ICI, BASF, Maag and Ciba documents the high interest of pesticide manufacterers in this new class of fungicides. Two photostable development candidates *(33)* and *(34)* were recently presented at the British Crop Protection Conference in Brighton. [36]

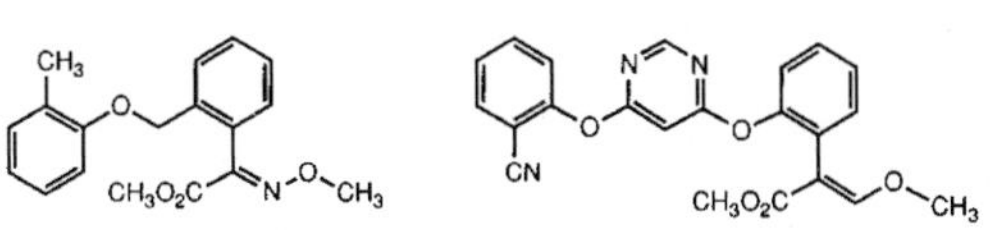

BAS 490 F (33)   ICI A 5504 (34)

Hydantocidin and Ribantoin

Additional synthetic work in the area of herbicides is exemplified with a metabolite called (+)-hydantocidin *(35)* which was isolated from *Streptomyces hygroscopicus*, SANK 63′584 by Sankyo. [37] The same structure was independently characterized from *S. hygroscopicus*, TÜ-2474, by Ciba and named (+)-ribantoin. [38]

(+)-Hydantocidin is the only one of 16 possible diastereoisomers with the four contiguous stereogenic centers that showed strong herbicidal activity. [39] In the retrosynthetic strategy of Mirza tribenzyl-oxyribose *(36)* was used as chiral starting material. [40] The first step was based on the Bucherer reaction *(36)* → *(37)* + *(38)*. The natural isomer *(35)* was made in six steps, as shown in Scheme 4. It was interesting to note that Sankyo's several synthetic strategies were distinct from Mirza's approach with exception of the final *cis*-dihydroyylation step *(42)* → *(35)* + lyxoisomer *(44)*. A completely different approach to the spironucleoside *(35)* by Chemla is based on the readily available 1,2:3,4-di-*O*-isopropylidene-D-psicofuranose as a chiral starter unit using a new oxygen-bridged intramolecular Vorbrüggen coupling. [41]

The production of ribantoin *(35)* by fermentation was inferior to the total synthesis due to very low fermentation yields and difficulties in separating ribantoin from other metabolites, such as desoxy-xylitol, 4-hydroxyphenyl-glyoxylic acid and homomycin. [11]

Avermectins/Emamectin – MK 244 and MK 243/Milbemycins

A great deal of attention was drawn by papers of Merck, Sharp and Dohme, in which 4″-deoxy - 4″ - *epi* - methylamino - avermectin B₁ hydrochloride, MK 244, emamectin *(48)* and MK 243 *(49)* were described. The semisynthetic derivative MK 244 *(48)* shows up to a 1600-fold improvement in potency against lepidopteran larvae. [41] With MK 243 *(49)* it was observed that 0,02 ppm led to 100 % mortality of *Spodoptera eridiana*. [42]

The successful synthesis of *(48)* and *(49)* started from avermectin B₁ *(45)* by the oxidation of the 4″-hydroxy group of the α-L-oleandrosyl -α - L - oleandrosyl - disaccharide-

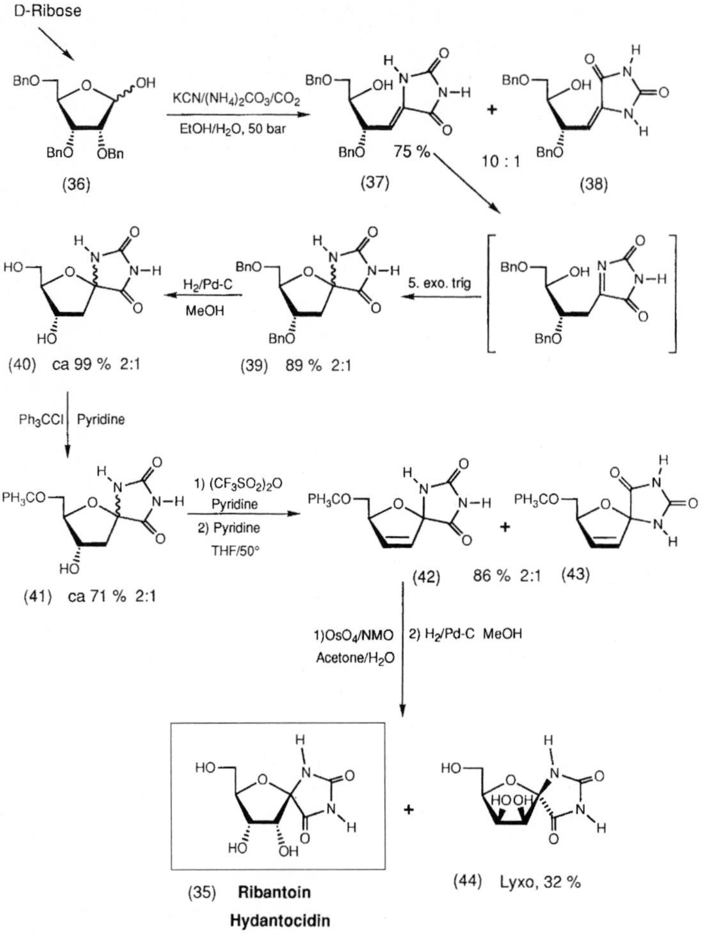

Scheme 4. Synthesis of ribantoin *(35)* by the Bucherer route [40].

component by Swern-oxidation, after the more reactive 5-hydroxy group was protected as the *O-tert*-butyl-dimethylsilylether *(46)*. Reductive amination of *(47)* with CH₃NH₂ and NaCNBH₃ and cleavage of the protecting group in the 5-position gave a mixture of the 4″-methylamino group, and MK 244, with the equatorial 4″-methylamino group, were isolated (Scheme 5).

The endoparasiticidal, acaricidal and insecticidal metabolites of the avermectin family are industrially produced by fermentation of optimized cultures of *Streptomyces avermitilis*,

for example, MA 4848 (ATCC 31′271). The mixture of avermectin B_{1a} and B_{1b}, abamectin, MK 936, which is less active in insect/mite control than the semisynthetic compounds *(48)* and *(49)* is formed in an 80:20 ratio. Ivermectin *(50)* is isolated by hydrogenation of the double bond in position 22–23 of *(45)*. This important product is a systemic endoparasiticide in animal health and can be used against tropical infections of humans with filaria, such as *Onchocerca volvulus*. [43]

The total synthesis of avermectins *(45)* or their aglycons *(51)* and of the related milbe-

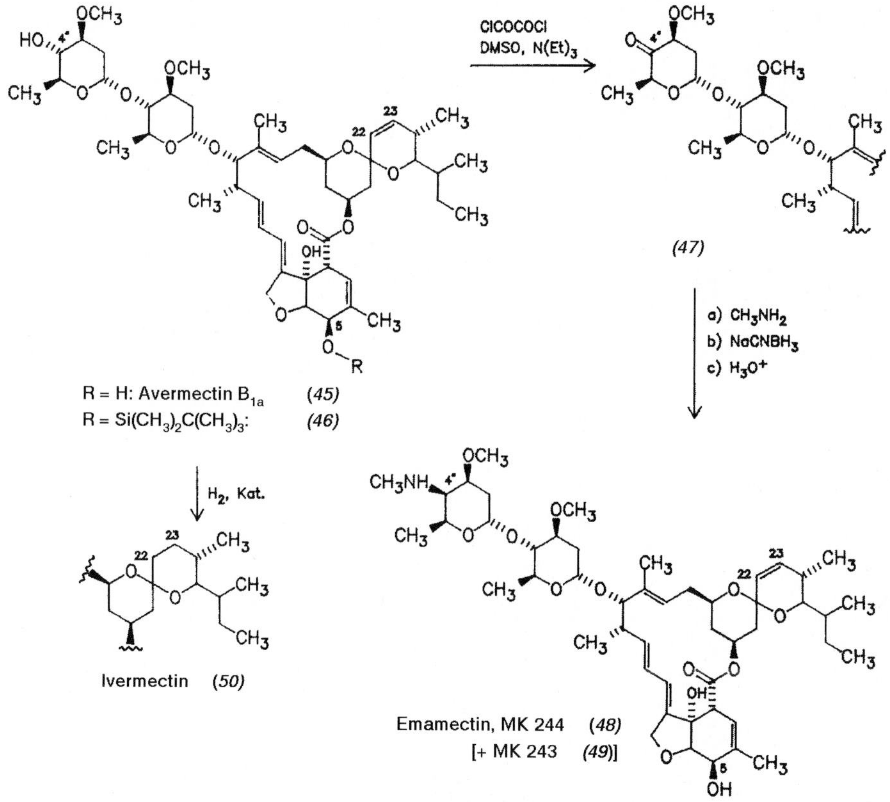

R = H: Avermectin B$_{1a}$ (45)
R = Si(CH$_3$)$_2$C(CH$_3$)$_3$:   (46)

(47)

a) CH$_3$NH$_2$
b) NaCNBH$_3$
c) H$_3$O$^+$

Ivermectin   (50)

Emamectin, MK 244   (48)
[+ MK 243   (49)]

Scheme 5. Modifications of avermectin B$_{1a}$ (45) to MK 244 (48) [41].

mycins, 13-deoxyavermectin-aglycons like (52), belong to the highlights of organic synthesis. New reactions were used and discovered in the course of these objectives. [44, 45] In this paper only the retrosynthetic strategy for the total synthesis of avermectin B$_{1a}$-aglycon by Hannesian's group is exemplified (Scheme 6). The stereochemical analysis of possible templates as chiral synthons was conceptually combined with enantionioselective synthetic steps. The synthesis was indeed started on the basis of the decoded chiral starter units, suggested by a computer program, and finalized successfully in several steps. [44, 46] In the course of these investigations a user-friedly computer program called "CHIRON" was developed, which assists che-

mists in the identification of useful chiral synthons for the synthesis of complex natural products. This software could be successfully applied to investigate alternative synthetic strategies for different industrial scale up processes of chiral pesticides.

Whereas the synthesis of milbemycins by degradation of the corresponding ivermectin-aglycons (56) → (57) and (52) is rather trivial, [46] it is more laborious to find methods for the reverse conversion, the regioselective oxidation of milbemycins (52) in the 13-position. [47] The direct diastereoselective microbial hydroxylation of milbemycin A$_4$ (52), for example with *Streptomyces violascens* (ATCC 31,560), to 13-β-hydroxymilbemycin (53), was successfully achieved in 92 % yield. [48] Com-

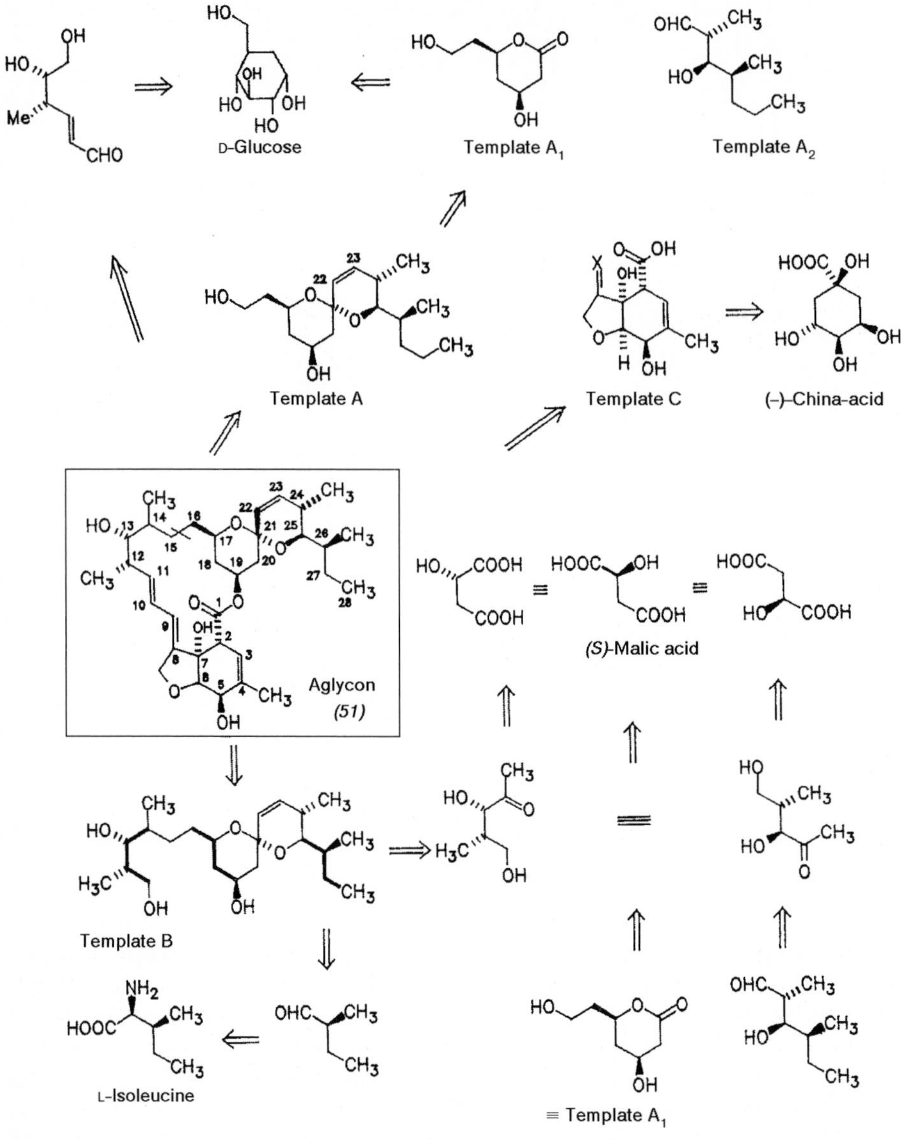

Scheme 6. Synthesic planning of avermectin B_{1a}-aglycon *(51)* from simple chiral building blocks by retrosynthetic analysis [44].

pound *(53)* can easily be transformed via *(55)* to *(56)* (Scheme 7). The β-epimer *(53)* can also be used to investigate biological activities of recently discovered 13-β-hydroxy-milbemycin derivatives. [49] The biotransformation process *(52)* → *(53)* also opened the possibility to make avermectin analogs from milbemycin A_4 producing microorganisms.

The avermectin and milbemycin [50] families find application as excellent parasiticides and insecticides in animal health, plant protection (Table 1) and, because of their activity

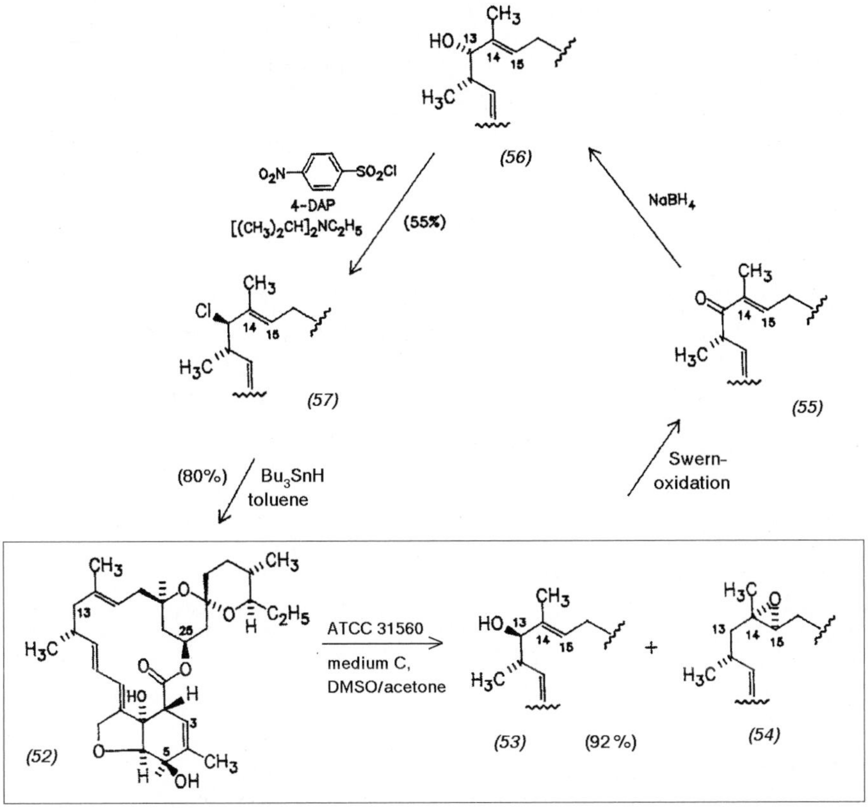

Scheme 7. Microbial oxidation of milbemycins *(52)* [47, 48].

against filaria, they contribute to the wellbeing of people with filarioses infections ("river blindness") in the Third World.

Additional Contributions to Plant Protection

Chemical synthesis and biotechnical processes offer a broad variety of alternatives for the production of useful secondary metabolites. This paper concentrates on a selection of important results within areas of potential commercial application to plant protection. Other recent synthesis results relating to pest control and regulators are impressive and con-

tribute important ideas to synthetic strategies and methodologies. They are focussed on synthesis, mimics and derivatives of antibiotics such as soraphen A (G. Höfle [51]), restrictici-din (S. Jendrzejewski [52]), validamycins (S. Ogawa [53]), β,γ-unsaturated amino acids (R. Duthaler et al. [54]), on plant growth regulators and herbicidal compounds like gibberellic acid GA$_3$ (Y. Yamada, L. N. Mander [55]), strigol (U. Berlage and O. D. Dailey [56]), herbimycin A (M. Nakata [57]), aniso-mycin (I. Felner, R. Ballini [58]), on insecticides and antifeedants like lepicidine A (D. A. Evans [59]), azadirachtin (S. V. Ley [60]), pyrrolomycin mimics (American Cyan-amid [61]), nikkomycins (H. Zähner,

W. A. König, A. G. M. Barret [62]), haedoxan (E. Taniguchi [63]), moncerin (K. Mori, M. Dillon [64]), philanthotoxins and andrimid (K. Nakanishi, A. V. R. Rao [65]), jasplakinolide (K. S. Chu [66]), allosamidin (A. Vasella, S. J. Danishefsky, S. Takahashi [67]), rocaglamid (R. J. K. Taylor, G. A. Kraus, B. M. Trost [68]) and the syntheses of numerous signal compounds, for example glycinoeclepin A (K. Mori [69]) and others. [70]

A large number of effective metabolites must be isolated before synthetic work for production processes, biological development, toxicity and ecology studies with a selected active substance can be justified in industry, where the aim to reach a optimal cost/benefit/ecology ratio is given priority. In addition, studies of the mechanism of action of new natural products and regulators are important. Investigations with new inhibitors of plant, fungi and insect/parasite metabolisms give biochemistry enormous impulses. They may advance the knowledge in metabolic processes that are *specific* to pest organisms. Results could possibly be used for the biorational design of new pesticides. [71] Primarily, generous support of isolation work of natural products for agriculture in industry and academia and studies of their role in ecology are of the utmost importance for the advancement of pest related biosciences, of synthetic methodology, and to ultimately achieve greater market penetration with new nature-based plant protection methods that are safer, far less damaging to the environment than many chemicals currently in use, and that are well accepted by society.

References

[1] D. Bellus, *Chimia* **1991**, *45*, 154; H. Geissbühler, P. Brenneisen, H.-P. Fischer, *Science* **1982**, *217*, 505; I. J. Graham-Bryce, *British Crop Protection Conference – Weeds* **1989**, *1*, 3.

[2] A. M. Agnello, J. R. Bradley, jr., in *Safer Insecticides* (eds.: E. Hodgson, R. J. Kuhr) Dekker, New York **1990**, Chapter 13, p. 509.

[3] F. Führ, W. Steffens, W. Mittelstaedt, B. Brumhard, *Jahresbericht der Kernforschungsanlage Jülich GmbH* **1988/89**, 11 and in *Pesticide Chemistry* (ed.: H. Frehse), VCH, Weinheim, **1990**, pp. 37–48.

[4] H. M. LeBaron, R. O. Mumma, R. C. Honeycutt, in ACS-Symposium Series no. 334 (ed.: J. H. Duesing) ACS, Washington, DC, **1987**; P. Eckes, G. Donn, F. Wengenmayer, *Angew. Chem.* **1987**, *99*, 392; Angew. Chem. Int. Ed. Engl. **1987**, *26*, 382.

[5] J. Landell Mils, D. Longman, D. D. Murray, *British Crop Protection Converence – Weeds* **1989**, *3*, 1005; M. J. Crawley, *British Crop Protection Converence – Weeds* **1989**, *3*, 969.

[6] H. Müller, *Nachr. Chem. Tech. Lab.* **1988**, *36*, 1011; K.-G. Malle, *Nachr. Chem. Tech. Lab.* **1988**, *36*, 396.

[7] S. Omura (ed.) in *The Search for Bioactive Compounds from Microorganisms*, Springer, New York **1992**, Part 5, pp. 213–262; E. A. Bell, L. E. Fellows, M. S. J. Simmonds in "Safer Insecticides" (eds.: E. Hodgson, R. J. Kuhr) Dekker, New York, **1990**, Chapter 9, p. 337.

[8] M. Elliot, *Pest. Sci.* **1989**, *27*, 337; J.-P. Demoute, *ibid.* **1989**, *27*, 375.

[9] B. N. Ames, M. Profet, L. S. Gold, *Proc. Natl. Acad. Sci. USA* **1990**, *87*, 7777–7786.; B. N. Ames, L. S. Gold, *Angew. Chem.* **1990**, *102*, 1233; Angew. Chem. Int. Ed. Engl. **1990**, *29*, 1197.

[10] B. Fugmann, F. Lieb, H. Moeschler, K. Naumann, U. Wachendorff, *Chem. unserer Zeit* **1991**, *25*, 317; and **1992**, *26*, 35; L. Lange in *Progress in Botany* (eds.: H.-D. Behnke et al.), Springer, Berlin, **1992**, pp. 252–270; S. Omura, *J. Industrial Microbiology* **1992**, *10*, 135; S. O. Duke, H. K. Abbas, C. D. Boyette in *Proc. Brighton Crop Protection Conference Weeds*, British Crop Protection Council, **1991**, pp. 155–164; H.-P. Fischer, *Nachr. Chem. Tech. Lab.* **1990**, *38*, 732; R. E. Hoagland (ed.) in ACS Symp. Ser. **439** (*Microbes and Microbial Products as Herbicides*), ACS, Washington, DC **1990**; A. L. Demain, G. A. Somkuti, J. C. Hunter-Cevera, H. W. Rossmoore (eds.)

in *Novel Microbial Products for Medicine and Agriculture*, Elsevier, Amsterdam **1989**; J. Davies, B. Brückner, E. Cundliffe in Ciba Foundation Symposium 171 *(Secondary Metabolites: Their Function and Evolution)*, Wiley, Chichester **1992**, pp. 1–2, 129–143, 199–214; S. W. Ayer, B. G. Isaac, D. M. Krupa, K. E. Crosby, L. J. Letendre, R. J. Stonard, *Pest. Sci.* **1989**, *27*, 221; H.-P. Fischer, D. Bellus, *Pest. Sci.* **1983**, *14*, 334; P. A. Worthington, *Natural Product Reports* **1988**, 47.

[11] H.-P. Fischer, R. Nyfeler, J. P. Pachlatko in *Proc. '92 Agric. BioTech. Symp. on New Bio-Pesticides* (ed.: S.-U. Kim), The Research Center for New Bio-Materials in Agriculture, Seoul – Suwon **1992**, pp. 17–54.

[12] L. Beerhues, *Deutsche Apotheker Zeitung* **1992**, *132*, 2486; A. J. Enyedi, N. Yalpani, P. Silverman, I. Raskin, *Cell* **1992**, *70*, 879; P. A. Hedin (Ed.) in *Naturally Occurring Pest Bioregulators*, ACS Symp. Ser. no. 449, ACS, Washington, DC, **1991**; D. J. Chadwick, J. Marsh (Eds.) in *Bioactive Compounds from Plants*, Ciba Foundation Symposium no. 154, Wiley, Chichester **1990**; A. C. Thompson (Ed.) in *The Chemistry of Allelopathy*, ACS Symp. Ser. no. 268, ACS, Washington, DC, **1984**; J. T. Arnason et al. (eds.), ACS Symposium Series no. 387, ACS, Washington **1989**; H. G. Cutler (Ed.), ACS Symposium Series no. 380, ACS, Washington, DC, **1988**; B. A. Leonhardt, M. Beroza (eds.) ACS Symposium Series no. 190, ACS, Washington, DC, **1982**; A. Nahrstedt, *Planta Medica* **1989**, *55*, 333.

[13] T. Misato in *Pesticide Chemistry*, Vol. 2 (Eds.: J. Miyamoto, P. C. Kearney), Pergamon Press, Oxford, **1983**, p. 241–252; T. Misato, I. Yamaguchi. *Outlook in Agriculture* **1984**, *13*, 136.

[14] G. M. Ramos Tombo, D. Bellus, *Angew. Chem.* **1991**, *103*, 1219; Angew. Chem. Int. Ed. Engl. **1991**, *30*, 1193.

[15] A. N. Collins, G. N. Sheldrake, J. Crosby (Eds.) in *Chirality in Industry*, Wiley, New York **1992**, Chaps. 2–4 and 15–18; J. Crosby, *Tetrahedron* **1991**, *47*, 4789; D. Seebach, *Angew. Chem.* **1990**, *102*, 1363; Angew. Chem. Int. Ed. Engl. **1990**, *29*, 1320.

[16] K. Mori, *Bull. Soc. Chim. Belg.* **1992**, *101*, 393; A. Akiyama, M. Bednarski, M.-J. Kim, E. S. Simon, H. Waldmann, G. M. Whitesides, CHEMTECH **1988**, 627; M. W. Fowler, *Plant Cell Culture Technology*. Botanical Monographs 23 (ed.: M. M. Yeoman), Blackwell, Oxford **1986**, p. 202.

[17] D. L. Illmann, *Chem. Eng. News* **1993**, July 12, p. 26.

[18] E. Bayer, K. H. Gugel, K. Hägele, H. Hagenmaier, S. Jessipow, W. A. König, H. Zähner, *Helv. Chim. Acta* **1972**, *55*, 224; Y. Kondo, T. Shomura, Y. Ogawa, T. Tsuruoka, H. Watanabe, K. Totsukawa, T. Suzuki, C. Moriyama, J. Yoshida, S. Inouye, T. Niida, *Sci. Reports of Meiji Seika Kaisha* **1973**, p. 34; S. Omura, K. Hinotozawa, N. Imamura, M. Murata, *J. Antibiot.* **1984**, *27*, 939.

[19] C. R. Worthing (Ed.), *The Pesticide Manual*, 8. edn. British Crop Protection Council **1987**, p. 448; K. Tachibana, in *Pesticide Science and Biotechnology* (Eds.: R. Greenhalgh, T. R. Roberts), Blackwell, Oxford, **1987**, p. 145.

[20] L. Maier, P. J. Lea, *Phosphorus Sulfur* **1983**, *17*, 1; L. Willms, *Pest. Sci.* **1989**, *27*, 219.

[21] H. J. Zeiss, *Tetrahedron Lett.* **1987**, *28*, 1255.

[22] N. Minowa, M. Hirayama, S. Fukatsu, *Bull. Chem. Soc. Jpn.* **1987**, *60*, 1761.

[23] M. G. Hoffmann, H. J. Zeiss, *Tetrahedron Lett.* **1992**, *33*, 2669.

[24] I. A. Natchev, *Bull. Chem. Soc. Japan* **1988**, *61*, 3699.

[25] A. Schulz, P. Taggeselle, D. Tripier, K. Bartsch, *Appl. Environ. Microbiol.* **1990**, *56*, 1; K. Bartsch, R. Dichmann, P. Schmitt, E. Uhlmann, A. Schulz, *Appl. Environ. Microbiol.* **1990**, *56*, 7.

[26] K. Kamigiri, T. Hidaka, S. Imai, T. Murakami, H. Seto, *J. Antibiot.* **1992**, *45*, 781.

[27] J. Laville, C. Voisard, C. Keel, M. Maurhofer, G. Défago, D. Haas, *Proc. Natl. Acad. Sci. USA* **1992**, *89*, 1562; S. Hasegawa, N. Kondo, F. Kodama in ACS Symp. Ser. no. 449 *(Naturally Occurring Pest Bioregulators)*, P. A. Hedin (ed.), ACS, Washington, DC **1991**, pp. 407–416; J. Ligon, D. S. Hill, J. I. Stein, C. R. Howell, J. O. Becker, S. T. Lam, Ciba-Geigy AG and USDA, EP 472494, **1990**; C. R. Howell, R. D. Stipanovic, *Phytopathology* **1980**, *70*, 172; C. R. Howell, R. D. Stipanovic, *Phytopathology* **1979**, *69*, 480.

[28] J. Gosteli, *Helv. Chim. Acta* **1972**, *55*, 451.

[29] A. Ueda, H. Nagasaki, Y. Takakura, S. Kojima, Nippon Soda Co. Ltd., EP 92890, **1982**.

[30] R. Nyfeler, P. Ackermann in *Synthetic Chemistry Agrochem III* (Eds.: D. R. Baker, J. G. Fenyes, J. J. Steffens), ACS Symp. Ser. no. 504, ACS, Washington, DC, **1992**, pp. 395–404.

[31] A. M. van Leusen, H. Siderius, B. E. Hoogenboom, D. van Leusen, *Tetrahedron Lett.* **1972**, *52*, 5337.

[32] D. Nevill, R. Nyfeler, D. Sozzi, *British Crop Protection Conference – Pests and Diseases* **1988**, *1*, 65.

[33] V. Musilek, J. Cerna, V. Sasek, M. Semerdzieva, M. Vondracek, *Folia Microbiol.* (Prague) **1969**, *14*, 377.

[34] T. Anke, W. Steglich in *Biologically Active Molecules* (ed.: U. P. Schlunegger), Springer, Berlin **1989**, p. 9; A. Fredenhagen, A. Kuhn, H. H. Peter, V. Cuomo, U. Giuliano, P. Hug, *J. Antibiot.* **1990**, *43*, 655; W. Weber, T. Anke, B. Steffan, W. Steglich, *J. Antibiot.* **1990**, *43*, 207.

[35] M. Sutter, *Tetrahedron Lett.* **1989**, *30*, 5417.

[36] K. Beautement, J. M. Clough, P. G. de Fraine, C. R. A. Godfrey, *Pestic. Sci.* **1991**, *31*, 499; E. Ammermann, G. Lorenz, K. Schelberger, p. 403, and J. R. Godwin, V. M. Anthony, J. M. Clough, C. R. A. Godfrey, *Brighton Crop Protection Conf. – Pest and Diseases*, p. 435; British Crop Protection Council **1992**.

[37] M. Mizukai, S. Mio, Sankyo Pat. JP-B 2085287, **1989**.

[38] J. P. Pachlatko, H. Zähner, Ciba-Geigy DE 4129616, **1990**.

[39] S. Mio, S. Sugai, *Sankyo Kenkyusho Nempo* **1991**, *43*, 133; S. Mio, R. Ichinose, K. Goto, S. Sugai, S. Sato, *Tetrahedron* **1991**, *47*, 2111; S. Mio, M. Shiraishi, S. Sugai, H. Haruyama, S. Sato, *Tetrahedron* **1991**, *47*, 2121; S. Mio, Y. Kumagawa, S. Sugai, *Tetrahedron* **1991**, *47*, 2133. S. Mio, M. Ueda, M. Hamura, J. Kitagawa, S. Sugai, *Tetrahedron* **1991**, *47*, 2145; S. Mio, S. Sano, M. Shindo, T. Honma, S. Sugai, *Agric. Biol. Chem.* **1991**, *55*, 1105; H. Haruyama, T. Takayama, T. Kinoshita, M. Kondo, M. Nakajima, T. Haneishi, *J. Chem. Soc. Perkin Trans. 1*, **1991**, 1637; M. Nakajima, K. Itoi, Y. Takamatsu, T. Okazaki, T. Kinoshita, M. Shindo, K. Kawakubo, N. Tohjigamori, T. Haneishi, *Nippon Nogeikagaku Kaishi* **1990**, *64*, 293; M. Nakajima, Y. Itoi, Y. Takamatsu, T. Kinoshita, T. Okazaki, K. Kawakubo,

M. Shindo, T. Honma, M. Tohjigamori, T. Haneishi, *J. Antibiot.* **1991**, *44*, 293.

[40] S. Mirza, Ciba-Geigy DE 4129728, **1990**; S. Mirza, R. Kolly, J. P. Pachlatko, G. Rihs, *Helv. Chim. Acta* **1995**, *78*; S. Mirza, R. Kolly, *Kurzfassungen der Vorträge und Poster*, SCG Herbstversammlung in Bern, October, p. 10, **1991**.

[41] P. Chemla, *Tetrahedron Lett.* **1993**, *34*, 7391. H.-P. Fischer, H.-P-Buser, P. Chemla, P. Huxley, W. Lutz, S. Mirza, G. M. Ramos Tombo. Bull. Soc. Chim. Belg. **1994**, 103, 565.

[42] M. H. Fisher, *Abstr. Pap. Am. Chem. Soc.* of 203 Meet. Pt. 1, AGRO No. 159, **1992**; H. Mrozik, P. Eskola, B. O. Linn, A. Lusi, T. L. Shih, M. Tischler, F. S. Waksmunski, M. J. Wyvratt, N. J. Hilton, T. E. Anderson, J. R. Babu, R. A. Dybas, F. A. Preiser, M. H. Fisher, *Experientia* **1989**, *45*, 315; R. A. Dybas, J. R. Babu, *British Crop Protection Conference – Pests and Diseases* **1988**, *1*, 57; R. A. Dybas, N. J. Hilton, J. R. Babu, F. A. Preiser, G. J. Dolce in *Novel Microbial Products for Medicine and Agriculture* (Eds.: A. L. Demain, G. A. Somkuti, J. C. Hunter-Cevera, H. W. Rossmoore), Elsevier, Amsterdam **1989**, pp. 203–212.

[43] H. G. Davies, R. H. Green, *Natural Product Reports* **1986**, 87; G. W. Benz, *Southwest Entomol. Suppl. 7* **1985**, 43; M. Lariviere, M. Aziz, D. Weinmann, J. Ginoux, P. Gaxotte, P. Vingtain, B. Beauvais, F. Derouin, H. Schulz-Key, D. Basset, C. Sarfati, *Lancet* **1985**, *2*, 174; M. H. Fisher, H. Mrozik, *Annu. Rev. Pharmacol. Toxicol.* **1992**, *32*, 537; J. A. Lasota, R. A. Dybas, *Annu. Rev. Entomol.* **1991**, *36*, 91; H. G. Davies, R. H. Green, *Chem. Soc. Rev.* **1991**, *20*, 211, 279.

[44] S. Hanessian, A. Ugolini, P. J. Hodges, P. Beaulieu, D. Dubé, C. André, *Pure Appl. Chem.* **1987**, *59*, 299; *J. Am. Chem. Soc.* **1986**, *108*, 2776.

[45] T. A. Blizzard, G. M. Margiatto, H. Mrozik, M. H. Fisher, *J. Org. Chem.* **1993**, *58*, 3201; S. V. Ley, A. Armstrong, D. Diez-Martin, M. J. Ford, P. Grice, J. G. Knight, H. C. Kolb, A. Madin, C. A. Marby, S. Mukherjee, A. N. Shaw, A. M. Z. Slawin, S. Vile, A. D. White, D. J. Williams, M. J. Woods, *J. Chem. Soc. Perkin Trans. 1* **1991**, 667; T. A. Blizzard, M. H. Fisher, H. Mrozik, T. Shih in *Recent*

Progress in the Chemical Synthesis of Antibiotics (Eds.: G. Lucacs, M. Ohno), Springer, Berlin **1990**, pp. 65–102; D. Dubé, Ph. D. Thesis, Department of Chemistry, University of Montreal, Canada, **1988**; J. D. White, G. L. Bolton, *J. Am. Chem. Soc.* **1990**, *112*, 1625; S. J. Danishefsky, D. M. Armistead, F. E. Wincott, H. G. Selnick, R. Hungate, *J. Am. Chem. Soc.* **1989**, *111*, 2967; S. V. Ley, N. J. Anthony, A. Armstrong, M. G. Brasca, T. Clarke, D. Culshaw, Ch. Greck, P. Grice, A. B. Jones, B. Lygo, A. Madin, R. N. Sheppard, A. M. Z. Slawin, D. J. Williams, *Tetrahedron* **1989**, *45*, 7162; D. R. Williams, B. A. Barner, K. Nishitani, J. G. Philips, *J. Am. Chem. Soc.* **1982**, *104*, 4708.

[46] H. Mrozik, J. C. Chabala, P. Eskola, A. Matzuk, F. Waksmunski, M. Woods, M. H. Fisher, *Tetrahedron Lett.* **1983**, *24*, 5333.

[47] B. Frei, P. Huxley, P. Maienfisch, H. B. Mereyala, G. Rist, A. C. O'Sullivan, *Helv. Chim. Acta* **1990**, *73*, 1905.

[48] G. M. Ramos Tombo, O. Ghisalba, H.-P. Schär, B. Frei, P. Maienfisch, A. C. O'Sullivan, *Agric. Biol. Chem.* **1989**, *53*, 1531; K. Nakagawa, K. Sato, T. Okazaki, A. Torikata, *J. Antibiot.* **1991**, *44*, 803.

[49] M. A. Haxell, B. F. Bishop, P. Bryce, K. A. F. Gration, H. Kara, R. A. Monday, M. S. Pacey, D. A. Perry, Y. Kojima, H. Maeda, S. Nishiyama, J. Tone, L. H. Huang, *J. Antibiot.* **1992**, *45*, 659.

[50] D. G. Stansfield, D. I. Hepler, *Canine Practice* **1991**, *16*, 11–16; A. C. O'Sullivan, B. Frei, ACS Symp. Ser. no. 504 *(Synth. Chem. Agrochem. III)* (Eds.: D. R. Baker, J. G. Fenyes, J. J. Steffens), ACS, Washington, DC, **1992**, pp. 239–257.

[51] M. Sutter, B. Boehlendorf, N. Bedorf, G. Höfle, EP 359706, **1990**; EP 358608, **1990**, and EP 358607, **1990**.

[52] S. Jendrzejewski, P. Erman, *Tetrahedron Lett.* **1993**, *34*, 615.

[53] Y. Miyamoto, S. Ogawa, *J. Chem. Soc. Perkin Trans. 1* **1991**, 2121; *ibid.* **1989**, 1013.

[54] R. O. Duthaler, *GIT Fachz. Lab.* **1992**, *36*, 479; M. Kugler, W. Loeffler, C. Rapp, A. Kern, G. Jung, *Arch. Microbiol.* **1990**, *153*, 276.

[55] H. Nagaoka, M. Shimano, Y. Yamada, *Tetrahedron Lett.* **1989**, *30*, 971; L. N. Mander, *Chem. Rev.* **1992**, *92*, 573.

[56] U. Berlage, J. Schmidt, U. Peters, P. Welzel, Z. Milkova, *Tetrahedron Lett.* **1987**, *28*, 3091; O. D. Dailey, A. B. Peppermann, ACS Symp. Ser. no. 355 *(Synth. Chem. Agrochem. I)*, ACS, Washington, DC, **1987**, p. 409.

[57] M. Nakata, T. Osumi, A. Ueno, T. Kimura, T. Tamai, K. Tatsuta, *Bull. Chem. Soc. Japan* **1992**, *65*, 2974.

[58] R. Ballini, E. Marcantoni, M. Petrini, *J. Org. Chem.* **1992**, *57*, 1316; H. H. Baer, M. Zamkanei, *J. Org. Chem.* **1988**, *53*, 4786; I. Felner, K. Schenker, *Helv. Chim. Acta* **1970**, *53*, 754.

[59] D. A. Evans, W. D. Black, *J. Am. Chem. Soc.* **1993**, *115*, 4497

[60] S. V. Ley, P. J. Lovell, A. M. Z. Slawin, S. C. Smith, D. J. Williams, A. Wood, *Tetrahedron* **1993**, *49*, 1675.

[61] R. W. Addor, T. J. Babcock, B. C. Black, D. G. Brown, R. E. Diehl, J. A. Furch, V. Kameswaran, V. M. Kamhi, K. A. Kremer, D. G. Kuhn, J. B. Lovell, G. T. Lowen, T. P. Miller, R. M. Peevey, J. K. Siddens, M. F. Treacy, S. H. Trotto, D. P. Wright in *Synth. Chem. Agrochem. III*, (Eds.: D. R. Baker, J. G. Fenyes, J. J. Steffens), ACS, Washington, DC, **1992**, pp. 283–312.

[62] H. Decker, H. Zähner, H. Heitsch, W. A. König, H. P. Fiedler, *J. Gen. Microbiol.* **1991**, *137*, 1805; A. G. M. Barrett, S. A. Lebold, *J. Org. Chem.* **1991**, *56*, 4875.

[63] S. Yamauchi, E. Taniguchi, *Biosci. Biotechnol. Biochem.* **1992**, *56*, 1744; F. Ishibashi, E. Taniguchi, *Agric. Biol. Chem.* **1989**, *53*, 1565.

[64] K. Mori, H. Takaishi, *Tetrahedron* **1989**, *45*, 1639; M. P. Dillon, T. J. Simpson, J. B. Sweeney, *Tetrahedron Lett.* **1992**, *33*, 7569.

[65] R. Goodnow, K. Konno, M. Niwa, T. Kallimopoulos, R. Bukownik, D. Lenares, K. Nakanishi, *Tetrahedron* **1990**, *46*, 326; A. V. R. Rao, A. K. Singh, C. V. N. S. Varaprasad, *Tetrahedron Lett.* **1991**, *32*, 4393.

[66] K. S. Chu, G. R. Negrete, J. P. Konopelski, *J. Org. Chem.* **1991**, *56*, 5196.

[67] J. L. Maloisel, A. Vasella, B. M. Trost, D. L. van Vranken, *Helv. Chim Acta* **1992**, *75*, 1515; S. Takahashi, H. Terayama, H. Kuzuhara, *Tetrahedron Lett.* **1992**, *33*, 7565; D. A. Griffith, S. J. Danishefsky, *J. Am. Chem. Soc.* **1991**, *113*, 5863.

[68] J. Janprasert, C. Satasook, P. Sukumalanand, D. E. Champagne, M. B. Isman, P. Wiriya-

chitra, *Phytochemistry* **1993**, *32*, 67; F. Ishibashi, C. Satasook, M. B. Isman, G. H. N. Towers, *Phytochemistry* **1993**, *32*, 307; A. E. Davey, M. J. Schaeffer, R. J. K. Taylor, *J. Chem. Soc. Perkin Trans. 1* **1992**, 2657; B. M. Trost, P. D. Greespan, B. V. Yang, M. J. Saulnier, *J. Am. Chem. Soc.* **1990**, *112*, 9022; G. A. Kraus, J. O. Sy, *J. Org. Chem.* **1989**, *54*, 77.

[69] H. Watanabe, K. Mori, *J. Chem. Soc. Perkin Trans. 1* **1991**, 2919.

[70] K. Mori, *Total Synthesis Natural Products*, Vol. 9 (Ed.: J. ApSimon), Wiley, New York, **1992**, pp. 1–521.

[71] J. R. Finney in *Pesticide Chemistry* (Ed.: H. Frehse), VCH, Weinheim, **1990**, pp. 555–576.

Penems:
A New Generation of β-Lactam Antibiotics

Gottfried Sedelmeier

Since the important class of β-lactam antibiotics was discovered by Fleming in 1929 (penicillin *(1)*), a large number of other very potent β-lactam antibiotics have been found in nature, for example, cephalosporin *(2)* (1945), olivanic acids *(3)* [1d] (1976), and thienamycin *(4)* (1976). Other β-lactam antibiotics, such as monobactams, clavulanic acid, or nocardicins will not be discussed in this paper. For information on these substances the reader is referred to two articles [1a] that have appeared in "Focus on Synthesis", to the reviews cited in these articles, and to more recent studies. [1b–g] Imipenem *(4a)*, a derivative of thienamycin *(4)*, which was isolated from a microorganism by a group from MSD [2] is the only carbapenem currently available on the market. No thiapenem of type *(5)* has yet reached the market. Penems, which unlike carbapenems have a sulphur atom in their five-membered ring, making them a "superimposition" of penicillin *(1)* and cephalosporin *(2)* to *(6)* [3], are synthetic products that have not yet been found to occur naturally. The 6α-[1-(R)-hydroxyethyl] side chain has been shown to be crucially important in achieving maximum potency and a broad spectrum of activity, in the case of both carbapenems and thiapenems. Since it is not economical to produce thienamycin *(4)* – as opposed to *(1)* and *(2)* – by biotechnology because its fermentation titer is too low, there has been a great challenge for many teams in industry and scientific institutes to find economically and ecologically acceptable methods of synthesis for the very similar structures *(4)*, *(5)*, and *(9)*.

Enantiomerically-Pure-Compound (EPC) Synthesis of a Common Intermediate Product

Each approach is more or less aimed at the synthesis of a key intermediate of type *(7)*, from which it should be possible to obtain *(4)*, the synthetically produced β-methylcarbapenem *(9)* via *(8)*, and type *(5)* penems.

For industrial production, asymmetric synthesis of *(7)* (R = H, SiR′$_3$; X = SO$_2$R, O–CO–R′) as an enantiomerically pure intermediate (EPC) [4] for a pharmaceutically active substance now requires synthetic techniques that take into account not just cost, but also environmental impact, in terms of process-integrated environmental protection, as well as by-products and degradation products (false diastereomers and enantiomers). The main problem with total synthesis of *(4)*, *(5)*, and *(9)*, and so most importantly with *(7)*, is controlling the relative and absolute stereochemistry of the three consecutive chiral centers and choosing a suitable and, if necessary, chiral starting material, at the most favorable cost possible.

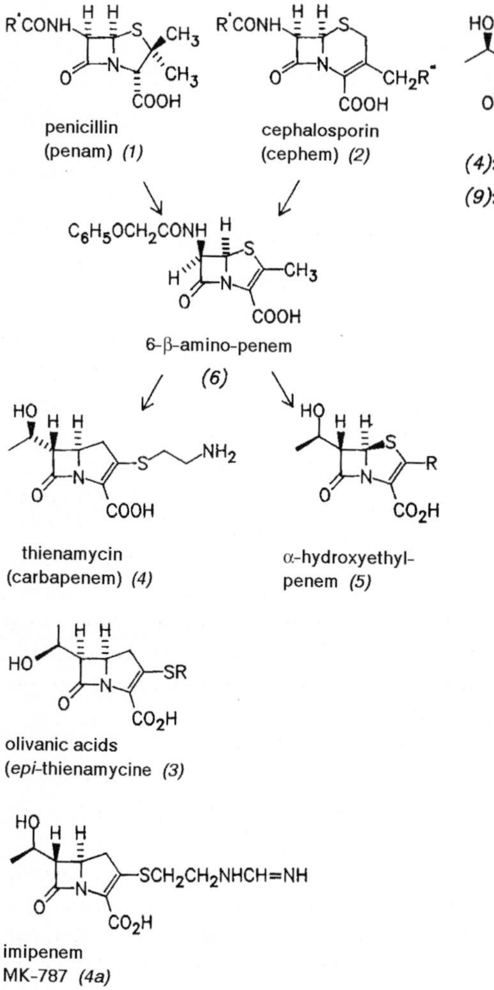

Scheme 1, Several very potent β-lactam antibiotics found in nature.

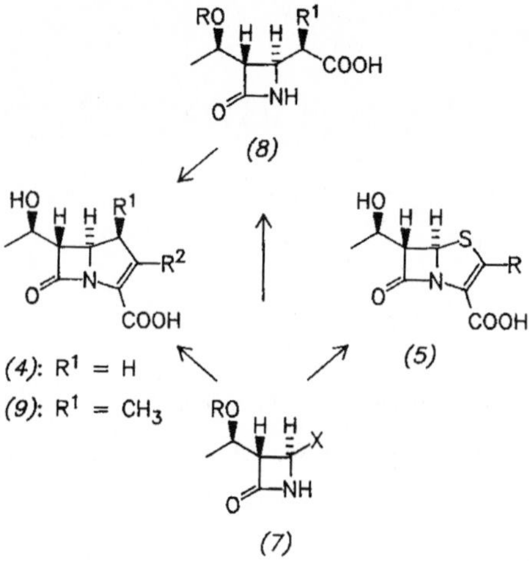

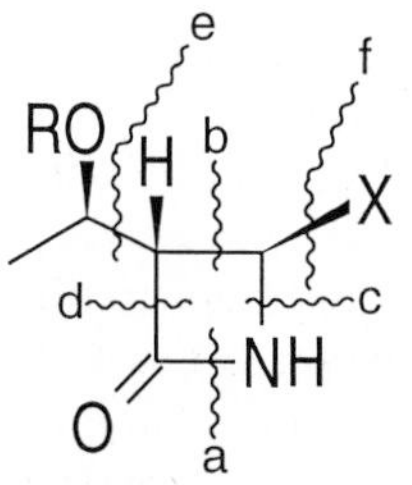

Scheme 2. Strategic bond formations that are (in principle) feasible in the synthesis of the chiral building block.

Of the three synthetic methods that might be used for this purpose (racemate separation, enantioselective conversion (stoichiometric or catalytic), and the "chiron approach") the latter [5] has been by far the most popular method with research groups (cf. Scheme 3).

This article therefore only discusses those syntheses that used the chiral pool, although remarkable results were also achieved with other methods, for example chemoenzymatic synthesis with pig liver esterase (PLE) [6a] and baker's yeast. [6b]

In planning the synthesis of the chiral penem building block *(7)*, the various possible strategic bond formations shown in Scheme 2 were used:

- β-lactam formation from a β-amino acid derivative (formation of bond a);
- cyclisation from nucleophile epoxide ring opening (bond b),

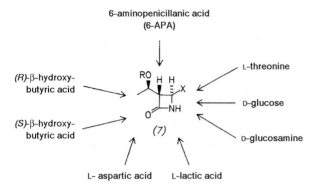

Scheme 3. Starting compounds for "chiron approach" syntheses of the chiral penem building block *(7)*.

- cyclisation of a β-substituted amide (bond c);
- [2 + 2]-cycloaddition of an isocyanate to an alkene (c and d);
- [2 + 2]-cycloaddition of a ketene to an imine (a and b);
- aldol addition of acetaldehyde or ethyl silyl ketone (e);
- functionalisation of a CH$_2$-group to the CHX-group or transformation of a C–C bond to a C–X bond (f).

Some of the most important "chiron approach" syntheses shown in Scheme 3 are discussed below, selected on the basis of their particular importance for the commercial synthesis of *(7)* or *(8)*. The yields from individual steps or total yields for each route were not a criterion for selection because these cannot always be accurately ascertained (patents) and also depend on the stage of development of the method.

Threonine Routes

Starting with enantiomerically pure (2*S*,3*R*)-threonine there are several variations. [7, 9, 10] The threonine routes of Ciba-Geigy (Scheme 4) [7b] and Sankyo [7a] have the advantage that both threonine stereocenters are converted without racemization to the

required absolute configuration at C-3 and C-1′, whereas the third stereocenter (C-4) is directed to the *trans*-position on the β-lactam ring. However, it must be pointed out that the necessary protecting-group technique, the use of mercaptans, and the relatively expensive reagents dicyclohexylcarbodiimide (DCC), Bu$_4$NF, m-chlorperoxybenzoic acid (MCP BA), and cerium (IV)-ammonium nitrate, tend to make these syntheses less economical and less ecologically acceptable. In addition, L-threonine is not one of the cheaper amino acids. [8]

Variants of the "threonine sulphone routes" are the "benzoate routes" of S. Hanessian [9] and Ciba-Geigy [10a] are shown in Schemes 5 and 6.

The same is true in respect of economy and environmental impact for the threonine variants shown in Schemes 5 and 6 as for the sulfone routes. However, the benzoate route (Scheme 5) uses the very cheap base K$_2$CO$_3$ in DMF for closing the β-lactam ring, whereas lithium-bis(trimethylsilyl)amide is employed with the dibenzoate route. Whereas amide formation costs about the same (using either mixed anhydride or acid chloride) for both methods, removal of the nitrogen protecting group in the case of the dibenzoate route takes place via selective enzymatic ester cleavage followed by an "environmentally problemat-

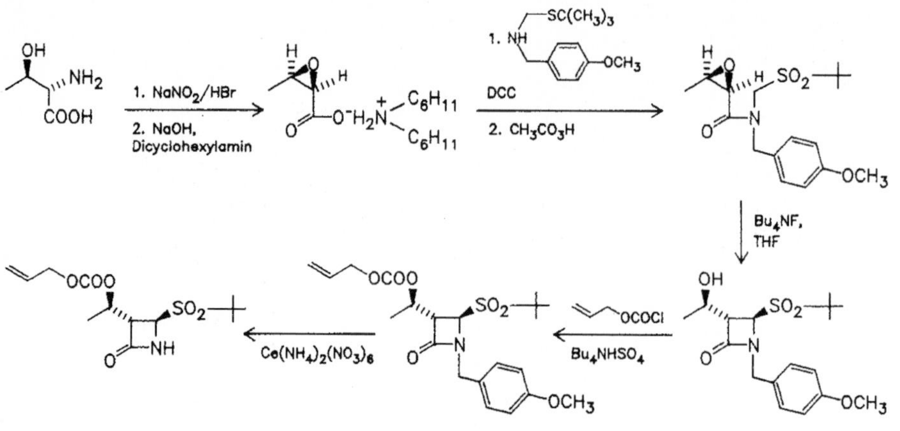

Scheme 4. Threonine route of Ciba-Geigy [7b].

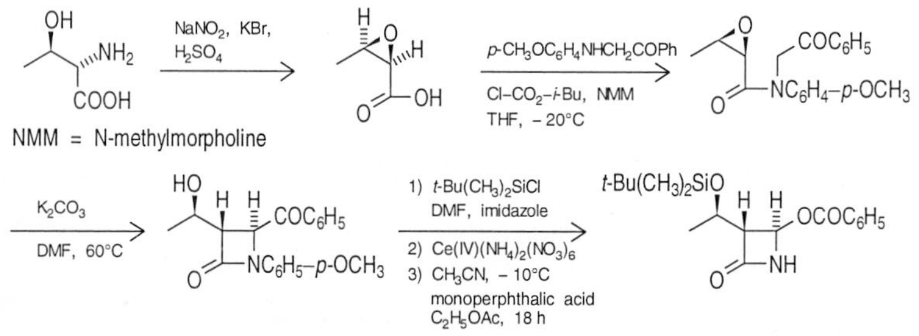

NMM = N-methylmorpholine

Scheme 5. Benzoate route of S. Hanessian [9].

ical" Jones oxidation. An interesting variation the of the Baeyer–Villiger oxidation has been reported more recently that avoids the use of organic peracids and does not require protection of the α-(R)-hydroxyethyl side chain. With this variant, $KHSO_5$ is employed under conditions of phase-transfer catalysis. [10b]

6-Aminopenicillanic Acid Routes

Another educt for a chiron approach to the synthesis of *(7)* is 6-aminopenicillanic acid (6-APA). These routes have been used in particular by Schering-Plough and Farmitalia to produce large quantities of various penem test preparations [1b, c] (Scheme 7), for example,

FEC 22101, FCE 22891, SCH 29482, and SCH 34343 (see also Schemes 8 and 16). The advantage of the 6-APA route is that the β-lactam ring does not have to be constructed first. But the attraction of this route is slightly offset by the fact that the hydroxyethyl side chain, which in the case of penems has to be added at C-6, is in a *trans* position to sulfur, which means that a series of often complex transformations with relatively expensive reagents (e.g., AgCl/1,8-diazobicyclo[5.4.0] undec-7-ene (DBU) or ozone in CH_2Cl_2, −75 °C) are necessary to reconstruct and extend the 6-APA skeleton.

However, in the synthesis of FCE 22891, [1c] a group from Farmitalia showed that it is not absolutely essential when using 6-APA as

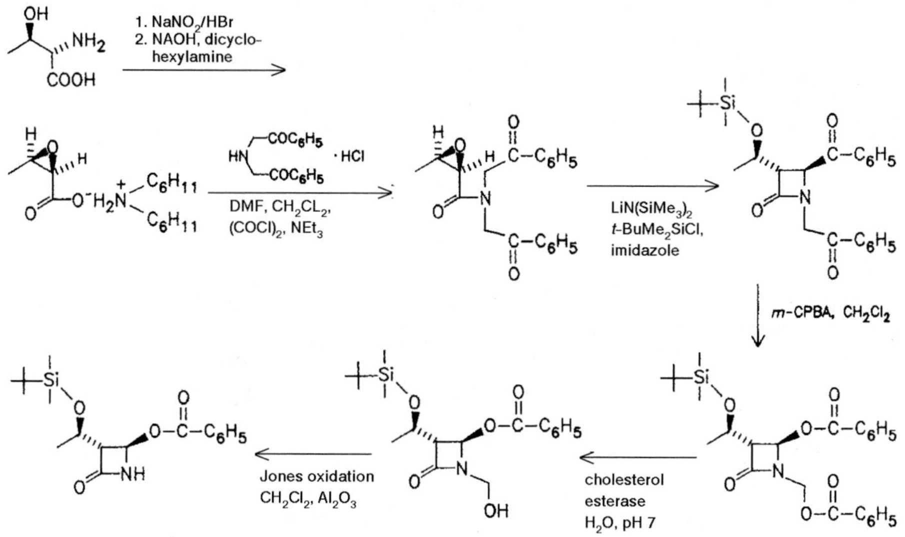

Scheme 6. Dibenzoate route of Ciba-Geigy [10a].

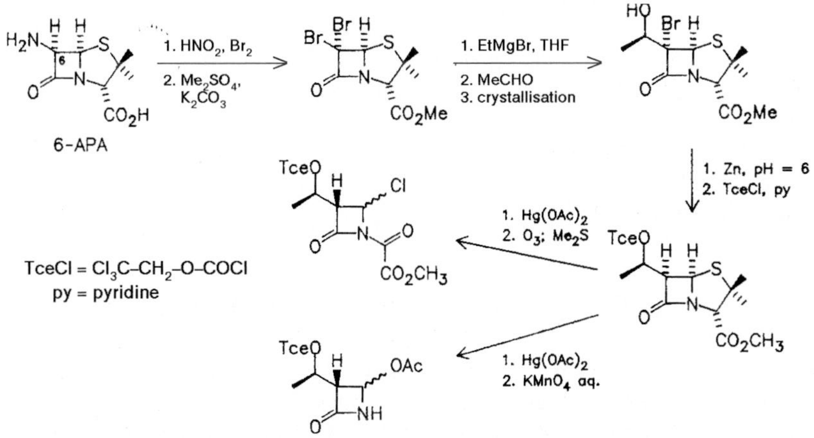

Scheme 7. 6-APA routes of Farmitalia/Schering-Plough [11].

a starting material to proceed via an intermediate product of type *(7)*, but that it is even possible to use primary structural components of the 6-APA five-membered ring to build the penem five-membered ring (Scheme 8). A group from Pfizer recently succeeded in improving the breakdown of the thiazoline ring and transformation to an intermediate of type *(7)* to make the procedure less expensive. [11c]

3-(R)-Hydroxybutyric Acid Routes

Both enantiomers of 3-hydroxybutyric acid are suitable for the synthesis of *(7)*. In order to produce the correct, absolute configuration by the enolate-imine approach [12] (Scheme 9) at C-3 of azetidinone, the *(S)*-enantiomer [13] must be used, which means

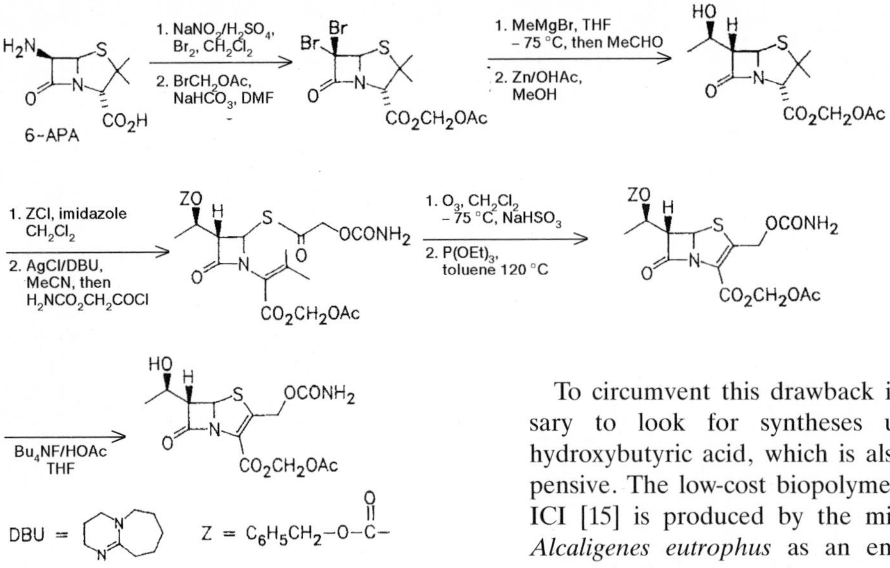

Scheme 8. FCE synthesis of FEC 22891 from 6-APA [1c].

that the configuration at the stereocenter of the hydroxy ethyl side chain must be inverted by the Mitsunobu reaction in a later step of the synthesis. [14]

To circumvent this drawback it was necessary to look for syntheses using 3-(R)-hydroxybutyric acid, which is also very inexpensive. The low-cost biopolymer PHB from ICI [15] is produced by the microorganism *Alcaligenes eutrophus* as an energy-storage material, glucose or molasses serving as nutrients. The price of PHB therefore depends on the world market price of glucose. The production costs of PHB should be around £ 1000–1500 per ton for bulk production. [16] Methylester or ethylester obtained by depolymerisation and simultaneous esterification was used for the synthesis of *(7)*. [17]

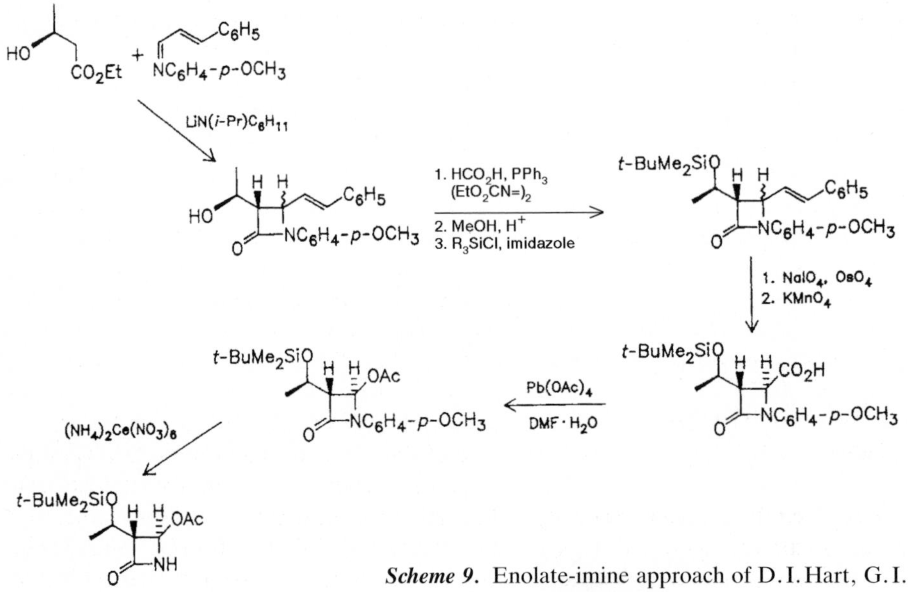

Scheme 9. Enolate-imine approach of D.I. Hart, G.I. Georg et al.

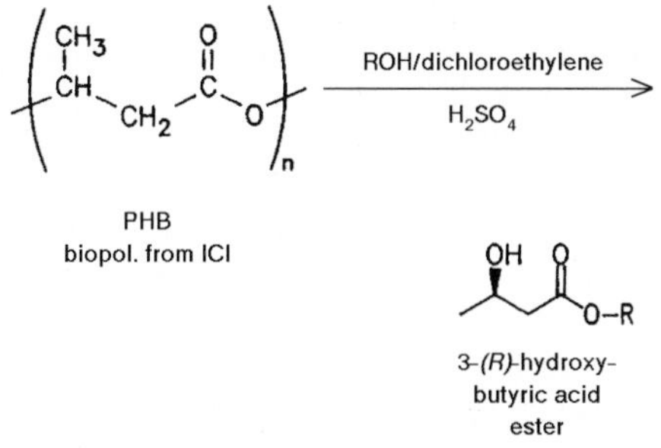

PHB
biopol. from ICI

ROH/dichloroethylene
H₂SO₄

3-(R)-hydroxy-
butyric acid
ester

2 LDA THF
– 60°C, (CO₂Me)₂

1) MesCl, NEt₃
2) NaN₃, acetone H₂O
3) H₂, Pd/C

H₂/Rh/alox.
then DBU

MeOH, NaOH,
then pH 3.9

(Me₃Si)₂NH/Me₃SiCl,
toluene, then
t-BuMgCl/THF

electrolysis:
AcOH, NEt₃,
Pt-anode

(9 : 1)

LDA = lithium diisopropylamide
DBU, cf. scheme 8

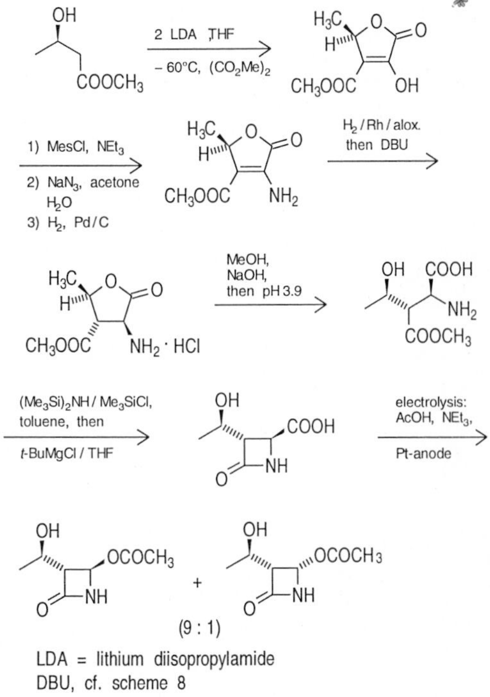

Scheme 10. Lactone approach of Ciba-Geigy [19].

Synthesis of β-Lactam *(7)* via the Lactone Approach

This method of synthesis (Scheme 10) is characterised by the use of simple transformations – with the exception of the first step (oxalic acid ester condensation at –60 °C) and the last step (electrochemistry) – and the fact that no protecting groups are needed apart from per-silylation (hexamethyldisilazane/trimethyl-chlorosilane) before Breckpot–Grignard cyclization to the β-lactam. [18] The correct absolute configuration at C-3/C-4 of the β-lactam is achieved by DBU-epimerization of the γ-lactone to the thermodynamically most stable all-*trans*-diastereoisomer.

The last step is a novel electrochemical decarboxylation/acetoxylation reaction; [19a, b] this replaces the standard reaction used to date, the Pb(OAc)₄-reaction, which was environmentally critical. [20] Very good yields are obtained with this new electrochemical reaction. [21] (See Ref. [19b] for electrochemistry and polarography of *(7)* and of penems).

Enol Ether (Enol Ester)-Chlorosulphonylisocyanate Cycloadditions

Another method of synthesis, using optically pure *(R)*- or *(S)*-β-hydroxybutyric acid, is provided by the so-called enol ether and enol ester/isocyanate cycloaddition (Scheme 11). The appropriate ester is reduced to aldehyde with DIBAL (at –70 °C) after silylation at C-3. This is then transferred to silyl enol ether with trimethylchlorosilane by the Kanekafuchi route; [22] the Ciba-Geigy route [23] via silylated enol ester and cycloaddition of chloro-sulfonylisocyanate (CSI) followed by reduction gives β-lactam. Whereas CSI-cyclo-addition to silyl enol ether at –70 °C gives a yield of 70 %, cycloaddition to enol ester gives a maximum yield of 17 %, which is unsatisfactory. The two low-temperature reactions (–70 °C) with DIBAL and CSI and the relatively expensive reagents, *tert*-butyl-dimethylsilylchloride, DIBAL, and CSI in the case of the Kaneka route, are a disadvantage from the production point of view. Roche described a third route via thioenol ether/isocyanate cycloaddition. [24] Other enol ether/isocyanate routes can be found in ref. [25]

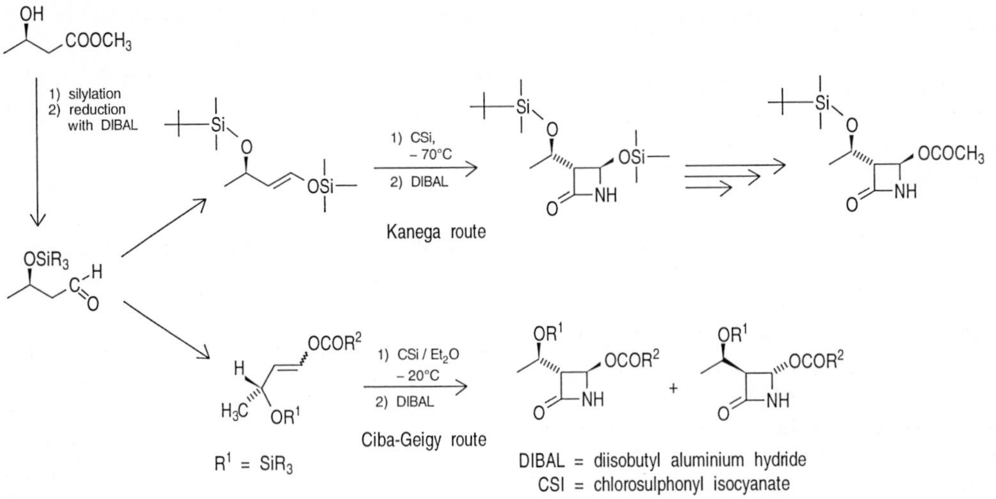

Scheme 11. Enol ether (enol ester)-chlorosulphonylisocyanate cycloadditions (Kanegafuchi [22] and Ciba-Geigy [23]).

Kanekafuchi and Takasago are the only companies so far to offer the β-lactam building block *(7)* (R = *t*BuMe₂Si, X = OAc) in bulk, but it is still very expensive.

Sagami Ltd. has a new, very promising, enol ether/CSI-cycloaddition route that uses double chiral induction of *(R)*-hydroxybutyric acid and protected *(S)*-lactic acid aldehyde via a *cis*-substituted 1,3-dioxene derivative with

high diastereoselectivity (98:2). [26] Unfortunately, this method of synthesis also requires the two low-temperature reactions mentioned above for the Kaneka route, as well as some relatively expensive reagents. Furthermore, the chiral auxiliary *(S)*-benzyloxypropanal is destroyed by oxidation (RuCl₃, HIO₄), which means that it cannot be restored to the cycle (Scheme 12).

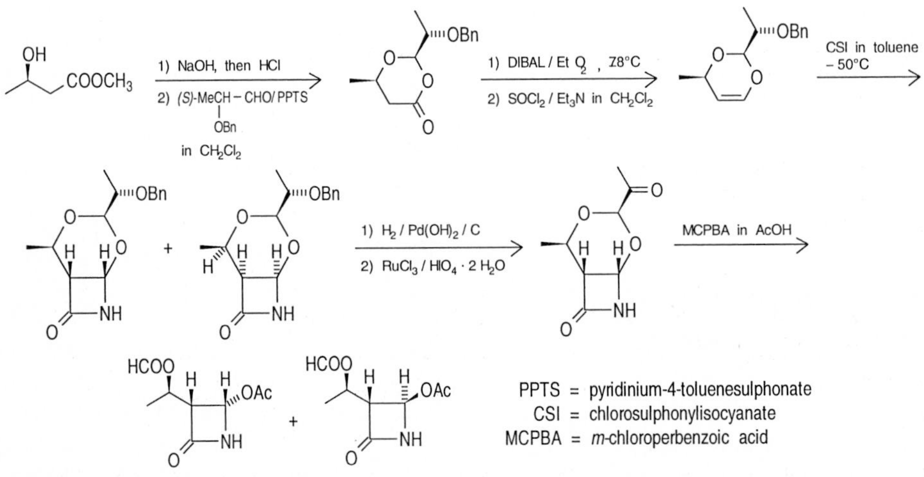

Scheme 12. Sagami route [26].

The L-Aspartic Acid Route

The cheap amino acid L-aspartic acid [8] was used by a group from Merck to synthesize *(S)*-azetidinone carbonic acid. [30a, b] This approach (Scheme 13) allows the α-configurated hydroxyethyl side chain to be introduced at C-3 with adequate diastereoselectivity. The two-step one-pot reaction (70–90 % yield) of Bouffard and Salzmann, which uses a sterically hindered silyl ketone as an acetaldehyde equivalent, gives the desired *trans*-1'-*(R)*-hydroxyethylazetidinonic acid (77 % isolated yield) [31] from *(S)*-azetidinone carboxylic acid, which can then be converted to *(7)* using the oxidative methods mentioned for the lactone approach (Scheme 10).

Routes via Other Molecules from the Chiral Pool

Types of sugar, for example, D-glucose, [27] D-glucosamine, [28] and mannitol in the form of isopropylidene glycerol aldehyde, [29] have so far been used, above all for synthesizing type *(8)* carbapenem building blocks. However, no industrial applications have been found for these approaches, presumably because they are lengthy and inefficient.

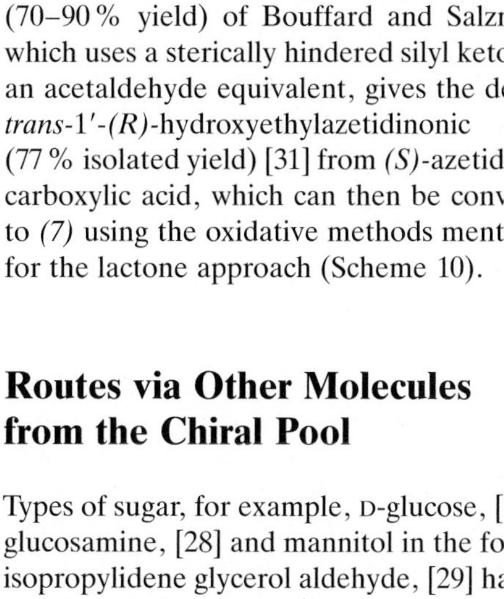

Synthesis and Biological Activity of Penems

Various strategies have been developed synthesizing the five-membered ring system of penem and carbapenem active substances via the intermediate products *(7)* and *(8)*, but these cannot be described in full here. In this context, therefore, only the frequently used Woodward route [3, 32] (intramolecular Wittig reaction with triphenylphosphine), substitution of acyloxy- or sulfonazetidinons of

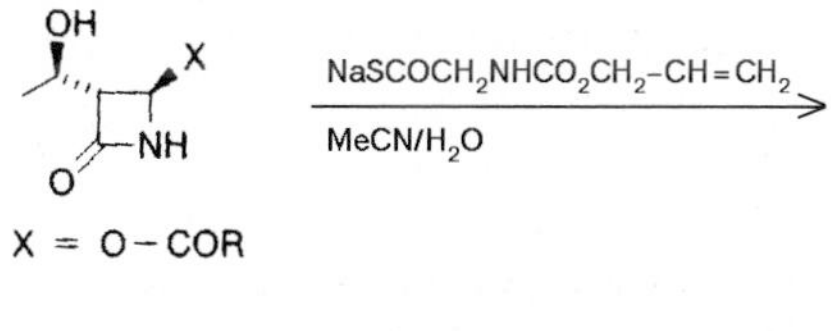

LDA = LiN(*i*-Pr)₂

Scheme 13. Aspartic acid route of Merck, Sharp and Dohme [30, 31].

CGP 31608

Scheme 14. Synthesis of penem CGP 31608 by Ciba-Geigy.

type *(7)* with thiocarbonic acids, [33] followed by intramolecular Wittig–Horner phospite cyclisation [34] are mentioned.

The last method was used at Ciba-Geigy for pilot synthesis of CGP 31 608 (Scheme 14).

The interesting intramolecular Michael reaction described by Hanessian should also be mentioned here. Carried out under mild conditions, this base-catalyzed addition links the C-2/C-3 bond to the thiazoline ring via a nitro-olefin. [9]

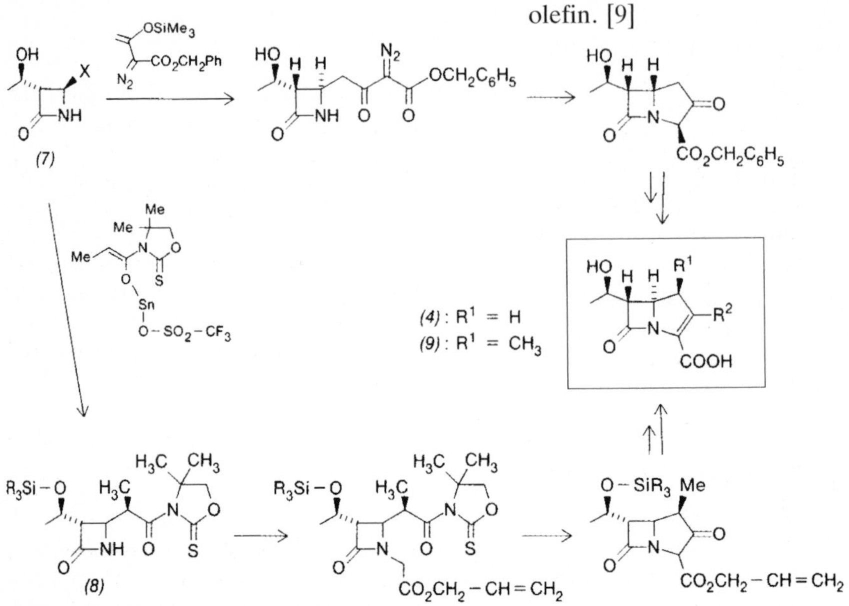

Scheme 15. Syntheses of carbapenems via *(7)* or *(8)*.

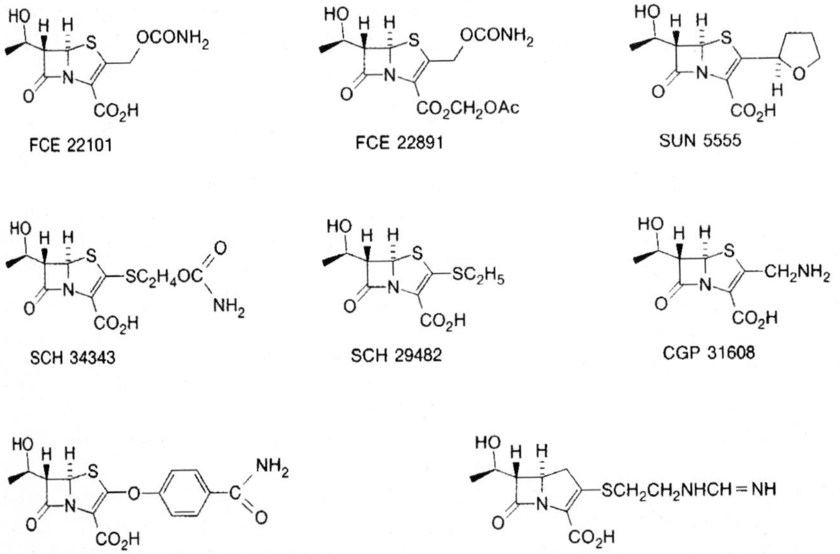

FCE 22101

FCE 22891

SUN 5555

SCH 34343

SCH 29482

CGP 31608

Scheme 16. Selected clinical penem test preparations.

The silyl enol ether, [35] or more recently the boron enolate [36] and tin enolate, [37] routes (Scheme 15), have been used to obtain type *(8)* intermediate products, which are then used to synthesize carbapenems and β-methylcarbapenems.

The main preclinical or clinical test preparations produced using the methods described in this article are shown in Scheme 16. SCH 34343, FCE 22101, CGP 31608, and HRE 664 are parenteral preparations, and SCH 29482, SUN 5555, and FCE 22891 are metabolisable prodrug esters that can be absorbed orally. A good comparison between in vitro and in vivo data on these preparations with respect to efficacy, pharmacokinetics, stability against β-lactamases, etc. can be found in refs. [38] and [1b, c]. Of the preparations in Figure 15, only CGP 31608 and imipenem are effective against the pyogenic microorganism *Pseudomonas aeruginosa*. An important advantage of penems over most penicillins and cephalosporins is their efficacy against a large number of anaerobic bacteria. [1b] Of the above-mentioned preparations, CGP 31608 and SCH 29482 have already been withdrawn.

Outlook

The new β-lactam generation consisting of penems and carbapenems, with their improved spectrum of activity and potentially lower dosage compared with cephalosporins, pose particular problems for development chemists in terms of both economy and environmental impact owing to their unusual β-lactam structure (three stereocenters). Technically feasible, inexpensive, ecologically acceptable, and stereoselective syntheses are therefore needed.

The future of penems as broad-spectrum antibiotics able to compete in price with the new-generation cephalosporins, which are already very effective, depends to a large extent on whether a synthesis is found – possibly via β-lactam building blocks of type *(7)* – that is not too expensive.

References

[1] a) D. Hoppe, *Nachr. Chem. Tech. Lab.* **1982**, *30*, 24 and **1979**, *27*, 127; b) S. W. McCombie, A. K. Ganguly, *Med. Res. Rev.* **1988**, *8*, 393; c) E. Perrone, *Farmaco, Ed. Sci.* **1988**, *43*, 1075; d) T. Nagahara, T. Kametani, *Heterocycles* **1987**, *25*, 729; e) R. Labia, C. Morin, *J. Antibiotics* **1984**, *37*, 1103; f) W. Dürkheimer, J. Blumbach, R. Lattrell, K. H. Scheunemann, *Angew. Chem.* **1985**, *97*, 183; Angew. Chem. Int. Ed. Engl. **1985**, *24*, 180; g) R. Kirstetter, W. Dürckheimer, *Pharmazie* **1989**, *44*, 177.

[2] a) B. G. Christensen et al., *J. Am. Chem. Soc.* **1978**, *100*, 6491; b) S. R. Norrby et al., *Antimicrob. Agents Chemother.* **1983**, *23*, 300.

[3] R. B. Woodward in *Recent Advances in Chemistry of β-Lactam Antibiotics*, Spec. Publ. No. 28. (Ed.: J. Elks) Chemical Society, London, **1977**.

[4] D. Seebach, E. Hungerbühler in *Modern Synthetic Methods 1980* (Ed.: R. Scheffold), Salle u. Sauerländer-Verlag, Frankfurt a. M. – Aarau, **1980**.

[5] S. Hanessian in *Total Synthesis of Natural Products: The Chiron Approach* (Ed.: J. E. Baldwin), Pergamon Press, Oxford, **1983**.

[6] a) M. Ohno et al., *J. Am. Chem. Soc.* **1981**, *103*, 2405 and *Tetrahedron Lett.* **1983**, *24*, 217; b) P. Schneider, G. Ramos, J. Bersier, Ciba-Geigy, E. P. A. 290385, **26.4.1988**.

[7] a) M. Shiozaki et al., *Tetrahedron Lett.* **1983**, *24*, 1037; **1981**, *22*, 5205; b) I. Ernest, J. Kalvoda, W. Froestl, E. P. A. 126709, **28.11.1984**; c) M. Shiozaki, N. Ishida, T. Hiraoka, H. Maruyama, *Bull. Chem. Soc. Jpn.* **1984**, *57*, 2135, and references cited therein.

[8] B. Hoppe, J. Martens, *Chem. unserer Zeit* **1984**, *18*, 73.

[9] S. Hanessian et al., *J. Am. Chem. Soc.* **1985**, *107*, 1438.

[10] a) J. Kalvoda, I. Ernest, M. Biollaz, E. Hungerbühler, E. P. A. 181831, **21.5.1986**; E. P. A. 221846, **13.5.1987**; b) M. Altamura, M. Ricci, *Synth. Commun.* **1988**, *18*, 2129.

[11] a) M. Alpegiani, A. Bedeschi, F. Gudici, E. Perrone, G. Franceschi, *J. Am. Chem. Soc.* **1985**, *107*, 6398; b) A. K. Ganguly et al., *J. Antimicrob. Chemother.* **1982**, *9*, Suppl. C, 1; c) G. J. Quallich et al., *J. Org. Chem.* **1990**, *55*, 367.

[12] D.-C. Ha, D. J. Hart, T.-K. Yang, *J. Am. Chem. Soc.* **1984**, *106*, 4819; G. I. Georg, *Tetrahedron Lett.* **1984**, *25*, 3779.

[13] D. Seebach, M. A. Sutter, R. H. Weber, M. F. Zueger, *Org. Synth.* **1985**, *63*, 1.

[14] G. I. Georg, J. Kant, H. S. Gill, *J. Am. Chem. Soc.* **1987**, *109*, 1129.

[15] a) ICI, Marlborough Biopolymers Ltd., Elta House, Yarm Road, GB-Stockton-on-Teese, Cleveland TS 18 3RX/Great Britain. Marlborough Biopolymers and Kanegafuchi Chem. Ind. Co. Ltd. also supply *(R)*-3-hydroxybutiric acid methylester in bulkquantities; b) B. Sonnleitner, E. Heinzle, G. Braunegg, R. M. Lafferty, *Europ. J. Appl. Microbiol. Biotechnol.* **1979**, *7*, 1; see also P. A. Holmes, L. F. Wright, S. H. Collins, ICI, E. P. A. 52 459, **26. 5. 1982**.

[16] P. A. Holmes, *Phys. Technol.* **1985**, *16*, 32; N. Uttley, *Manufacturing Chemist*, October **1985**, p. 63.

[17] D. Seebach, M. F. Zueger, *Helv. Chim. Acta* **1982**, *65*, 495.

[18] R. Breckpot, *Bull. Soc. Chim. Belg.* **1923**, *32*, 412.

[19] a) G. Sedelmeier, J. Bersier, E. P. A. 279 781, **24. 8. 1988**; b) P. M. Bersier, J. Bersier, G. Sedelmeier, E. Hungerbühler, *Electroanalysis*, in print.

[20] P. J. Reider, E. J. J. Grabowski, *Tetrahedron Lett.* **1982**, *23*, 2293; see also [14].

[21] M. Mori et al., *Chem. Express* **1989**, *4*, 89; M. Mori et al., *Tetrahedron Lett.* **1988**, *29*, 1409.

[22] T. Ohashi, K. Kazunori, I. Soda, A. Miyama, K. Watanabe, E. P. A. 167 154/167 155, **8. 1. 1986**; E. P. A. 247 378, **2. 12. 1987**.

[23] E. Hungerbühler, E. P. A. 259 268, **9. 3. 1988**.

[24] G. Schmid, E. P. A. 179 318, **30. 4. 1986**.

[25] Hungerbühler, E.; Biollaz, M.; Ernest, I.; Kalvoda, J.; Lang, M.; Schneider, P.; Sedelmeier, G. in *New Aspects of Organic ChemistryI;* Yoshida, Z.; Shiba, T., Ohshiro, Y., Eds.; VCH: Tokyo, 1989; p 419.

[26] Y. Ito, Y. Kobayashi, S. Terashima, *Tetrahedron Lett.* **1989**, *30*, 5631.

[27] see Ref. 1 e) and ref. 128–132 cited therein.

[28] M. Miyashita, N. Chida, A. Yoshikoshi, *J. Chem. Soc., Chem. Commun.* **1982**, 1354.

[29] H. Matsunaga, T. Sakamaki, H. Nagaoka, Y. Yamada, *Tetrahedron Lett.* **1983**, *24*, 3009.

[30] a) T. N. Salzmann, R. W. Ratcliffe, B. G. Christensen, F. A. Bouffard, *J. Am. Chem. Soc.* **1980**, *102*, 6163, and E. P. A. 7973; b) R. Labia, C. Morin, *Chem. Lett.* **1984**, 1007.

[31] F. A. Bouffard, T. N. Salzmann, *Tetrahedron Lett.* **1985**, *26*, 6285.

[32] R. B. Woodward et al., *J. Am. Chem. Soc.* **1978**, *100*, 8214.

[33] M. Lang, P. Schneider, R. Scartazzini, W. Tosch, E. A. Konopka, O. Zak, *J. Antibiot.* **1987**, *40*, 217; *Helv. Chim. Acta* **1986**, *69*, 1576; see also R. M. Cozens, M. Lang, *Drugs Future* **1988**, *13*, 19.

[34] A. Afonso, F. Hon, J. Weinstein, A. K. Ganguly, *J. Am. Chem. Soc.* **1982**, *104*, 6138.

[35] Y. Sugimura et al., *Tetrahedron Lett.* **1985**, *26*, 4739; A. G. M. Barrett, P. Quayle, *J. Chem. Soc., Chem. Commun.* **1981**, 1076.

[36] L. M. Fuentes, I. Shinkai, T. N. Salzmann, *J. Am. Chem. Soc.* **1986**, *108*, 4675; T. Shibata, Y. Sugimura, *J. Antibiot.* **1989**, *42*, 374.

[37] Y. Nagao et al., *J. Am. Chem. Soc.* **1986**, *108*, 4673; R. Déziel, D. Favreau, *Tetrahedron Lett.* **1986**, *27*, 5687; **1989**, *30*, 1345.

[38] O. Zak, M. Lang, R. Cozens, E. A. Konopka, H. Mett, P. Schneider, W. Tosch, R. Scartazzini, *J. Clin. Pharmacol.* **1988**, *28*, 128.

Synthesis of *O*-Glycosides

Herbert Waldmann

Carbohydrates play central roles in numerous biological processes. Therefore, the development of efficient methods for the construction of complex oligosaccharides is of increasing interest. [1, 2] For the formation of a glycoside bond a leaving group X, which is attached to the anomeric center of the glycosyl donor **A,** must be activated to leave the molecule giving rise to the oxonium ion **B**. This intermediate is then attacked by a hydroxyl group of the glycosyl acceptor **C**. For an extension of the carbohydrate chain in both directions the formed disaccharide **D** must fulfill two demands: (1) it must be deprotectable selectively at a single OH group and (2) it must carry a leaving group Y that can be activated in a further glycosylation reaction without affecting the already formed glycoside bond. The Koenigs–Knorr reaction and its numerous variants, have proven to be efficient methods for this purpose. [1a, 2a] However, the glycosyl chlorides and -bromides (**A** : X = Cl, Br) employed in these transformations are often generated under harsh reaction conditions, they are sensitive to hydrolysis, and are frequently thermally unstable. Furthermore, in most cases expensive and often also toxic silver and mercury salts are needed for their activation. To overcome these disadvantages various leaving groups have been developed and, in particular, the trichloroacetimidates (**A** : X = O–C(=NH)CCl$_3$) introduced by Schmidt et al. [1b, 2a] proved to be a viable alternative to the classical techniques. Recently, further new and promising methods for the synthesis of *O*-glycosides have been introduced.

X = e.g. Cl, Br: promotor: Ag$^+$, Hg^{2+}

X = O–C(=NH)CCl$_3$: promotor: BF$_3$, OEt$_2$, TMSOTf

Glycosyl Fluorides

Glycosyl fluorides, which have already been known for some time, were considered to be too stable to function as glycosyl donors. However, Mukaiyama et al. demonstrated that they can be activated with the catalyst system SnCl$_2$/AgClO$_4$ and they synthesized, for example, the *gluco* disaccharide *(4)* in high

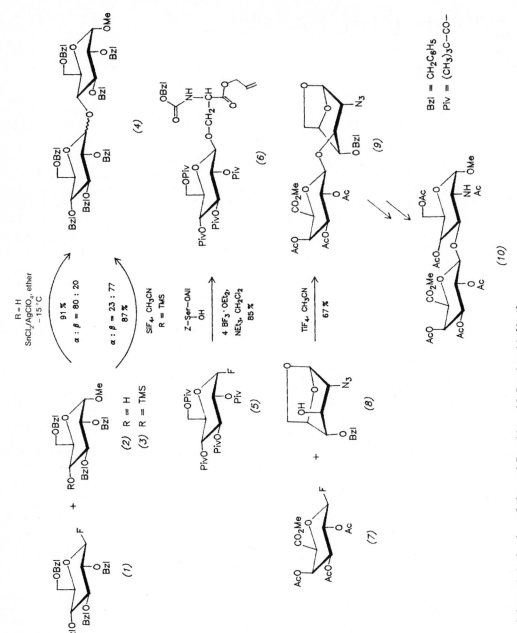

Scheme 1. Activation of glycosyl fluorides with Lewis acids [3, 4].

yield (Scheme 1). [3] Subsequently several groups studied the application of glycosyl fluorides and investigated the use of various Lewis acid promotors. Noyori et al. showed that *(4)* can also be obtained from the silyl ether *(3)* in the presence of gaseous SiF$_4$. [4a] In this and further similar transformations the anomeric configuration of the fluoride *(1)* does not influence the product ratio, but in ether and acetonitrile complementary α/β-selectivity is obtained (Scheme 1). Nicolaou et al. [4b] and Kunz et al. [4c] employed BF$_3$·OEt$_2$, which is a liquid and is, therefore,

easier to handle in the generation of the intermediary oxonium ions of type **B**. By this technique, for example, the serine glycoside *(6)* could be constructed stereospecifically from the pivaloyl-protected carbohydrate *(5)*. Thiem et al. [4d] reported that the use of the heterogeneous catalyst TiF$_4$ in acetonitrile is particularly advantageous. By means of this promotor, for example, the repeating core unit *(10)* of isohyaluronic acid could be synthesized. To this end, the diacchacide *(9)* was built up with complete β-selectivity and then transformed into *(10)* (Scheme 1). In particu-

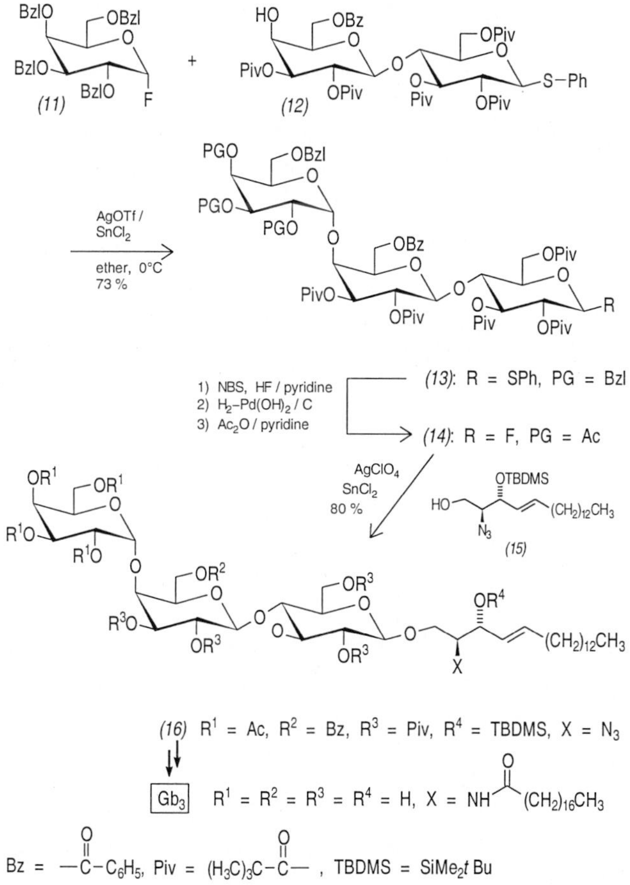

Scheme 2. Synthesis of the globotriaosylceramide by means of glycosyl fluorides according to Nicolaou et al. [5c].

lar, Nicolaou et al. [4b, 5] and Ogawa et al. [6] employed glycosyl fluorides for the construction of complex glycoconjugates. Thus, to synthesize the biologically relevant globotriaosyl ceramide (Gb$_3$) the perbenzylated galactosyl fluoride *(11)* and the selectively deblocked acceptor *(12)* were condensed under Mukaiyama conditions with complete α-selectivity to give the trisaccharide *(13)* (Scheme 2). [5c] Next, the thioglycoside *(13)* was converted to the β-fluoride *(14)* by treatment with NBS and the HF pyridine complex. [4b] After exchange of the benzyl protecting groups in the terminal galactose unit for acetates ($\rightarrow$ *(14)*) activation of the fluoride with SnCl$_2$/AgClO$_4$ delivers the desired β-glycoside *(16)* in 80 % yield. On the one hand, in this step the pivaloyl groups guarantee that only the β-anomer is formed and that competing orthoester formation is suppressed. [7a] On the other hand, the azido-substituted sphingosine equivalent *(15)* is the complex alcohol of choice to achieve a high

yield. [7b] Finally, conversion of the azide to the octadecylamide and the removal of all protecting groups completed the synthesis of Gb$_3$. This glycosylation method also opened routes to the tumor associated Le^X-glycosphingolipids [5d] and the complex oligosaccharide part of calicheamicin γ_1^1. [5e] Suzuki et al. described that the hard Lewis acids Cp$_2$XCl$_2$(X = Zr, Hf) are efficient promotors for the activation of glycosyl fluorides. [8] In combination with AgClO$_4$ these metallocenes rapidly induce glycosylation. They proved themselves, for instance, in the synthesis of the macrolide mycinomycine IV (Scheme 3). To this end the benzoylated mycinolide IV *(17)* is first coupled with the desosaminyl fluoride *(18)* to give the glycoside *(19)*. The glycosylation of the second OH-group with 1-fluoro-D-mycinose *(20)* in benzene proceeds with high β-selectivity. Removal of the acyl protecting functions finally delivers the desired natural product *(21)*.

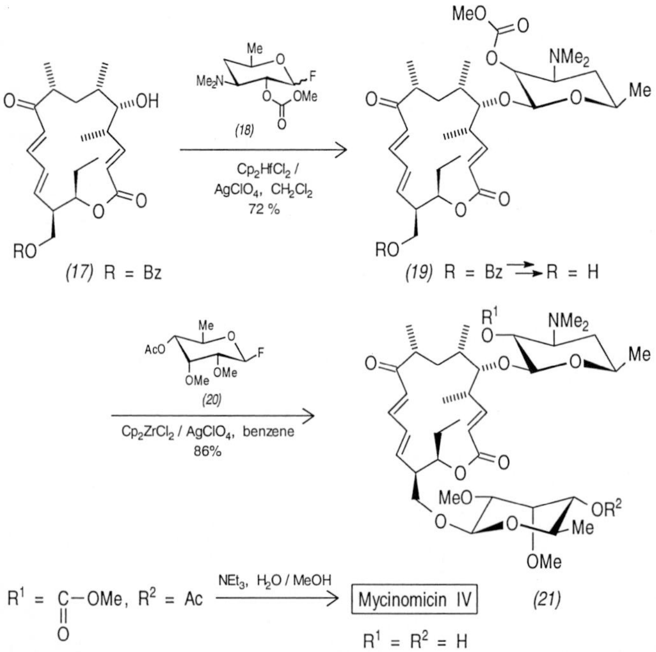

Scheme 3. Activation of glycosyl fluorides with Hf- and Zr-metallocenes according to K. Suzuki et al. [8].

Thioglycosides

Recently, thioglycosides have been applied in numerous cases as glycosyl donors. After initial attempts of several groups to activate these thioacetals with thiophilic metal salts (carbohydrates were attached to the aglycon by activation of thioglycosides as early as the synthesis of erythromycin A by Woodward et al.), Lönn [10] achieved the decisive breakthrough. In the synthesis of the trisaccharide *(22)*, a complex hexasaccharide, and a nonasaccharide the 1-ethylthioethers were alkylated with methyl triflate to give sulfonium salts that were then attacked by the *O*-nucleophiles (Scheme 4). The high stereocontrol over the anomeric center is guaranteed by the active neighbouring phthaloyl group at C-

2. In subsequent developments Fügedi et at, [11a, b] Paulsen et al. [11c] and Lönn et al. [11d] introduced dimethyl(methylthio)sulfonium triflate (DMTST) *(23)* and the corresponding tetrafluoroborate (DMTSB) as superior promotors. Methylsulfenyl bromide and triflate [12a, b] and phenylselenyl triflate [12c] are also advantageous. Furthermore, NOBF₄, [12d] SO₂Cl₂/HCF₃SO₃H [12e] and iodonium compounds [12f, g] have been used successfully for this purpose. In addition, thioglycosides can be activated by means of the stable radical cation tris(4-bromophenyl) ammoniumyl hexachloroantimonate [13a] *(24)* and electrochemically. [13b, c] Particularly unreactive alcohols and even amides can be glycosylated if the phenylthioglycosides are first converted to sulfoxides that are then acti-

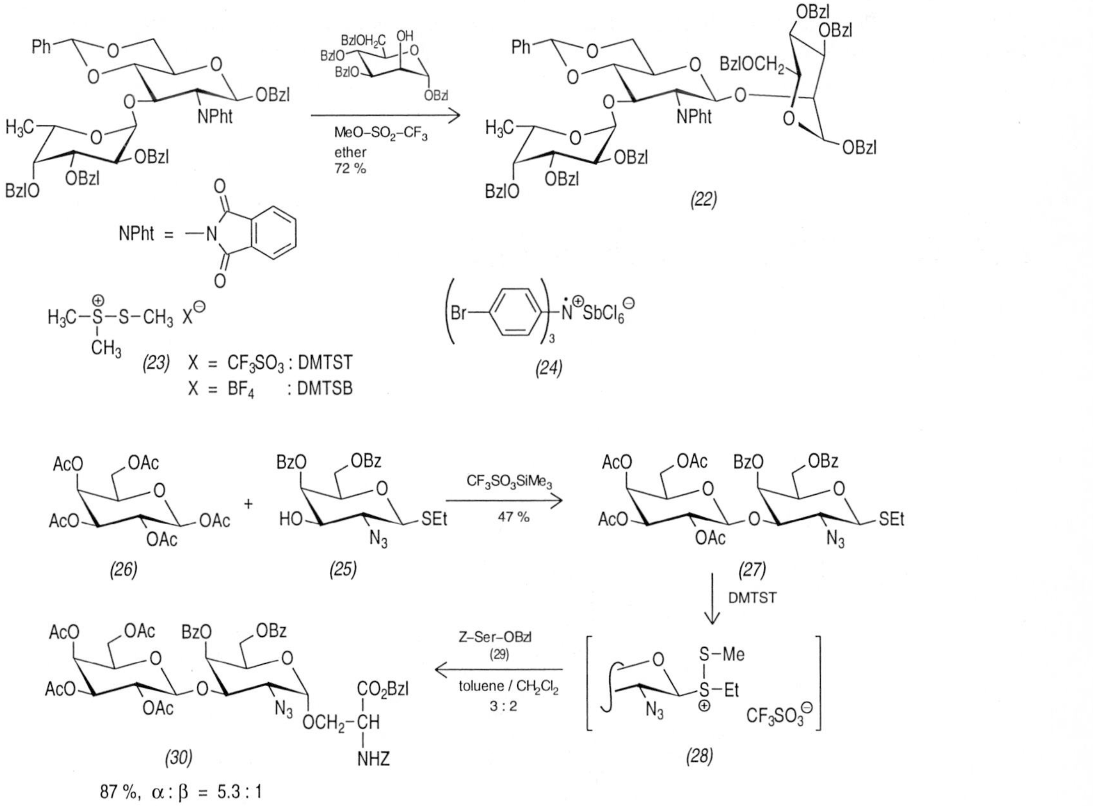

Scheme 4. Activation of thioglycosides according to Lönn [10], Fügedi et al. [11a] and Paulsen et al. [11c].

vated with trifluoromethanesulfonic acid anhydride. [13d] Finally, the use of anomeric xanthogenates together with the radical cation *(24)* [13a] and the glycosylation employing 2-pyridyl-thioglycosides together with methyl iodide as promotor [14a] should be mentioned. These methods have been applied in a multitude of oligosaccharide syntheses. Two representative examples of the use of DMTST and phenylselenyl triflate involve construction of the glycopeptides *(30)* and *(35)*. In a synthesis of the core T structure of the mucine *O*-glycoproteins Paulsen et al. [11c] first attached the peracetylated galactose *(26)* to the 2-azidogalactose thioglycoside *(25)* (Scheme 4). In this reaction the thioether remained intact.

On treatment with DMTST the sulfonium intermediate *(28)* is formed, which reacts with *Z*-serine benzyl ester within one hour to give the glycoside *(30)* in high yield. If methyl triflate is used as promotor after seven days only 30% product is isolated. Kunz et al. [15a] employed the disaccharide thioglycoside *(32)* in a synthesis of the β-mannosylchitobiosyl core structure of *N*-glycoproteins. In the presence of phenylselenyl triflate *(32)* was coupled in 76% yield to the selectively deprotected glucosamine derivative *(33)* to give the β-mannosyltrisaccharide *(34)*, which was subsequently converted to the desired asparagine *(35)* (Scheme 5). In this route the installation of the β-configured mannosidic bond, which is

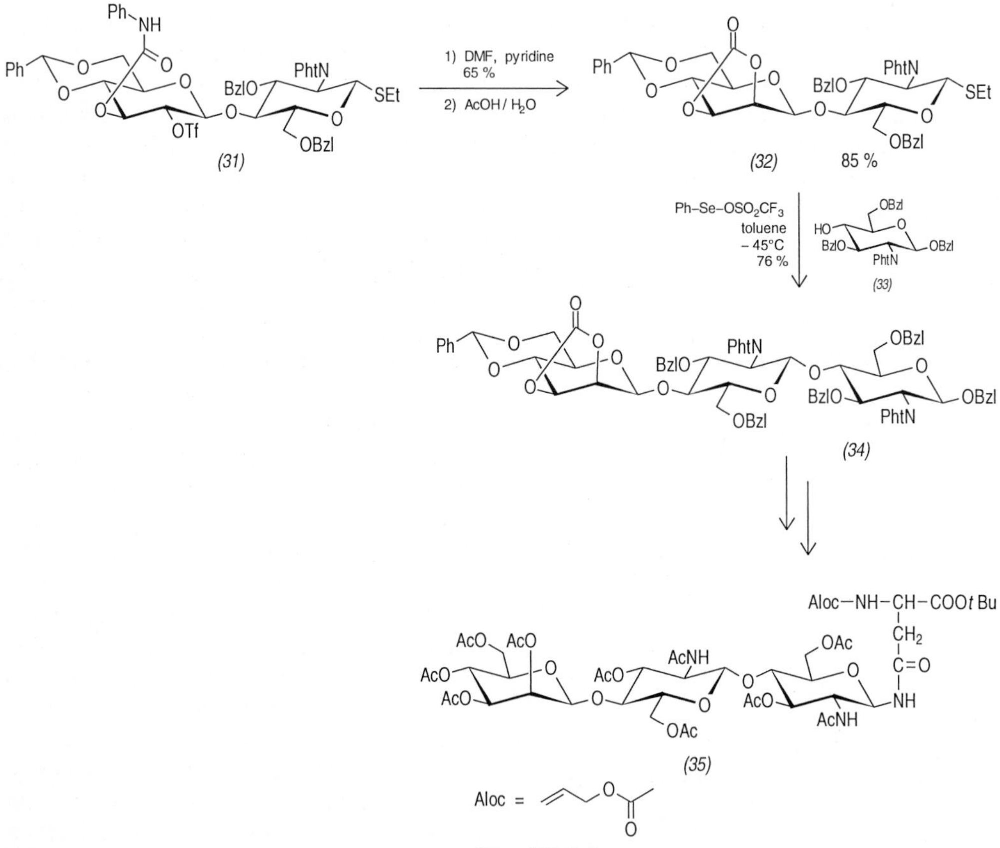

Scheme 5. Synthesis of the β-mannosidic core unit of the *N*-glycoproteins according to Kunz et al. [15].

regarded as being particularly difficult, [15c, d] was elegantly achieved by first building up the β-glucoside *(31)* and then inverting its configuration at C-2' via intramolecular nucleophilic substitution (→ *(32)*) [15b]

The use of thioglycosides is particularly advantageous in the chemistry of *N*-acetylneuraminic acid glycosides. [16] For instance, if the methylthioglycoside *(36)* is activated with DMTST, α-glycosides of this complex carbohydrate can be built up in high yield and with complete selectivity. [16b] This was demonstrated impressively in the synthesis of the α-(2-3)-linked galactoside *(37)*, which can be obtained by other methods only with great difficulty.

Overall, thioglycosides are valuable and advantageous reagents for *O*-glycoside synthesis [9] that can be readily synthesized, are stable under a variety of reaction conditions, and that can be easily activated to become reactive glycosyl donors or, alternatively, be converted to glycosyl fluorides [4b] and -bromides. [9, 17]

Electrophile-initiated Cyclizations

Glycosyl cations **B** can also be generated if the leaving group X participates in a cyclization. Fraser-Reid et al. [18, 19] discovered this principle and developed the use of pentenyl glycosides to a method by which complex glycosides also become available. [18b, d] If such unsaturated alkyl glycosides, for example *(38)*, are treated with I^+-donors like iodonium di-*sym*-collidine perchlorate *(39)* or *N*-iodosuccinimide/CF_3SO_3H [18c] in the presence of glycosylacceptors, for example, *(41)*, via intermediate iodonium ions (e. g. *(40)*) 2-iodomethyltetrahydrofuran *(43)* (with an enantiomeric excess up to 85 % [20]) and the desired saccharides (e. g. *(42)*) are formed (Scheme 6). If solvents containing ethers are used in the activation of glycosyl donors carrying alkyl protecting groups the α-anomers predominate, whereas in acetonitrile β-glycosides are formed in excess. If acyl substituents that are active as participating neighbouring groups are present, 1,2-*trans*-glycosides are generated as expected. Of particular interest is the observation that in the coupling of *(38)* and *(41)* only *(38)* is activated: the self-condensation of *(41)* does not occur. To explain this result the authors assume that ester groups at the 2-position of the carbohydrates reduce the nucleophilicity of anomeric oxygen atoms more than do ether groups. Therefore, 2-acyl substituted glycosyl donors are considered to be "disarmed" towards activation by electrophiles whereas 2-alkyl substituted donors are "armed". Furthermore, in **F** the generation of a positive charge at the anomeric center is less favorable than in **E**. The validity of this rationalization is supported by the observation that "disarmed" pentenyl glycosides can be "armed". Thus, if in *(42)* the acetates are replaced by benzyl ethers the monosaccharide unit that could not be activated in the construction of *(42)* functions as glycosyl donor so that the trisaccharide *(44)* is readily formed. The "armed/dis-

(36)

DMTST
47 %

(37)

$Bz = $ (benzoyl structure) ; $TMSE = (CH_2)_2SiMe_3$

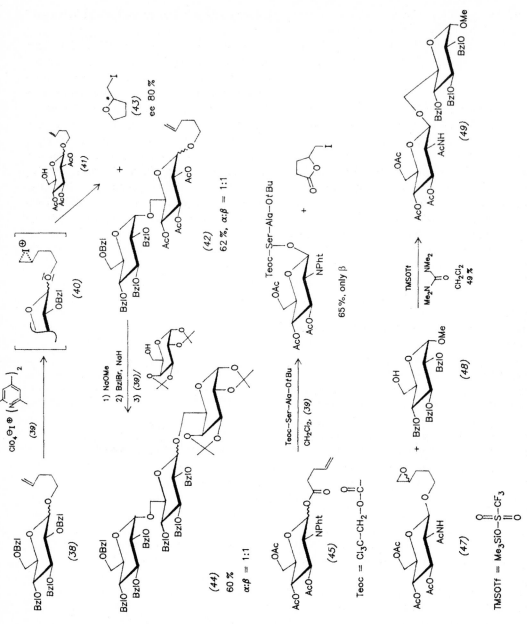

Scheme 6. Glycosylation via electrophile-initiated cyclizations according to Fraser-Reid et al. [18, 19]. Kunz et al. [21a] and Peter et al. [21d].

E   F   (46)

armed" concept does not only account for the reactivity of pentenyl glycosides. The concept can also be applied to different glycosyl donors (**E** and **F**: X = OR, SR; Br, OPent) and electrophiles [12f, 12g, 19b] (**E** and **F**: E = H$^+$, I$^+$, Me$^+$, Ag$^+$, Hal$^+$) and may, thus, prove to be a general principle.

The generation of glycosyl cations by electrophile initiated cyclizations can also be realized with other glycosyl donors. Thus, Kunz et al. [21a] and Fraser-Reid et al. [21b] investigated the pentenoic acid ester as leaving group at the anomeric center and found that in the presence of glycosyl acceptors and various electrophiles, for example, iodonium compounds and the 1,3-dithianium salt *(46)*, 1-pentenoates form iodolactones and glycosides (Scheme 6). Similarly, anomeric allylic urethanes can be employed for the same purpose. [21c] Under the mild reaction conditions various protecting groups that are widely used in peptide and carbohydrate chemistry remain intact. Peter et al. [21b] reported that *N*-acetylglucosaminyl glycosides cannot be synthesized from *n*-pentenyl glycosides, but that 4′,5′-epoxypentenyl glycosides are suitable glycosyl donors to reach this goal. By treatment of the epoxide *(47)* with TMS-triflate in the presence of the pyranose *(48)* the GlcNAc glycoside *(49)* could be synthesized in 49 % yield.

In addition to these methods a variety of new techniques for the formation of *O*-glycoside bonds has been developed. [2b] Among these the application of phosphoric acid derivatives and related glycosyl donors [22] as well as heterocycles [23] as leaving groups appear to be particularly promising. Moreover, reagents that are used in peptide chemistry for the activation of carboxylic acids

are being applied at the anomeric center of carbohydrates. For instance, Ley et al. [24] developed 1-imidazolylcarbonyl- and thiocarbonyl- mono- and disaccharides, for example *(50)*, which were generated from carbohydrates deprotected in the 1-position and carbonyldiimidazole or thiocarbonyldiimidazole. In the presence of ZnBr$_2$ or AgClO$_4$ as promotors they prove to be efficient glycosyl donors. [24a] This method was applied with great success in the total synthesis of avermectin B$_{1a}$ and allowed the coupling of the bis-oleandrose disaccharide *(50)* and the aglycon *(51)* to be carried out in 80 % yield.

(50)

ROH | AgClO$_4$, K$_2$CO$_3$
THF, toluene
80 %

α : β = 4 : 1

ROH =

(51)

References

[1]   a) H. Paulsen, *Chem. Soc. Rev.* **1984**, *13*, 15; b) R. R. Schmidt, *Angew. Chem.* **1986**, *98*, 213; Angew. Chem. Int. Ed. Engl. **1986**, *25*, 212; c) H. Kunz, *Angew. Chem.* **1987**, *99*, 297; Angew. Chem. Int. Ed. Engl. **1987**, *26*, 294.

[2]   a) K. Krohn, *Nachr. Chem. Tech. Lab.* **1987**, *35*, 930; b) K. Toshima, K. Tatsuta, *Chem. Rev.* **1993**, *93*, 1503.

[3]   T. Mukaiyama, Y. Murai, S. Shoda, *Chem. Lett.* **1981**, 431.

[4]   a) S. Hashimoto, M. Hayashi, R. Noyori, *Tetrahedron Lett.* **1984**, *25*, 1379; b) K. C. Nicolaou, A. Chucholowski, R. E. Dolle, J. L. Randall, *J. Chem. Soc., Chem. Commun.* **1984**, 1155; K. C. Nicolaou, R. E. Dolle, D. P. Papahatjis, J. E. Randall, *J. Am. Chem. Soc.* **1984**, *106*, 4189; c) H. Kunz, W. Sager, *Helv. Chim. Acta* **1985**, *68*, 283; H, Kunz, H. Waldmann, *J. Chem. Soc., Chem. Commun.* **1985**, 638; d) M. Kreuzer, J. Thiem, *Carbohydr. Res.* **1986**, *149*, 347.

[5]   a) R. E. Dolle, K. C. Nicolaou, *J. Am. Chem. Soc.* **1985**, *107*, 1695; b) K. C. Nicolaou, J. L. Randall, G. T. Furst, *J. Am. Chem. Soc.* **1985**, *107*, 5556; c) K. C. Nicolaou, T. J. Caulfield, H. Kataoka, *Carbohydr. Res.* **1991**, *202*, 177; d) K. C. Nicolaou, T. J. Caulfield, H. Kataoka, N. A. Stylianides, *J. Am. Chem. Soc.* **1990**, *112*, 3693; e) R. D. Groneberg, T. Miyazaki, N. A. Stylianides, T. J. Schulze, W. Stahl, E. P. Schreiner, T. Suzuki, Y. Iwabuchi, A. L. Smith, K. C. Nicolaou, *J. Am. Chem. Soc.* **1993**, *115*, 7593.

[6]   a) Y. Takahashi, T. Ogawa, *Carbohydr. Res.* **1987**, *169*, 127; b) Y. Nakahara, H. Ijima, S. Sibayama, T. Ogawa, *Tetrahedron Lett.* **1990**, *31*, 6897.

[7]   a) H. Kunz, A. Harreus, *Liebigs Ann. Chem.* **1982**, 41; b) P. Zimmermann, R. Bommer, T. Bär, R. R. Schmidt, *J. Carbohydr. Chem.* **1988**, *7*, 435.

[8]   T. Matsumo, H. Maeta, K. Suzuki, G. Tsuchihashi, *Tetrahedron Lett.* **1988**, *29*, 3567, 3571, 3575.

[9]   Short review: P. Fügedi, P. Garegg, H. Lönn, T. Norberg, *Glycoconjugate J.* **1987**, *4*, 97.

[10] H. Lönn, *Carbophydr. Res.* **1985**, *139*, 105 and 115.

[11] a) P. Fügedi, P. Garegg, *Carbohydr. Res.* **1986**, *149*, C9; b) F. Andersson, P. Fügedi, P. Garegg, M. Nashed, *Tetrahedron Lett.* **1986**, *27*, 3919; c) H. Paulsen, W. Rauwald, U. Weichert, *Liebigs Ann. Chem.* **1988**, 75, d) S. Nilsson, H. Lönn, T. Norberg, *Glycoconjugate J.* **1989**, *6*, 21.

[12] a) F. Dasgupta, P. Garegg, *Carbohydr. Res.* **1988**, *177*, C13; b) F. Dasgupta, P. Garegg, *Carbohydr. Res.* **1990**, *202*, 225; c) Y. Ito, T. Ogawa, *Carbohydr. Res.* **1990**, *202*, *165*; d) V. Pozsgay, H. Jennings, *J. Am. Chem. Soc.* **1987**, *52*, 4635; e) E. Kallin, H. Lönn, T. Norberg, *Glycoconjugate J.* **1988**, *5*, 3; f) G. H. Veeneman, S. H. van Leeuwen, J. H. van Boom, *Tetrahedron Lett.* **1990**, *31*, 1331; g) P. Konradsson, U. E. Udodong, B. Fraser-Reid, *Tetrahedron Lett.* **1990**, *31*, 4313.

[13] a) A. Marra, J.-M. Mallet, C. Amatore, P. Sinay, *Synlett* **1990**, 572; b) C. Amatore, A. Jutand, J.-M. Mallet, G. Meyer, P. Sinay, *J. Chem. Soc., Chem. Commun.* **1990**, 718; c) G. Balavoine, A. Gref, J.-C. Fischer, A. Lubineau, *Tetrahedron Lett.* **1990**, *31*, 5761 d) D. Kahne, S. Walker, Y. Cheng, D. Van Engen, *J. Am. Chem. Soc.* **1989**, *111*, 6881.

[14] G. V. Redy, V. R. Kulkarni, H. B. Mereyala, *Tetrahedron Lett.* **1989**, *30*, 4283.

[15] a) W. Günther, H. Kunz, *Angew. Chem.* **1990**, *102*, 1068; Angew. Chem. Int. Ed. Engl. **1990**, *29*, 1050; b) H. Kunz, W. Günther, *Angew. Chem.* **1988**, *100*, 1118; Angew. Chem. Int. Ed. Engl. **1988**, *27*, 1086; For new methods for the synthesis of β-mannosides see: c) F. Baresi, O. Hindsgaul, *J. Am. Chem. Soc.* **1991**, *113*, 9376; *Synlett* **1992**, 759; d) G. Stork, H. Kim, *J. Am. Chem. Soc.* **1992**, *114*, 1087.

[16] a) T. Murase, H. Ishida, M. Kiso, A. Hasegawa, *Carbohydr. Res.* **1988**, *184*, C1; b) Review: K. Okamoto, T. Goto, *Tetrahedron* **1990**, *46*, 5835.

[17] J. O. Kihlberg, D. A. Leigh, D. R. Bundle, *J. Org. Chem.* **1990**, *55*, 2860.

[18] a) B. Fraser-Reid, P. Konradsson, D. R. Mootoo, U. Udodong, *J. Chem. Soc. Chem. Commun.* **1988**, 823; b) D. R. Mootoo, P. Konradsson, B. Fraser-Reid, *J. Am. Chem. Soc.* **1989**, *111*, 8540; c) P. Konradsson, D. R. Mootoo, R. E. McDevitt, B. Fraser-Reid, *J. Chem. Soc., Chem. Commun.* **1990**, 270; d) A. J. Ratcliffe, P. Konradsson, B. Frasser-Reid, *J. Am. Chem.*

Soc. **1990,** *112,* 5665; e) Review: B. Fraser-Reid, U. E. Udodong, Z. Wu, H. Ottosson, J. R. Merritt, C. S. Rao, C. Roberts, R, Madsen, *Synlett* **1992,** 927.

[19] a) D. R. Mootoo, P. Konradsson, U. Udodong, B. Fraser-Reid, *J. Am. Chem. Soc.* **1988,** *110,* 5583; b) B. Fraser-Reid, Z. Wu, U. E. Udodong, H. Ottoson, *J. Org. Chem.* **1990,** *55,* 6068.

[20] J. M. Llera, J. C. Lopez and B. Fraser-Reid, *J. Org. Chem.* **1990** *55,* 2997.

[21] a) H. Kunz, P. Wernig, M. Schultz, *Synlett* **1990,** 631; b) J. C. Lopez, B. Fraser-Reid, *J. Chem. Soc., Chem. Commun.* **1991,** 159; c) H. Kunz, J. Zimmer, *Tetrahedron Lett.* **1993,** *34,* 2907; d) P.-C. Boldt, M. Schumacher-Wandersleb, M. Peter, *Tetrahedron Lett.* **1991,** *32,* 1413.

[22] a) S. Hashimoto, T. Honda, S. Ikegami, *J. Chem. Soc., Chem. Comun* **1989,** 685; b) T. Yamanoi, T. Inazu, *Chem. Lett.* **1990,** 849; c) S. Hashimoto, T. Honda, S. Ikegami, *Chem. Pharm. Bull* **1990,** *30,* 775; d) T. J. Martin, R. R. Schmidt, *Tetrahedron Lett.* **1992,** *33,* 6123; e) M. M. Sim, H. Kondo, C.-H. Wong, *J. Am. Chem. Soc.* **1993,** *115,* 2260.

[23] W. Bröder, H. Kunz, *Synlett* **1990,** 251.

[24] a) M. J. Ford, S. V. Ley, *Synlett* **1990,** 255, b) M. J. Ford, J. G. Knight, S. V. Ley, S. Vile, *Synlett* **1990,** 331.

Carbacyclines:
Stable Analogs of Prostacyclines

Dieter Schinzer

After the discovery of the prostaglandins in the thirties [1] it took nearly another thirty years until the literature almost exploded and many papers appeared concerning their biolo- gical activity and approaches to synthesize these targets. This increase in activity can be explained by solution to the biosynthesis been found in the meantime. The key reaction in

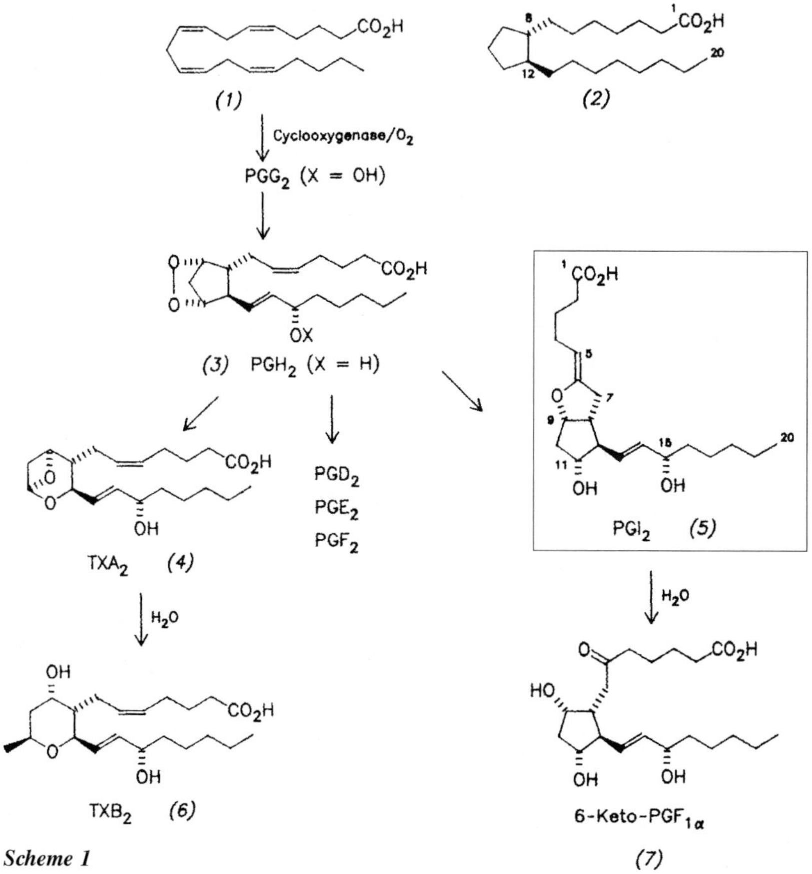

Scheme 1

the biosynthesis is the physiological oxidation of arachidonic acid, the so-called arachidonic acid cascade (Scheme 1), with its two enzyme systems. [2]

The extraordinary interest in these molecules is explained by their high biological activity. A major drawback for potential medical uses is the high sensitivity to hydrolytic destruction under physiological conditions. Therefore, more stable acyclic molecules like leucotrienes were studied intensively. [3]

A further group of related natural products are the prostacyclines because of their remarkable antihypertensive and inhibiting platelet aggregation properties. [4]

In order to overcome the hydrolytic instability analogs were required with potent physiological activity and promising therapeutic properties for treatment of cardiovascular and circulatory disease.

Carbacycline, isocarbacycline and various carbacyclic analogs received particular attention as promising therapeutic agents. The following chapter will focus on efficient syntheses for the controlled construction of the bicyclo-octene framework and the regiodefined synthesis of the double bond.

An entry to optically active isocarbacycline was published by Korozumi et al. [5] (Scheme 2), who used "Noyori-enone" *(8)* as starting material. [6]

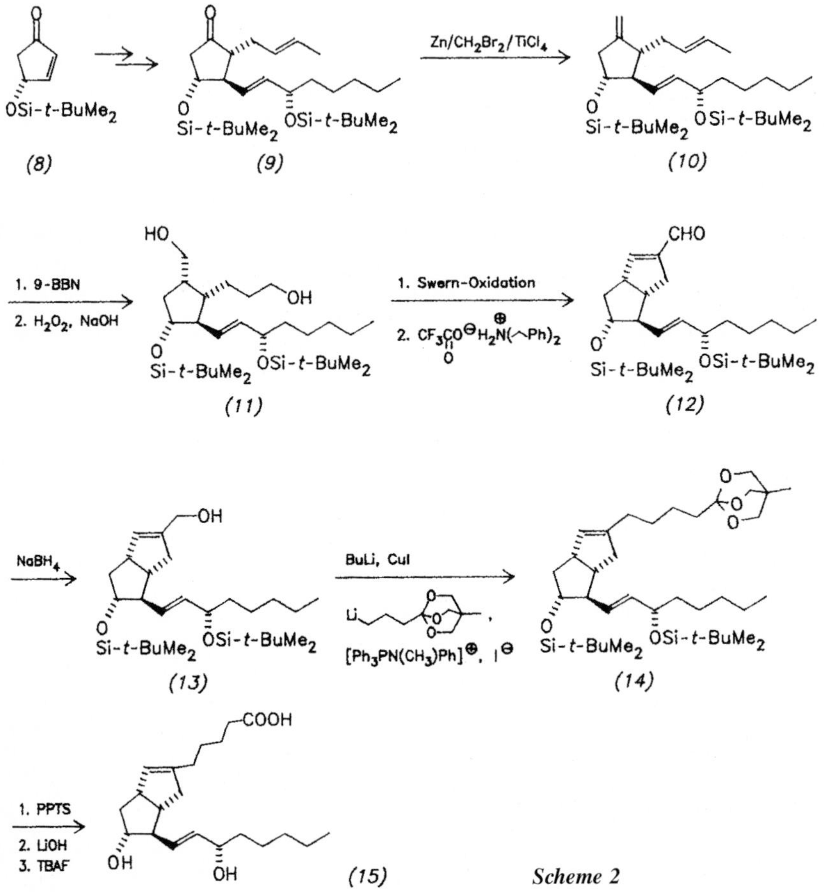

Scheme 2

Enone *(8)* was transformed in a short sequence using Shibasaki's method into the allylic alcohol *(13)*. [7] The *exo*-methylene group was introduced by the Lombardo reaction [8] to give compund *(10)*, followed by hydroboration with 9-borabicyclo[3.3.1] nonane to obtain diol *(11)*. The latter was oxidized under Swern conditions and directly transformed by an intramolecular aldol reaction to yield *(12)*. Boronate reduction provided allylic alcohol *(13)*. Key reaction of the synthesis was a regioselective, direct alkylation of *(13)* to give *(14)*.

Noyori et al. published a short and very efficient synthesis of isocarbacycline *(15)* in which a radical cyclization followed by a regioselect-

ive protodesilylation of an allylsilane were performed as the key steps (Scheme 3). [9]

The precursor *(20)* for the radical cyclization was synthesized by the use of a tandem reaction with the optically active enone *(8)*. Both functionalized side chains were introduced in a one-pot procedure. Ketone *(16)* was transformed into the *exo*-methylene derivative, again by a Lombardo reaction. [8] Hydroboration followed by oxidation yielded the sensitive aldehyde *(18)*, which gave the α-silylated alcohol *(19)* after addition of a silyl cuprate. [10] The OH-group in alcohol *(19)* was transformed into a leaving group and photolyzed to give a *Z/E*-mixture of the desired allylsilane *(21)*. Compound *(21)* was regiose-

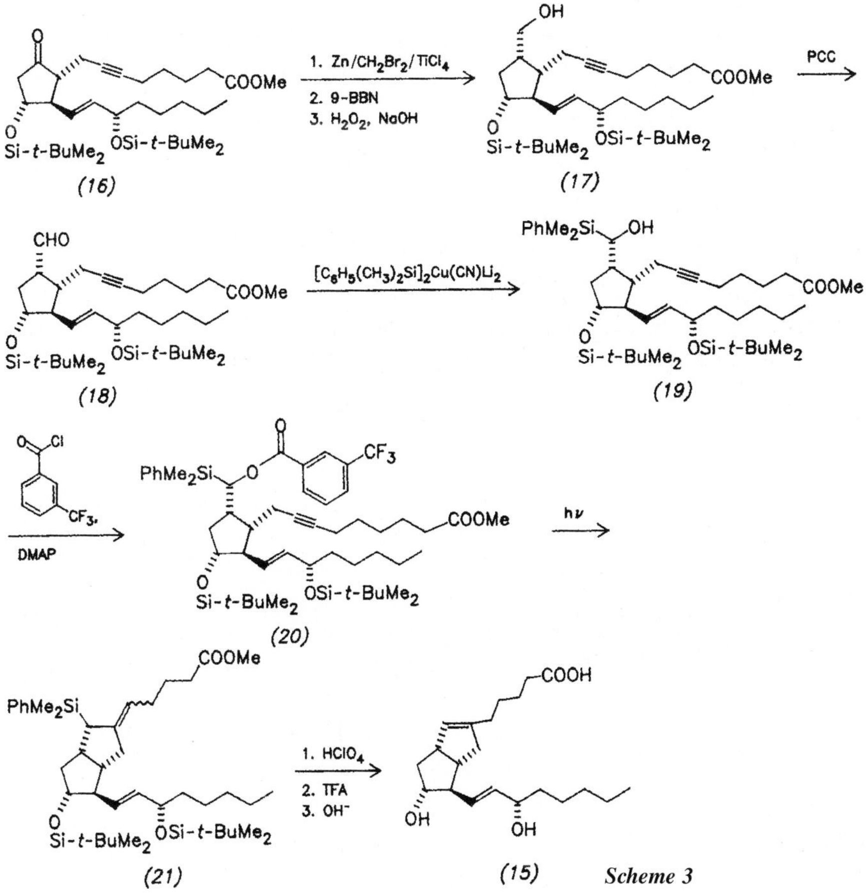

Scheme 3

lectively protodesilylated to yield isocarbacycline *(15)* in optically active form. The regioselectivity of the *endo*-double bond was controlled by the β-effect of the silyl group.

Nagao et al. described an asymmetric synthesis of carbacycline using a novel chiral induction by an asymmetric cleavage of a symmetric cyclohexene diester *(22)* with thiophenolate (Scheme 4). [11] The product mixture was separated by chromatography to yield pure asymmetric diester *(23)*.

Compound *(25)* was transformed by conventional procedures (Dieckmann condensation) into *(26)*. After protecting group manipulation and oxidative cleavage of the double bond a second ester condensation was used to construct the bicyclo-octane framework. The exocyclic double bond with the functionalized side chain was introduced by a non-selective Wittig reaction to obtain compound *(31)* as a *Z/E*-mixture of diastereomers.

Ikegami et al. approached *(33)* by a metal-catalyzed C-H-insertion to give bicyclo-octane system *(34)* (Scheme 5). [12]

The correctly placed ester group allowed regioselective construction of the lower side chain using a lead reagent. Finally, isocarbacycline *(15)* was approached via ketone *(37)*. Sulfenylation of the kinetically controlled enolate, followed by oxidation, alkylation, and reductive elimination, provided *(15)*.

In a series of papers Gais et al. published strategies to synthesize regiodefined the *exo*- or *endo*-cyclic double bond. [13–15] Dilithiosulfones were used in a stereoselective cycloalkylation with functionalized cyclopentanes *(40)* (Scheme 6).

The subsequent cuprate addition yielded the *exo*-cyclic fragment as a mixture of diastereomers (*E:Z* = 2:1), which is similar to that of other methods used. Considerable progress was achieved with transition metal-catalyzed

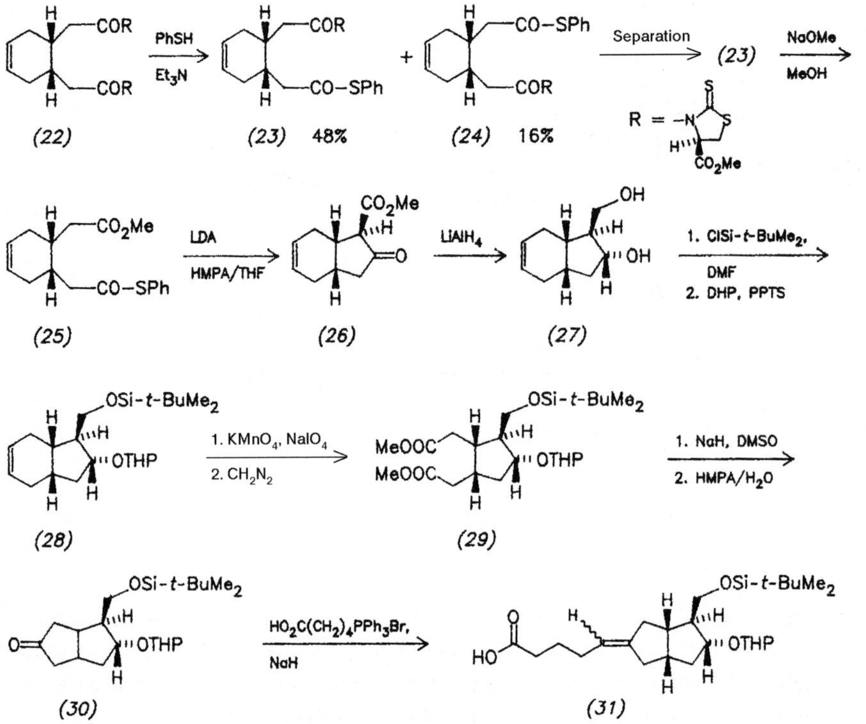

Scheme 4

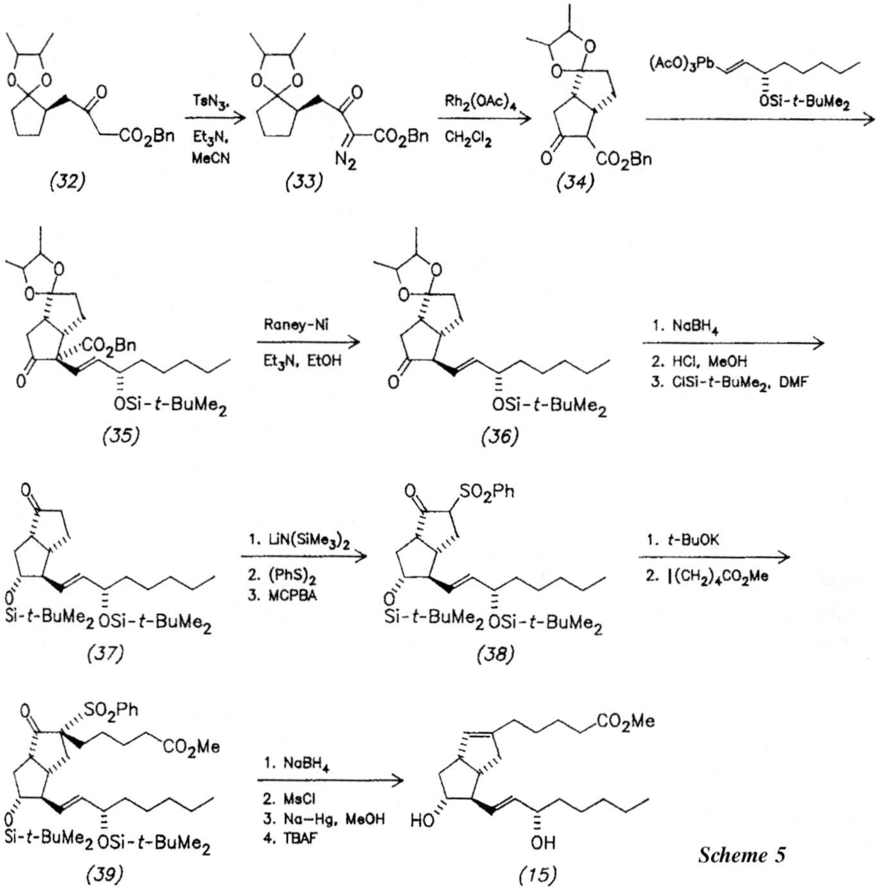

(32)   *(33)*   *(34)*

(35)   *(36)*

(37)   *(38)*

(39)   *(15)*   *Scheme 5*

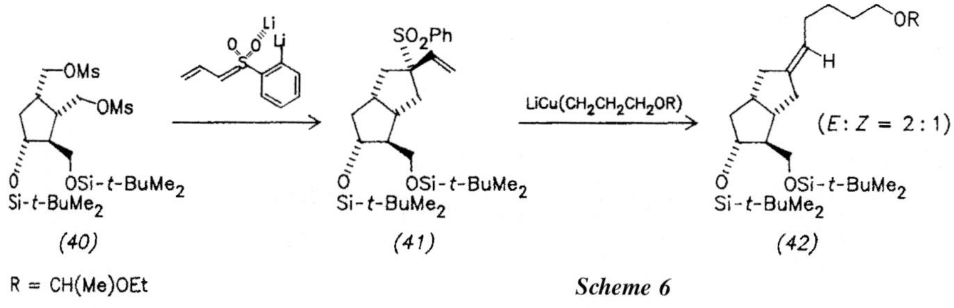

(40)   *(41)*   *(42)*

R = CH(Me)OEt   *Scheme 6*

coupling of optically active sulfoximines *(44)* (Scheme 7). [14]

The starting materials required were obtained by an asymmetric elimination *(43)* → *(44)*. The latter was further transformed in

excellent selectivity (99:1) with the catalyst system MgBr$_2$/NiCl$_2$/Ph$_2$P (CH$_2$)$_3$PPh$_2$ and the zinc reagent shown in the scheme into the carbacycline derivative.

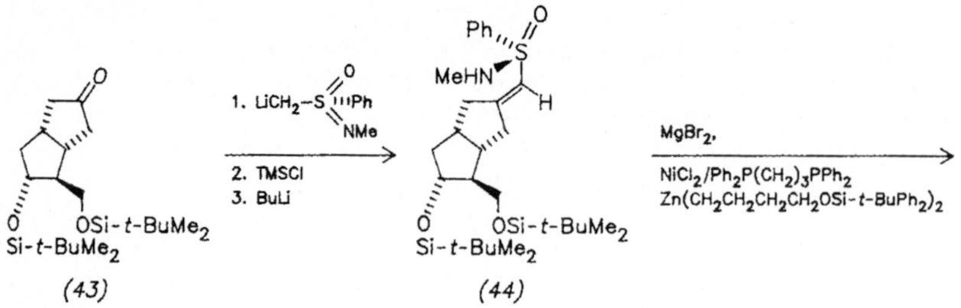

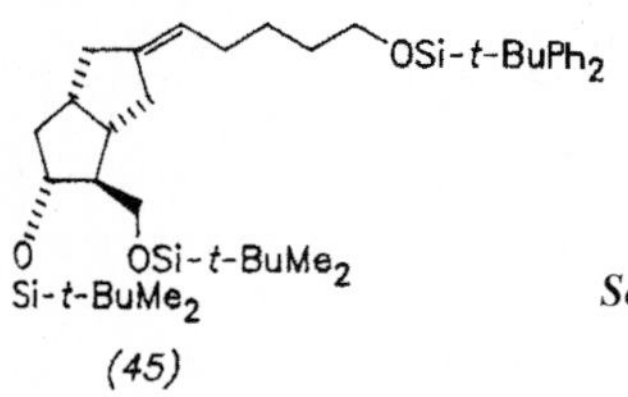

(43)   (44)

(45)   *Scheme 7*

(46)   (47)

(48)

(49)

Finally, Gais described an entry – similar to Noyori's approach – to isocarbacyclines by a radical-initiated 5-*exo-dig*-cyclization. [16] The allylester *(50)* reacted with a cuprate to the isocarbacycline precursor *(51)*. The –CH₂–OH group in the 5-membered ring will be needed later to attach the lower side chain. Gais has developed a microbiological reaction for the asymmetric construction of the 5-membered ring *(46)*. [17]

The lower side chain was introduced by standard procedures and some protecting group manipulations. The configuration at C-15 was controlled by the use of Yamamoto's reagent (9:1). [18] The epimeric allylic alcohols can be separated by chromotography and (+)-isocarbacycline *(15)* was isolated after hydrolysis of the methyl ester (Scheme 8). [16]

The Schering group used the commercially available Corey lactone *(52)* as the key intermediate. After reaction with lithiated acetic ester and some additional maneuvers a retro-Michael and subsequent Michael addition yielded the desired bicyclo octane *(55)*. The problem of the stereoselective construction of the exocyclic double bond was solved by the

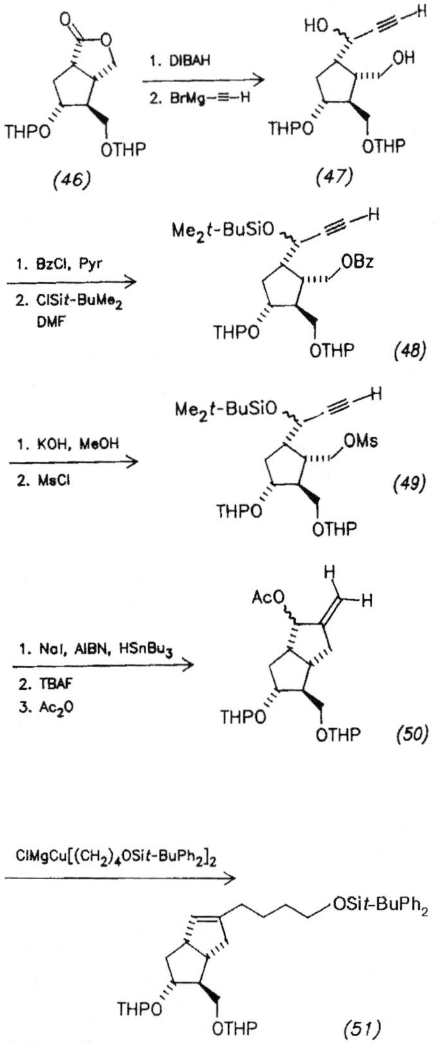

(50)

(51)

Scheme 8

use of optically active phosphine ligands in the Wittig reaction (Scheme 9). In addition, a microbiological process has been developed by the same group. [4a]

The last synthesis in particular has demonstrated that industrial laboratories are using organometallic reactions and newly developed methods to synthesize physiological important molecules by efficient routes.

The different strategies to construct the bicyclo-octane framework and the methods to solve the selectivity problems in the endo- or exocyclic double bond were presented with priority. The construction of the lower side chain is almost a routine operation.

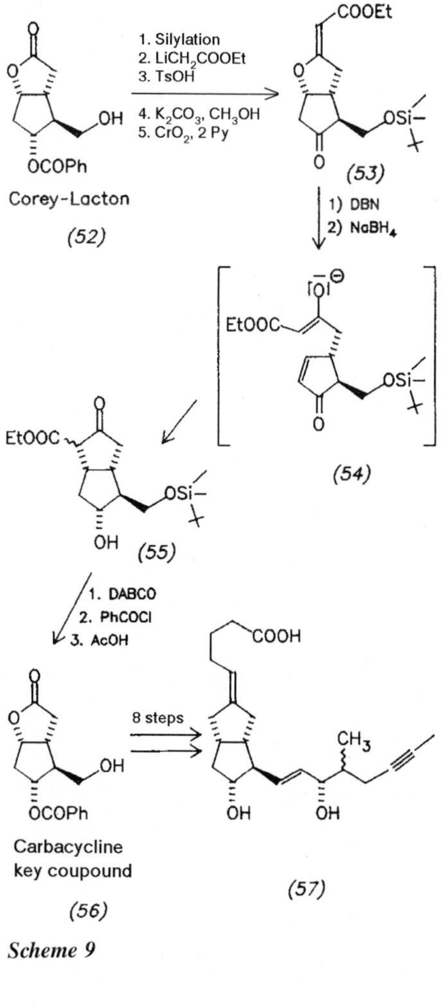

References

[1]   M. W. Goldblatt, *Chem. Ind. (London)* **1933**, *52*, 1056.

[2]   S. M. F. Lai, P. W. Manley, *Nat. Prod. Reports* **1984**, *1*, 409.

[3]   B. Samuelsson, P. Borgeat, S. Hammarström, R. C. Murphy, *Adv. Prostaglandine Thromboxane Res.* **1980**, *6*, 1.

[4]   W. Bartmann, G. Beck, *Angew. Chem.* **1982**, *94*, 767. Angew. Chem. Int. Ed. Engl. **1982**, *21*, 751.

[4a] W. Skuballa, M. Schäfer, *Nachr. Chem. Tech. Lab.* **1989**, *37*, 584.

[5]   K. Bannai, T. Tanaka, N. Okamura, A. Hazato, S. Sugiura, K. Manabe, K. Tomimori, S. Kurozumi, *Tetrahedron Lett.* **1986**, *27*, 6353.

[6]   R. Noyori, M. Suzuki, *Angew. Chem.* **1984**, *96*, 854.

[7]   T. Mase, M. Sodeoka, M. Shibasaki, *Tetrahedron Lett.* **1984**, *25*, 5087.

[8]   L. Lombardo, *Org. Synth.* **1987**, *65*, 81.

[9]   M. Suzuki, H. Koyano, R. Noyori, *J. Org. Chem.* **1987**, *52*, 5583.

[10] I. Fleming, N. K. Terret, *J. Organomet. Chem.* **1984**, *99*, 264.

[11] Y. Nagao, T. Nakamura, M. Ochiai, K. Fuji, E. Fujita, *J. Chem. Soc. Chem. Commun.* **1987**, 267.

[12] S. Hashimoto, T. Shinoda, S. Ikegami, *J. Chem. Soc. Chem. Commun.* **1988**, 1137.

[13] H.-J. Gais, W. A. Ball, J. Bund, *Tetrahedron Lett.* **1988**, *29*, 781.

[14] I. Erdelmeier, H.-J. Gais, *J. Am. chem. Soc.* **1989**, *111*, 1125.

[15] H. Hemmerle, H.-J. Gais, *Angew. Chem.* **1989**, *101*, 362. Angew. Chem. Int. Ed. Engl. **1989**, *28*, 349.

[16] G. Stork, R. Mock, Jr., *J. Am. Chem. Soc.* **1983**, *105*, 3720.

[17] H. Hemmerle, H.-J. Gais, *Tetrahedron Lett.* **1987**, *30*, 3471.

[18] S. Iguchi, H. Nakai, M. Hayashi, H. Yamamoto, K. Maaruoka, *Bull Chem. Soc. Japan.* **1981**, *54*, 3033.

Scheme 9

Synthesis of Mitomycins

Herbert Waldmann

Mitomycin A *(1)*, mitomycin C *(2)* and porfiromycin *(3)* are the most important representatives of the mitomycins [1] a class of antibiotics found in *streptomyces* species, which are active against gram positive and gram negative bacteria as well as mycobacteria. Of particular interest is the pronounced activity of mitomycin C *(2)* against a broad variety of tumors, which renders it one of the most efficient drugs in the clinical chemotherapy of cancer. The cell-destroying activity of mitomycin C is most probably initiated by reductive activation of its quinone subunit. In the course of this reaction an intermediate quinone methide is formed, which then reacts with the amino groups of two deoxyguanidins (→ *(5)*), thereby crosslinking the DNA double strands *(4)*. [2] Due to their clinical importance and their unusual structure the mitomycins have been the subject of numerous synthetic studies. However, all these endeavors were complicated by the high reactivity of the quinone

Mitomycin C
(2)

1) reductive activation
2) DNA¹
DNA²

(4)

(5)

and the aziridine structures present, as well as the facile elimination of the methoxy groups at C-9a, so that hitherto only two successful total syntheses could be reported.

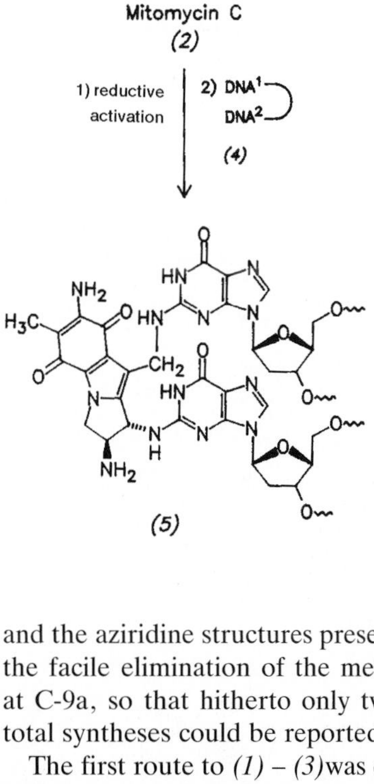

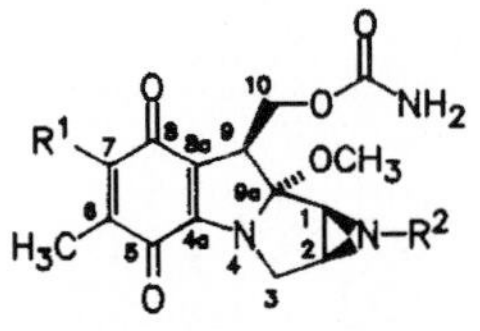

The first route to *(1)* – *(3)* was developed by Kishi et al. [3] (Scheme 1). The synthesis starts with the conversion of the allyl phenyl ether *(6)* to the alcohol *(7)* by Claisen rearrangement, epoxidation of the double bond and opening of the oxirane with the anion of acetonitrile. After oxidation of the alcohol to the corresponding ketone a crossed aldol reaction with formaldehyde was carried out, resulting in an intermediate that was con-

1): R¹ = OCH₃, R² = H Mitomycin A

2): R¹ = NH₂, R² = H Mitomycin C

3): R¹ = NH₂, R² = CH₃ Porfiromycin

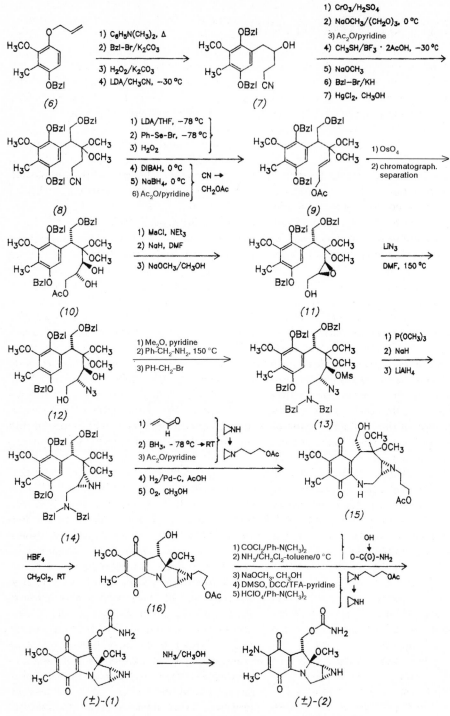

Scheme 1. Synthesis of the mitomycins *(1)–(3)* according to Kishi et al. [3].

verted to the dimethylacetal *(8)* by means of established methods. Next, Kishi et al. introduced a double bond by phenylselenation, oxidation of the selenide and elimination. Then the nitrile was transformed into the acetylated alcohol *(9)*, which was converted to a mixture of two diastereomeric diols by *cis*-hydroxylation with OsO$_4$. After chromatographic separation the desired isomer *(10)* was converted to the aziridine *(14)* in ten steps. In the course of this synthetic sequence the regioselective methylation of the less hindered OH-group in *(10)* paved the way to the stereoselective ring closure to the oxirane *(11)* and its subsequent regioselective opening to the diol *(12)* via attack of azide at the more accessible carbon atom. Next, both alcohols were mesylated and the primary mesylate was displaced to give the amine *(13)*. On treatment with trimethylphosphite, the azido function in *(13)* was converted to an amino group, which then attacked the neighbouring carbon atom to the desired aziridine *(14)*. To finish the synthesis of the mitomycins successfully, the aziridine nitrogen had to be protected by a blocking group, which could finally be removed without destruction of the product. The use of acyl functions did not meet this demand, whereas the introduction of the 3-acetoxypropyl group, which can be removed by oxidation/retro Michael addition, turned out to be the method of choice. After hydrogenolytic cleavage of the benzyl ethers and subsequent oxidation of the aromatic ring to the quinone, conjugate addition of the liberated amino group to the α,β-unsaturated system occurred immediately resulting in formation of the 8-membered quinone *(15)*. On careful treatment with HBF$_4$ this compound underwent a transannular cyclization to give the mitomycin A congener *(16)*, which was formed as a single stereoisomer. The OH group was transformed to the urethane and finally the *N*-protecting group was removed to liberate mitomycin A *(1)*. On treatment with ammonia [4], this compound can be converted

to mitomycin C *(2)*, which delivers porfiromycin *(3)* on *N*-methylation with methyl iodide. [4] Overall, the synthetic sequence developed by Kishi et al. includes ca. 40 individual operations and, therefore, does not seem to offer straightforward access to mitomycin analogs with modified physiological activity.

A much more practical route to the mitomycins was reported by Fukuyama et al. [5a, b] (Scheme 2). As the central intermediate in this synthesis the chalcone *(18)* is employed, which was constructed in 13 steps, with an overall yield of 64%, from 2,6-dimethoxytoluene *(17)*. [5c] Compound *(18)* is then coupled to the silyloxy-substituted furan *(19)* to give the silylenol ether *(20)* which, on heating in toluene, undergoes an azide-olefin cycloaddition resulting in the formation of the aziridine *(21)*. This lactone is then reduced to the acetal and the double bond degraded with RuO$_4$ to the aldehyde with simultaneous oxidation of the thioether to the corresponding sulfone. After reduction of the carbonyl group and acylation of the alcohol formed thereby to the activated urethane *(23)*, subsequent treatment with ammonia in methanol and NaBH$_4$ initiates a reaction cascade that passes throuth the aldehyde *(24)* and then finally terminates in the formation of the bridgehead semiaminal *(25)*. On treatment with acid in methanol *(25)* is transformed to the methyl ether (OH → OMe), which is then deprotected in the hydroquinone part and oxidized to the quinone *(26)* (= isomitomycin A). Finally, isomitomycin A was treated with ammonia in methanol. In the course of the occurring reaction sequence isomitomycin C *(27)* is first formed, from which the desired mitomycin C is generated by intramolecular conjugate addition of the pyrrolidine nitrogen to give *(28)* and a subsequent retro-Michael reaction resulting in the cleavage of the C–N bond to the aziridine nitrogen. By means of this "mitomycin rearrangement" [6] *(26)* can also be converted to mitomycin A. Although this total synthesis also requires numerous chemical

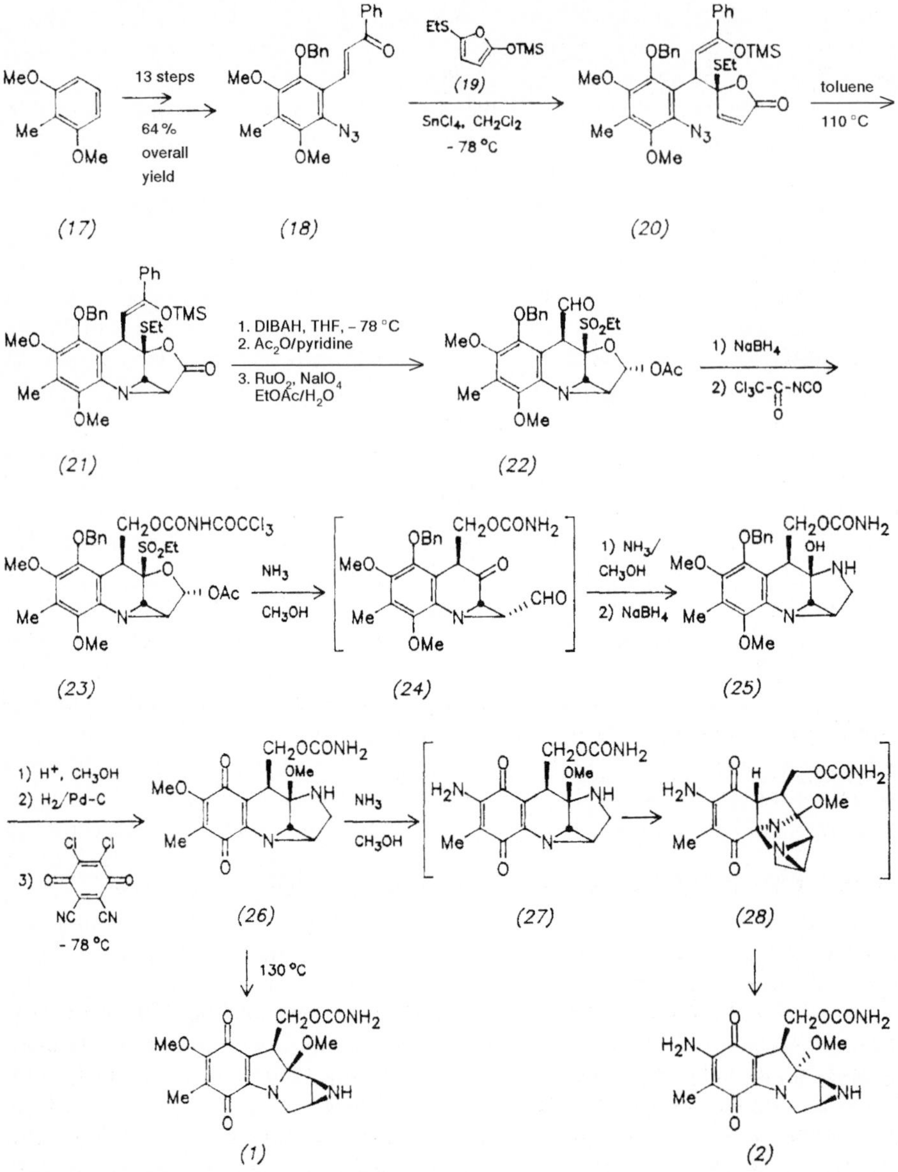

Scheme 2. Synthesis of the mitomycins *(1)–(3)* according to Fukuyama et al. [5].

transformations overall it proceeds with high
efficiency and makes mitomycin C available
from 2,6-dimethoxytoluene with an overall
yield of 10%.

Rappoport et al. [7a] succeeded in the construction of a non-racemic aziridinomitosene

(37) (Scheme 3). The key steps in the route
followed are the addition of the aziridinopyrrolidine *(32)* to the dibromoquinone *(33)* and
a Pd-mediated cyclization resulting in the
formation of the complete carbon framework.
To form *(32)* in enantiomerically pure form

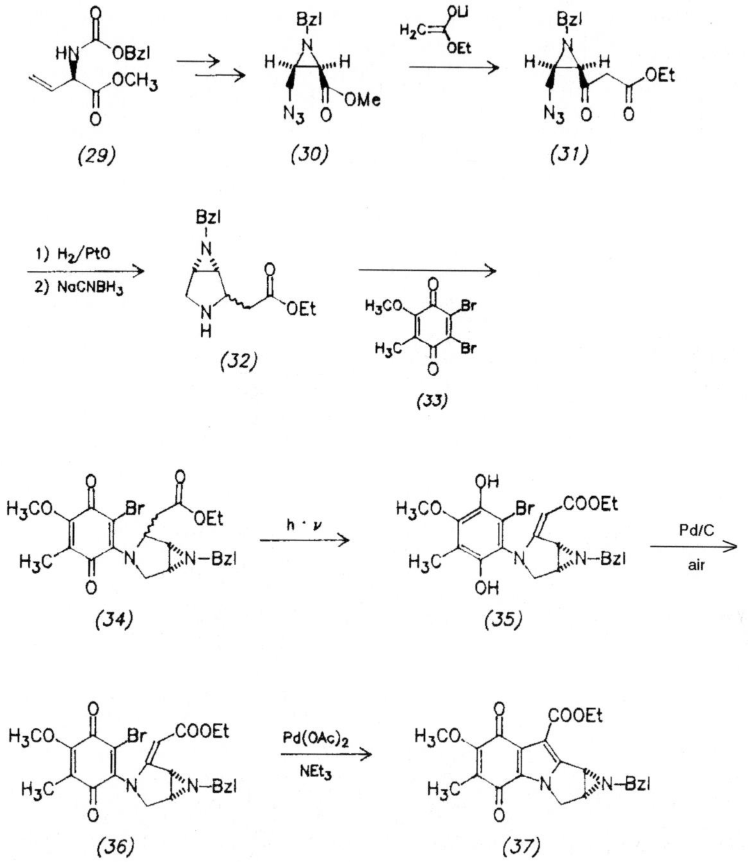

Scheme 3. Synthesis of the enantiomerically pure aziridinomitosene *(37)* according to Rappoport et al. [7a, b].

the D-vinylglycine ester *(29)* (obtained from D-methionine) was converted to the aziridine *(30)* by a sequence of reactions that included an epoxidation and suitable inter- and intra-molecular nucleophilic substitutions. Homo-logisation of *(30)* to the β-keto ester *(31)*, followed by selective reduction of the azide and subsequent cyclization, delivers an unsaturated aziridinopyrrolidine, which is reduced with NaCNBH$_3$ to the nitrogen base *(32)*. In the following steps *(32)* is condensed with the dibromoquinone *(33)* to the vinylogous amide *(34)* which, on irradiation with light, is converted to the hydroquinone *(35)* in a photo-

chemical oxidation/reduction. [7b] After reoxidation of *(35)* to the quinone *(36)* the planned ring closure to the enantiomerically pure aziridinomitosene *(37)* can be effected by employing a Pd-mediated cyclization. The absolute configuration at C-1 and C-2 of *(37)* is opposite to the structure of the naturally occurring mitomycins, however, starting from L-methionine, this synthetic route of course would make the respective enantiomer available.

In addition to the syntheses highlighted, the chemistry of the mitomycins has provided a rich and rewarding source of interesting problems for numerous research groups. [1,8–10] However, the unusual structure and the pronounced reactivity of these antitumor antibiotics continue to present challenges that by no means are easily met.

References

[1] Review: R. W. Franck, *Fortschr. Chem. Org. Naturst.* **1979**, *38*, 1.
[2] a) M. Tomasz, R. Lipman, D. Chowdary, J. Pawlak, G. L. Verdinge, K. Nakanishi, *Science* **1987**, *235*, 1204; b) J. T. Millard, M. F. Weidner, S. Raucher, P. B. Hopkins, *J. Am. Chem. Soc.* **1990**, *112*, 3637; c) H. Kohn, Y. P. Hong, *J. Am. Chem. Soc.* **1990**, *112*, 4596; d) V. Li, H. Kohn, *J. Am. Chem. Soc.* **1991**, *113*, 275–238.
[3] a) F. Nakatsubo, T. Fukuyama, A. J. Cocuzzo, Y. Kishi, *J. Am. Chem. Soc.* **1977**, *99*, 8115; b) T. Fukuyama, F. Nakatsubo, A. J. Cocuzzo, Y. Kishi, *Tetrahedron Lett.* **1977**, *49*, 4295.
[4] J. S. Webb, D. B. Cosulich, J. H. Mowat, J. B. Patrick, R. W. Broschard, W. E. Meyer, R. P. Williams, C. F. Wolf, W. Fulmor, C. Pidachs, J. E. Lancaster, *J. Am. Chem. Soc.* **1962**, *84*, 3185.
[5] a) T. Fukuyama, L. Yang, *J. Am. Chem. Soc.* **1989**, *111*, 8303; b) T. Fukuyama, L. Yang, *J. Am. Chem. Soc.* **1987**, *109*, 7881; c) T. Fukuyama, L. Yang, *Tetrahedron Lett.* **1986**, 6299.
[6] Y. Arai and S. Ishii, *J. Am. Chem. Soc.* **1987**, *109*, 7224.
[7] a) K. J. Shaw, J. R. Luly, H. Rappoport, *J. Org. Chem.* **1985**, *50*, 4515; b) J. R. Luly, H. Rappoport, *J. Am. Chem. Soc.* **1983**, *105*, 2859.
[8] Aziridinomitosenes and -mitosanes: a) S. Danishefsky, E. M. Bermann, M. Ciufolini, S. J. Etheredge, B. S. Segmuller, *J. Am. Chem. Soc.* **1985**, *107*, 3891; b) S. J. Danishefsky, M. Egbertson, *J. Am. Chem. Soc.* **1986**, *108*, 4648; c) S. Nakajima, K. Yoshida, M. Mori, Y. Ban, M. Shibasaki, *J. Chem. Soc. Chem. Commun.* **1990**, 468.
[9] Mitosenes: a) P. A. Wender, C. B. Cooper, *Tetrahedron Lett.* **1987**, *49*, 6125; b) S. Nakatsuka, O. Asano, T. Goto, *Chem. Lett.* **1987**, 1225; c) J. Rebek, Jr., S. H. Shaber, Y. Shue, J. Gehret, S. Zimmerman, *J. Org. Chem.* **1984**, *49*, 5164; d) R. M. Coates, P. A. MacManus, *J. Org. Chem.* **1982**, *47*, 4822.
[10] a) A. P. Kozikowski, B. J. Mugrage, *J. Chem. Soc., Chem. Commun.* **1988**, 198; b) W. Verboom, E. O. M. Oremans, H. J. Berga, M. W. Scheltinga, D. N. Reinhoudt, *Tetrahedron*, **1986**, *42*, 5053; c) Y. Naruta, N. Agai, T. Yokota K. Maruyama, *Chem. Lett.* **1986**, 1185; d) W. Flitsch, P. Rußkamp, *Liebigs Ann. Chem.*, **1985**, 1398 and 1422; e) W. Flitsch, P. Rußkamp, W. Langer, *Liebigs Ann. Chem.*, **1985**, 1413; f) T. Kametani, T. Ohsawa, M. Ihara, *J. Chem. Soc. Perkin Trans. I*, **1981**, 290; g) K. F. McClure, J. W. Benbow, S. J. Danishefsky, G. K. Schulte, *J. Am. Chem. Soc.* **1991**, *113*, 8185.

Syntheses of Ergot Alkaloids

Thomas Brumby

Ergot alkaloids are pharmacologically important molecules, the most widely known of which is the diethylamide of lysergic acid (LSD). They owe their interesting properties to interaction with the receptors of the neurotransmitters noradrenaline, dopamine and serotonin. The interest of the medicinal chemist is attracted by the fact that selectivity for these receptors can be influenced markedly by structural variation. Thus, it is not surprising that many groups deal with total synthesis as well as variation of ergot alkaloids.

The synthesis of lysergic acid was reviewed in this column five years ago [1]. Since then two new total syntheses have been published [2, 3].

Kurihara et al. [2] started with Kornfeld's ketone *(1)* (Scheme 1), which was treated with diethyl phosphorocyanidate (DEPC) to yield phosphorylated cyanohydrin *(2)*. Boron trifluoride diethyl etherate catalysed elimination then gave unsaturated nitrile *(3)*. This sequence is generally applicable to aromatic ketones [4]. DiBAH reduction with subsequent benzoylation gave aldehyde *(4)*, which has been used in previous syntheses. The reaction with the enolate of Boc-protected ethyl 3-(methylamino)propionate *(5)* to *(6)* completed the C-frame of lysergic acid. The following steps were carried out without isolation of intermediates: Mesylation and cleavage of the Boc-group gave *(7)*. Unfortunately, the Boc-group was not inert in the mesylation

step: 30 % of the material is lost as oxazinone *(8)*. The diene *(9)*, generated by treatment of *(7)* with DBU, cyclized in situ. The isomerisation in position 8 under the basic reaction conditions was expected and *(10)* was accordingly obtained as mixture of diastereomers. The overall yield from *(1)* was 22 % nevertheless. The conversion to racemic lysergic acid *(11)* is known.

The second synthesis, which allows 14-substitutents to be incorporated [5] follows an unusually interesting strategy [3] (Scheme 2). Haefliger generates the indole-system by his own method at the *end of the synthesis,* thus avoiding many problems usually associated with the presence of the reactive aromatic system. The synthesis commenced with construction of the D-ring from 5-nitro-tetralone *(12)* in one step. This reaction was developed by Grob and Renk [6] for ergoline synthesis and allows easy incorporation of 14-substituents into the ergoline via 6-substituted tetralones. Reduction of enamine *(13)* with sodium cyanoborohydride gave *trans*-compound *(14)* in 53 % yield; 20 % is lost as the unwanted *cis*-isomer. Catalytic hydrogenation of the nitro-group, formylation of the resulting amine with formylimidazole and dehydration with phosphorus oxychloride/diisopropylethylamine gave isonitrile *(15)* in excellent yield. LDA-treatment generates a benzyl anion which reacts with the isonitrile to form the indole ring. As in the former synthesis the basic con-

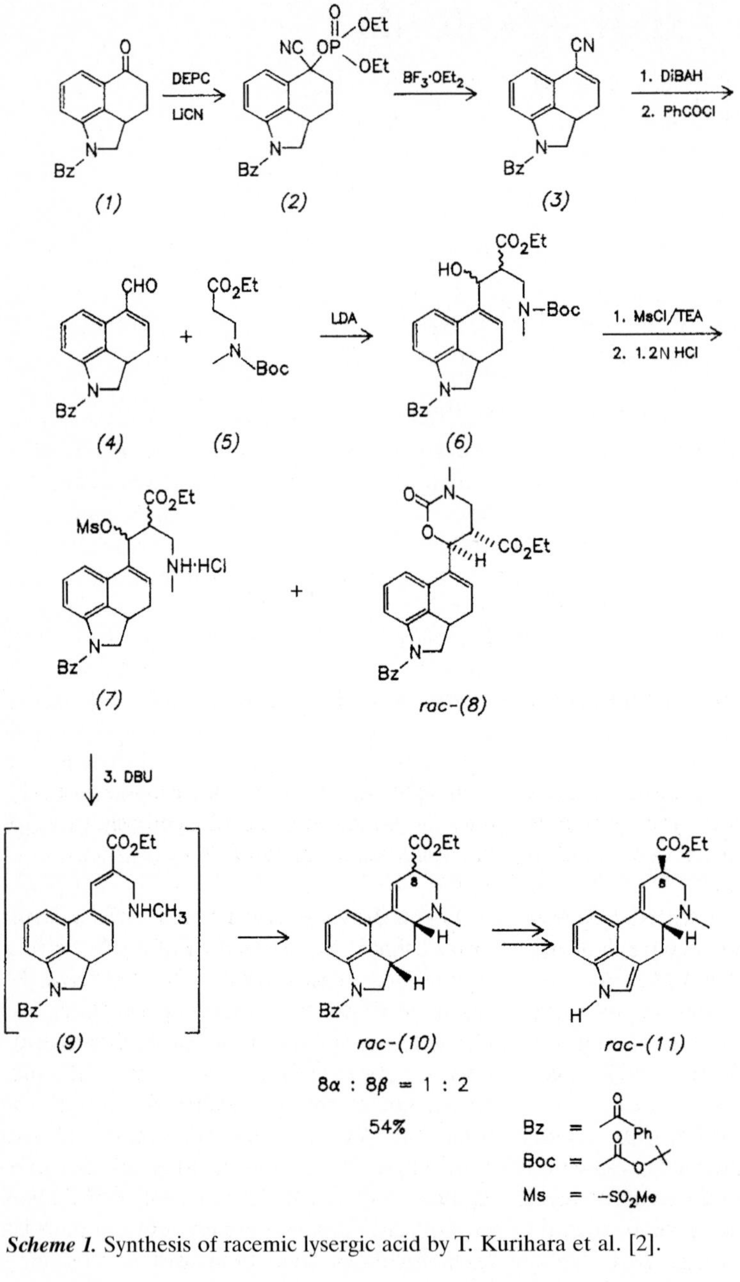

Scheme 1. Synthesis of racemic lysergic acid by T. Kurihara et al. [2].

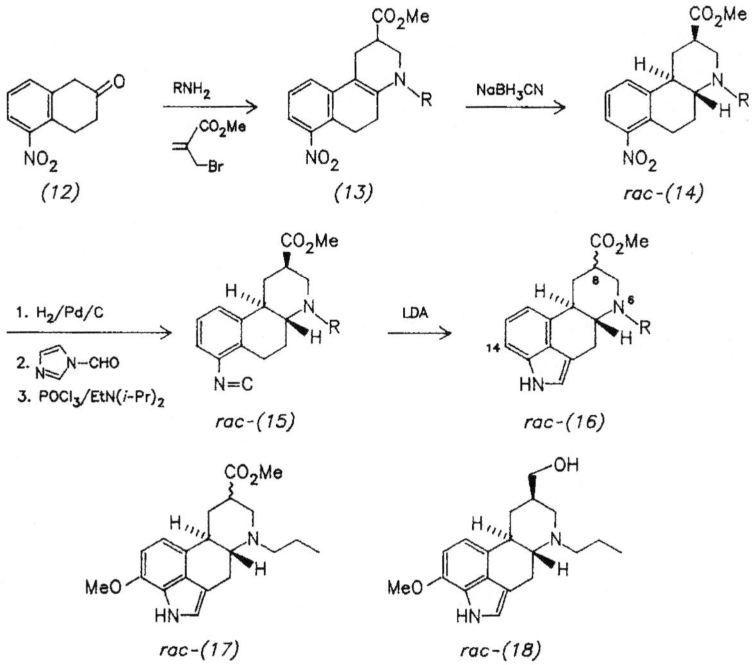

Scheme 2. Synthesis of racemic dihydro lysergic acid derivatives by W. E. Haefliger [3].

ditions cause equilibration in position 8 and a 1:1 mixture of the $8\alpha/8\beta$-dihydrolysergates *(16)* (R = Me) is obtained. If one wants to give up the ester functionality (which cannot be regenerated after indol formation) it is possible to reduce *(14)* and take the THP-protected alcohol through the following reactions. The 14-methoxy-derivatives *(17)* and *(18)* were synthesised in overall yields of 11 and 14 %, respectively; in the case of *(18)* pure 8β-diastereomer was obtained.

An atypical ergot alkaloid is clavicipitic acid *(19)*, the structure of which was determined by X-ray analysis in 1980. [7] Since then the total synthesis and the biosynthesis of clavicipitic acid have interested several groups. [8–13] Some *Claviceps*-strains produce clavicipitic acid as a diastereomeric mixture (at C-10 [14]), the *cis*-isomer being predominant. It is believed that a derailment of the normal metabolism after the prenylation of trypto-

phan leads to *(19)*. Ring closure does not take place as in the normal biosynthesis of ergolines between C-10 und C-5, but between C-10 and N-6.

The published syntheses of *(19)* also rely on this strategy. Matsumoto et al. [12] started with 4-cyanomethylindole *(20)* (Scheme 3). Alkylation with methallyl tosylate gave *(21)*, which was subjected to aerobic oxidation after deprotonation with potassium *tert*-butoxide. This reaction gave a mixture of ketones *(22)* and *(23)*. Subsequent deprotection of the indole by basic hydrolysis afforded only conjugated ketone *(24)*, the deconjugated double bond isomerised completely under the reaction conditions. Functionalisation of position 3 by Mannich reaction and condensation with diethyl *N*-formylamidomalonate gave *(26)*, which contains all necessary carbon atoms for ring closure. The critical step was the removal of the *N*-formyl group with consecutive

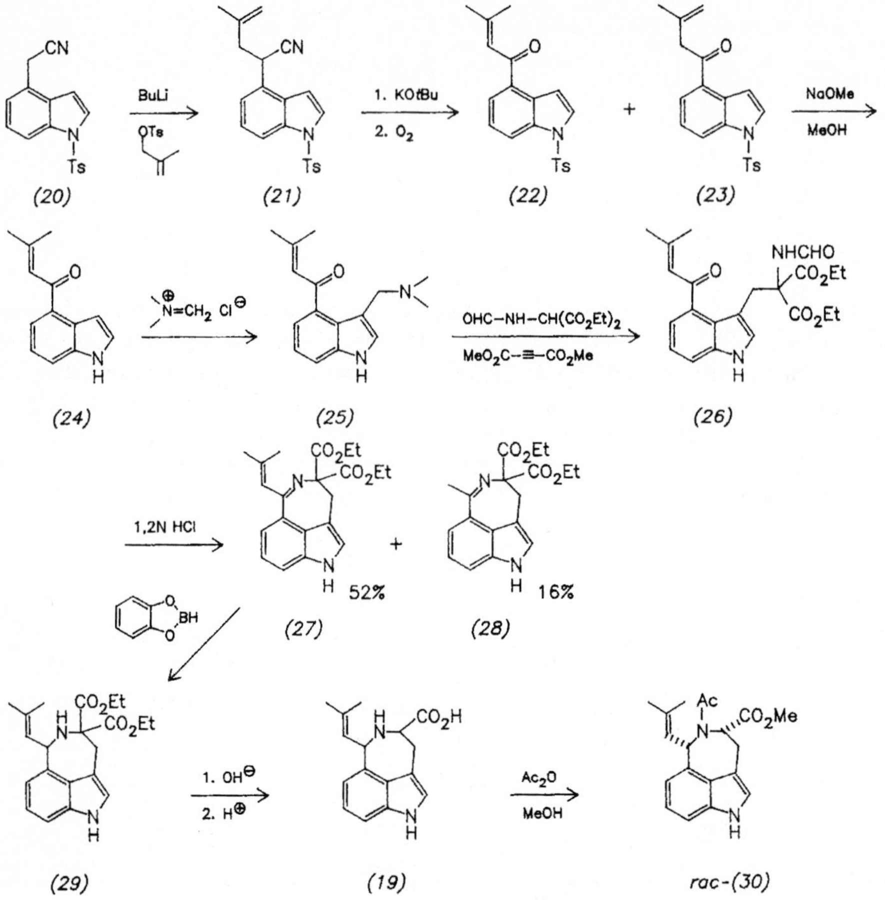

Scheme 3. Synthesis of clavicipitic acid by M. Matsumoto et al. [12].

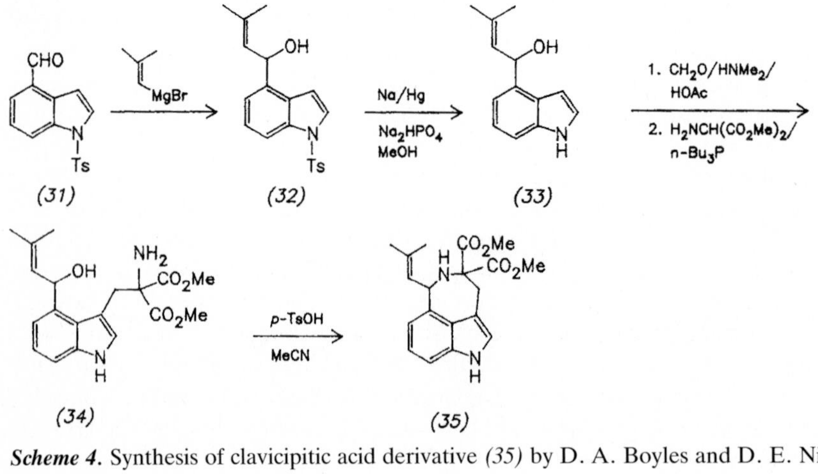

Scheme 4. Synthesis of clavicipitic acid derivative *(35)* by D. A. Boyles and D. E. Nichols [13].

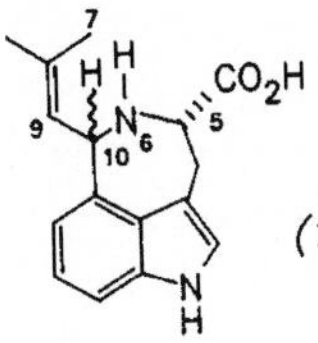

formation of Schiff base *(27)*. Usually conjug-ated addition of primary amines to α,β-unsaturated ketones is observed. In this case, however, only 1,2-addition was seen, the formation of a nine-membered ring probably being disfavored for steric reasons. The only isolated byproduct *(28)* is likely formed via a retro aldol reaction. The selective reduction of the imine is possible with catecholborane in

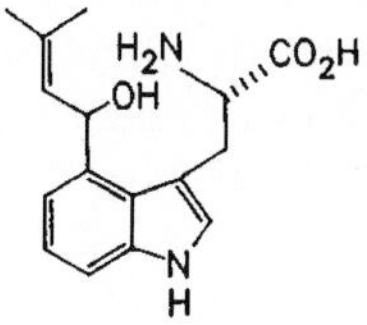

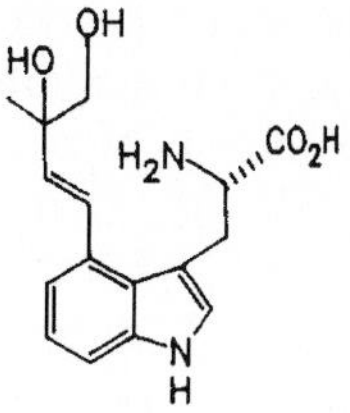

good yield. Saponification of *(29)* and decarb-oxylation completed the synthesis of racemic clavicipitic acid as a *cis/trans* mixture. To cha-racterise *(19)*, it was converted to the *cis-N*-acetyl methyl ester *(30)*. The isomerisation of *trans-(19)* in this reaction was also found by other authors. [8, 9]

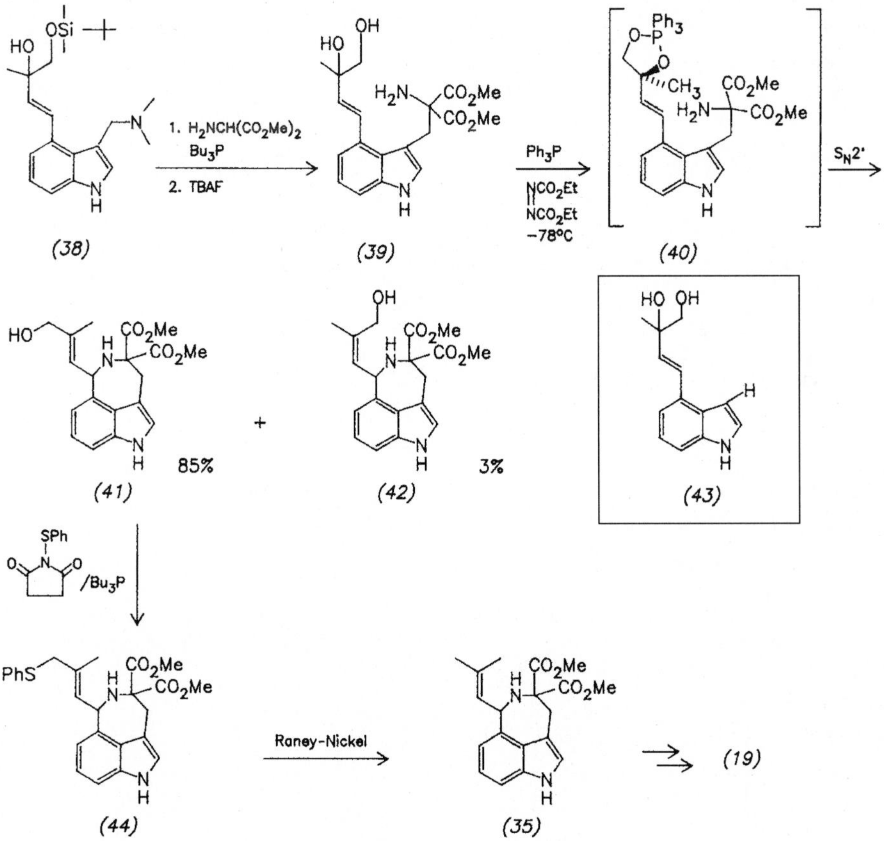

Scheme 5. Synthesis of clavicipitic acid derivative *(35)* by A. P. Kozikowski and M. Okita [10].

This somewhat lengthy sequence can be shortened considerably, as demonstrated by Boyles and Nichols [13] (Scheme 4). Grignard reaction of 2-methyl-1-propenylmagnesium bromide with formylindole (31) gave the hydroxylated prenylindole-derivative (32). The protecting group was removed quantitatively with sodium amalgam in buffered solution. The side chain was introduced as in the previous synthesis. However, the formyl protecting group proved unnecessary – the Mannich compound directly condensed with dimethyl aminomalonate and tri-n-butylphosphine to give (34). The final cyclisation with p-toluenesulfonic acid in acetonitrile afforded (35) in 48%, the overall yield over five steps amounted to an excellent 26%!

In this reaction an analogue of the postulated biosynthetic intermediate (36) was cy-clised. The cyclization of an analogue of diol (37), another possible precursor of (19) was done in the synthesis of Kozikowski et al. [10] (Scheme 5). Protected diol (38), which was synthesized in several steps from 4-ethinyl-indole, was treated with dimethyl aminomalonate/tri-n-butylphosphine. Subsequent removal of the silyl group with tetrabutylammonium fluoride (TBAF) gave (39), the analog to (37). Cyclization under Mitsunobu conditions is highly stereoselective: in addition to 85% of E-allylalcohol (41), only 3% Z-isomer (42) was obtained. Probable intermediate in this reaction is compound (40). Subjecting model diol (43) to Mitsunobu conditions, the ^{31}P-resonance of the postulated cyclic phosphorane was observed. The intermediacy of (40) could be an explanation for the high stereoselectivity. The sterically less

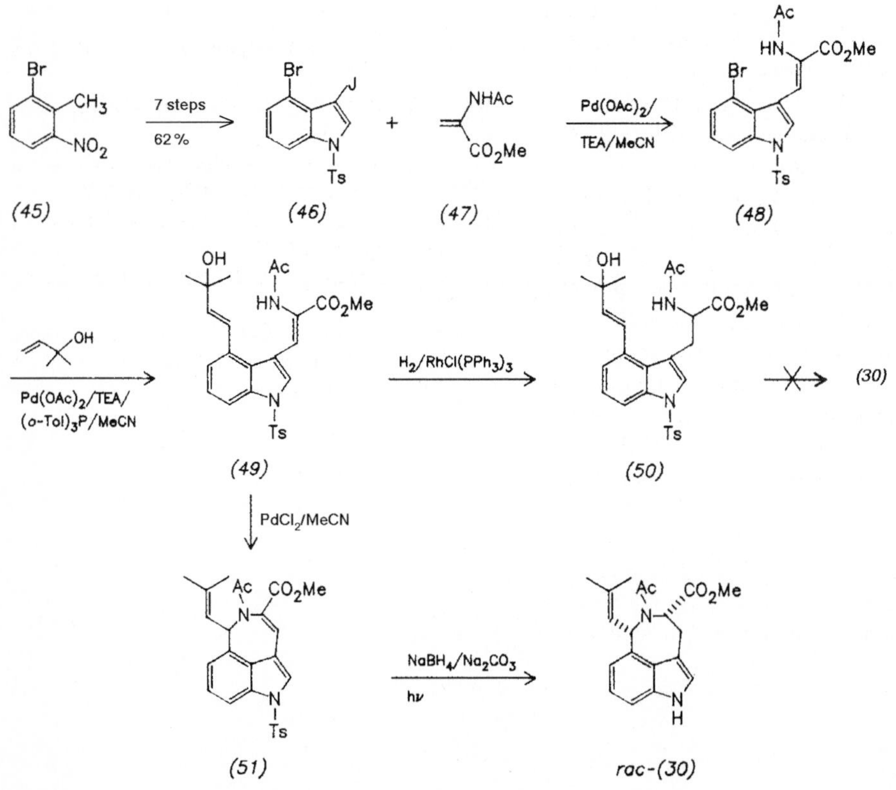

Scheme 6. Synthesis of racemic N-acetyl clavicipitic acid methyl ester (30) by L. S. Hegedus et al. [11].

demanding of the two possible *syn*-SN2′ transition states leads to the observed product. The formal synthesis was completed by conversion of *(41)* to the phenyl sulfide *(44)* and desulfurisation with Raney nickel in DMSO.

The last synthesis of clavicipitic acid by Hegedus et al. [11] reviewed here is remarkable in the consistent use of palladium catalyzed reactions.

Bromo-iodo-indole *(46)* was made available by the authors through an efficient 7 step synthesis in 62 % overall yield from 2-bromo-6-nitro-toluene *(45)*. Only three palladium catalyzed steps are then necessary to assemble the frame of clavicipitic acid. The first step is the Heck reaction of methyl α-acetamidoacrylate *(47)* with *(46)*. Only Z-isomer *(48)* is formed in 60 % yield along with 20 % deiodinated indole, which can be recycled to *(46)*. The side chain in position 4 is introduced similarly by palladium catalyzed reaction with 2-methyl-3-buten-2-ol in 89 % yield. The final cyclization of *(49)* to *(51)* is carried out with palladium dichloride in acetonitrile in 95 %. However, similar yields can be realised with *p*-toluenesulfonic acid. The originally planned cyclization of saturated compound *(50)*, which was obtained by homogeneous hydrogenation in 98 % using Wilkinson's catalyst, to *(30)*, succeeded neither under acid or basic conditions nor under transition metal catalysis. The authors explain the easy cyclization of *(49)* by the favorable positioning of the amide function by the Z-geometry of the acetamidoacrylate side chain. The known [9] photoreduction of *(51)* with simultaneous removal of the tosyl protection group was optimized to give 60 % pure *cis*-isomer *(30)*. The overall yield of this elegant 12-step synthesis is 18 %, starting from *(45)*.

References

[1] J. Mulzer, *Nachr. Chem. Tech. Lab.* **1984,** *32,* 721.

[2] T. Kurihara, T. Terada, S. Harusawa, R. Yoneda, *Chem. Pharm. Bull.* **1987,** *35,* 4793.

[3] W. E. Haefliger, *Helv. Chim. Acta* **1984,** *67,* 1942.

[4] T. Kurihara, Y. Hamada, T. Shioiri, S. Harusawa, R. Yoneda, *Tetrahedron Lett.* **1984,** *25,* 427.

[5] Syntheses of 14-substituted lysergic acid derivatives from natural ergolines are not known to date.

[6] C. A. Grob, E. Renk, *Helv. Chim. Acta* **1961,** *44,* 1531.

[7] J. E. Robbers, H. Otsuka, H. G. Floss, E. V. Arnold, J. Clardy, *J. Org. Chem.* **1980,** *45,* 1117.

[8] A. P. Kozikowski, M. Greco, *Heterocycles* **1982,** *19,* 2269 and *J. Org. Chem.* **1984,** *49,* 2310.

[9] H. Muratake, T. Takahashi, M. Natsume, *Heterocycles* **1983,** *20,* 1963.

[10] A. P. Kozikowski, M. Okita *Tetrahedron Lett.* **1985,** *26,* 4043.

[11] P. J. Harrington, L. S. Hegedus, K. F. McDaniel, *J. Am. Chem. Soc.* **1987,** *109,* 4335.

[12] M. Matsumoto, H. Kobayashi, N. Watanabe, *Heterocycles* **1987,** *26,* 1197.

[13] D. A. Boyles, D. E. Nichols, *J. Org. Chem.* **1988,** *53,* 5128.

[14] Stereochemistry at C-5 was not experimentally determined but postulated from investigation of the biosynthesis [7].

Enantioselective Synthesis of Piperidine Alkaloids

Peter Hammann

Alkaloids represent a large group of plant- and animal compounds with a wide variety of activity. The discovery of alkaloids in the last century led to a great surge of interest in natural product chemistry. Alkaloids are not only important pharmacological products, they are also interesting models for the design of new drugs. In addition, as a result of their specific affinity for receptors, they possess great importance for biochemical investigations. Piperidines and pyrrolidines are integral components of countless alkaloids; construction of these fragments is thus frequently a central step in the synthesis of alkaloids. Moreover, simple substituted piperidines and pyrrolidines are often biologically active. Example of piperidines include the pseudodistamines A and B from *Pseudodistoma kanoko* (cytotoxic), [1] derivatives of desoxyfuconojirimicin (antiviral agent against HIV viruses), [2] the synthetic etoxadrol (phencyclidine agonist), [3] and the 2- or 2,6–substituted compounds (neurotoxic, insecticidal) [4] derived from coniin, pinidine, or solenopsine. In the following account, exemplary examples of piperidine synthesis, which have been developed between 1983 and 1989, will be presented which, in slightly modified form, could also be used for the synthesis of the pyrrolidines. The main objective of the synthesis of piperidine alkaloids directed towards the investigation of their biological activities is the preparation of enantiomerically pure com-

pounds in high yield whereby with modern synthetic concepts both enantiomers should be synthesized, so that these can be individually examined. If the construction of a stereocenter is predictable, then the absolute configuration of a natural product can be clarified by a total synthesis. In particular when X-ray structure analysis cannot be used, because of a lack of suitable crystals of a natural product isolated in only trace amounts, synthesis is obviously the only way to complete the structure determination.

To obtain both enantiomers separately, both enantiomers of a chiral reagent, of a chiral building block, or a chiral auxiliary are usually required for an asymmetric synthesis. In many cases one of the two enantiomers is either not available from natural sources or is extremely expensive. An alternative is therefore the selective formation of both enantiomers starting from a single enantiomeric, chiral building block.

Stereocontrolled reactions can be performed, for example, by substrate control. In this case an achiral reagent will be converted with a chiral substrate into diasteromers, whereby the chirality is already present in the substrate or can be introduced by connection to a chiral auxilliary. L-Lysine, which is also the starting product for the biosynthesis of several of these compounds, is an ideal chiral substrate for the synthesis of piperidine alkaloids. An efficient method for the cyclization

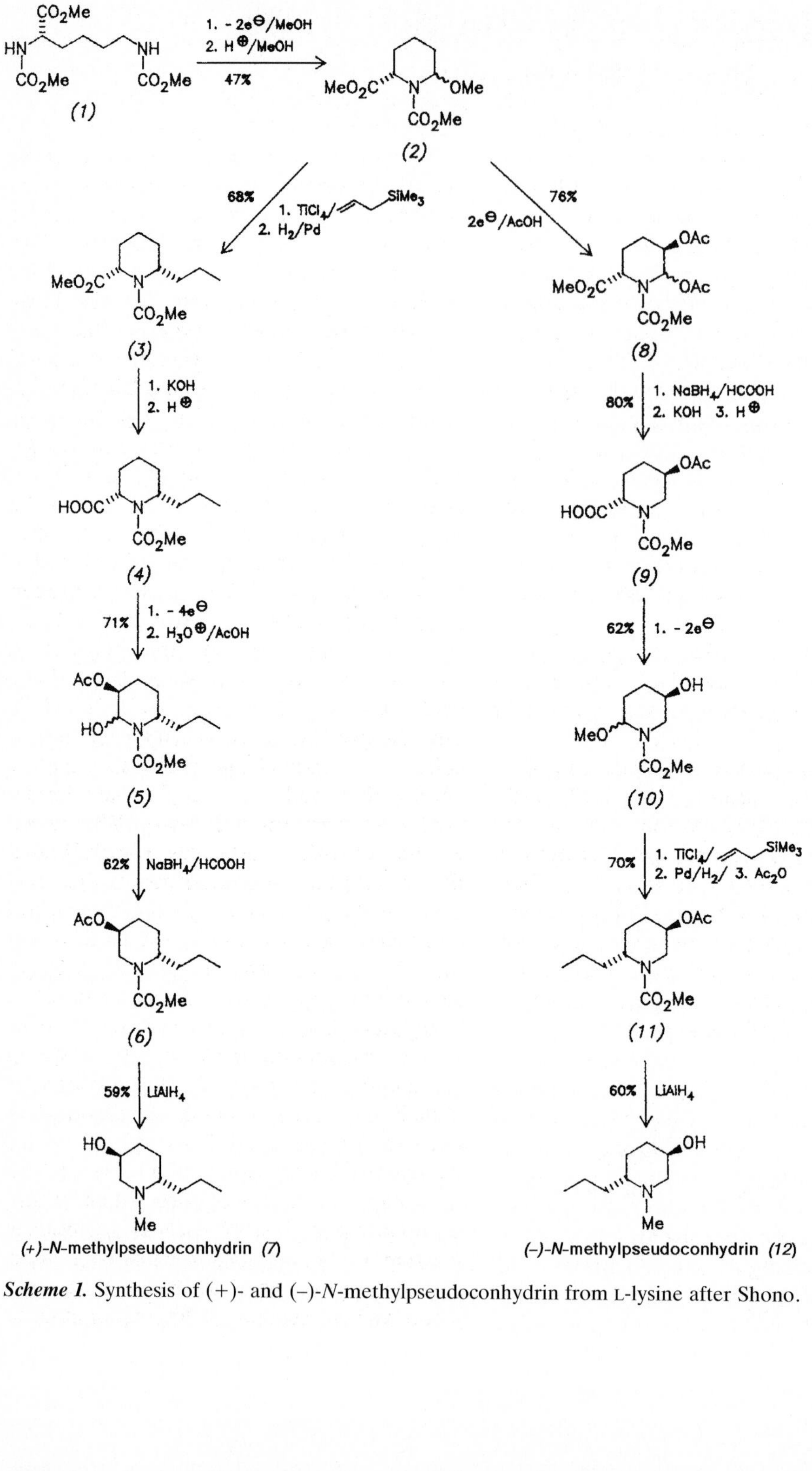

Scheme 1. Synthesis of (+)- and (−)-N-methylpseudoconhydrin from L-lysine after Shono.

of L-lysine to the piperidine skeleton involves electrochemical anodic oxidation of $N\alpha,N\varepsilon$-bis(methoxycarbonyl)-L-lysine methyl ester *(1)* (Scheme 1). [5] The intermediate open-chain ε-N,O-acetal undergoes cyclization to the chiral, cyclic N,O-acetal *(2)* with retention of the stereocenter of L-lysine. The intermediate *(2)* is a valuable building block for the synthesis of a multitude of alkaloids. [6, 7] Thus, for example, a propyl group can be introduced diastereoselectively into *(3)* under the control of the methoxycarbonyl group in the 6-position by amidoalkylation [8, 9] of *(2)* with TiCl$_4$/allylsilane via a reactive acyl-iminium ion and subsequent hydrogenation. After hydrolysis of the ester the N,O-hemiacetal *(5)* is formed from *(4)* by anodic oxidation and hydrolysis. The reaction probably takes place by way of an anodic decarboxylation and subsequent elimination to give an enamine, which can be anodically diacetoxylated diastereoselectively under the influence of the propyl group. Subsequent reduction of *(5)* with NaBH$_4$ to *(6)* and then LiAlH$_4$ led to the natural $(+)$–$(2S,5S)$-N-methylpseudoconhydrin *(7)*.

The synthesis of the $(-)$-enantiomer *(12)* starts with anodic diacetoxylation of *(2)* to *(8)*. The stereochemistry is controlled by the center at C-6. After reduction and hydrolysis to *(9)* compound *(10)* is obtained by anodic decarboxylation. Amidoalkylation with allyltrimethylsilane/TiCl$_4$, hydrogenation, and acetylation gave *(11)*, which was converted into the unnatural enantiomer $(-)(2R,5R))$-N-methylpseudocon-hydrin *(12)*.

The strategy for the synthesis of both enantiomers of pinidine *(24)* and *(28)* according to Kibayashi is based on the construction of a new stereocenter by diastereoselective hydride addition to a chiral α,β-bis[(methoxymethyl)oxy]ketone with subsequent cleavage of the original chirality-inducing group. [10] The starting material is L-tartaric acid diethyl ester *(13)*, [11] which is converted into the chiral ketone *(20)* via compounds *(14–19)* (Scheme 2). [12] Reduction with zinc borohydride gave almost exclusively the *anti*-diastereomer *(26)* in good yield. Formation of the *anti*-compound can be explained by α-coordination of the zinc atom in the transition state *(25)*, in which the nucleophile attacks the *re*-side of the carbonyl group (1,2-asymmetric induction). The preference of the α- over the β-chelate formation can probably be traced to a "crown ether" effect, through which a stable coordination structure is formed from the carbonyl oxygen and the MOM groups with the zinc as central atom. On the other hand, reduction with L-selectride leads predominantly to the *syn*-diastereomer *(22)* (91:9). The course of this reaction can be understood either via an open-chain transition state *(21a)* (Anh-Felkin model) or via β-chelate formation (1,3-asymmetric induction) *(21b)*. As trialkyl borohydrides show only a weak inclination to coordination, as a result of their low Lewis acidity, it is assumed that the reaction takes place via the transitiion state *(21a)* and that the *syn*-selectivity is only additionally strengthened via β-chelate formation in *(21b)*. Compound *(27)*, which can be converted over six further steps into the natural $(-)$-$(2R,6R)$-pinidine *(28)*, can be obtained from *(26)* with phthalimide under Mitsunobu conditions. The exact reaction sequence from *(27)* starts with the hydrazinolysis of the phthalimide protecting group and subsequent benzylation of the amino group to give *(29)* (Scheme 3). Wacker reaction to ketone *(30)* and hydrogenation gave, after cleavage of the benzyl protecting group and subsequent reductive amination,

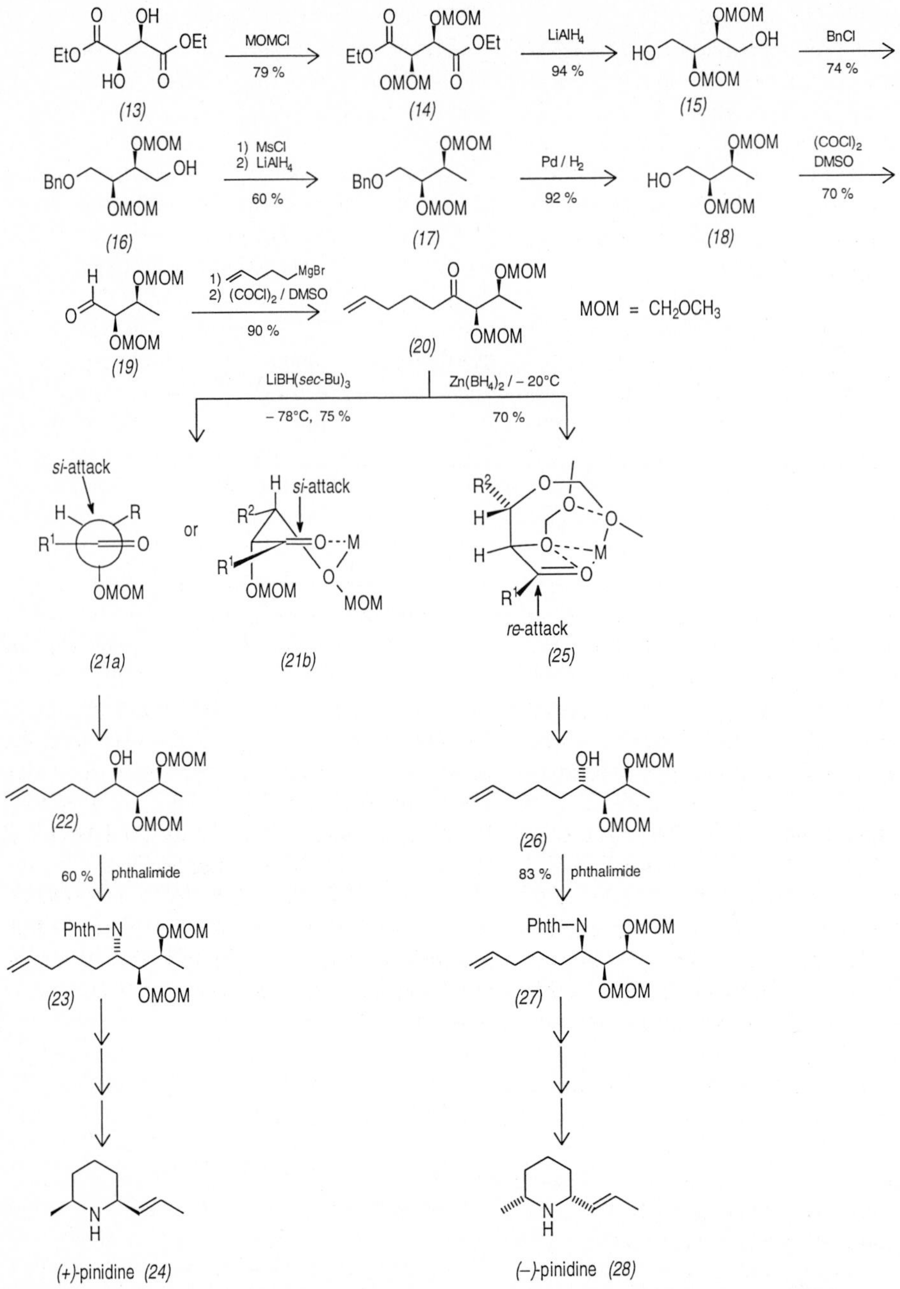

Scheme 2. Synthesis of (+)- and (−)-pinidine from L-diethyl tartrate after Kibayashi.

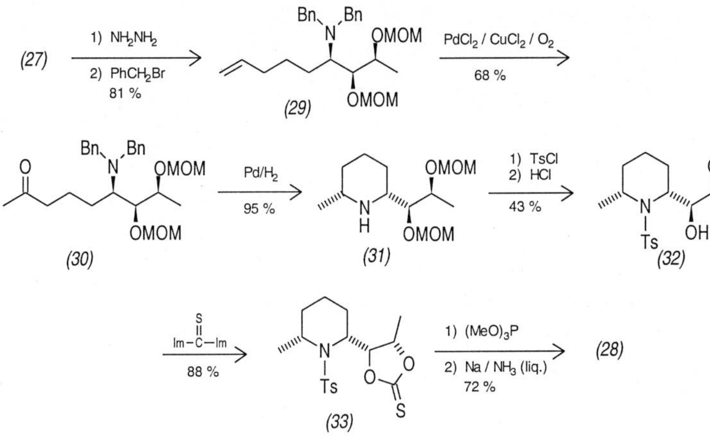

(27) 1) NH$_2$NH$_2$ 2) PhCH$_2$Br 81 %

Bn, Bn
N   OMOM

(29)   OMOM

PdCl$_2$ / CuCl$_2$ / O$_2$ 68 %

Bn, Bn
N   OMOM

O

OMOM

(30)

Pd/H$_2$ 95 %

OMOM

N
H   OMOM

(31)

1) TsCl 2) HCl   43 %

OH

N
Ts   OH

(32)

Im–C(=S)–Im   88 %

N
Ts

(33)

O   O
S

1) (MeO)$_3$P 2) Na / NH$_3$ (liq.)   72 %

(28)

Scheme 3. Conversion of the key compound *(27)* into (–)-pinidine *(28)*.

piperidine *(31)*. Thus, the asymmetric center at C-2 controls the diastereomeric course of hydrogenation of the intermediate ketimine. After tosylation at the nitrogen atom and removal of the protecting group to give *(32)*, both original chiral centers in the side chain were converted via the cyclic thiocarbonate *(33)* to a *trans*-double bond. Subsequent cleavage of the nitrogen protecting group gave (–)-pinidine *(28)*. (+)-(2*S*,6*S*)-Pinidine *(24)* is obtained in analogous fashion from *(22)*. Although the starting material in this synthesis is cheap, the number of steps is relatively high and the eventual moderate overall yield is only about 2 %. In the meantime a widely used method (CN(RS) method) for the preparation of 2- and 2,6-substituted piperidines has been developed by Husson, [13] that, by way of a further cyclization, is also very suitable for the synthesis of indolizidine- and quinolizidine alkaloids. [14] The key compound is a chiral 2-cyano-6-oxazolopiperidine *(37)*, the synthesis of which was achieved by Robinson–Schöpf condensation of glutardialdehyde *(35)* with the chiral auxiliary (+)-norephedrin *(34)* in the presence of KCN (Scheme 4). The (2*S*,6*R*)-configuration was determined by NMR measurements and

energy calculations. Regioselective and diastereoselective reactions are possible, contingent on the α-aminonitrile and the α-aminoether groups on the one hand and on the chiral auxiliary on the other. If *(37)* is treated with propylmagnesium bromide in the presence of AgBF$_4$, the α-aminonitrile group reacts regioselectively. As the intermediate iminimum ion *(38)* allows only axial attack of the nucleophile under stereoelectronic control, the reaction runs strictly diastereoselectively to *(39)*. Reduction of the α-aminoether group with NaBH$_4$ to *(40)* and cleavage of the substituents on the piperidine nitrogen atom gave (–)-(*R*)-coniin *(41)*.

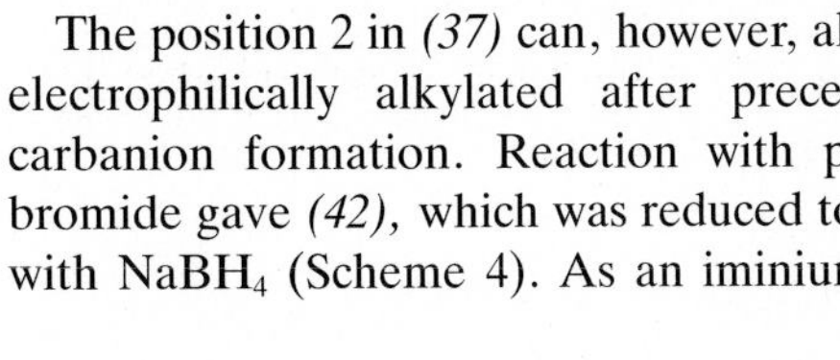

(37) →

Nu$^{\ominus}$

CN$^{\ominus}$
O   N=

R^2 R^1

(38)

→ (39)

The position 2 in *(37)* can, however, also be electrophilically alkylated after preceeding carbanion formation. Reaction with propyl bromide gave *(42)*, which was reduced to *(44)* with NaBH$_4$ (Scheme 4). As an iminium ion

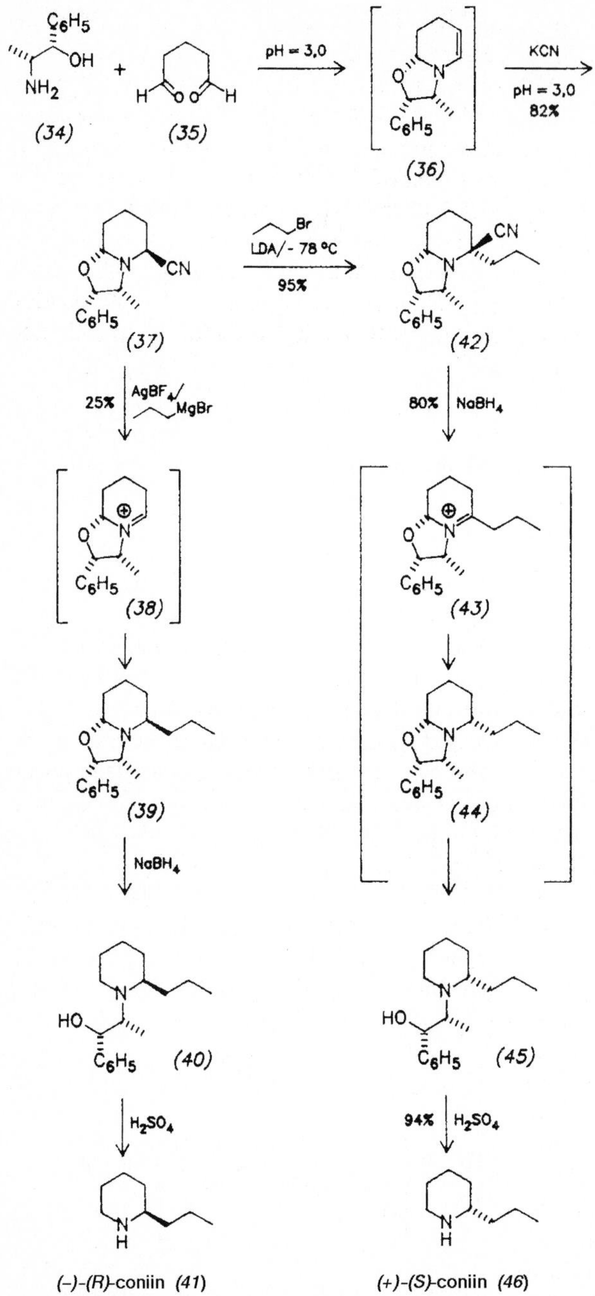

Scheme 4. Synthesis of (+)- and (−)-coniin using (+)-norephedrin *(34)* as chiral auxiliary, according to Husson (CN(RS) method).

transition state *(43)* is also involved here and the nucleophilic hydride anion attacks at the axial side, one obtains at C-2 in *(44)* the inverted configuration to that in *(39)*. Simultaneously, the α-aminoether group is cleaved under these reaction conditions to give *(45)*. Removal of the substituents at the piperidine nitrogen atom gave (+)-(*S*)-coniin *(46)*.

The potential of this strategy lies in the possibility to attack the electrophilic 6-position regio- and stereoselectively with similar ease and thus to form 2,6-disubstituted piperidines (Scheme 5). If *(47)*, the condensation product from glutaraldehyde *(35)*, (–)-phenylglycinol, and KCN, is alkylated at C-2 to give *(48)* and subsequently reduced diastereoselectively with $AgBF_4/ZnBH_4$ to *(49)*, nucleophilic attack of the Grignard compound at C-6 of the α-aminoether group leads stereoselectively to the *cis*-product *(50)*. Hydrogenation gave (+)-(2*S*,6R)-dihydropinidine *(51)*. Alkylation of *(47)* with methyl bromide gave *(52)* and

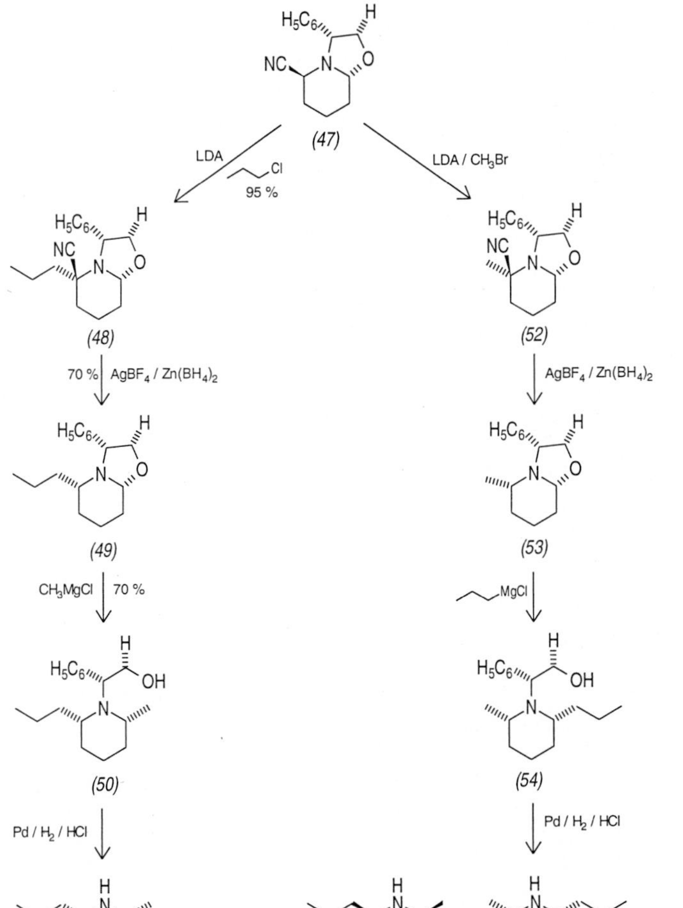

(+)-(2S, 6R)-dihydropinidine *(51)* *(–)-(2R, 6S)*-dihydropinidine *(55)*

Scheme 5. Synthesis of (+)- and (–)-dihydropinidine with (–)-norephedrin as chiral auxiliary (Husson CN(RS) method).

subsequent catalytic reduction gave *(53)*. Grignard reaction to *(54)* and hydrogenation led to the (–)-(2R,6S)-dihydropinidine *(55)*. One recognizes the variability of this method from the number of alkaloids that have been synthesized by Husson hitherto. [14] Nevertheless, the chiral auxiliary cannot be recovered.

The use of *O*-acyl-glycosylamines as chiral auxiliaries in the Strecker- and Ugi syntheses of α-amino acids has been demonstrated impressively by Kunz. [15] This strategy has been extended to the preparation of piperidines.[16] The conversion of 2,3,4,6-tetra-*O*-pivaloyl-α-D-galactopyranosylamine *(56)* with butanal gave Schiff base *(57)*. The latter could be converted via a stereoselective tandem Mannich-Michael reaction with 1-methoxy-3-trimethylsilyloxybutadiene *(58)*

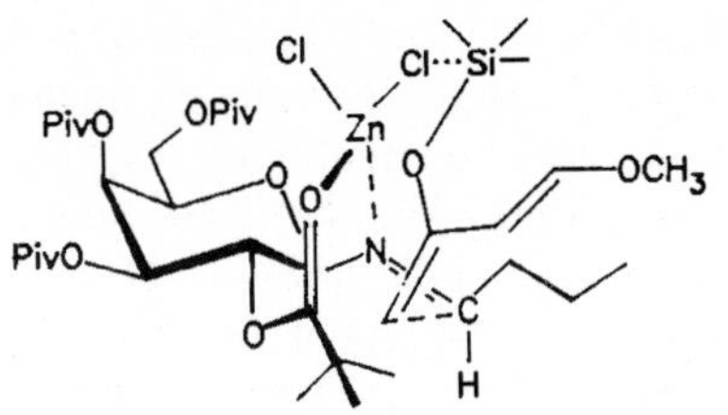

under catalysis by zinc chloride into *(60)* (Scheme 6). The high diastereoselectivity (97.5:2.5) is of particular significance in that hitherto only a few examples of stereoselective Mannich reactions have been described. As delocalization of the C=N-π-electrons into the σ* orbital of the ring C–O bond exists, a conformation is preferred in *(57)* in which the C–O bond of the ring is oriented almost orthogonally to the plane of the double bond. The Lewis acid, zinc chloride coordinates with

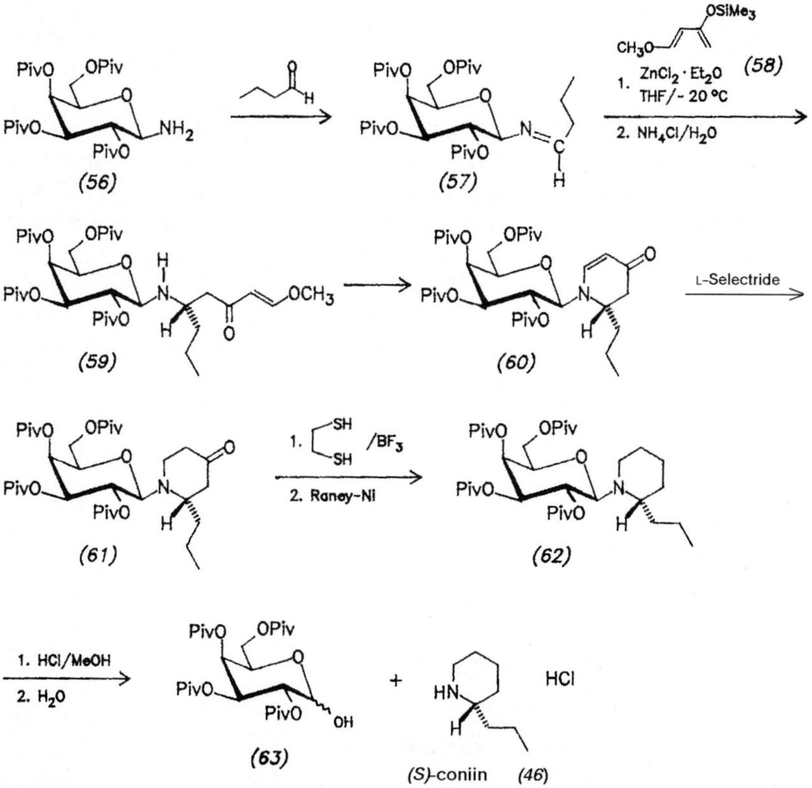

Scheme 6. Synthesis of (*S*)-coniin with an *O*-acetyl-protected glycosylamine as chiral auxiliary (after Kunz).

the nitrogen and the carbonyl oxygen of the 2-pivaloyl group. The initial step in the Mannich reaction is release of the latent nucleophile of the silyl dienol ether *(58)*. As the required interaction between the chloride ligands on the zinc with the silyl group takes place in front of the plane of the C=N double bond, the *(S)*-diastereomer *(59)* is formed, which subsequently undergoes cyclization to the piperidone *(60)*.

(S)-Coniin *(46)* is obtained after reduction of the double bond with L-Selectride to give *(61)*, conversion of the piperidone via the dithiolane, Raney nickel reduction to piperidine *(62)*, and removal of the carbohydrate residue. The cleaved carbohydrate *(63)* (yield 90 %) can be reconverted to the starting auxiliary *(56)*. Using this synthetic sequence one obtains, starting from *(56)* and pyridine carbaldehyde *(64)*, the natural *(S)*-anabasin *(68)*, which possesses the opposite stereochemistry at C-2 to that of *(S)*-coniin *(46)* (Scheme 7). Two equivalents of zinc chloride are required, however, for the diastereoselective Mannich reaction. The first equivalent coordinates with the pyridine nitrogen, whilst the second equivalent is responsible for activation of the aldimine bond. The activation of the silyl dienol ether *(58)* probably results from the Lewis acid at the pyridine nitrogen, whereby attack from the free rear side to give *(66)* is possible.

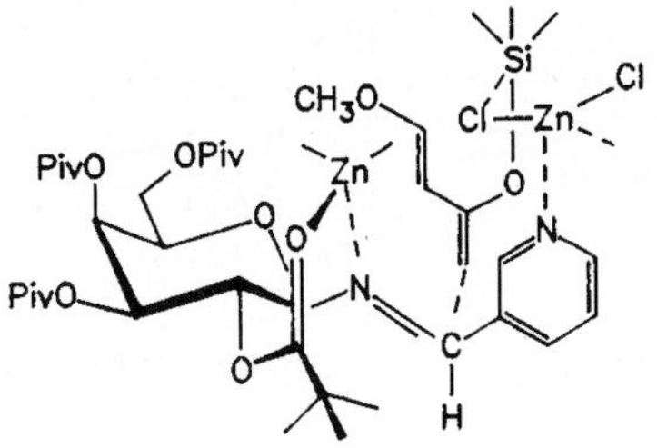

This hypothesis would explain why the Mannich reaction takes place with the opposite stereochemistry. In analogous fashion to that described for *(S)*-coniin *(46)*, the compound *(66)* is converted to *(S)*-anabasin *(68)* via *(67)*. In contrast to the results of Husson, Kunz managed to recover the chiral auxiliary. Nevertheless, only one enantiomer is accessible with this route.

Of course, stereocontrolled reactions can also be carried out under reagent control, not only under substrate control, as just discussed. In this manner, one obtains enantiomers by reaction of prochiral substrates with chiral reagents. Thus, Tanne [17] used the Sharpless epoxidation with (−)-tartaric acid diethyl ester as the key step in the synthesis of (+)-nitramine *(74)* (Scheme 8). Although hitherto only the synthesis of the natural (+)-nitramine *(74)* has been reported, in principle synthesis of the unnatural enantiomer

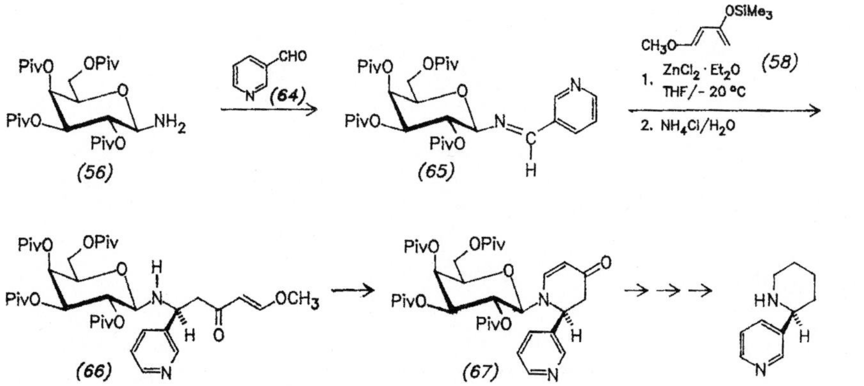

Scheme 7. Synthesis of (S)-anabasine using the Kunz method.

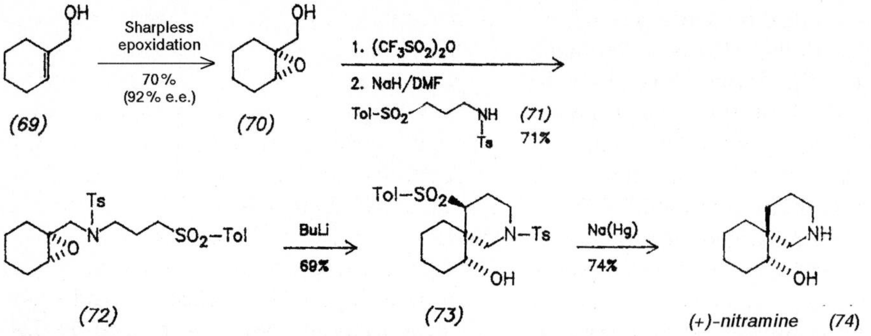

Scheme 8. Preparation of (+)-nitramine according to Tanne, using Sharpless epoxidation as the key step.

could also be considered by the use of (+)-diethyl tartrate. Starting from 1-cyclohexene methanol *(69)* two chiral centers are introduced to *(70)* under Sharpless conditions with (–)-tartaric acid diethyl ester. The optical purity was determined by ^{1}H NMR measurements with the use of chiral shift reagents as 90–94 % *ee*. Exchange of the hydroxyl group for an amino group to *(72)* was achieved via the triflate of *(70)* and reaction with the sulfonamide *(71)*. The latter can be prepared from acrolein according to Pinnick. [18] The spirocyclization by nucleophilic attack of the carbanion formed from attack of butyl lithium on the oxirane gave *(73)* which can be further converted into (+)-nitramine *(74)* with an optical purity of 93 %. The product is diastereomerically but not enantiomerically pure,

which represents a problem for the pharmacological evaluation.

Carbohydrates offer themselves as chiral building blocks for an economic synthesis of polyhydroxylated piperidines. In terms of a potential therapy for AIDS, one is currently primarily interested in derivatives of the natural glycosidase inhibitors and less in the unnatural enantiomers. Chiral pool syntheses, starting from carbohydrates, offers the advantage, of starting with the corresponding number of correctly configured chiral centers. The synthetic problem in this case is not the formation of new stereogenic centers, rather more to guarantee regioselective reactions at the hydroxyl groups using appropriate protecting groups. This was nevertheless hitherto possible only in isolated cases with minimal effort.

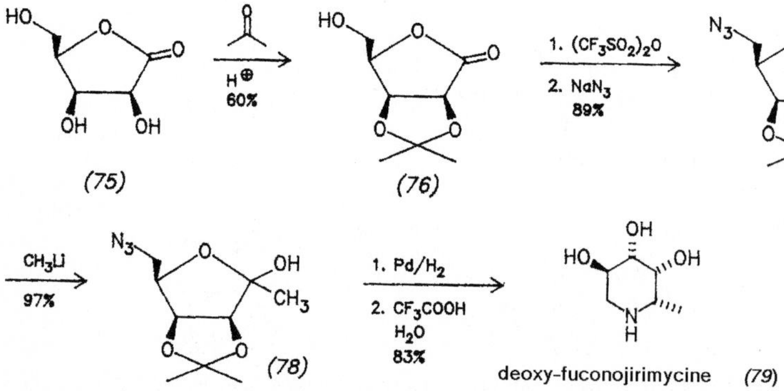

Scheme 9. Chiral pool synthesis of desoxyfuconojirimycin, after Fleet.

The synthesis of desoxyfuconojirimycin *(79)* by Fleet [19] can be regarded as an example (Scheme 9). Starting from lyxonolactone *(75)* [20] the 2,3-*O*-isopropylidene-D-lyxono-lactone *(76)* is obtained by ketalization with acetone. Formation of the azidolactone *(77)* via the triflate and addition of methyl lithium gave the hemiketal *(78)*. Subsequent hydrogenation led to reduction of the azide and finally to intramolecular reductive amination. Compound *(79)* was obtained after removal of the protecting groups. A very elegant route to the synthesis of polyhydroxylated piperidines was described first by Effenberger [21] and a little later by Wong. [22] First, a suitable carbohydrate precursor is constructed with an asymmetric synthesis. The asymmetric induction is carried out enzymatically. Both groups used the aldolase-catalyzed aldol addition of dihydroxyacetone phosphate *(80)* with 3-

azido-2-hydroxypropanal *(81)* or *(84)* to *(82)* or *(85)*, respectively (Scheme 10). It is not even necessary to use the pure enantiomers *(81)* or *(84)*, as the exclusively formed enantiomerically pure diastereomers *(82)* or *(85)*, respectively, can be separated easily. Effenberger used rabbit muscle aldolase, whereas Wong employed a recombinant bacterial aldolase. Hydrogenation of *(82)* gave (+)-1-desoxynojirimycin *(83)* exclusively, and (–)-1-desoxynojirimycin *(86)* was formed exclusively from *(85)*. If the *N*-protected 3-aminopropanal *(87)* is converted enzymatically with *(80)*, one obtains *(88)*, which is converted to fagomin *(89)* by hydrogenation. Similarly, amine *(91)* is formed from *(80)* and 3-(*N*-butyl)-2-hydroxypropanal *(90)*. Compound *(91)* can be hydrogenated to *N*-butyl-1-desoxynojirimycin *(92)*, [22] which is active against the AIDS virus. Nevertheless, in the

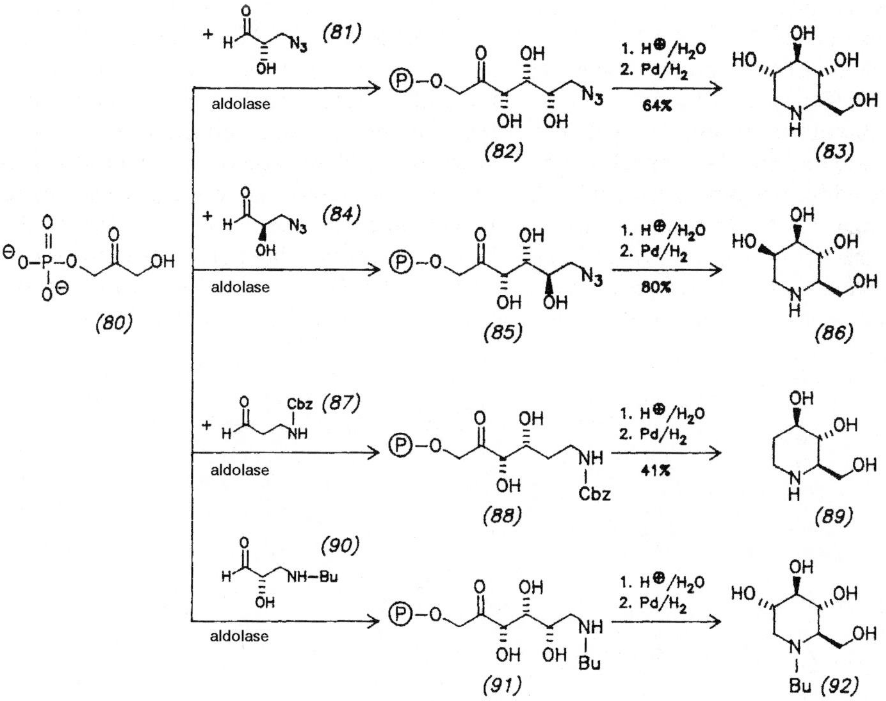

Scheme 10. Synthesis of polyhydroxylated piperidines after Wong and Effenberger with an enzymatic Aldol reaction as the key step.

long term a synthesis of the corresponding enantiomers will also be required for this series of compounds. The unnatural (–)-nojirimycin has proven itself to be a potent inhibitor of β-glucosidases and α-mannosidases. [23]

With the methods described herein even complex alkaloids that contain a piperidine or pyrrolidine structure can be prepared enantioselectively. Most advanced here are the investigations of Husson with the CN(RS) method. As adequate amounts of these compounds for biological investigations are now available through synthesis, interesting impulses could be expected for pharmacological research.

References

[1] M. Ishibashi, Y. Ohizumi, T. Sasaki, H. Nakamura, Y. Hirata, J. Kobayashi, *J. Org. Chem.* **1987,** *52,* 450.

[2] G. W. Fleet, A. Karpas, R. A. Dwek, L. E. Fellows, A. S. Tyms, S. Petursson, S. K. Namgoong, N. R. Ramsden, P. W. Smith, J. C. Son, F. Wilson, D. R. Witty, G. S. Jacob, T. W. Rademacher, *FEBS Lett.* **1988,** *237,* 128; see also V. A. Johnson, B. D. Walker, M. A. Barlow, T. J. Paradis, T.-C. Chou und M. S. Hirsch, *Antimicrob. Agents Chemother.* **1989,** *33,* 53; D. A. Winkler, G. Holan, *J. Med. Chem.* **1989,** *32,* 2084.

[3] A. Thurkauf, P. C. Zenk, R. L. Balster, E. L. May, C. George, F. I. Carroll, S. W. Mascarella, K. C. Rice, A. E. Jacobson, M. V. Mattson, *J. Med. Chem.* **1988,** *31,* 2257.

[4] M. S. Blum, J. R. Walker, P. S. Callahan, A. F. Novak, *Science* **1958,** *128,* 306.

[5] T. Shono, Y. Matsumura, K. Inoue, *J. Chem. Soc. Chem. Commun.* **1983,** 1169.

[6] K. Irie, K. Aoe, T. Tanaka, S. Saito, *J. Chem. Soc. Chem. Commun.* **1985,** 633.

[7] T. Shono, Y. Matsumura, K. Tsubata, K. Uchida, *J. Org. Chem.* **1986,** *51,* 2590.

[8] T. Shono, Y. Matsumura, O. Onomura, M. Sato, *J. Org. Chem.* **1988,** *53,* 4118.

[9] H.-U. Reißig, *Nachr. Chem. Tech. Lab.* **1986,** *34,* 656.

[10] N. Yamazaki, C. Kibayashi, *J. Am. Chem. Soc.* **1989,** *111,* 1396.

[11] H. Iida, N. Yamazaki, C.Kibayashi, *J. Org. Chem.* **1986,** *51,* 1069.

[12] H. Iida, N. Yamazaki, C. Kibayashi, *J. Org. Chem.* **1986,** *51,* 4245.

[13] L. Guerrier, J. Royer, D. S. Grierson, H.-P. Husson, *J. Am. Chem. Soc.* **1983,** *105,* 7754.

[14] S. Arsenyadis, P. Q. Huang, H.-P. Husson, *Tetrahedron Lett.* **1988,** *29,* 1391.

[15] H. Kunz, W. Sager, *Angew. Chem.* **1987,** *99,* 595; *Angew. Chem. Int. Ed. Engl.* **1987,** *26,* 557.

[16] H. Kunz, W. Pfrengle, *Angew. Chem.* **1989,** *101,* 1041; *Angew. Chem. Int. Ed. Engl.* **1989,** *28,* 1067.

[17] D. Tanne, H. M. He, *Tetrahedron* **1989,** *45,* 4309.

[18] Y. H. Chang und H. W. Pinnick, *J. Org. Chem.* **1978,** *43,* 373.

[19] G. W. Fleet, S. Peturson, A. L. Campbell, R. A. Mueller, J. R. Behling, K. A. Bablak, J. S. Ng und M. G. Scaros, *J. Chem. Soc. Perkin Trans. I* **1989,** 665.

[20] W. J. Humphlett, *Carbohydr. Res.* **1967,** *4,* 157.

[21] T. Ziegler, A. Straub und F. Effenberger, *Angew. Chem.* **1988,** *100,* 737; *Angew. Chem. Int. Ed. Engl.* **1988,** *27,* 716.

[22] C. H. von der Osten, A. J. Sinskey, C. F. Barbas, R. L. Pederson, Y.-F. Wong, C.-H. Wong, *J. Am. Chem. Soc.* **1989,** *111,* 3924.

[23] N. Chida, Y. Furuno und S. Ogawa, *J. Chem. Soc. Chem. Commun.* **1989,** 1230.

Taxanes:
An Unusual Class of Natural Products

Dieter Schinzer

The natural product class of taxanes belongs to an unusual type of terpene: the 6-8-6 ring combination with the rare 1,3-fusion of the six-and eight-membered ring and – synthetically quite difficult – a geminal dimethyl group in the A ring and numerous oxygen functions spread over the skeleton (Scheme 1). The enormous variety of challanges required to synthesize the target molecule explains the worldwide interest in this fascinating molecule. Even more important is the high biological activity, especially as an anti-tumor compound. [1] To satisfy the actual clinical need of taxol (ca. 2.5 kg) 12.000 Califorian yew trees had to be felled. [2] These figures show that a flexible approach to taxol is needed: not an easy task, knowing all the problems with this complex structure.

The following account will focus on the most important strategies for synthesizing the complex 6-8-6 ring system. The article cannot present all details but will show the principles of the most important approaches. Worldwide more than 20 research groups involved in taxol synthesis. The most striking problem in all these syntheses is the construction of the 1,3-annulated ring system containing the

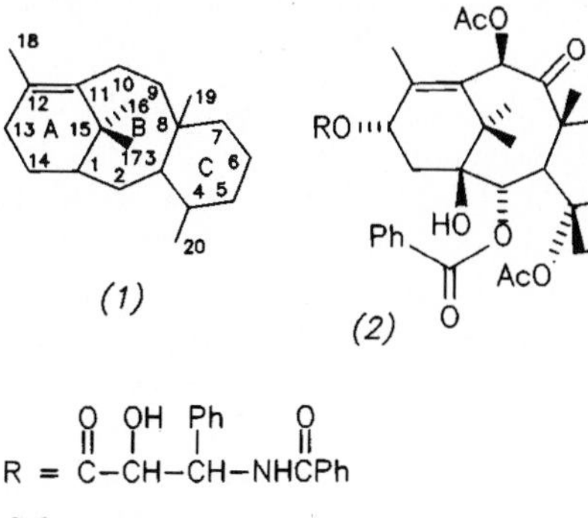

Scheme 1

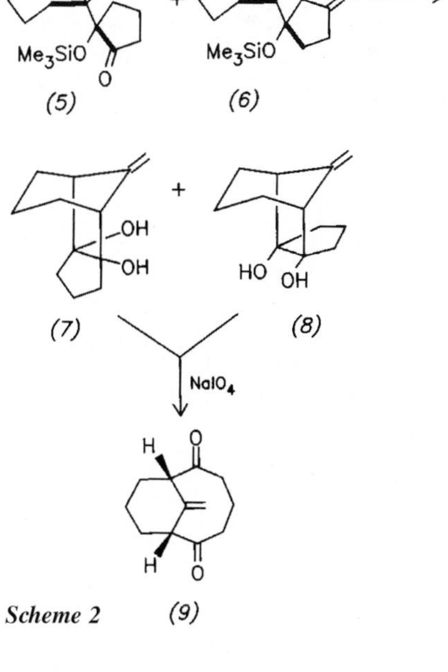

Scheme 2

geminal dimethyl group. This will be a good test for potential total syntheses.

Trost's group started quite early with the basic construction of a bicyclic 6-8-membered ring system, in which an elegant three-carbon ring expansion with a bifunctional reagent is performed (Scheme 2). [3, 4] The geminal dimethyl group is already masked in the exocyclic double bond and the two carbonyl groups contain the required oxygen functionality.

Wender's group uses a Ni-catalyzed bicyclization of compound (12) of a [4+4]-cycloaddition. Compound (12) is available from commercial myrcene (10) in four steps. [5, 6] This model system does not contain the geminal dimethyl group. The usefulness of this strategy is the additional approach for the B-C-ring system (Scheme 3). Compound (19) was obtained in 51 % yield by Kraus et al. with

a 2-carbon ring expansion (Scheme 4). [7] The precursor (16) is obtained in 3 steps from cyclohexenone by an in situ reaction with electron-rich olefins. Compound (18) can be reductively transformed into bicyclo[5.3.1]undecanone (19). Again, this approach only produced a basic model system.

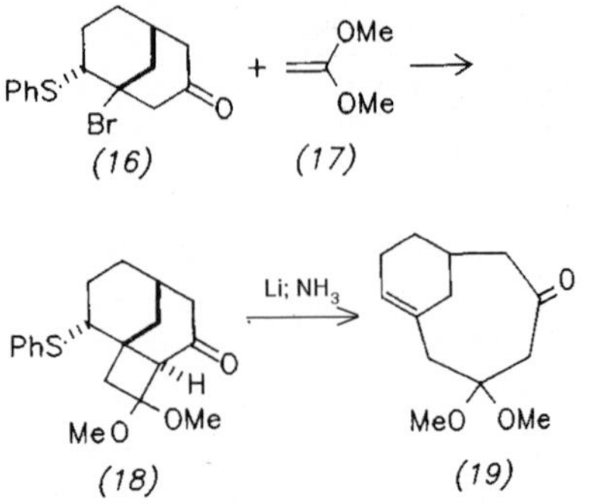

Scheme 4

Scheme 3

Considerable progress was made with Blechert's deMayo approach. [8, 9] The key reaction is a stereoselective [2+2]-photocycloaddition of the bicyclic compound (20) with cyclohexene. The stereochemistry in the cycloaddition can be controlled by the ketal function: attack from the β-face controls the stereochemistry at carbon 8. The benzyl carbonate can be reductively cleaved to the alcohol, which undergoes a retro-aldol reaction to give compound (23) (Scheme 5). This sequence provides, for the first time, the tricyclic skeleton containing the critical geminal dimethyl group.

An additional, direct entry to the tricyclo[9.3.1.0$_{3,8}$]pentadecane system uses an intramolecular Diels-Alder reaction with an aromatic C-ring (Scheme 6). [10] Intermediate (24) was synthesized from a simple aromatic dibromide, which was transformed under thermal conditions in xylene into the cycloadduct (25) (70 %).

A new paper from Shea has now solved the problem with the aromatic C-ring. [11] The

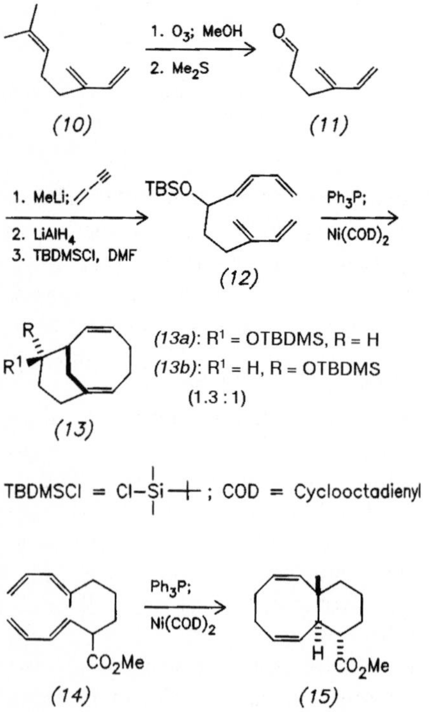

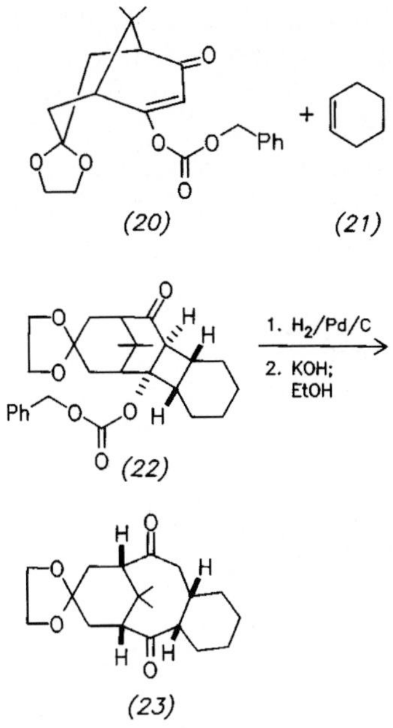

(20)   (21)

(22)

1. H₂/Pd/C
2. KOH; EtOH

(23)

Scheme 5

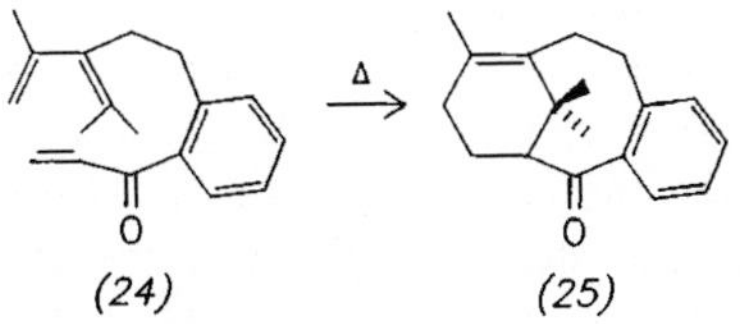

(24)   (25)

Scheme 6

(26)

1. BrMg
2. H⁺

(27)

1. HS
2. t-BuLi
3. DMF

SH; BF₃

(28)   (29)

EtAlCl₂

CHO

Scheme 7

(30)   (31)

BF₃·OEt₂

Scheme 8

new approach again uses an intramolecular Diels-Alder reaction yielding compound *(29)*, which allows the addition of the missing methyl group at carbon 8. The conditions for the cycloaddition are much milder now because Lewis acid catalysis is used. Trienone *(28)* is easily obtained in four steps from a cyclic vinylogous ester (Scheme 7).

A similar construction of the a tricyclic compound has been published by Jenkins et al. Compound *(31)* already contains the carbonyl group at carbon 2 and the angular methyl group at carbon 8 (Scheme 8). [12]

The first synthesis of a taxane-triene of type *(38)* was recently discovered by Kende et al. [13] Compound *(38)* represents an excellent

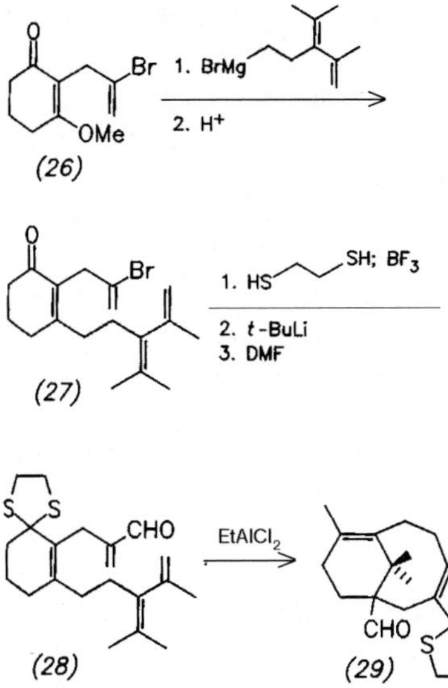

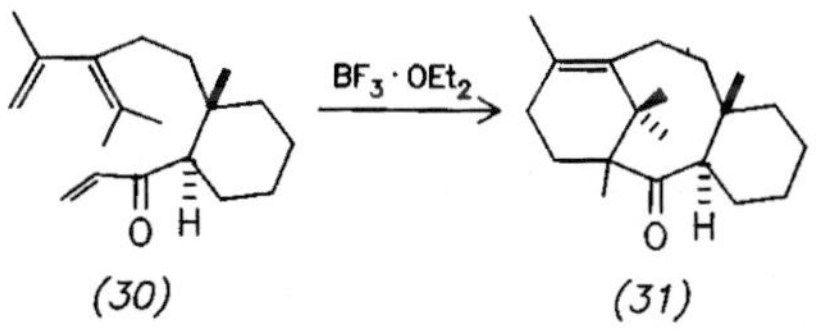

precursor for the construction of the missing oxygens. Key reaction in this synthesis is a McMurry reaction of dialdehyde *(36)* in the presence of Ti*(0)*; so far only in low yield *(20%)* (Scheme 9). The cyclization precursor can be obtained by a Lewis acid-catalyzed aldol reaction: *(32)* + *(33)* → *(34)*.

The stereocontrolled synthesis of the 8-membered ring system by a Claisen rearrangement of the macrolide *(43)* was described by Funk et al. [14] The required lactone *(42)* was synthesized by a short sequence of reactions. The ketene acetal was stereoselectively rearranged into compound *(45)* in 82% yield (Scheme 10).

The first total synthesis of a member of the taxane family, taxusine *(59)*, has been reported by Holton et al. [15, 16] The key

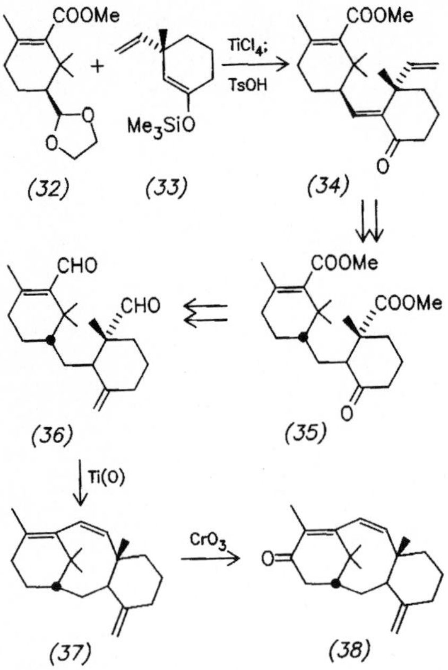

COOMe

(32)

TiCl₄;
TsOH

Me₃SiO

(33)

COOMe

(34)

CHO

CHO

(36)

COOMe

COOMe

(35)

Ti(O)

CrO₃

(37)

(38)

Scheme 9

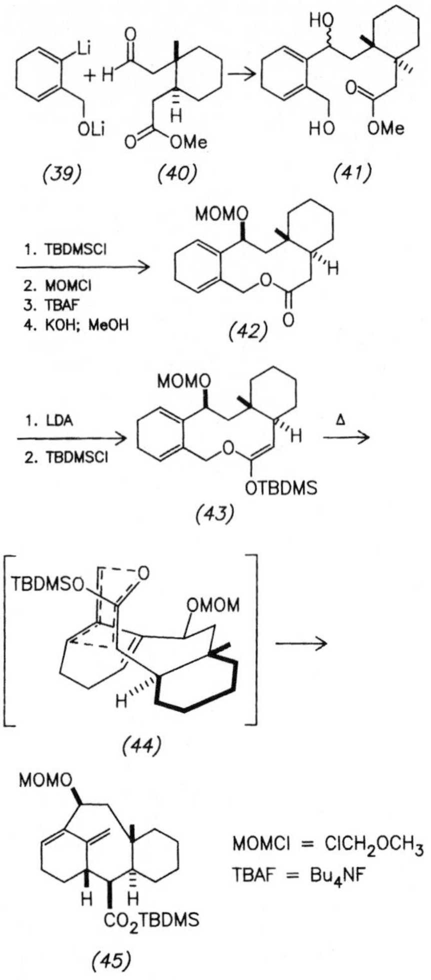

t-BuOOH;
Ti(O-i-Pr)₄

OH

(46)

O-H

(47)

HO

(48)

1. MOMCl
2. BMDA/TMSCl
3. TMS

MOMO

(49)

BMDA

MOMO

OH

(50)

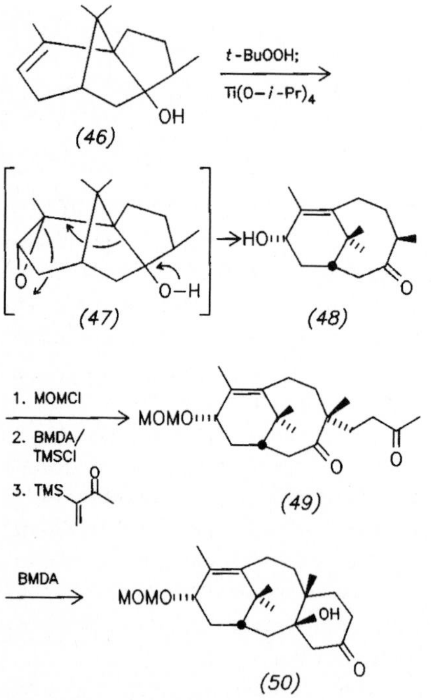

BMDA = BrMg—N(iPr)₂

Scheme 11

HO

(39) + (40)

Li

OLi

O

OMe

HO

HO   OMe

(41)

1. TBDMSCl
2. MOMCl
3. TBAF
4. KOH; MeOH

MOMO

H

(42)

1. LDA
2. TBDMSCl

MOMO

H

OTBDMS

(43)

Δ

TBDMSO

OMOM

H

(44)

MOMO

H   H

CO₂TBDMS

(45)

MOMCl = ClCH₂OCH₃
TBAF = Bu₄NF

Scheme 10

reaction in this synthesis is the fragmentation of epoxy alcohol *(46)*. [15] The missing C-ring is introduced by an intramolecular aldol reaction *(48)* → *(49)* (Scheme 11).

This simple model has been expanded to the total synthesis of taxusine *(59)*, starting with the commercially available optically active *β-patchulene oxide (51)*. An eight-step synthesis leads to compound *(52)*, which is transformed *in situ* after epoxidation to the 8-membered ring *(53)*. The C-ring was built via

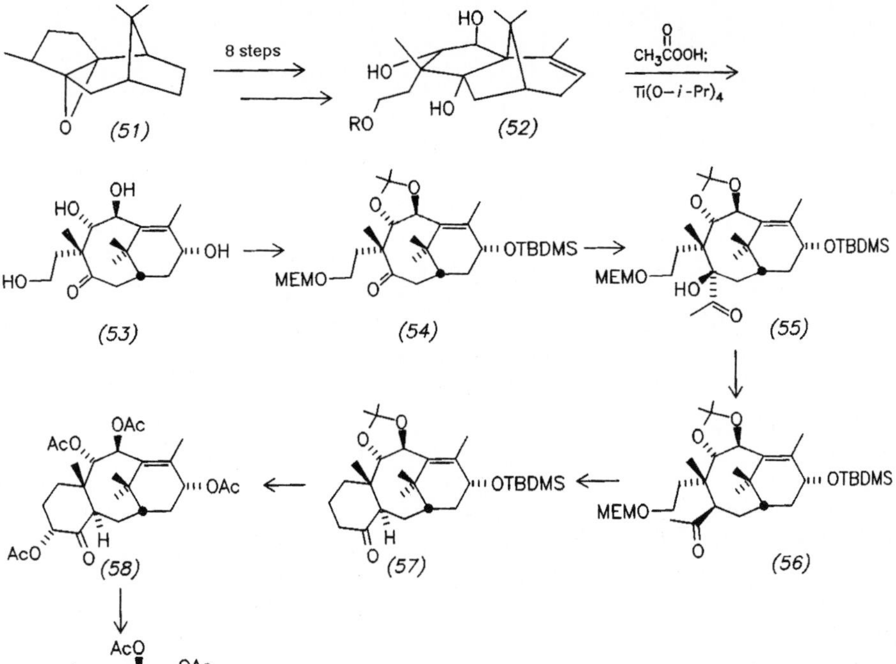

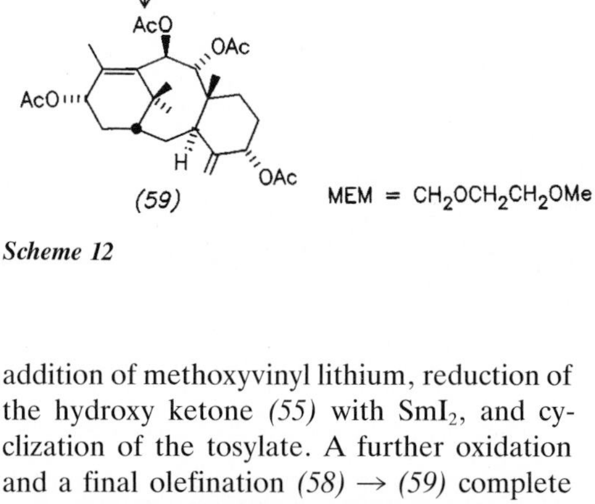

Scheme 12

MEM = CH₂OCH₂CH₂OMe

addition of methoxyvinyl lithium, reduction of the hydroxy ketone *(55)* with SmI₂, and cyclization of the tosylate. A further oxidation and a final olefination *(58)* → *(59)* complete the synthesis yielding the natural product in optically active form (Scheme 12).

In the meantime taxol *(1)* has been synthesized by Holton et al. using the strategy described above. [17, 18] In this total synthesis terpene alcohol *(60)* was transformed via an epoxy-alcohol fragmentation into the 1,3-annulated bicyclic ketone *(61)*, followed by an aldol addition and closure to *(63)*, which was rearranged to *(64)* and further transformed to the tricyclic lactone *(66)*. The side chain was ozonized and the lactone cleaved by an intra-

molecular Dieckmann condensation to yield *(68)*, which was oxidized to *(71)*. Addition of methyl Grignard, followed by a Burgess elimination yielded *(72)*, which allowed the construction of the sensitive oxetane ring after a short sequence. After several protecting group manipulations an interesting carbonyl shift-oxidation sequence was used to establish the missing oxygen atom in the eight-membered ring *(77)* → *(78)*. Finally, the side chain in the A Ring was introduced using the β-lactam route, followed by deprotection of oxygen functions to give taxol *(1)* in optically active form (Scheme 13).

In addition, Nicolaou et al. have synthesized taxol by a flexible strategy using a Shapiro reaction and a McMurry coupling as the key operations to construct the basic skeleton. [19–21] In a straightforward sequence all the oxygens required are maneuvered around the molecule yielding taxol in a 14-step sequence. Nicolaou's synthesis started with compound *(81)*, which is easily available by a Diels-Alder approach. Subsequent transformation into

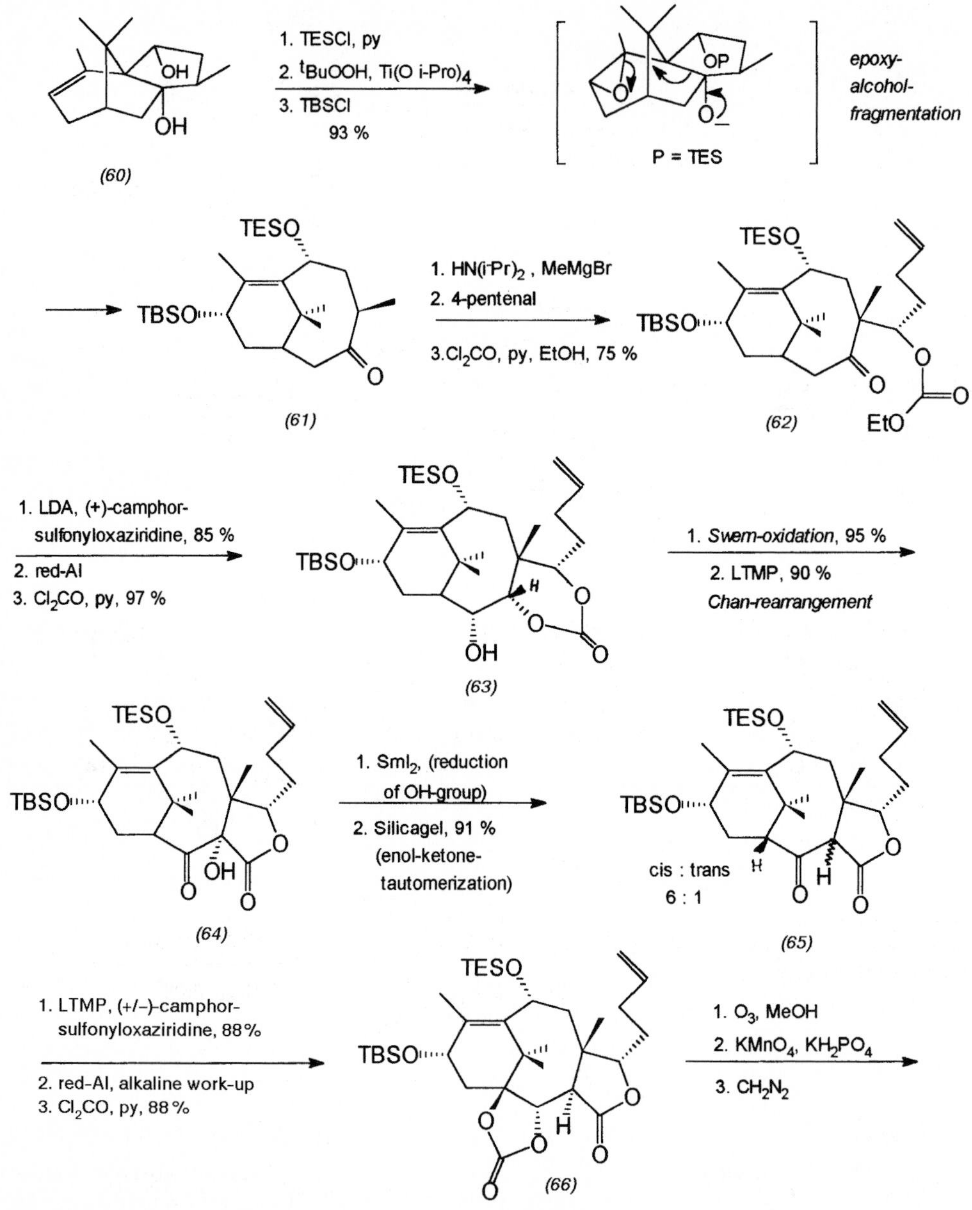

1. TESCl, py
2. ^{t}BuOOH, Ti(O i-Pro)$_4$
3. TBSCl
93 %

*epoxy-
alcohol-
fragmentation*

P = TES

(60)

1. HN(i-Pr)$_2$, MeMgBr
2. 4-pentenal
3. Cl$_2$CO, py, EtOH, 75 %

(61)

(62)   EtO

1. LDA, (+)-camphor-
   sulfonyloxaziridine, 85 %
2. red-Al
3. Cl$_2$CO, py, 97 %

1. *Swern-oxidation*, 95 %
2. LTMP, 90 %
   Chan-rearrangement

(63)

1. SmI$_2$, (reduction
   of OH-group)
2. Silicagel, 91 %
   (enol-ketone-
   tautomerization)

cis : trans
6 : 1

(64)

(65)

1. LTMP, (+/–)-camphor-
   sulfonyloxaziridine, 88%
2. red-Al, alkaline work-up
3. Cl$_2$CO, py, 88 %

1. O$_3$, MeOH
2. KMnO$_4$, KH$_2$PO$_4$
3. CH$_2$N$_2$

(66)

LTMP = lithiumtetramethylpiperidide
Red-Al = Na$^+$ AlH$_2$(OC$_2$H$_4$OCH$_3$)$_2^-$

TES = triethylsilyl
TBS = ^{t}butyldimethylsilyl

Scheme 13

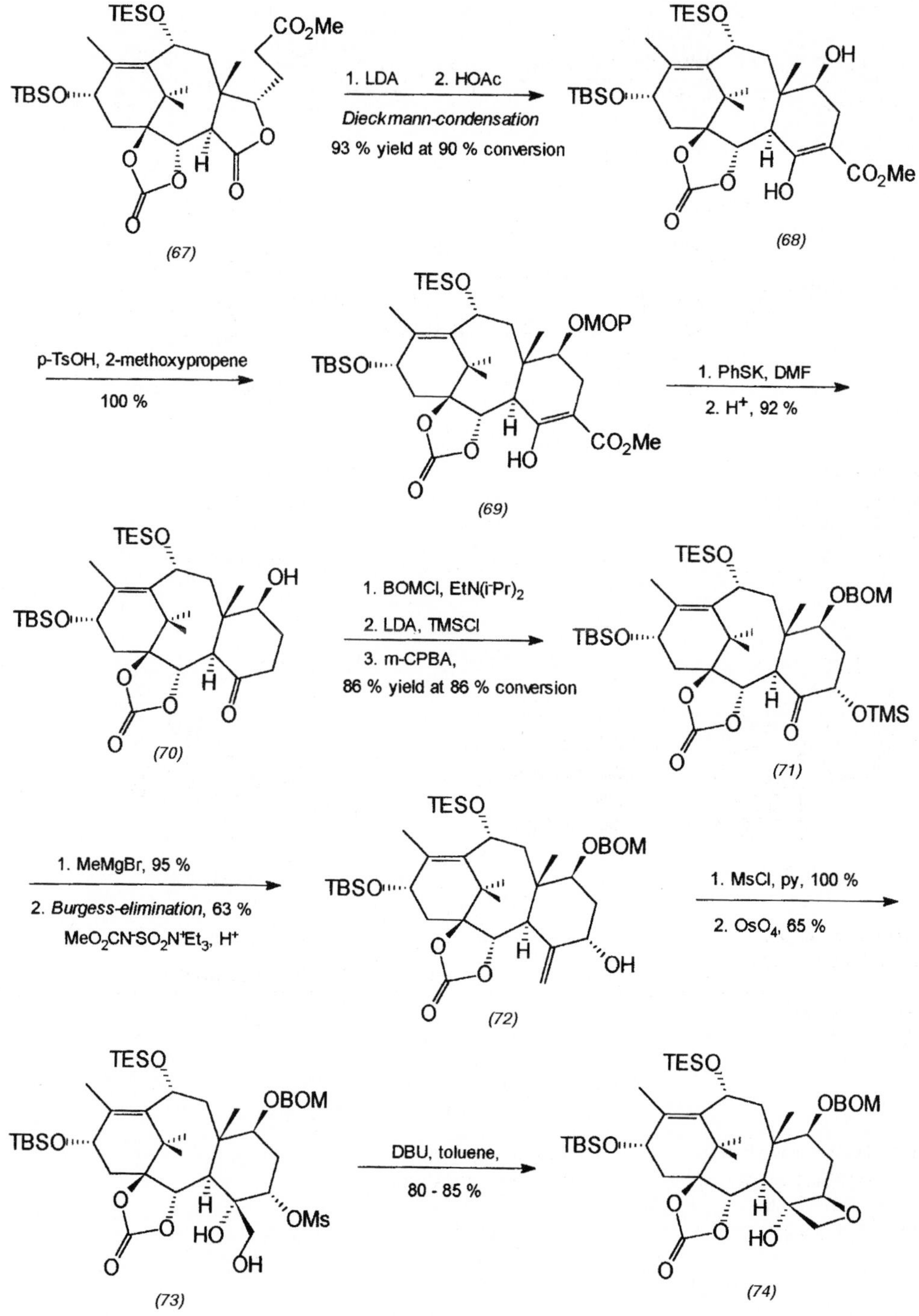

1. LDA   2. HOAc

Dieckmann-condensation

93 % yield at 90 % conversion

(67)

(68)

p-TsOH, 2-methoxypropene

100 %

(69)

1. PhSK, DMF

2. H⁺, 92 %

1. BOMCl, EtN(i-Pr)₂

2. LDA, TMSCl

3. m-CPBA,

86 % yield at 86 % conversion

(70)

(71)

1. MeMgBr, 95 %

2. *Burgess-elimination*, 63 %

MeO₂CN⁻SO₂N⁺Et₃, H⁺

(72)

1. MsCl, py, 100 %

2. OsO₄, 65 %

DBU, toluene,

80 - 85 %

(73)

(74)

Scheme 13.2 cont.

BOM = benzyloxymethyl   DBU = diazabicyclo[5.4.0]undecene-7

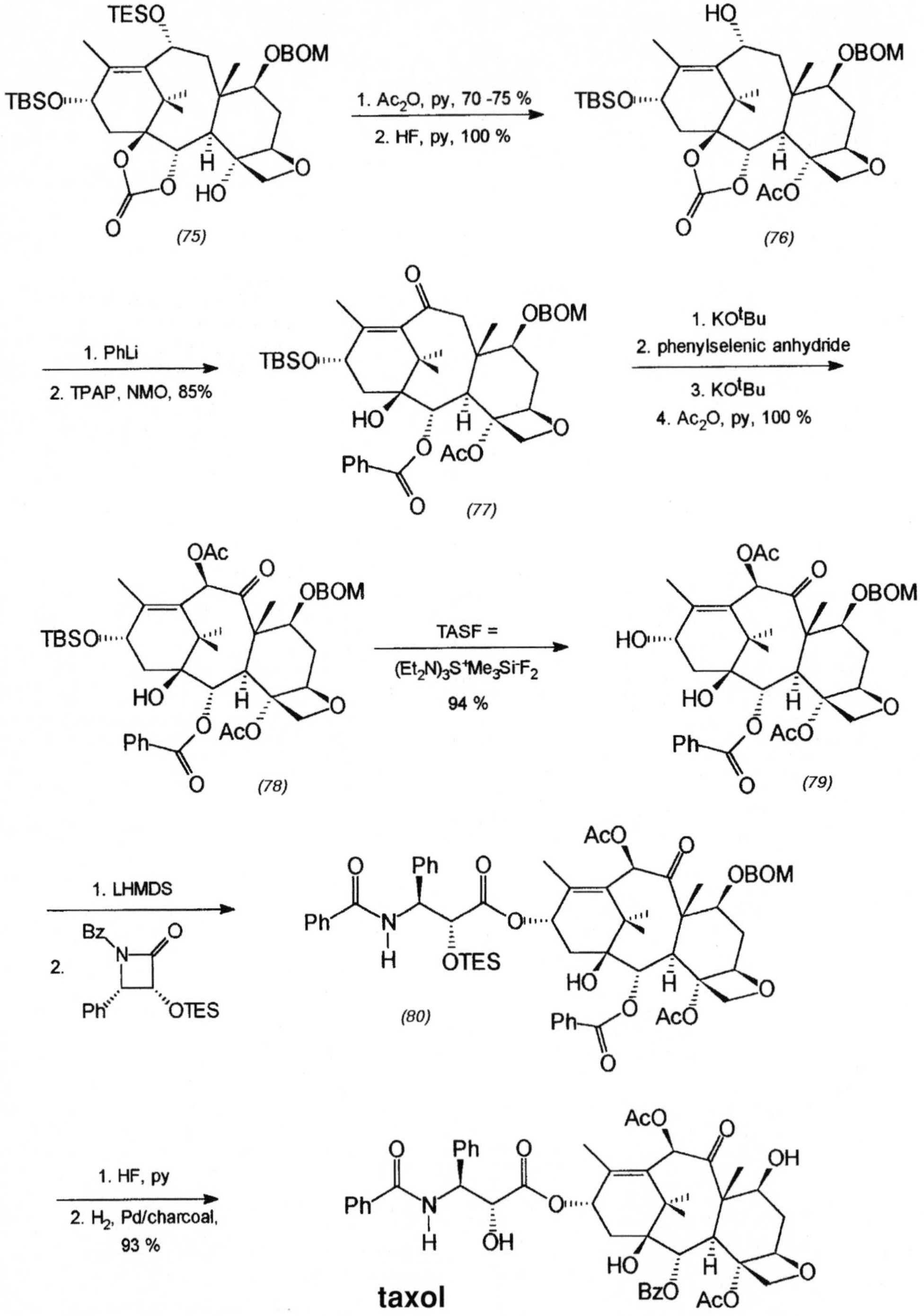

taxol

TPAP = tetrapropylammoniumperruthenate   LHMDS = lithiumhexamethyldisilazide
NMO = N-methylmorpholine N-oxide

Scheme 13.3 cont.

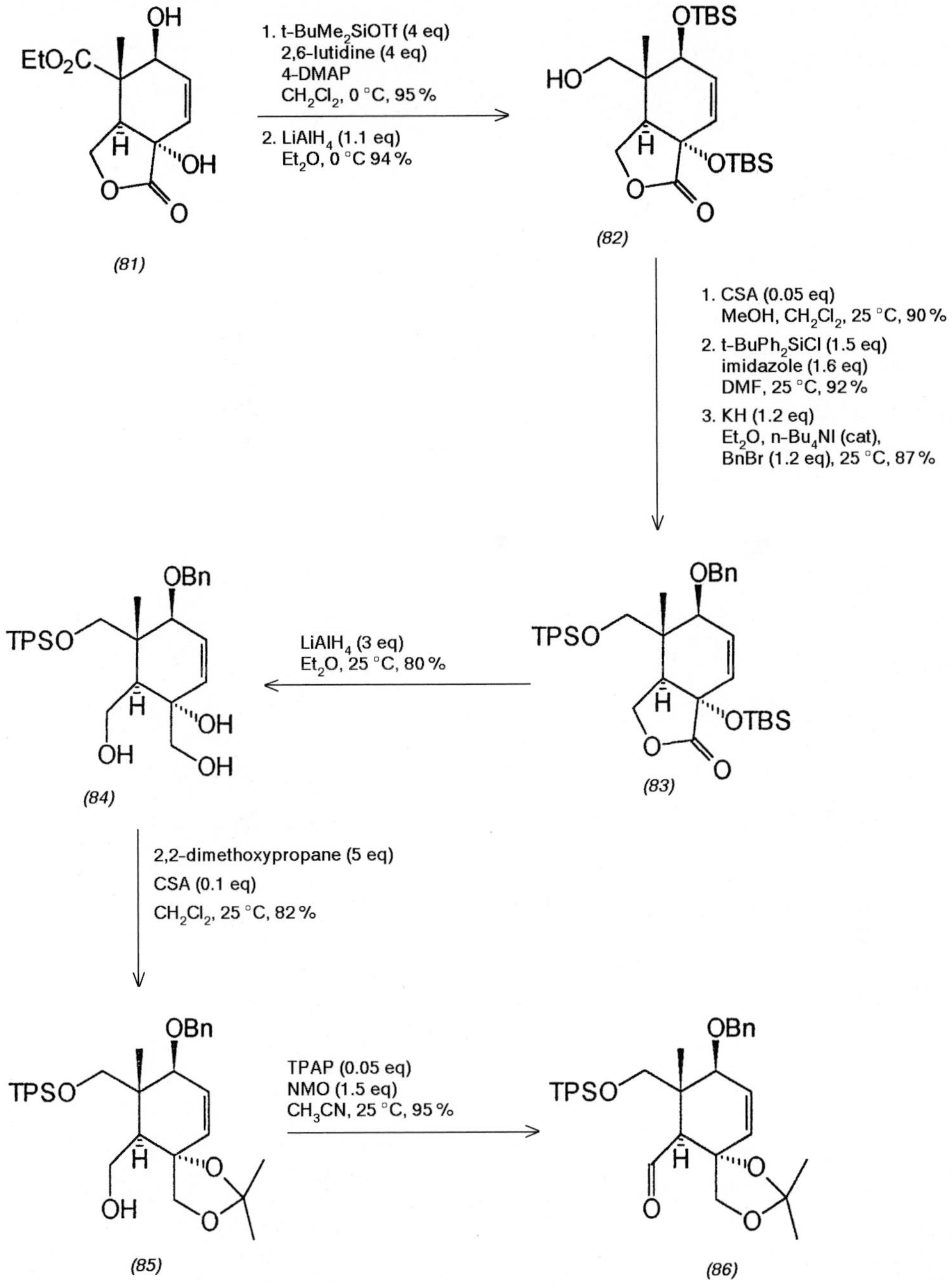

Scheme 14

Scheme 14.2 cont.

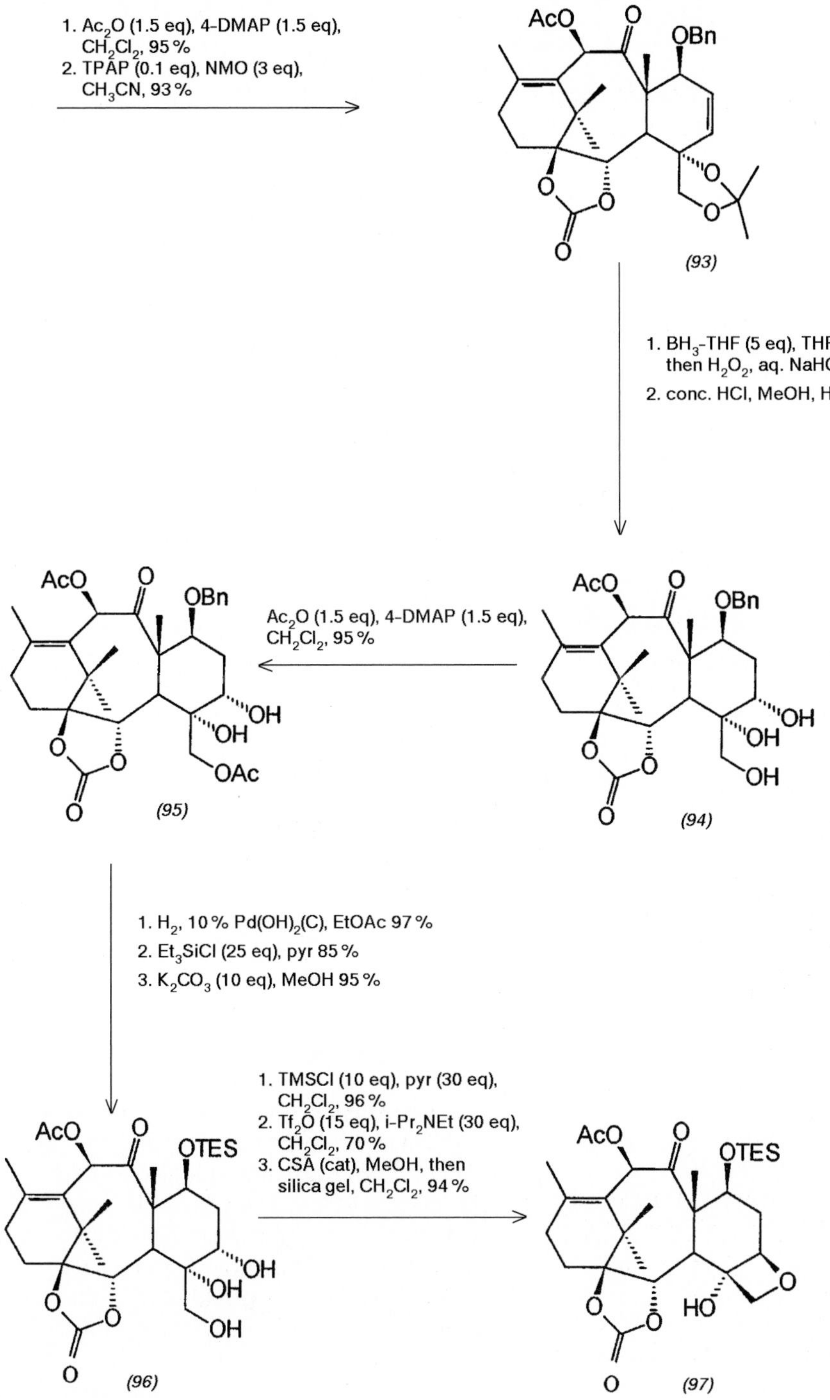

1. Ac$_2$O (1.5 eq), 4-DMAP (1.5 eq),
   CH$_2$Cl$_2$, 95%
2. TPAP (0.1 eq), NMO (3 eq),
   CH$_3$CN, 93%

(93)

1. BH$_3$-THF (5 eq), THF,
   then H$_2$O$_2$, aq. NaHCO$_3$, 55%
2. conc. HCl, MeOH, H$_2$O, 80%

Ac$_2$O (1.5 eq), 4-DMAP (1.5 eq),
CH$_2$Cl$_2$, 95%

(95)

(94)

1. H$_2$, 10% Pd(OH)$_2$(C), EtOAc 97%
2. Et$_3$SiCl (25 eq), pyr 85%
3. K$_2$CO$_3$ (10 eq), MeOH 95%

1. TMSCl (10 eq), pyr (30 eq),
   CH$_2$Cl$_2$, 96%
2. Tf$_2$O (15 eq), i-Pr$_2$NEt (30 eq),
   CH$_2$Cl$_2$, 70%
3. CSA (cat), MeOH, then
   silica gel, CH$_2$Cl$_2$, 94%

(96)

(97)

Scheme 14.3 cont.

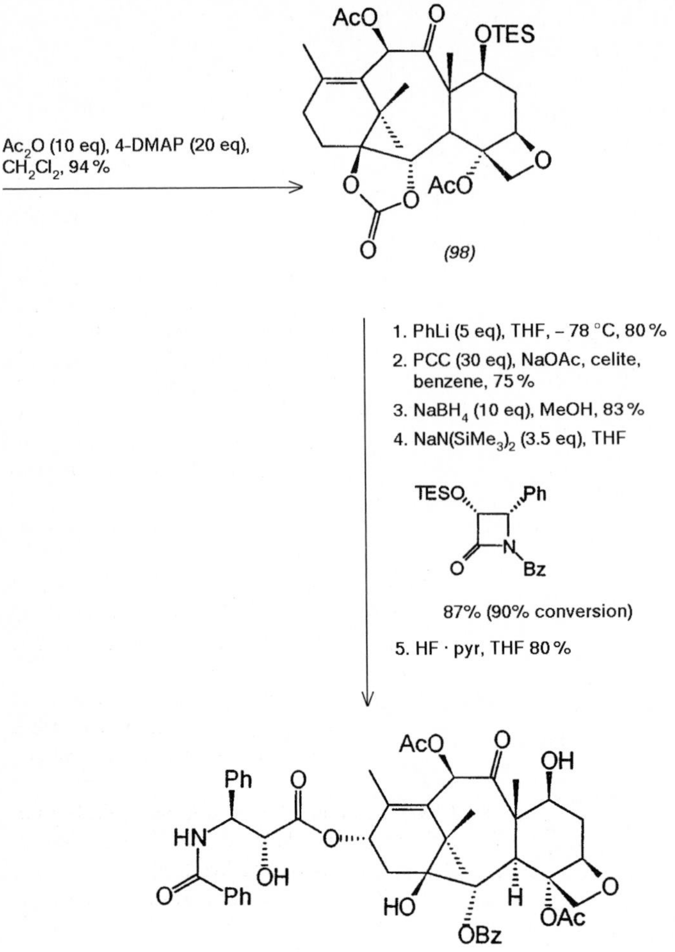

Ac$_2$O (10 eq), 4-DMAP (20 eq),
CH$_2$Cl$_2$, 94 %

AcO O OTES

(98)

1. PhLi (5 eq), THF, – 78 °C, 80 %
2. PCC (30 eq), NaOAc, celite,
   benzene, 75 %
3. NaBH$_4$ (10 eq), MeOH, 83 %
4. NaN(SiMe$_3$)$_2$ (3.5 eq), THF

TESO,   Ph

O   N
       Bz

87% (90% conversion)

5. HF · pyr, THF 80 %

AcO O OH

Ph O

HN

O   Ph OH

O

HO H OAc

ŌBz

taxol

4-DMAP	4-dimethylaminopyridine
CSA	camphorsulfonic acid
TPAP	tetra-n-propylammonium perruthenat
NMO	N-methylmorphonilin-N-oxide
TES	Et$_3$Si
TBS	t-BuMe$_2$Si
TPS	t-BuPh$_2$Si

Scheme 14.4 cont.

aldehyde *(86)* provided the coupling partner
in the following Shapiro reaction with compound *(87)*. The adduct *(88)* was further trans-

formed into epoxide *(89)*, which was reductively ring-opened to *(90)*. After several
protecting group manipulations and oxidation
to the dialdehyde, *(91)* was ready for the key
transformation: McMurry coupling provided
compound *(92)* with almost all oxygen atoms.
This intermediate was taken on to construct
the oxetane ring using a similar strategy to
that used in Holton's synthesis to give *(97)*. A
final allyl oxidation and addition of the side
chain in the A ring by the β-lactam route provided taxol *(1)* (Scheme 14).

In summary, after many investigations and
employment of different strategies taxol has

finally been synthesized by two groups using completely different strategies. In addition, success has been made in the isolation of natural taxol from yew trees without felling them. One can only hope that an anti-tumor drug will appear in the near future.

References

[1]   R. W. Miller, *J. Nat. Prod.* **1980,** *43,* 425.

[2]   R. W. Miller, R. G. Powell, C. R. Smith, Jr., *J. Org. Chem.* **1981,** *46,* 1469.

[3]   B. M. Trost, H. Hiemstra, *J. Am. Chem. Soc.* **1982,** *104,* 886.

[4]   B. M. Trost, M. J. Fray, *Tetrahedron Lett.* **1984,** *25,* 4605.

[5]   P. A. Wender, N. C. Ihle, *J. Am. Chem. Soc.* **1986,** *108,* 4678.

[6]   P. A. Wender, M. L. Snapper, *Tetrahedron Lett.* **1987,** *28,* 2221.

[7]   G. A. Kraus, P. J. Thomas, Y.-S. Hon, *J. Chem. Soc., Chem. Commun.* **1987,** 1849.

8]   H. Neh, S. Blechert, W. Schnick, M. Jansen, *Angew. Chem.* **1984,** *96,* 903. Angew. Chem. Int. Ed. Engl. **1984,** *23,* 905.

[9]   R. Kaczmarek, S. Blechert, *Tetrahedron Lett.* **1986,** *27,* 2845; see also H. Cervantes, D. D. Khac, M. Fetizon, F. Guir, *Tetrahedron* **1986,** *42,* 3491.

[10]   K. J. Shea, P. D. Davis, *Angew. Chem.* **1983,** *95,* 422; Angew. Chem. Int. Ed. Engl. **1983,** *22,* 419.

[11]   K. J. Shea, C. D. Haffner, *Tetrahedron Lett.* **1988,** *29,* 1367.

[12]   R. V. Bonnert, P. R. Jenkins, *J. Chem. Soc., Chem. Commun.* **1987,** 1540.

[13]   A. S. Kende, S. Johnson, P. Sanfilippo, J. C. Hodges, L. N. Jungheim. *J. Am. Chem. Soc.* **1986,** *108,* 3513.

[14]   R. L. Funk, W. J. Daily, M. Parvez, *J. Org. Chem.* **1988,** *53,* 4143.

[15]   R. A. Holton, *J. Am. Chem. Soc.* **1986,** *106,* 5731.

[16]   R. A. Holton, R. R. Juo, H. B. Kim, A. D. Williams, S. Harusawa, R. E. Lowenthal, S. Yogai, *J. Am. Chem. Soc.* **1988,** *110,* 6558.

[17]   R. A. Holton, C. Somoza, H.-B. Kim, F. Liang, R. J. Biediger, P. D. Boatman, M. Shindo, C. C. Smith, S. Kim, H. Nadizadeh, Y. Suzuki, C. Tao, P. Vu, S. Tang, P. Zhang, K. K. Murthi, L. N. Gentile, J. H. Liu, *J. Am. Chem. Soc* **1994,** *116,* 1597.

[18]   R. A. Holton, C. Somoza, H.-B. Kim, F. Liang, R. J. Biediger, P. D. Boatman, M. Shindo, C. C. Smith, S. Kim, H. Nadizadeh, Y. Suzuki, C. Tao, P. Vu, S. Tang, P. Zhang, K. K. Murthi, L. N. Gentile, J. H. Liu, *J. Am. Chem. Soc* **1994,** *116,* 1599.

[19]   For an excellent review see: K. C. Nicolaou, W.-M. Dai, R. K. Guy, *Angew. Chem.* **1994,** *106,* 38. Angew. Chem. Int. Ed. Engl. **1994,** *33,* 45.

[20]   K. C. Nicolaou, C. F. Claiborne, P. G. Nantermet, E. A. Couladouros, E. J. Sorensen, *J. Am. Chem. Soc.* **1994,** *116,* 1591.

[21]   K. C. Nicolaou, Z. Yang, J. J. Liu, H. Ueno, P. G. Nantermet, R. K. Guy, C. F. Claiborne, J. Renaud, E. A. Couladouros, K. Paulyannan, E. J. Sorensen, *Nature* **1994,** *367,* 630.

B. Synthesis of Individual Natural products

CC-1065: One of the Most Powerful Anti-Tumor Compounds

Dieter Schinzer

Apart from taxanes [1] and antibiotics of the calicheamicin- and esperamicin-type [2], a third class of highly potent anti-tumor compounds exists: the antibiotic CC-1065 [3] and some related compounds, which were isolated by the Upjohn group in the U.S. [4]. At the time of its isolation CC-1065 was the most active anti-tumor compound in the world. No name – just a number – already indicates an industrial laboratory, and is probably a good idea: it would not be easy to name this complex molecule using IUPAC nomenclature rules. In this chapter some of the known approaches and the first total synthesis of the natural product *(1)* will be presented (Scheme 1).

Like other interesting target molecules a race was started around the world with several well known groups, recently finished by Boger et al. with a report of the first total synthesis. [5, 6] The antibiotic CC-1065 is twice as active in the L-1210 test with leucemia cells than maytansine.

Analysis shows three pyrroloindol units that are connected with amide bonds. From the stereochemical point of view two asymmetric centers (cyclopropane ring in the fragment A) must be created. Most of the research groups involved have first studied the synthesis of the subunits, which will be discussed first. The

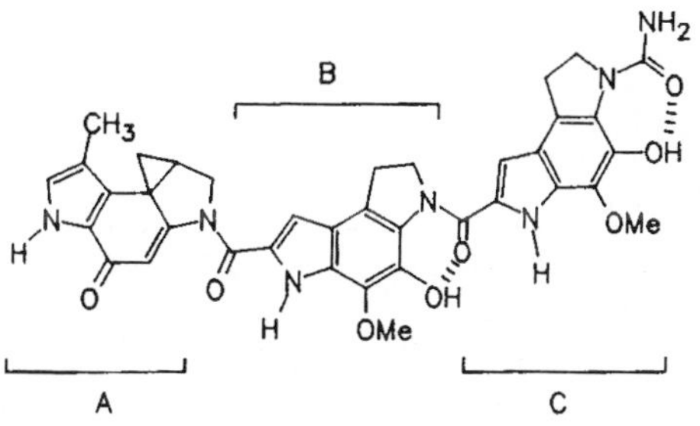

Scheme 1   *(1)*

first synthesis of fragment A was reported by Wierenga et al. (from the Upjohn group), who started from a functionalized aromatic compound *(2)* and constructed the indol fragment later (Scheme 2). [8]

Compound *(2)* was transformed into a malonic ester derivative. The bis mesylate *(4)* cyclized after reduction of the nitro group to form the indoline *(5)*. After protecting group manipulation and transformation to the acetate, a modified oxindole synthesis was used to synthesize *(8)*. Demethylation gave *(9)*, the acetate was cleaved, the alcohol transformed into the bromide, and final cyclization with Hünig base yielded cyclopropane derivative *(10)*. Fragment A of the natural product *(1)*

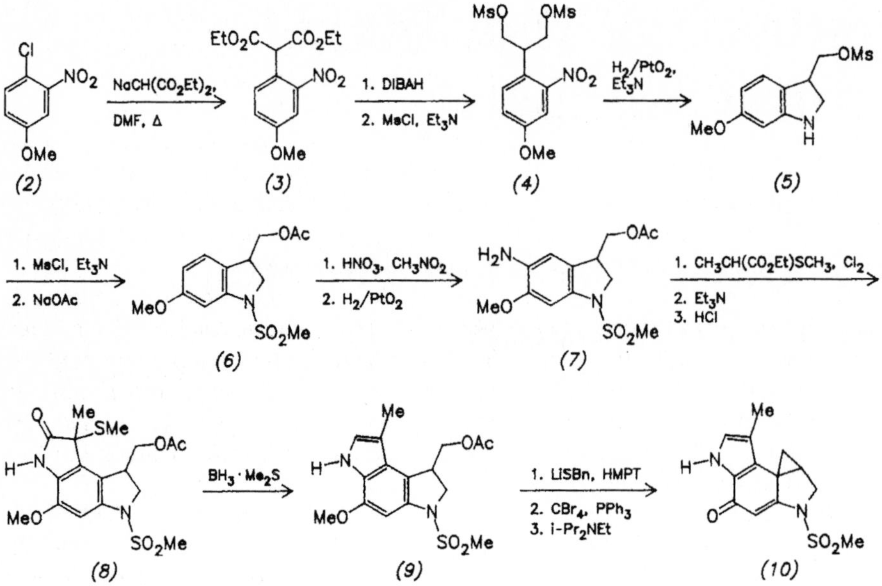

Scheme 2

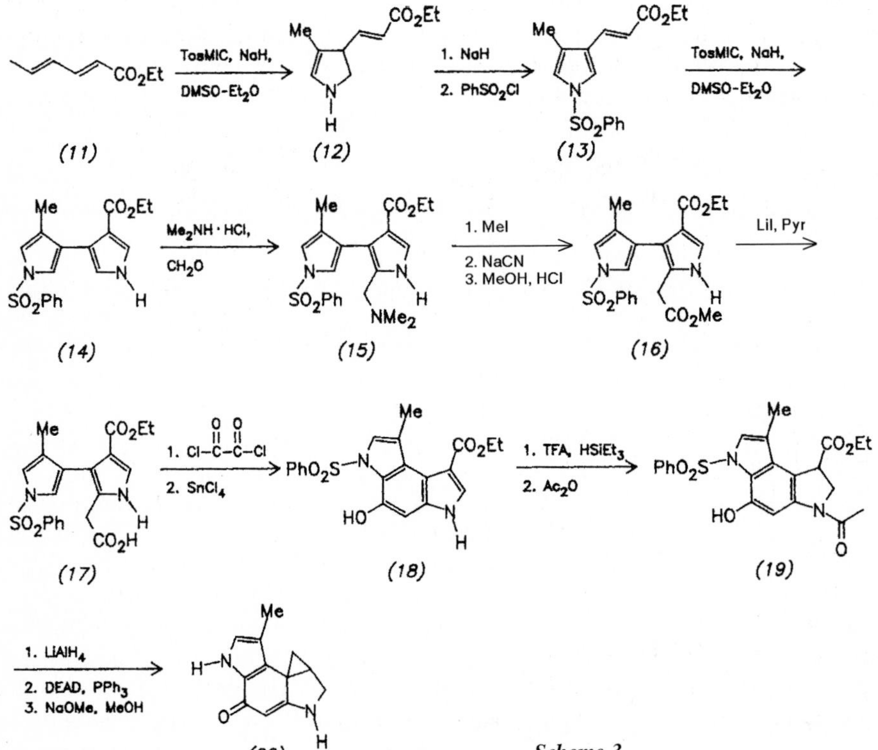

Scheme 3

was synthesized over 14 steps with an overall yield of 3%.

A better result concerning the overall yield was achieved by Magnus, who used a completely different strategy (Scheme 3).[9–11]

This synthesis began with an acceptor diene of type *(11)* as starting material, which was transformed as Michael acceptor with *p*-toluenemethylsulfonyl isocyanate (TOSMIC) to pyrrole *(12)*. It was not possible to use two equivalents, directly thereby forming two rings simultaneously in a one-pot procedure. The strong electron donating effect of the nitrogenatom had to be decreased by transformation into the *N*-tosylate. This compound generated the desired second pyrrole ring to give *(14)*. Under Mannich conditions *(15)* was isolated and transformed into the ester *(16)* after addition of methyl iodide, NaCN, and subsequent hydrolysis of the nitrile in the presence of methanol. Selective ester hydrolysis, followed by the synthesis of the acid chloride, gave the expected phenol *(18)* under Lewis acid conditions. Reduction of the indol portion with TFA/HSiEt₃ gave *(19)*, which was

N-acetylated, reduced and transformed under Mitsunobu conditions to give the cyclopropane *(20)*, which represents part A of the natural product *(1)*.

A very elegant approach by Kraus et al. used a Diels–Alder reaction as the key transformation. [12] Reaction of the diene *(22)* and imino quinone *(21)* gave the Diels–Alder adduct *(23)* (Scheme 4).

After a few transformations the tricycle *(26)* was generated from the Diels–Alder adduct. The construction of the cyclopropane ring was already designed in this sequence. The mesylate *(27)* was treated solely with base and *(28)* was obtained as a protected building block A of *(1)*.

A formal total synthesis of CC-1065 was recently described by Rees et al. [13] All subunits of the natural product (the cyclopropane part A and the dimeric pyrrole units B + C) have been synthesized by vinyl azide chemistry. All six nitrogen atoms of the natural product are formally coming from azide groups, basically from NaN₃ (Scheme 5)!

Scheme 4

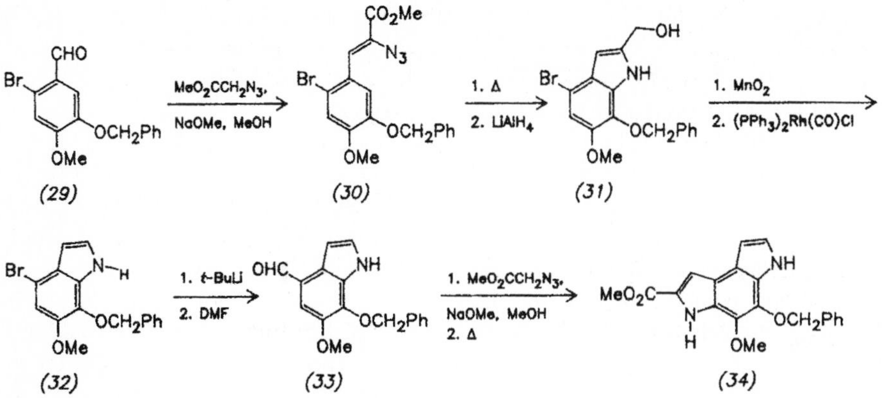

Scheme 5

This approach is somewhat similar to Wierenga's approach described earlier. Both use quite accessible aromatic precursors. The bromobenzaldehyde *(29)* can be prepared on a large scale from isovanillin. A simple condensation with methylazido acetate yielded the desired acyl azide *(30)*, which was thermolyzed to the indole derivative *(31)*. A drawback in the sequence was the transformation of the indole *(32)* from the ester by reduction and decarbonylation of the aldehyde. The second pyrrole ring was synthesized by a similar strategy: condensation of the aldehyde *(33)* with methylazido acetate, followed by thermolysis of the product to obtain tricycle *(34)*. The intermediate *(34)* was used in two ways: 1) Reduction of the indoline system *(35)*; 2) Hydrolysis of the ester *(34)* to the acid, and, finally, coupling of the pieces to the dimer *(37)* (Scheme 6).

The cyclopropane *(45)* was synthesized in a straightforward way from ketone *(38)*, which was further transformed via the epoxide into the azide *(40)*. Thermolysis gave the indole, which was coupled to the tricycle *(43)*. After a few further operations *(45)* was obtained (Scheme 7).

Both subunits of the natural product were synthesized by this route and can be coupled to the natural product, as already described by Kelly et al. [14]

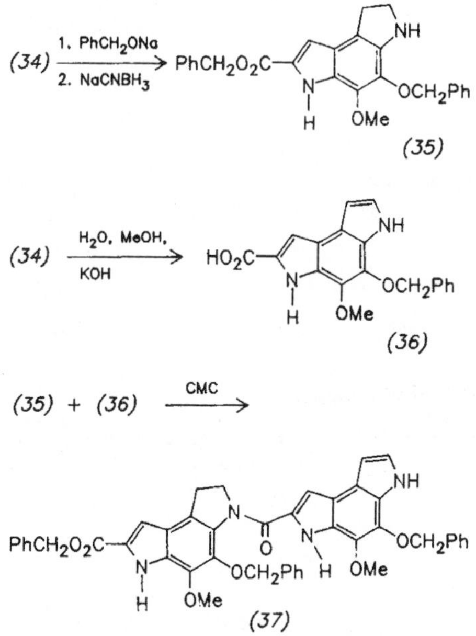

CMC = 1-Cyclohexyl-3-[2-(4-methylmorpholinoethyl]carbodiiminiumtoluene-4-sulfonate

Scheme 6

The first total synthesis of CC-1065 was reported by Boger et al. [15] He used two different strategies with quite interesting new reactions to synthesize the subunits (Scheme 8).

As the key reaction in the synthesis of the dimer an intramolecular Diels–Alder reaction

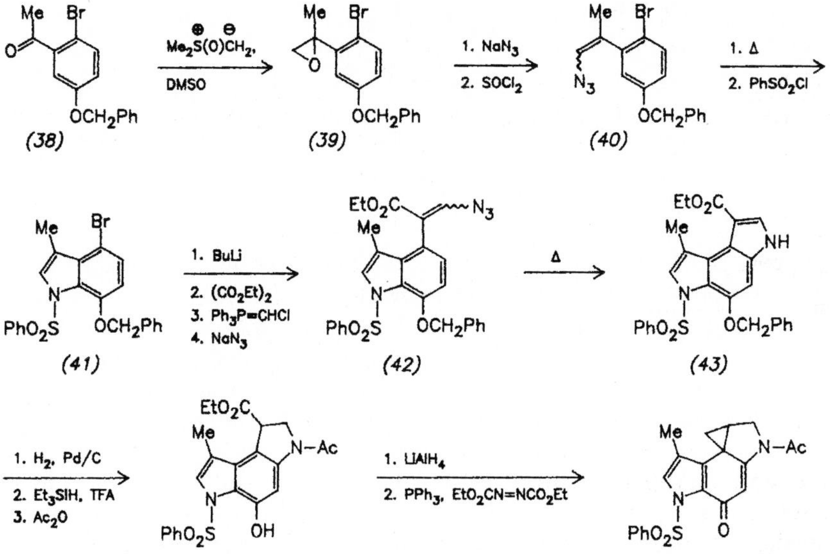

Scheme 7

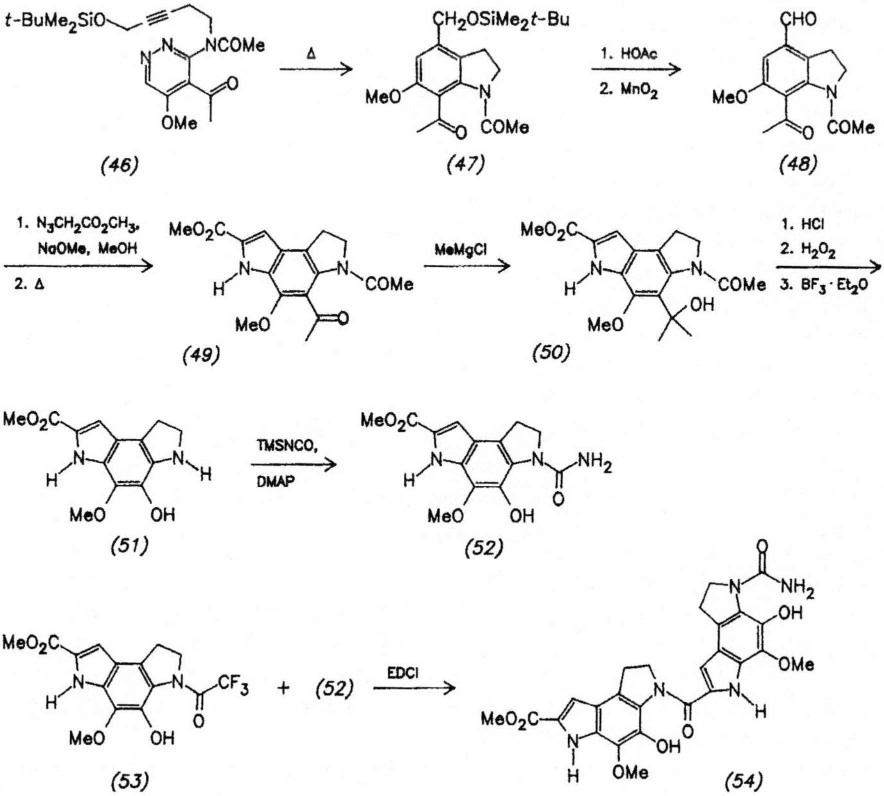

Scheme 8

(46) → (47) – which resulted directly in the indoline (47) – was used. The required precursor was quite easily obtained – alkylation under Mitsunobu conditions – but had to be optimized by a conformational study in order to increase the yield in the Diels–Alder reaction. [15] The missing ring was synthesized by the route of Rees (via the nitrene) yielding the required tricycle (49). The hydroxyl group of the aromatic ring was introduced by a Lewis acid catalyzed hydroperoxide-rearrangement of the tertiary alcohol (50). The originally planned Baeyer–Villiger rearrangement could not be realized. After a few protecting group manipulations the fragments could be coupled by EDCl. The next operation introduced part A by a radical-induced cyclization as the key step. In order to manage this step an indole with the required functionality had to be synthesized. Boger used an unusual nucleophilic addition of an enamine (56) to an activated diimido quinone (55) (Scheme 9).

The additional elimination of piperidine gave the required indole derivative (58). Bromination at the 4-position took place at low temperature, and smooth N-alkylation with propargyl bromide followed to yield (60). The latter was cyclized to the tricyclic compound (61) by a radical initiated exo-dig reaction in the presence of AIBN and Bu₃SnH. The desired alcohol (62) was obtained, after hydroboration, in racemic form and was separated using a procedure developed by Boger [16] yielding the two enantiomers (65a) and (65b) (Scheme 10). The cyclopropanation was the last step after coupling of the fragments, which was carried out using techniques developed by Kelly with a derivative of carbodiimide. The coupled molecules (66a) and (66b) could be cyclopropanated under basic conditions to give (1) in both antipodes (Scheme 11).

The syntheses discussed to obtain CC-1065 clearly show that know-how developed by different groups could be elegantly combined to make this project a success.

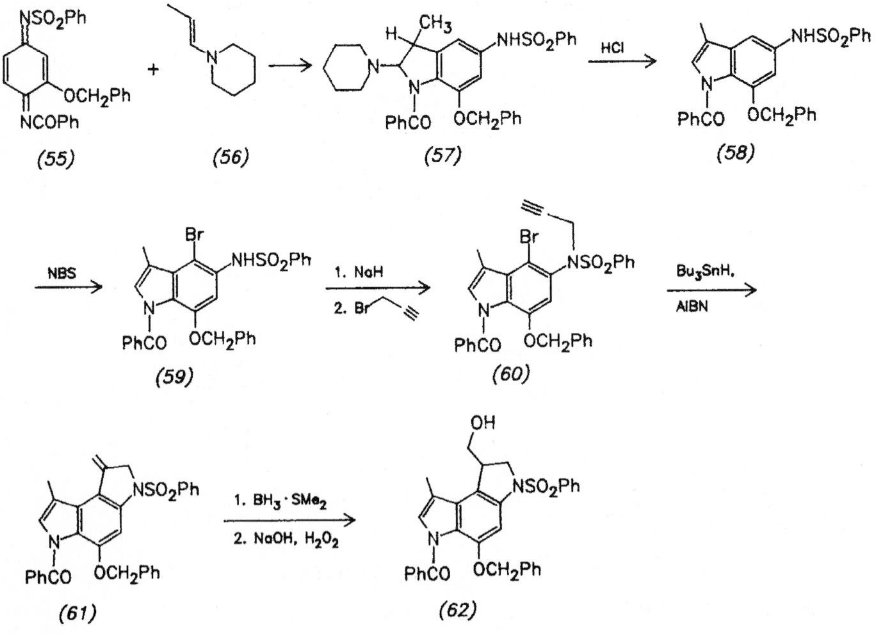

Scheme 9

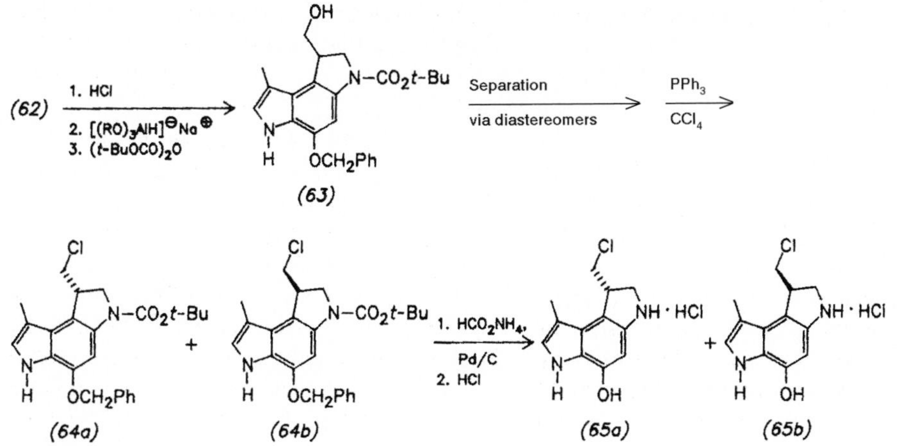

Scheme 10

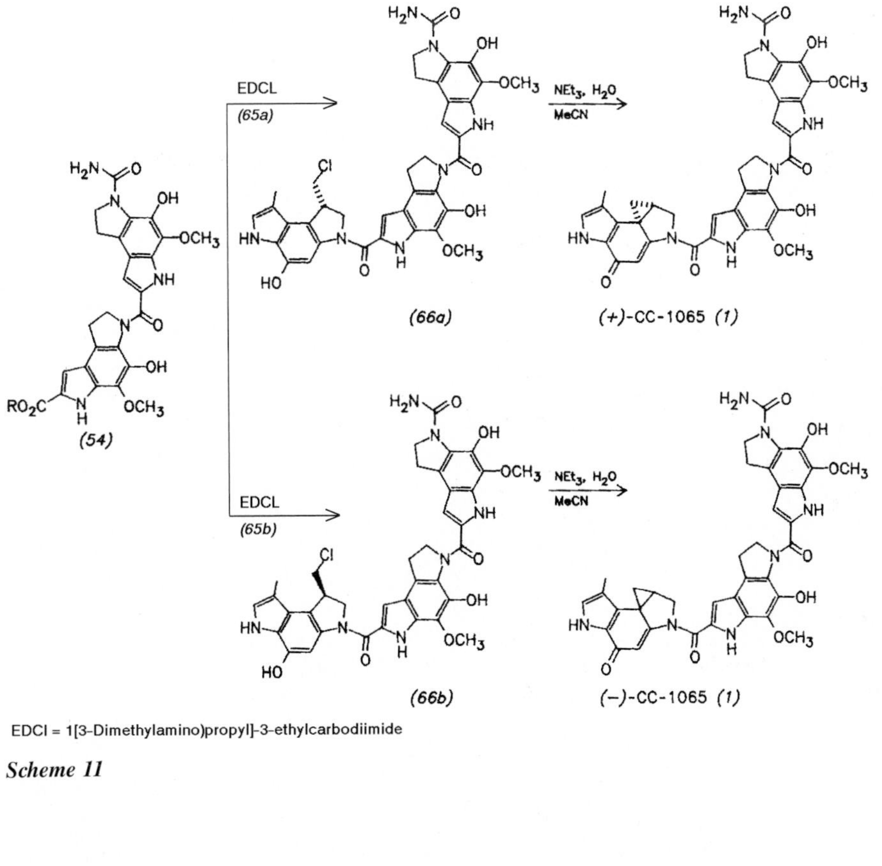

EDCI = 1[3-Dimethylamino)propyl]-3-ethylcarbodiimide

Scheme 11

References

[1]   D. Schinzer, *Nachr. Chem. Tech. Lab.* **1989,** *37,* 172.

[2]   J. Golik, J. Clardy, G. Dubay, G. Groenewold, H. Kawaguchi, M. Konishi, B. Krishnan, H. Okuma, K. Sitloh, T. W. Doyle, *J. Am. Chem. Soc.* **1987,** *109,* 3462.

[3]   Review: V. H. Rawal, R. J. Jones, M. P. Cava, *Heterocycles* **1987,** *25,* 701.

[4]   L. J. Hanka, A. Dietz, S. A. Gerpheide, S. L. Kuentzel, D. G. Martin, *J. Antibiot.* **1978,** *31,* 1211.

[5]   D. L. Boger, R. S. Coleman, *J. Am. Chem. Soc.* **1988,** *110,* 4796.

[6]   D. L. Boger, T. Ishizaki, R. J. Wysocki, S. A. Munk, *J. Am. Chem. Soc.* **1989,** *111,* 6461.

[7]   D. G. Martin, L. J. Hanka, G. L. Neil, *Proc. Am. Assoc. Cancer Res.* **1978,** *19,* 99.

[8]   W. Wierenga, *J. Am. Chem. Soc.* **1978,** *103,* 5621.

[9]   P. Magnus, Y.-S. Or, *J. Chem. Soc. Chem. Commun.,* **1983,** 26.

[10] P. Magnus, T. Gallager, *J. Chem. Soc. Chem. Commun.* **1984,** 389.

[11] P. Magnus, S. Halazy, *Tetrahedron Lett.* **1985,** *26,* 2985.

[12] G. A. Kraus, S. Yue, J. Sy, *J. Org. Chem.* **1984,** *50,* 284.

[13] R. E. Bolton, C. J. Moody, M. Pass, C. W. Rees, G. Tojo, *J. Chem. Soc. Perkin Trans. I,* **1988,** 2491.

[14] R. C. Kelly, I. Gebhard, N. Wicnienski, P. A. Aristoff, P. D. Johnston, D. G. Martin, *J. Am. Chem. Soc.* **1987,** *109,* 6837.

[15] D. L. Boger, R. S. Coleman, *J. Am. Chem. Soc.* **1987,** *109,* 2717.

[16] D. L. Boger, R. S. Coleman, *J. Org. Chem.* **1988,** *53,* 695.

Syntheses of Morphine

Martin Maier

Since drugs have been known, they have been contested in society. This is quite understandable in the light of the misery and the crime caused as a result of unresponsible drug consumption. Nevertheless, it is illusory to dream of a drug-free society and it is questionable whether a repressive drug policy can diminish the problems. [1] For science, however, drugs have proven to be very useful, because they represent important tools for studying the function of the brain. Particularly promising in this regard are marihuana and heroin, or morphine, since in the brain they interact with specific receptors. Thus, they are different from cocaine and other drugs, which merely disrupt certain processes in the brain. In the meantime the receptors for morphine and for tetrahydrocannabinol (THC) have been isolated and cloned. Moreover, for both receptor types the natural (endogenous) ligands are known. While the encephalins and endorphines, which bind to the opiate receptor, are peptides, the anandamides, the ligand for the THC-receptor, is derived from arachidonic acid. [2] An important function of the endogenous ligands is to block pain signals in certain situations. Simultaneously, the endogenous ligands cause a feeling of well-being. Some questions that come to mind are how the corresponding receptors change on consumption of drugs or how they might be manipulated with therapeutics. With this knowledge one hopes to separate the potent analgesic properties from the other, undesired, effects. In both cases the enormous structural differences between endogenous and exogenous ligands are amazing, with structural overlap only in small areas. Possibly, the addicting properties of the exogenous ligands may be explained by a "gain of function" model. [3] This means that only after binding to the receptor is an interaction with other receptors possible via the formation of ternary complexes. Morphine *(1)* is also an attractive synthetic target because of its complex structure. Despite the fact that the first synthesis dates back almost forty years, its attraction as a synthetic target hardly has diminished. A compar-

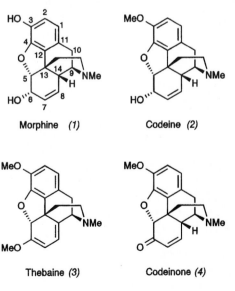

Morphine *(1)*

Codeine *(2)*

Thebaine *(3)*

Codeinone *(4)*

ison of the syntheses also demonstrates to some degree the evolution of synthetic strategies and methodologies.

Morphine *(1)* belongs to the class of the morphinanedienone alkaloids; [4] further important representatives of this class of compounds are codeine *(2)*, thebaine *(3)* and codeinone *(4)*. The diacetyl derivative of morphine is known under the name heroin. The potency of heroin is about three to five times that of morphine and because of its higher lipophilicity it is able to cross the blood-brainbarrier more easily.

Retrosynthesis

The polycyclic ring system of morphine represents a great challenge from a synthetic point of view. In addition, there are the five stereo

centers that must be established in the course of the synthesis, although in cyclic systems the correct stereochemistry often establishes itself automatically or can be easily controlled or corrected. In the retrosynthetic analysis it is important, in the first place, to cut the right bonds, whereby the aim is to get to simpler structures, the syntheses of which are obvious. If one does not want to rely on intuition or trial-and-error, a procedure developed by Corey is available that allows the determina-

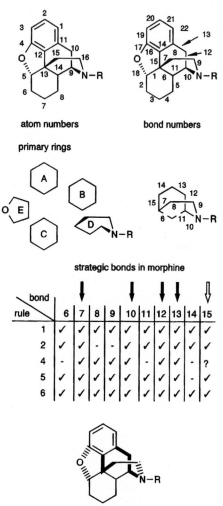

Scheme 1. Retrosynthetic analysis of the morphine skeleton.

tion of strategic bonds (SB) in polycyclic systems. [5] Strategic bonds are those, the breaking of which leads to a reduction a) in the number of side chains, b) the number of side chains with chiral centers, c) of the number of medium-sized rings or large rings, and d) of the number of bridged rings. The SB can be ascertained with the help of six rules, whereby *all* presuppositions must be fulfilled. In principle an SB is a bond that can be formed very easily.

Rule 1: An SB must be part of a primary, four- to six-membered ring. This is connected to the ease of formation of such rings. A pri-

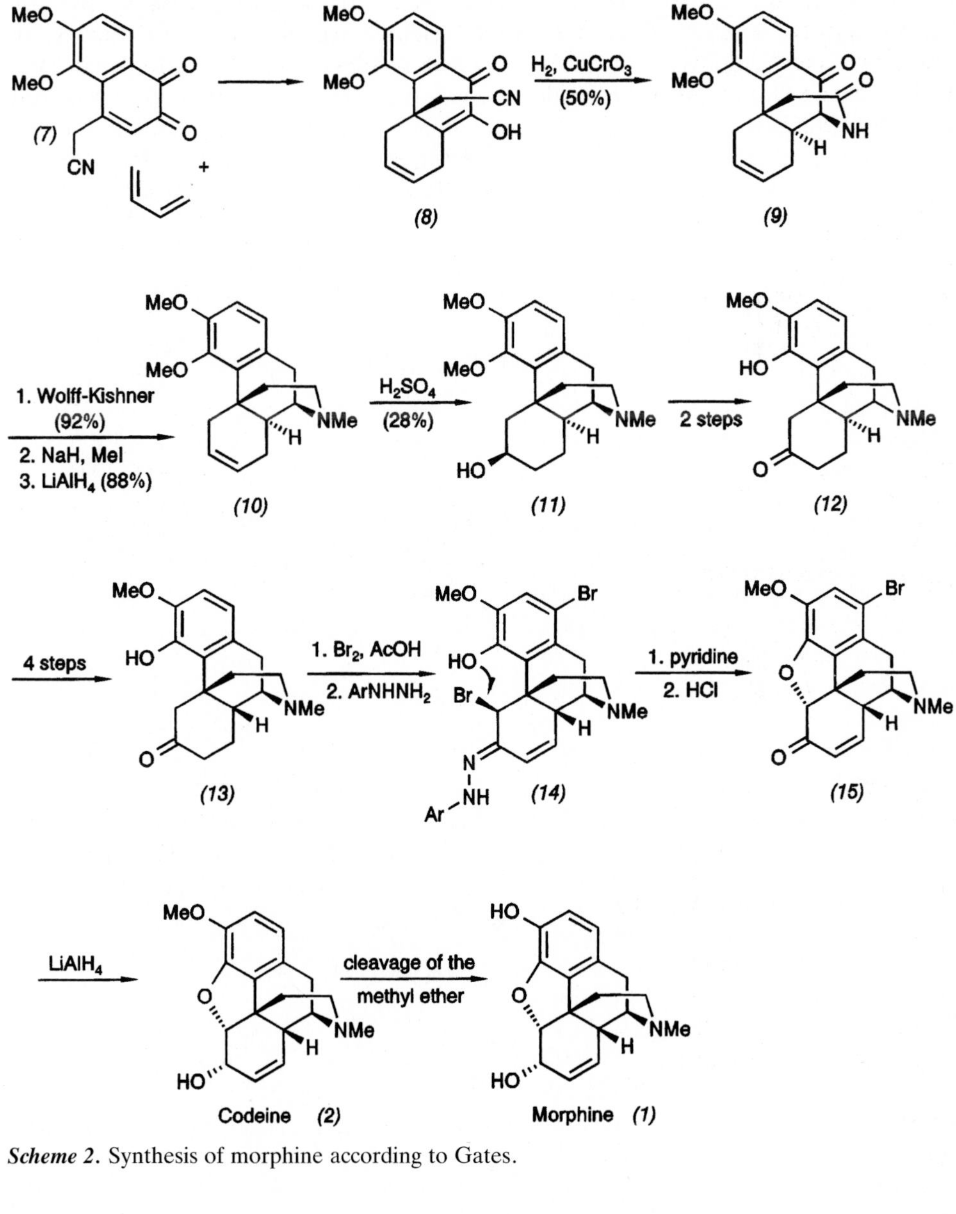

Scheme 2. Synthesis of morphine according to Gates.

mary ring cannot be represented as a sum (envelope) of two or more rings. The envelope of two annulated or bridged rings is termed a secondary ring.

Rule 2: An SB must be *exo* to another ring. If the bond is, however, *exo* to a three-membered ring, it is not considered to be strategic (Rule 2B).

Rule 3: An SB must be in the ring that is bridged at the highest number of positions (highest degree of bridging). This or these rings, respectively, is (are) termed the maximum bridging ring (MBR). In addition, the MBR must belong to the set of synthetically significant rings (SSR). The SSR's include primary rings and secondary rings with a size of less than eight atoms. For example, in the ring system *(5)* there are two secondary rings (D and F, $n < 8$), besides the three primaries. Ring A is bridged at positions 1 and 3 with B. It should be noted that although atom 2 is a bridgehead atom and is part of ring A, this ring is not bridged to another ring with this atom. Ring C is also bridged at two positions (2 and 4) with other rings. In contrast, ring B is bridged at four positions (1–4) with other rings and is, therefore, the MBR.

Rule 4: Bonds common to bridged or fused rings cleavage of which leads to a ring with more than seven atoms are not considered to be strategic. Such bonds are also termed core-bonds. If, however, two bridged or fused rings are connected to each other via another bond, even a core-bond is strategic.

Rule 5: Bonds in aromatic systems are not strategic, except in a heteroaromatic ring.

Rule 6: If rings contain a pair of common atoms (a spiro center, annulated or fused rings), bonds that connect the common atoms are only strategic if the bond path is free of stereocenters. By this rule structures that contain chiral atoms in the side chain are avoided. If, however, only one chiral center is within the chain connecting the two atoms, then the bonds to the chiral center are candidates for strategic bonds because their disconnection

leads to a reduction of chirality (e. g. bonds 1 to 3 in *(6)*).

Moreover, for heterocyclic systems carbon-heteroatom bonds are considered to be strategic if they satisfy the requirements of the rules 2B, 4 and 6. This is due to the ease of formation of such bonds.

In practice it is recommended to proceed according to the following steps: 1. Identify the primary rings of a polycycle, that is, all rings of size four to seven. 2. Identify the secondary rings. In order to find these rings, all possible ring combinations that share common atoms must be screened. At the same time one can determine whether a secondary ring is synthetically significant and which of the bonds are core bonds. 3. Determine whether

Reticuline *(16)*

Salutaridine *(17)*

Isoboldine *(19)*

Isosalutaridine *(18)*

an SSR is bridged with another ring and at which positions this is the case. This analysis leads to the MBR. Because this is the most restrictive rule, in the following it is sufficient to check the bonds of the MBR for the fulfillment of the other five rules. 4. Finally, one looks for the carbon-heteroatom bonds that are strategic.

If the morphine skeleton is analyzed according to this procedure, one finds five primary rings (Scheme 1). Because the smallest primary ring is a five-membered ring, a synthetically significant secondary ring ($n < 8$) cannot exist. That is, the analysis is restricted to the bicyclic ring system consisting of the rings B and D because only these two are bridged. [6] The envelope (i.e. $(R_i \cup R_j) - (R_i \cap R_j)$) [7] of the rings B and D yields an

eight-membered ring, that is bonds 6 and 11 are core bonds and are therefore not strategic.

It is appropriate to lay on the bond numbers against the rules, whereby one finds that bonds 7, 10, 12, and 13 are strategic. Actually, one also has to mark bond 15 as strategic, although it is a core bond of rings B and E. Quite often, however, the heterocyclic ring is formed only at the end of the synthesis. That means, in these immediate precursors, bond 15 is not a core bond. Further analysis shows that all of the C-X bonds are strategic.

Of course, the whole procedure has its limitations and according to the principle "the exception makes the rule" connections of bonds that officially are not strategic can also lead to the target. In particular, synthetic strategies that rely on intramolecular cycloadditions or pericyclic reactions in general are not well recognized. For example, in decalin systems the central bond is a core bond and therefore not strategic, despite the fact that this bond can be formed very efficiently by an intramolecular Diels–Alder reaction. [8]

Syntheses

Interestingly, in the first morphine synthesis [9] (Scheme 2) a cycloaddition strategy was chosen in order to construct the tricyclic system *(8)*. The dienophile *(7)* was obtained in ten steps from 2,6-dihydroxynaphtalene. The attachment of the cyanomethyl side chain by a Michael addition is in accordance with the retrosynthetic analysis. A chemoselective reaction with H_2/copper chromite resulted in a reductive cyclization with formation of the ketone amide *(9)*. The unsaturated amide *(10)* which was obtained by Wolff–Kishner reduction, *N*-methylation and amide reduction could be transformed to the alcohol *(11)* by regioselective hydration of the double bond. Two further steps, which included a selective cleavage of the methyl ether and an oxidation, provided the ketone *(12)*. After introduction of a double bond, the stereochemistry at C-14 could be corrected through the sequence double bromination, dinitrophenyl hydrazide formation, acid treatment and hydrogenation. Through a triple bromination of compound *(13)* and subsequent treatment with dinitrophenyl hydrazine, leading through the intermediate *(14)*, the ring system was completed. Altogether 28 steps were necessary. Quite complicated is the preparation of the dienophile *(7)* and the epimerization of the stereo-

center at C-14. Certainly, for many of these transformations one would use other reagents or conditions. Still, it is remarkable what can be achieved without using lithium diisopropylamide (LDA), silyl protecting groups, or organometallic chemistry. It should be noted, however, that in those days abundant use was already made of relay syntheses, which was admitted by the authors. Thus, many intermediates, for example *(10)*, were ultimately prepared from morphine itself. By this chemical correlation, however, the structure of morphine could be established unambiguously. As an aside, most of the subsequent syntheses of morphine lead to intermediates of the Gates synthesis.

An attractive possibility to approach a synthesis of a natural product is to follow the biosynthetic pathway – particularly because the retrosynthetic analysis is avoided. The key step of the biosynthesis is the oxidative coupling of reticuline *(16)* to salutaridine *(17)* *(ortho-para)*, whereby bond 15 between the positions that are marked with an arrow, is formed. However, it is not easy to realize this pathway in an enzyme-free manner, because isosalutaridine *(18)* *(para-para)* and isoboline *(19)* *(ortho-ortho)* are formed as major products. Nevertheless, there have been many attempts to direct this reaction in the desired sense. The best result so far was achieved by Schwartz et al., who succeeded in the cyclization of *N*-ethoxycarbonylnorreticuline to the corresponding salutaridine derivative in a yield of 23 % by using thallium trifluoroacetate. [10]

Instead by an oxidative coupling reaction, bond 15 can also be formed in an acid-catalyzed, intramolecular, electrophilic aromatic substitution. [11] Although now bond formation occurs unambiguously to C-13, the wrong regioisomer (OH group points to the top) is still favored. The solution to the regiochemical problem consists of using a symmetrical aromatic ring for the electrophilic substitution or to block the position that would lead to the wrong isomer. Thus, Beyermann et al. started with compound *(20)*, which by the way is very easily obtained from the corresponding 1-benzyl-1,2,3,4-tetrahydryisochinolin by Birch reduction (85 %) and *N*-formylation. Cyclization of *(20)* with 80 % sulfuric acid furnished the morphinane derivative *(21)*. Removal of the superfluous OH group to give dihydrothebainone *(23)* proceeded after selective etherification with 5-chloro-1-phenyl-tetrazole and subsequent hydrogenation. [12] In a variation of this strategy, position 1 of the aromatic ring A is brominated, which directs the cyclization to the correct isomer. [13]

Yet another way to construct bond 15 is illustrated in a synthesis of (–)-dihydrocodeinone *(29)* by the group of Overman. [14] They use an intramolecular Heck reaction to build the critical quarternary center of the morphinane skeleton. In a sequence of five steps, 2-allylcyclohexenone was converted to the chiral allylsilane *(25)*. The optical activity of *(25)* was secured by an enantioselective reduction of the cyclohexenone derivative with catecholborane in the presence of the (*R*)-oxazaborolidine catalyst. Condensation of the allylsilane *(25)* and the aldehyde *(24)* gave the crystalline *(27)* via the intermediate *(26)*. The stereochemistry of compound *(27)* results from the preferential formation of an (*E*)-iminium ion *(26)* that enters into a stereoselective iminium ion-allylsilane cyclization. With the iodine atom already in place, the problem of regioisomers does not exist in the subsequent Heck cyclization of *(27)* to compound *(28)*. The unsaturated morphinane *(28)* was transformed in four steps to

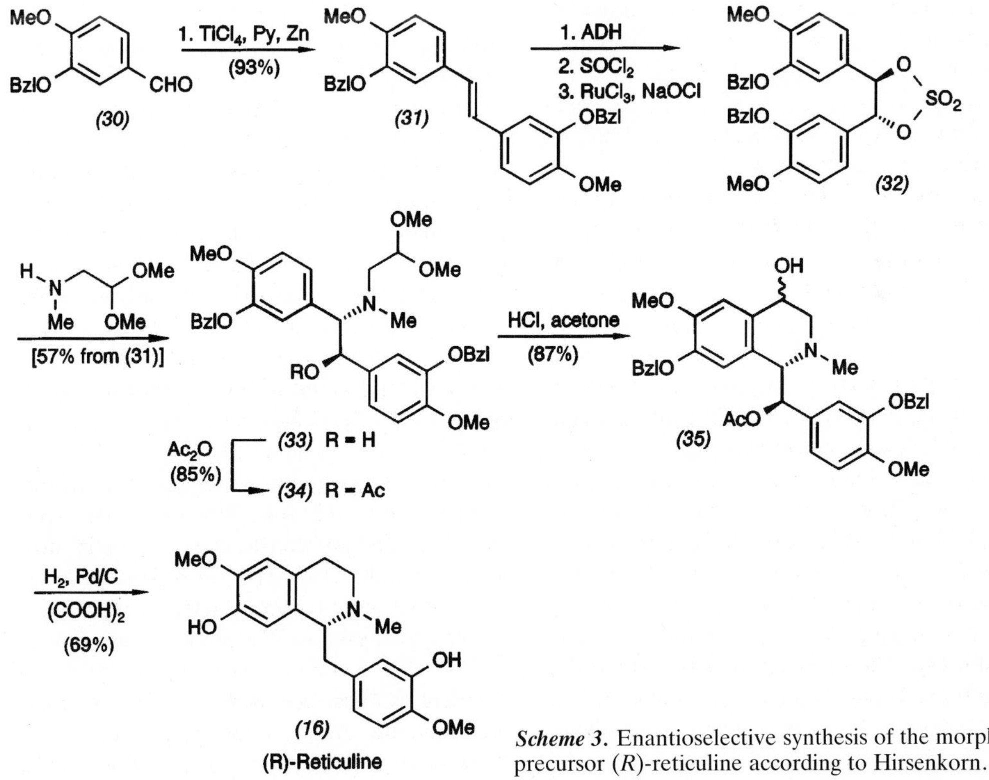

Scheme 3. Enantioselective synthesis of the morphine precursor (*R*)-reticuline according to Hirsenkorn.

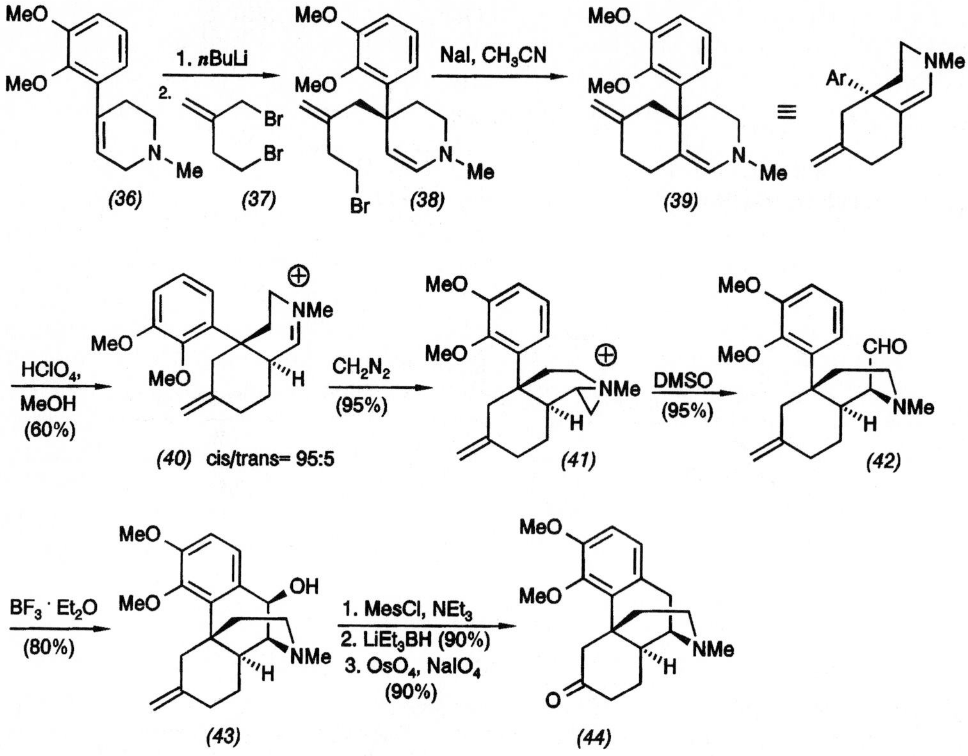

Scheme 4. Morphine synthesis of Evans.

(–)-dihydrocodeinone *(29)*. From this, in turn, (–)-morphine could be obtained in another five steps.

Despite the difficulties of controlling the regioselective cyclization of reticuline derivatives, the biomimetic approach is of particular interest, because the cyclization to tetracycles of type *(17)* necessarily proceeds in a diastereospecific manner. Through asymmetric syntheses of 1-benzyl-1,2-dihydroisoquinolines, morphine, or derivatives thereof, would be accessible in enantiomerically pure form. In a very elegant strategy (Scheme 3) stilbene *(31)* served as starting material, which itself could be prepared very easily by McMurry coupling. A key step is the enantioselective dihydroxylation (ADH) of the stilbene *(31)* according to the Sharpless procedure [15] providing, after

functionalization of the diol, the cyclic sulfate *(32)* in 88% *ee*. [16] Reaction of *(32)* with methylaminoacetaldehyde dimethyl acetal furnished the amino alcohol *(33)* which (as its acetate) cyclized under Pomeranz–Fritsch conditions to compound *(35)*. Hydrogenation of *(35)* led to (R)-reticuline *(16)*.

Although in the synthesis of Evans (Scheme 4) five C–C bonds (three of them are strategic) are formed in the course of the synthesis of two rings, the number of steps is nevertheless rather small. [17] By addition of 2,3-dimethoxylithium to N-methyl-4-piperidone and acid-catalyzed elimination, the heterocyclic olefin *(36)* is formed. Compound *(39)* could be constructed in a two-step procedure whereby the allylic anion, which was generated by treatment with *n*BuLi, was alkylated

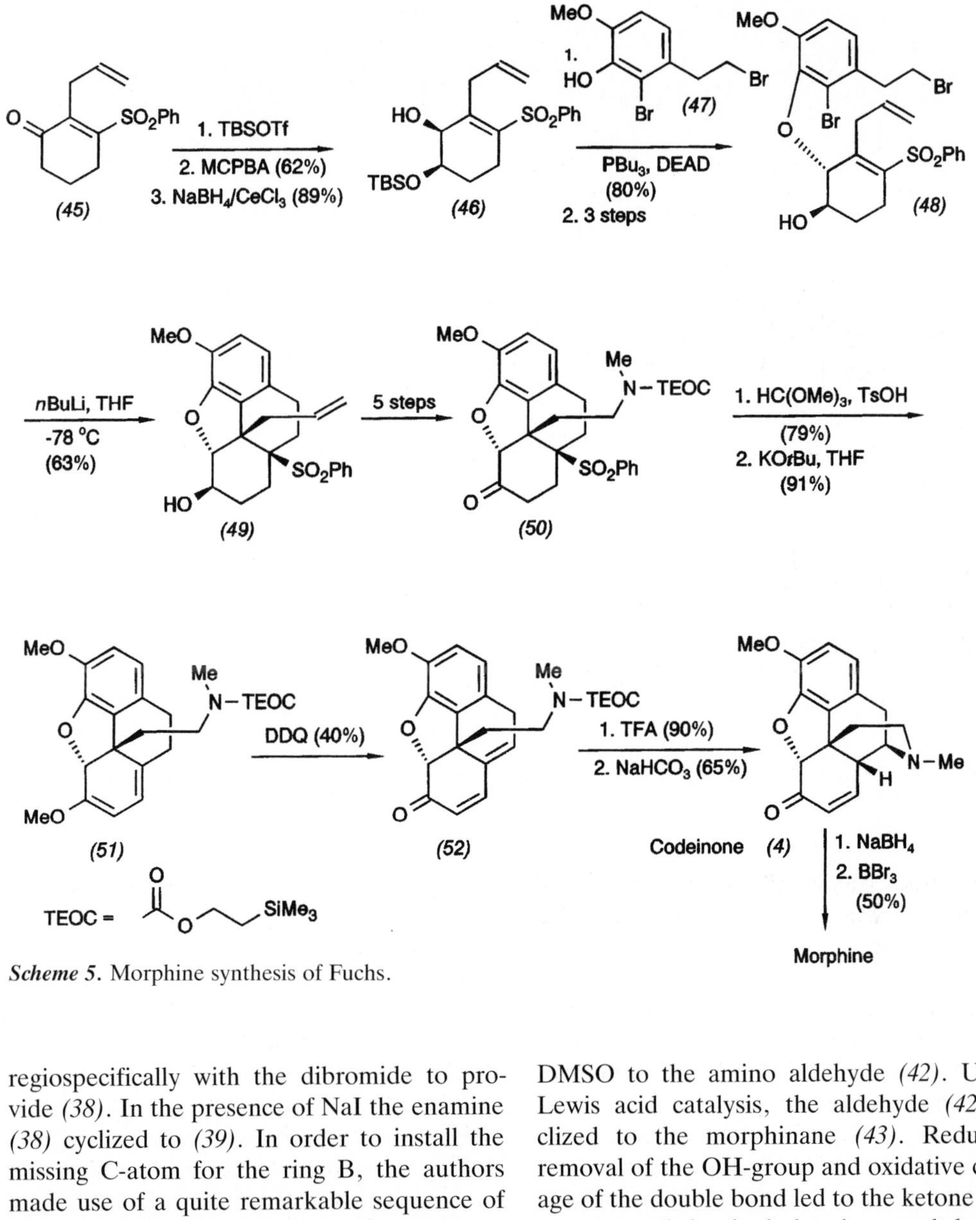

Scheme 5. Morphine synthesis of Fuchs.

regiospecifically with the dibromide to provide *(38)*. In the presence of NaI the enamine *(38)* cyclized to *(39)*. In order to install the missing C-atom for the ring B, the authors made use of a quite remarkable sequence of reactions. The *trans*-iminium salt, which was formed by acid treatment from *(39)*, can be equilibrated to the desired *cis*-isomer *(40)*. If compound *(40)* was treated with diazomethane, the aziridinium salt *(41)* was generated in a diastereospecific manner, which then reacted in a Swern-analogous reaction with

DMSO to the amino aldehyde *(42)*. Under Lewis acid catalysis, the aldehyde *(42)* cyclized to the morphinane *(43)*. Reductive removal of the OH-group and oxidative cleavage of the double bond led to the ketone *(44)*, a compound that had already passed through in the Gates synthesis (as the 4-OH-free derivative).

While the oxygen-containing ring is formed only at the end in the above-mentioned syntheses, Fuchs used this ether bond in a very clever way (Scheme 5) to guarantee an intra-

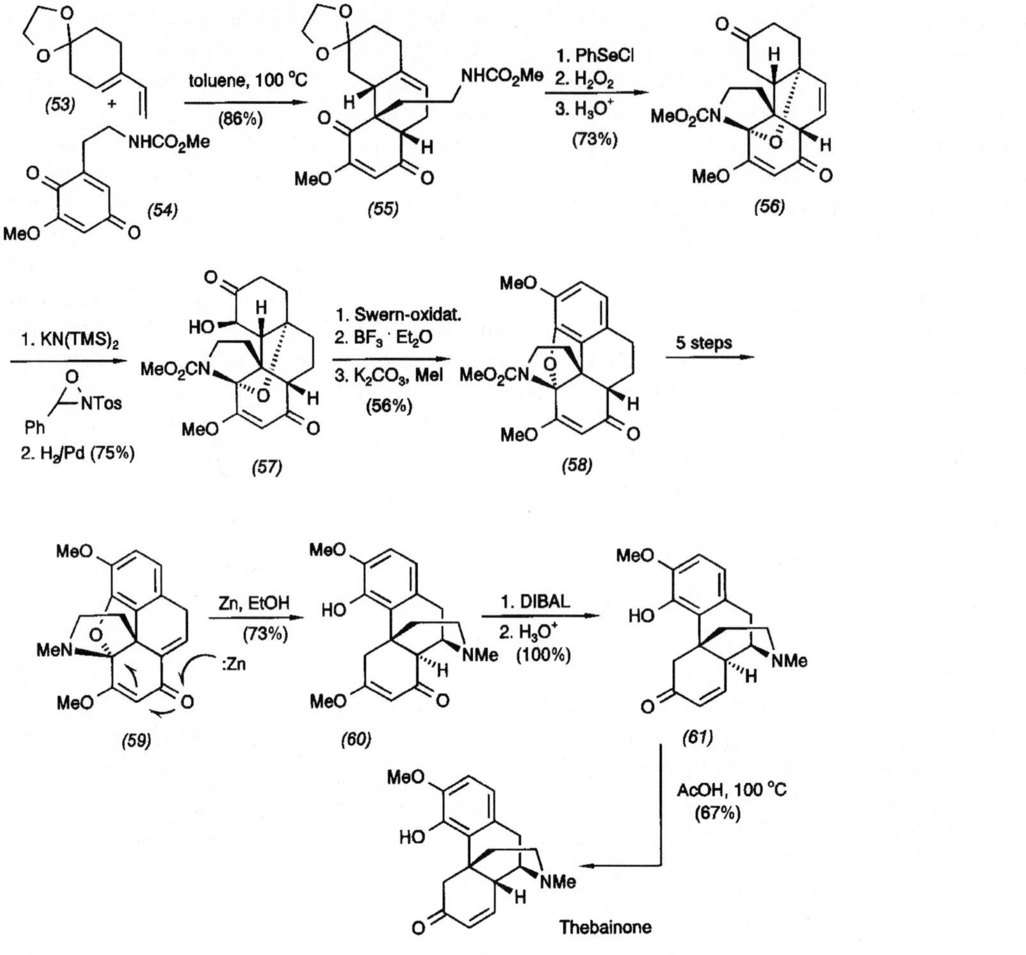

Scheme 6. Morphine synthesis of Tius.   *(62)*

molecular reaction during the formation of the strategic bond 15. [18] The alcohol *(46)* was prepared from 1,3-cyclohexanedione by standard reactions by passing through the intermediate *(45)*. Subsequent Mitsunobu reaction with the functionalized phenol *(47)* led with inversion of configuration to the *trans*-ether *(48)*. The key step of this synthesis is a tandem reaction, consisting of a nucleophilic addition of an aryl carbanion (prepared from *(48)* by treatment with *n*BuLi) to the vinyl sulfone function. The intermediate car-

banion now reacts in an intramolecular S_N2 reaction to the tricycle *(49)*. It took five further steps to convert the allyl group into a carbamate. Attempts to carry the amino function through the whole synthesis, in the form of an tosyl amide, led to a dead end, because deprotection was not possible without destroying the whole molecule. After enol ether formation the diene *(51)* was generated by base-induced elimination of benzene sulfinic acid. Oxidation of the diene (formal abstraction of H⁻) furnished the dienone *(52)*. Removal of the nitrogen protecting group was followed in the basic medium by addition of the amino function to the dienone with formation of codeinone *(4)*. Besides that, the β, γ-unsaturated ketone was formed, which, however, could be isomerized to compound *(4)* without a problem. Reduction of the carbonyl group and cleavage of the methyl ether provided morphine *(1)*.

Actually, it is quite logical to incorporate ring A from an aromatic precursor into the target molecule. That it can be done differently was demonstrated in a recent synthesis by Tius et al. (Scheme 6). [19] This synthesis is remarkable because the aromatic ring A is derived from an alicyclic precursor while on the other hand ring C was developed from an aromatic ring! The bonds 11 and 15 are constructed in a Diels–Alder reaction between the diene *(53)* and the quinone *(54)*. The next steps serve to aromatize ring A. Treatment of *(55)* with phenylselenyl chloride in methanol induced a selenocyclization with concomitant formation of a hemiaminal structure. Subsequent oxidative elimination and hydrolysis of the ketal yielded compound *(56)*. The introduction of the hydroxyfunction at C-17 was performed with the Davis reagent whereby the rigid ring system, in combination with conformative effects, enable regioselective formation of the required enolate. After hydrogenation of the superfluous double bond and Swern oxidation to the 1,2-diketone, treatment with Lewis acid caused aromatization by elimination of water. Because of the spatial vicinity of the functional groups a hemiaminal was again formed. Introduction of a double bond, reduction of the carbonyl function, cleavage of the carbamate and regeneration of the carbonyl group led to compound *(59)*. With the use of zinc the dihydrofuran ring could be opened reductively with simultaneous liberation of the amino group, which in one process undergoes a Michael addition to the α,β-unsaturated carbonyl system furnishing *(60)*. Epimerization

of β-thebainone *(61)* under acidic conditions finally provided thebainone *(62)*. Although this synthesis seems quite original, it requires, in a addition to the preparation of the starting materials *(53)* and *(54)*, a large number of steps (24 altogether) many for the aromatization of ring A. In chemistry the direct path is also usually the shortest!

Another recent synthesis of morphine by Parker et al. [20] is conceptually identical to the synthesis of Fuchs. The same bonds are constructed in the same order (Scheme 7).

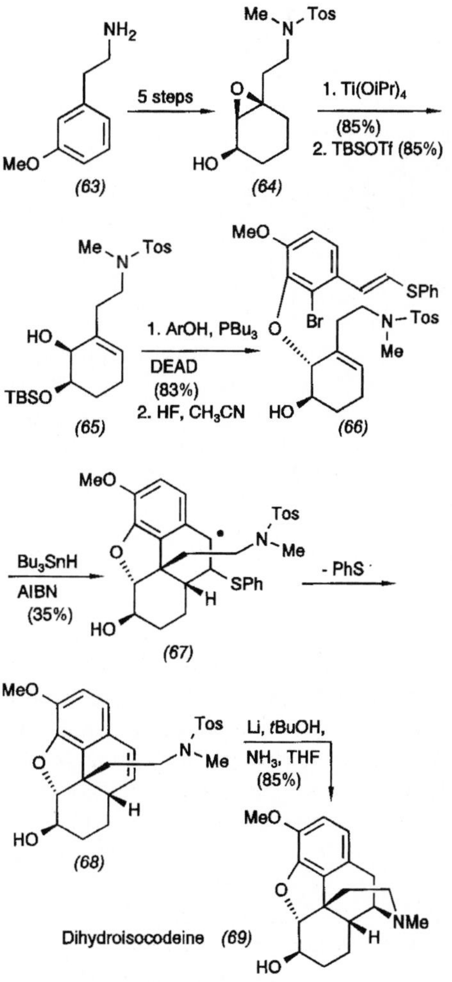

Scheme 7. Morphine synthesis of Parker.

Starting from amine *(63)* the *cis*-epoxide *(64)* was constructed in a conventional and transparent way (Birch-reduction, etc.). Opening of the epoxide provided the *cis*-diol *(65)*, the sterically less hindered OH group of which could be selectively protected. Subsequent Mitsunobu reaction provided the ether *(66)*. Treatment of *(66)* with AIBN/Bu₃SnH generated a radical that initiates a reaction cascade that ultimately leads to the formation of two C–C bonds. The following ring closure of *(68)* to give *(69)* was clearly a stroke of luck. In an attempt to reductively remove the tosyl group, the intermediate nitrogencentered radical or anion added directly to the double bond with the formation of dihydroisocodeine *(69)*. It is evident that with only 11 steps this synthesis defeats that of Fuchs. This synthesis underscores the fact that, besides the strategic concept the experimental execution, that is, the proper choice of reactions, is very important. Moreover, it represents an elegant application of a tandem radical reaction.

The syntheses presented here demonstrate that the retrosynthetic analysis of polycyclic systems according to Corey is absolutely useful, although the procedure seems rather complicated at first sight. By implementation of this or similar heuristic procedures using a computer any polycyclic molecule can be retrosynthetically dissected more or less easily. [21] Nevertheless, synthetic chemistry is still an experimental science. A large number of incomplete or failed syntheses provide clear evidence for this fact. [22] The sometimes expressed opinion, that the synthesis of organic molecules is not a problem or a challenge any longer, seems definitely to be wrong. Even today, only a few groups are able to synthesize complex natural products.

References

[1] See for example, W. Neskovic, *Ohne Drogen geht es nicht,* in *Die Zeit,* **1993,** no. 25, 40.

[2] a) M. Baringa, *Science* **1992,** *258,* 1882–1884; b) R. Mestel, *New Scientist* **1993,** July 31, 21–23.

[3] M. K. Rosen, S. L. Schreiber, *Angew. Chem.* **1992,** *104,* 413–430; *Angw. Chem. Int. Ed. Engl.* **1992,** *31,* 384–400.

[4] G. A. Cordell, *Introduction to Alkaloids,* Wiley, New York, **1981,** 422–450.

[5] E. J. Corey, W. J. Howe, H. W. Orf, D. A. Pensak, G. Petersson, *J. Am. Chem. Soc.* **1975,** *97,* 6116–6124. b) E. J. Corey, X. Cheng, *The Logic of Chemical Synthesis,* Wiley, New York, **1989;** c) F. Serratosa, *Studies in Organic Chemistry* 41, *Organic Chemistry in Action. The Design of Organic Synthesis,* Elsevier, Amsterdam-Oxford-New-York-Tokio, **1990,** 168–184.

[6] The numbering of the bonds is arbitrary.

[7] ∪ = Element either of the group 1 or 2; ∩ = Elements that belong to both groups.

[8] D. Craig, *Chem. Rev.* **1987,** *16,* 187.

[9] M. Gates, G. Tschudi, *J. Am. Chem. Soc.* **1956,** *78,* 1380–1393; see also: N. Anand, J. S. Bindra, S. Ranganathan, *Art in Organic Synthesis,* Holden-Day, San Francisco-London-Amsterdam, **1970,** 251–256.

[10] M. A. Schwartz, I. S. Mami, *J. Am. Chem. Soc.* **1975,** *97,* 1239–1240.

[11] R. Grewe, W. Friedrichsen, *Chem. Ber.* **1967,** *100,* 1550.

[12] H. C. Beyerman, T. S. Lie, L. Maat, H. H. Bosman, E. Buurman, E. J. M. Bijsterveld, H. J. M. Sinnige, *Recl. Trav. Chim. Pays-Bas* **1976,** *95,* 24–25.

[13] K. C. Rice, *J. Org. Chem.* **1980,** *45,* 3135–3137.

[14] C. Y. Hong, N. Kado, L. E. Overman, *J. Am. Chem. Soc.* **1993,** *115,* 11028–11029.

[15] a) Preparation of ligands for ADH: W. Amberg, Y. L. Bennani, R. K. Chadha, g. A. Crispino, W. D. Davis, J. Hartung, K.-S. Jeong, Y. Ogino, T. Shibata, K. B. Sharpless, *J. Org. Chem.* **1993,** *58,* 844–849; b) review, H. C. Kolb, M. S. Van Nieuwenhze, K. B. Sharpless, *Chem. Rev.* **1994,** *94,* 2483-2547.

[16] a) R. Hirsenkorn, *Tetrahedron Lett.* **1990,** *31,* 7591–7594; b) R. Hirsenkorn, *Tetrahedron Lett.* **1991**, *32,* 1775–1778; for further asymmetric syntheses of 1-benzyl-1,2-dihydro isoquinolines, see: a) A. I. Myers, J. Guiles, *Heterocycles,* **1989,** *23,* 295–301; b) M. Kitamura, Y. Hsiao, M. Ohta, M. Tsukamoto, T. Ohta, H. Takaya, R. Noyori, *J. Org. Chem.* **1994,** *59,* 297–310.

[17] D. A. Evans, C. H. Mitch, *Tetrahedron Lett.* **1982,** *23,* 285–288; for an alternative synthesis of *(39)* see: W. H. Moss, R. D. Gless, H. Rapoport, *J. Org. Chem.* **1993,** *48,* 227–238.

[18] J. E. Toth, P. R. Hamann, P. L. Fuchs, *J. Org. Chem.* **1988** *53,* 4694–4708.

[19] M. Tius, M. A. Kerr, *J. Am. Chem. Soc.* **1992,** *114,* 5959–5966.

[20] K. A. Parker, D. Fokas, *J. Am. Chem. Soc.* **1992,** *114,* 9688–9689.

[21] I. Ugi, J. Bauer, K. Bley, A. Dengler, A. Dietz E. Fontain, B. Gruber, R. Herges, M. Knauer, K. Reitsam, N. Stein, *Angew. Chem.* **1993,** *105,* 210–239; *Angew. Chem. Int. Ed. Engl.* **1993,** *32, 201.*

[22] See for example: a) M. Chandler, P. J. Parsons., *J. Chem. Soc., Chem. Commun.* **1984,** 322–323; b) J. D. White, R. J. Butlin, H.-G. Hahn, A. T. Johnson, *J. Am. Chem. Soc.* **1990,** *112,* 8595–8596; c) T. Hudlicky, C. H. Boros, E. E. Boros, *Synthesis* **1992,** 174–178; d) P. Magnus, I. Coldham, *J. Am. Chem. Soc.* **1991,** *113,* 672–673.

Synthesis of Calicheamicin γ_1^I

Herbert Waldmann

Calicheamicin γ_1^I *(1)*, esperamicin A_{1b} *(2)*, the chromophor *(3)* of neocarcinostatin and dynemicin A *(4)* are representatives of the enediyne antitumor antibiotics, which are characterized by a pronounced physiological activity even at very low concentration. The cell-destroying activity of these compounds is caused by a Bergmann cyclization that occurs after an initiation step. In the course of this reaction diradicals like *(29)* are formed (see Scheme 3) which abstract hydrogen atoms from the deoxyribose units of DNA and thereby initiate strand scissions. [1] In the case of the calicheamicins and the esperamicins this process is triggered by a bioreductive cleavage of the allylic trisulfide (see Scheme 3: *(28)* → *(29)* → *(30))*. The thiolate thus formed adds intramolecularly to the enone system. Thereby the two acetylenes present are brought into closer contact so that the Bergmann cycliza-

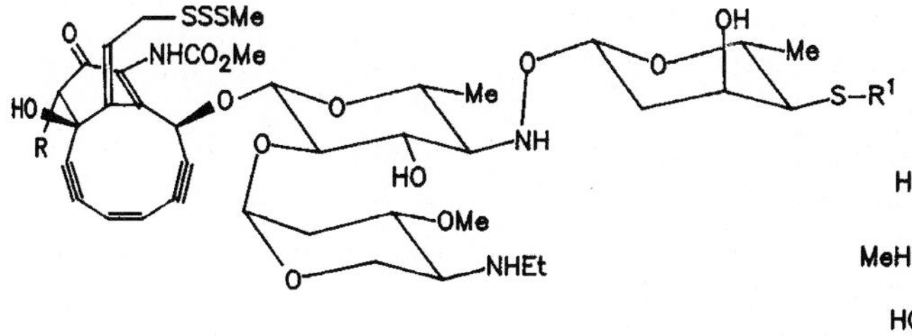

(1) calicheamycin γ_1^I:

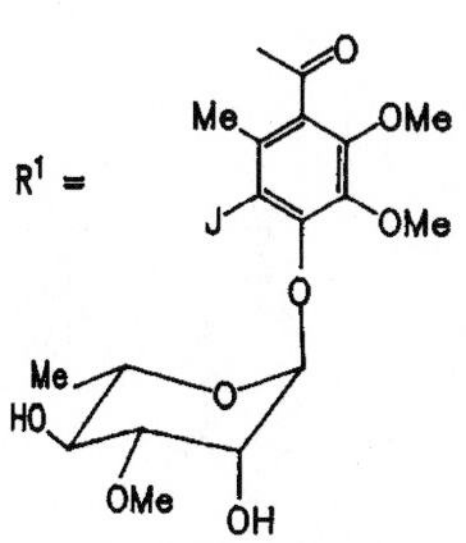

(2) esperamicin A_{1b}:

R = H;

R =

R^1 =

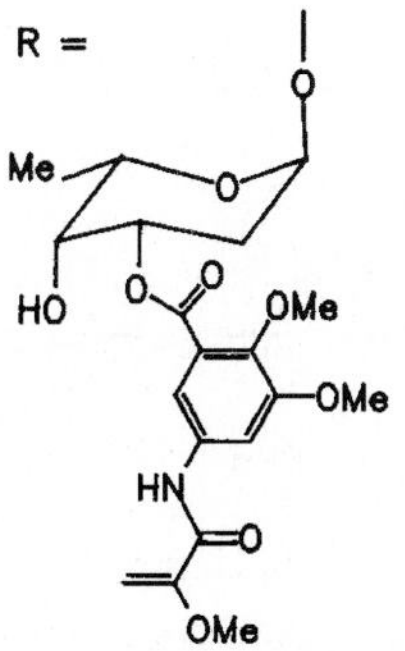

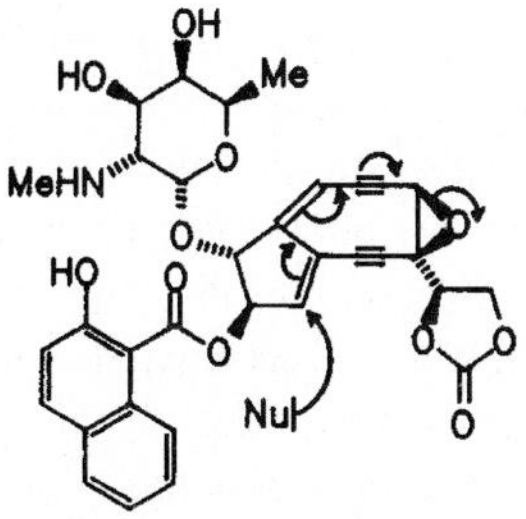

(3) chromophor of neocarcinostatin

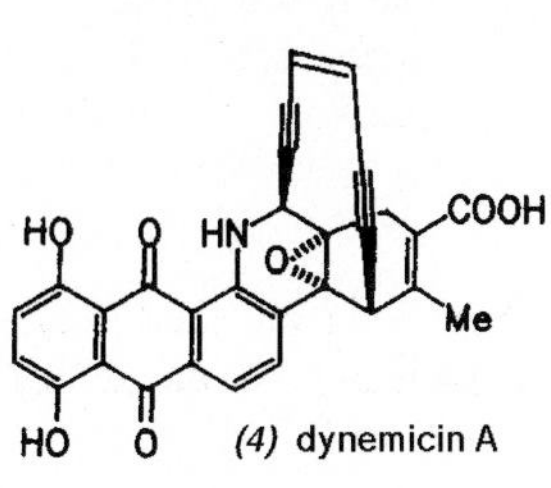

$R^1 = CH_3$

(4) dynemicin A

tion proceeds even at room temperature. In the case of dynemicin A the opening of the epoxide initiates the respective reaction cascade and in the case of neocarcinostatin the vinylogous addition of an S- or O-nucleophile to the epoxide (see (3)) leads to the formation of an enynecumulene intermediate, which reacts further to give an indacene diradical. The hope and expectation to develop less complex and consequently more accessible analogs with similar physiological activity, [1] which might open up new routes for cancer chemotherapy and the challenge to synthesize these complex antibiotics were driving forces initiating numerous research activities. [1] These efforts have recently culminated in the successful total synthesis of enantiomerically pure calicheamicin γ₁ᴵ. [2, 7]

For the construction of the bicyclo[7.3.1]tridecaenediyne core structure of the aglycon of the calicheamicins three synthetic routes were developed. [1] In each case initially the enediyne was built up from cis-1,2-dichloroethylene (5) via Pd(0)-mediated coupling with suitably functionalized copper acetylide [3] in the sense of a Heck reaction. By means of this technique Schreiber et al. [4] synthesized the aldehyde (6), to which a further diene unit was added by nucleophilic addition of a vinyl lithium compound (Scheme 1). After the removal of the silyl protecting group from the terminal acetylene, in a further Pd(0)-catalyzed coupling reaction the authors attached a vinylene carbonate to (7) to give (8), thereby installing the functionality required for an intramolecular Diels–Alder reaction that delivers the bicyclic compound (9). To build up the hydrocarbon framework (10) of the esperamicins the protecting groups were cleaved and after regioselective mesyl-

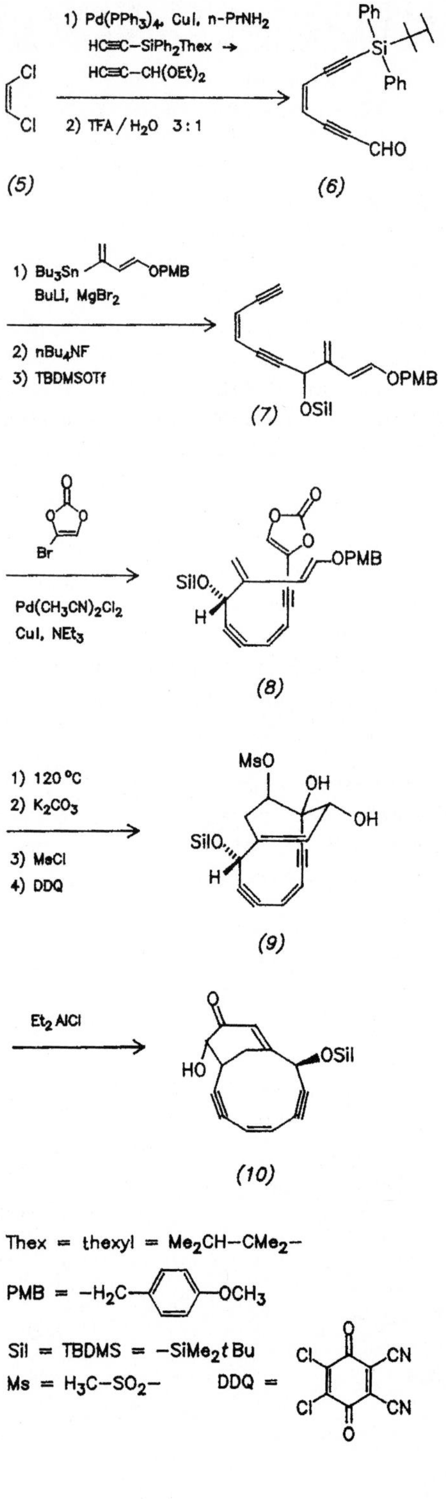

Scheme 1. Construction of the bicyclco[7.3.1]tridecaenediyne core structure of the calicheamicins according to Schreiber et al. [4].

ation an Et₂AlCl-induced pinacol rearrangement was initiated, that was immediately followed by an acyloin shift.

Magnus et al. [5a,b] and Tomioka et al. [5c]
employed the Pettit–Nicholas reaction and an
intramolecular aldol addition, respectively, to
effect the decisive ring closure leading to the
bicyclic system. The precursors *(13)* and *(14)*
used for this purpose were built up according
to the above mentioned method of successive
acetylene couplings (Scheme 2). In the
enediyne *(11)* first the enol ether protecting
group was exchanged (MEM → TBDMS) and
the sterically less hindered alkyne was converted to the dicobalthexacarbonyl cluster

(13). The cyclization product *(15)* was formed
from *(13)* in the presence of TiCl₄/DABCO via
an intermediate propargyl cation. In the case
of the aldehyde *(14)* the Nicholas reaction was
ineffective, although an aldol addition via a
boron enolate turned out to be successful. On
treatment of the ketoaldehyde *(14)* with dibu-
tylboron triflate/DABCO the aldol adduct
(16) was formed exclusively from which the
desired enediyne *(17)*, having the correct relative stereochemistry, was liberated by oxidative decomplexation.

The first total synthesis of the racemic aglycon of the calicheamicins (±)-calicheami-
cinone *(28)* was reported by Danishefsky et al.
[6] For the construction of the bicyclic framework the lithium salt of the enediyne *(20)* was
added chemoselectively to the keto group of
the ketoaldehyde *(18)*, which was obtained
from 3,5-dimethoxybenzoic acid. In the coupling of *(18)* and *(20)* the aldehyde group was
protected in situ by preceeding conversion to
the lithium salt *(19)* (Scheme 3). Subsequent

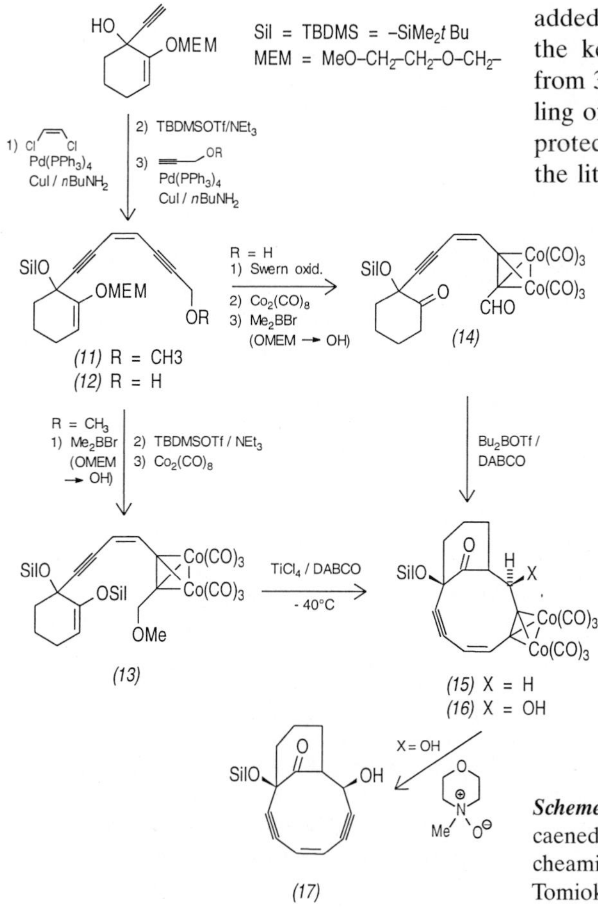

Scheme 2. Construction of the bicyclo[7.3.1]tride-
caenediyne core structure of the aglycon of the calicheamicins according to Magnus et al. [5a] and
Tomioka et al. [5b].

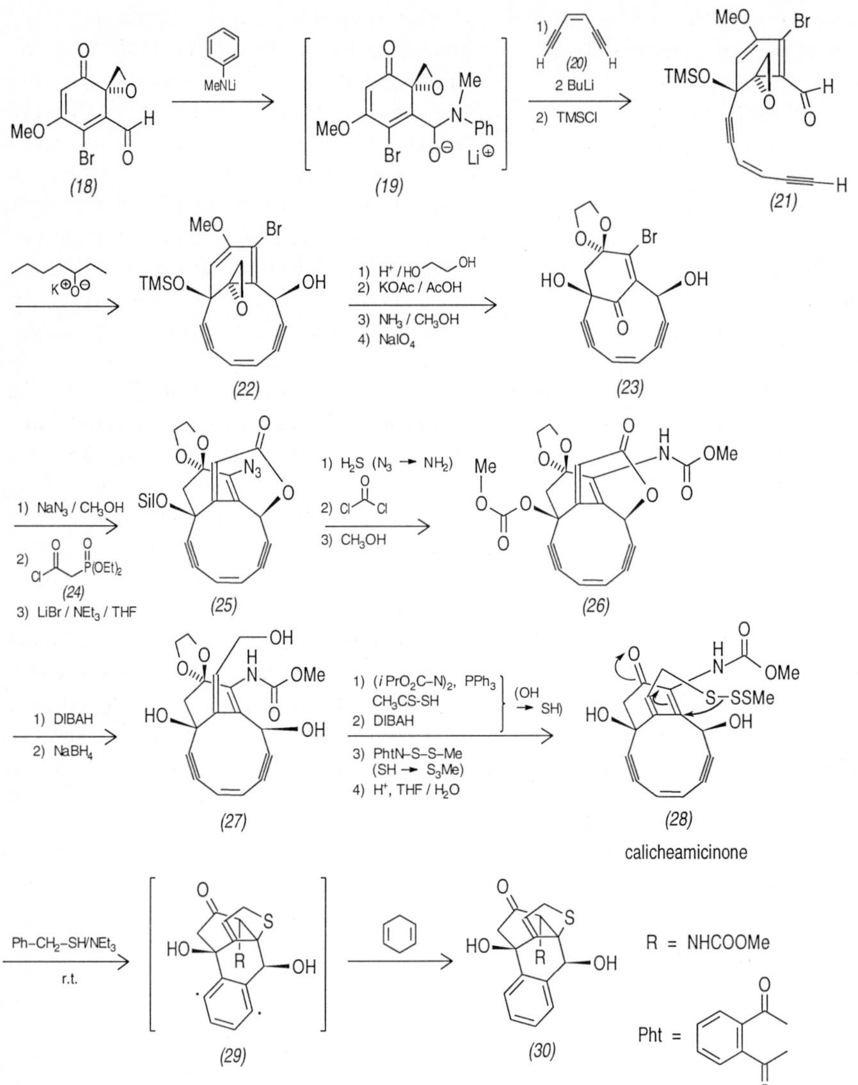

Scheme 3. Synthesis of racemic calicheamicinone according to Danishefsky et al. [6].

nucleophilic addition of the potassium acetylide to the aldehyde delivers the desired bicyclic framework. The functionalities introduced thereby then allowed for the installation of the structure of the natural products. After conversion of the enol ether to a ketal the epoxide could be opened by acetolysis and

the respective diol was subsequently oxidized to the ketone *(23)*. In the following steps the vinyl bromide was transformed to the required urethane. After exchange of the bromide by azide via a conjugate addition/elimination sequence the authors attached the olefination reagent *(24)* regioselectively to the secondary alcohol and then elongated the double bond system by an intramolecular Horner–Emmons reaction to give the lactone *(25)*. The extension of the conjugated system was necessary to guarantee the configurational stability of the enamine, which was generated by liberation of the amine function from the azide. After formation of the urethane *(26)* subsequent treatment with diisobutylaluminum hydride (DIBAH) resulted in the simultaneous cleavage of the carbonate present in *(26)* and the reduction of the lactone to the allyl alcohol *(27)*. By means of a Mitsunobu reaction the alcohol could then be converted to the thiol, which, upon treatment with methylphthalimidodisulfide, formed the trisulfide structure present in the natural product. Finally, calicheamicinone *(28)* was isolated after cleavage of the ketal. For this aglycon of the calicheamicins and related model compounds the authors demonstrated that the cycloaromatization of the enediyne may be triggered by a thiol-mediated reductive cleavage of the trisulfide. On treatment of *(28)* with benzyl mercaptan and NEt$_3$ the diradical *(29)* is formed,

which is converted to the tetracyclic compound *(30)* in the presence of 1,4-cyclohexadiene.

The first enantioselective total synthesis of enantiomerically pure (–)-calicheamicinone was reported by Nicolaou et al. [7b]. The decisive steps are highlighted in Scheme 4. By means of the diisopinocampheyl borane *(32)* the lactol *(31)* was converted to the diol *(33)*, which was formed with >98 % *de*. This optically active compound was then transformed to the enantiomerically pure isoxazoline *(34)* employing an intramolecular nitrile oxide cycloaddition as the key step. The stereocenters present in *(34)* then direct the steric course of the attack of a lithium acetylide to the carbonyl group to give *(35)*, which was further converted to the protected acyclic enediyne *(36)* in six steps. Subsequently the isoxazole moiety present in *(36)* was transformed to the protected amino aldehyde *(37)* via reductive opening of the heterocycle by means of molybdenum hexacarbonyl. Next the ring closure was effected by intramolecular addition of the acetylide generated from *(37)* to the

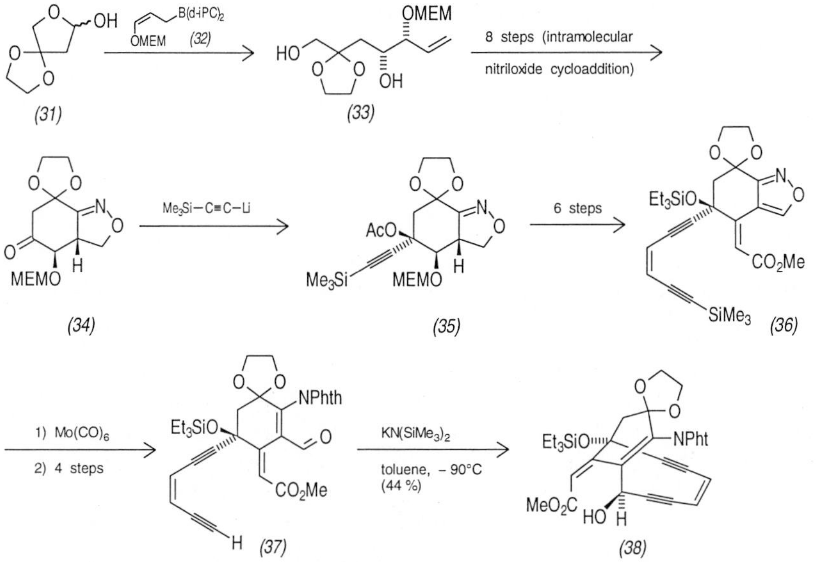

Scheme 4. Synthesis of enantiomerically pure (–)-calicheamicinone according to Nicolaou et al. [7b].

aldehyde function. Finally, *(38)* was transformed to calicheamicinone *(28)* in 11 steps.

Whereas the unusual reactivity of the aglycon is responsible for the DNA cleaving activity of calicheamicin γ_1^I, the carbohydrate part seems to recognize TCCT tetrades of DNA, thereby allowing for a sequence specific mode of action. [1] Nicolaou et al. [7a] succeeded in

the first total synthesis of this complex oligosaccharide (Scheme 5) and Danishefsky et al. [8a] subsequently described the construction of this aryl tetrasaccharide in a protected form but with reducing terminus.

First, Nicolaou et al. linked the saccharide units **A** and **B** (see *(49)*) by coupling of the selectively deprotected fucose *(39)* and the

1-fluorodeoxypentopyranose *(40)* obtained from L-serine to give the disaccharide *(41)* (Scheme 5). After removal of the carbonate protecting group the liberated 1,2-diol was oxidized regioselectively to give the α-hydroxy ketone *(42)*, which underwent condensation with the glycosylhydroxylamine *(43)* to deliver the trisaccharide *(44)*. The correct substitution pattern in the **C** part was installed by employing a [3,3]-sigmatropic rearrangement as the key step. After exchange of the *m*-chlorobenzoyl group in *(44)* into a thionoimidazolide warming in toluene resulted in the formation of the 4-thioester, which was cleaved to give the thiol *(46)*. Nicolaou et al. chose this elegant reaction sequence although an efficient synthesis of the required 4-thiohexopyranose had been developed by Scharf et al. [9] After liberation of the thiol the extension of the aryl oligosaccharide by the

units **D** and **E** was achieved by employing the acid chloride *(47)*. After having connected all subunits required, the silyl enol ether in ring **C** was cleaved selectively with fluoride ion and the ketone generated thereby was reduced stereoselectively with K-selectride. Next the silyl ethers in ring **E** and the Fmoc group in ring **A** were removed by treatment with HF/pyridine and HNEt₂/THF, respectively. Finally, reduction of the oxime to give the desired hydroxylamine using NaCNBH₃ in the presence of BF₃·OEt₂ completed the synthesis of the complex oligosaccharide, which was obtained as the methyl glycoside *(49)*.

For the construction of the **D–E** subunit *(47)* the glycosyl fluoride *(50)* was used (Scheme 6a). Compound *(50)* was activated by a method developed by Mukaiyama et al. and reacted with the hexasubstituted phenol *(51)* to give the phenyl glycoside *(52)* in high

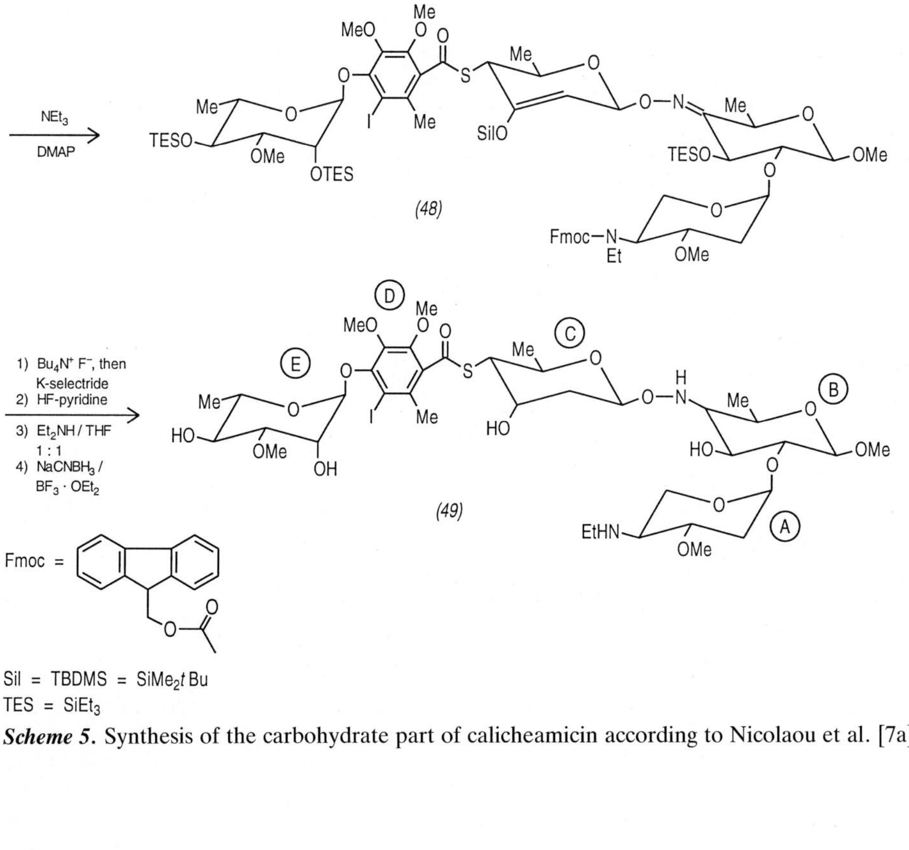

Scheme 5. Synthesis of the carbohydrate part of calicheamicin according to Nicolaou et al. [7a].

378   B. *Synthesis of Individual Natural Products*

yield and with complete stereoselectivity. The glycosylhydroxylamine *(43)* was built up from the selectively deblocked glycal *(53)* (Scheme 6b). Directed by the 3-OH group the enol ether was stereoselectively converted into the epoxide, which was immediately attacked by the formed *m*-chlorobenzoic acid to deliver the *trans*-product *(54)*. After silylation of the 3-OH group the alcohol in the 2-position was subjected to Swern oxidation, during which the 4-benzoate was simultaneously eliminated (→ *(55)*). Stereoselective reduction of the ketone generated the equatorial alcohol, which immediately captured the protecting group at C-1 in the course of an acyl migration. Thereby, the anomeric center in *(56)* was deblocked and could be used for introduction of the hydroxylamine. For this purpose a Mitsunobu reaction with a hydroxamic acid as monofunctional nucleophile turned out to be particularly suitable.

Having gained an access to both the complex oligosaccharide and the aglycon of calicheamicin the Nicolaou group finally succeeded in the first total synthesis of the entire drug [7c] by coupling of the two fragments (Scheme 7). To this end, the enantiomerically pure precursor *(57)* to calicheaminocone was constructed from the intermediate *(38)* (see Scheme 4) in seven steps. By analogy to the synthesis of the aryl tetrasaccharide

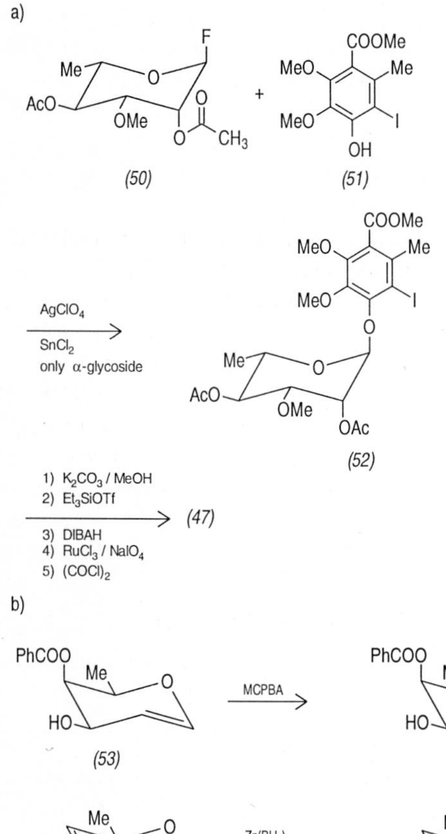

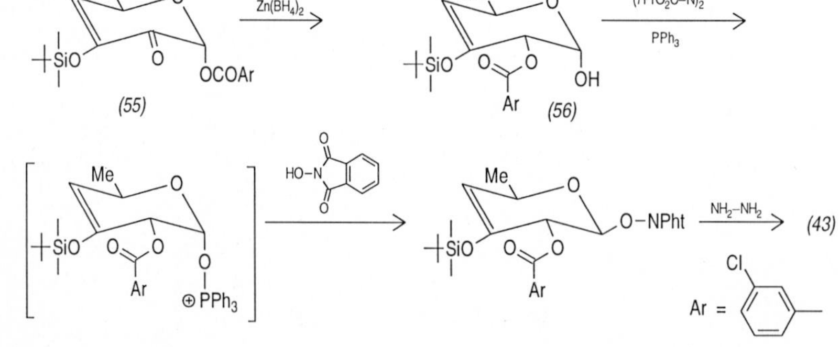

Scheme 6. Synthesis of the **D–E** subunit *(47)* and the glycosylhydroxylamine *(43)* according to Nicolaou et al. [7a].

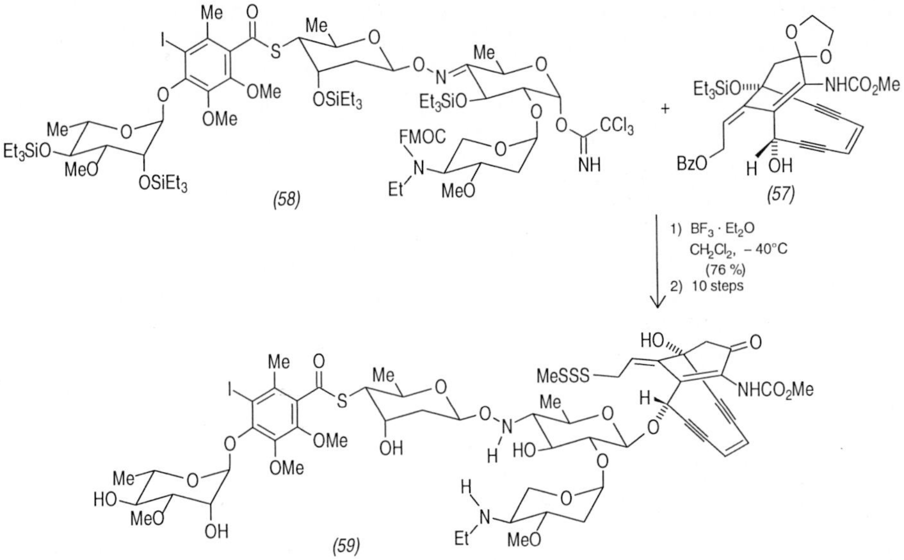

Scheme 7. Final steps in the total synthesis of enantiomerically pure calicheamicin according to Nicolaou et al. [7c].

(40) (see Scheme 5) the respective ortho-nitrobenzyl glycoside was built up, from which the blocking group at the anomeric center of glycosyl unit **B** could be removed selectively by photodeprotection. The two complex fragments were then coupled by making use of the trichloroacetimidate method developed by Schmidt. Thus, the oligosaccharide was converted into the glycosyl donor *(58)*, which reacted with the glycosyl acceptor *(57)* in the presence of BF₃ · OEt₂ to give the desired glycoside in 76% yield. Finally, the complete natural product *(59)* was constructed in 10 further steps, which were required to install the trisulfide moiety, for the reduction of the oxime bond and for the final deprotections. By this impressive achievement Nicolaou et al. have certainly set a landmark in the total synthesis of complex compounds. It should, however, be noted that Danishefsky et al. also independently succeeded in the coupling of an advanced precursor to calicheamicinone to a protected derivative of the saccharide [8a] and that this group very recently reported the sec-

ond total synthesis of enantiomerically pure calicheamicin by a shorter and more efficient synthetic route. [8b]

References

[1] Review: K. C. Nicolaou, W.-M. Dai. *Angew. Chem.* **1991**, *103*, 1453; *Angew. Chem. Int. Ed. Engl.* **1991**, *30*, 1387.

[2] Review: K. C. Nicolaou, *Angew. Chem.* **1993**, *105*, 1462; *Angew. Chem. Int. Ed. Engl.* **1993**, *32*, 1377.

[3] a) R. D. Stephans, C. E. Castro, *J. Org. Chem.* **1963**, *28*, 3313; b) K. Sonogashira, Y. Tohda, N. Hagihara, *Tetrahedron Lett* **1975**, 4467.

[4] F. J. Schoenen, J. A. Porco, Jr., S. L. Schreiber, *Tetrahedron Lett.* **1989**, *30*, 3765 and references given therein.

[5] a) P. Magnus, H. Annoura, J. Harling, *J. Org. Chem.* **1990**, *55*, 1709 and references given therein; b) P. Magnus, R. T. Lewis, J. C. Huffmann, *J. Am. Chem. Soc.* **1988**, *110*, 6921; c) K. Tomioka, H. Fujita, K. Koga, *Tetrahedron Lett.* **1989**, *30*, 851.

[6] J. N. Haseltine, M. Paz Cabal, N. B. Mantlo, N. Iwasawa, D. S. Yamashita, R. S. Coleman, S. J. Danishefsky, G. K. Schulte, *J. Am. Chem. Soc.* **1191,** *113,* 3850 and references given therein.

[7] R. D. Groneberg, T. Miyazaki, N. A. Stylianides, T. J. Schulze, W. Stahl, E. P. Schreiner, T. Suzuki, A. L. Smith, K. C. Nicolaou, *J. Am. Chem. Soc.* **1993,** *115,* 7593; b) A. L. Smith, E. N. Pitsinos, C.-K. Hwang, Y. Mizuno, H. Saimoto, G. R. Scarlato, T. Suzuki, K. C. Nicolaou, *J. Am. Chem. Soc.* **1993,** *115,* 7612; c) K. C. Nicolaou, C. W. Hummel, M. Nakada, K.

Shibayama, E. N. Pitsinos, H. Saimoto, Y. Mizuno, K.-U. Baldenius, A. L. Smith, *J. Am. Chem. Soc.* **1993,** *115,* 7625.

[8] R. L. Halcomb, S. H. Boyer, S. J. Danishefsky, *Angew. Chem.* **1992,** *104,* 314; *Angew. Chem. Int. Ed. Engl.* **1992,** *31.* 338; b) S. A. Hitchcock, S. H. Boyer, M. Y. Chu-Moyer, S. H. Olson, S. J. Danishefsky, *Angew. Chem.* **1994,** *106,* 928; *Angew. Chem. Int. Ed. Engl.* **1994,** *33,* 858.

[9] K. van Laack, H.-D. Scharf, *Tetrahedron Lett.* **1989,** *30,* 4505.

Total Synthesis of Rapamycin

Martin Maier

It is sufficiently known that the discovery of new classes of natural products can influence chemistry, biology and medicine. If natural products interact selectively with biological systems, they can serve as important probes for the discovery of hitherto unknown biological processes. Recent prominent examples for such compounds are the ene-diyne antitumor antibiotics [1] and the immunosuppressives cyclosporin (Csa), FK506 *(2)*, and rapamycin *(1)*. Biochemical studies with these immunosuppressives enable one to determine how signals from receptors in the cell wall pass through the cytoplasm to certain gene segments in the nucleus. [2] Normally, the binding of an antigen to the T-cell receptor causes

a transition of the cell from the G_0 to the G_1-phase. Within the cell, a kinase cascade is started (kinases are enzymes that catalyze the phosphorylation of substrates), which, among other things, induces an increase in the Ca^{2+}-concentration in the cytoplasm. Furthermore, the calmodulin-dependent phosphatase calcineurin (CaN) is activated. It is assumed that one target of CaN is the cytoplasmatic subunit of the transcription factor NF-AT. This factor, which regulates the transcription of the IL2-gene, can, after cleavage of the phosphate groups by CaN, migrate into the nucleus and take over its regulatory duties. Now the activated cell produces IL2 and, in addition, exprimes IL-2 receptors (IL-2R) at the cell

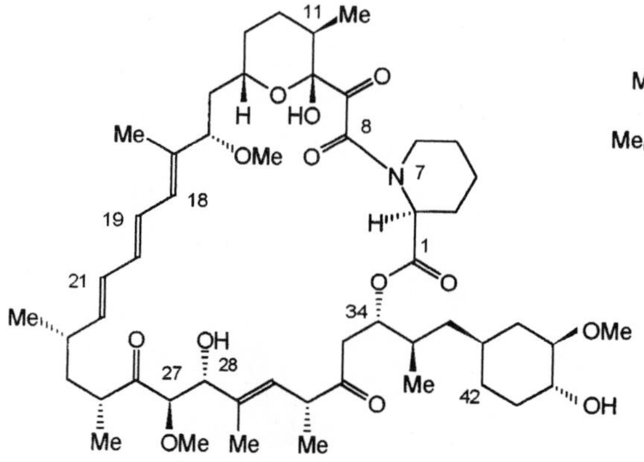

Rapamycin *(1)*

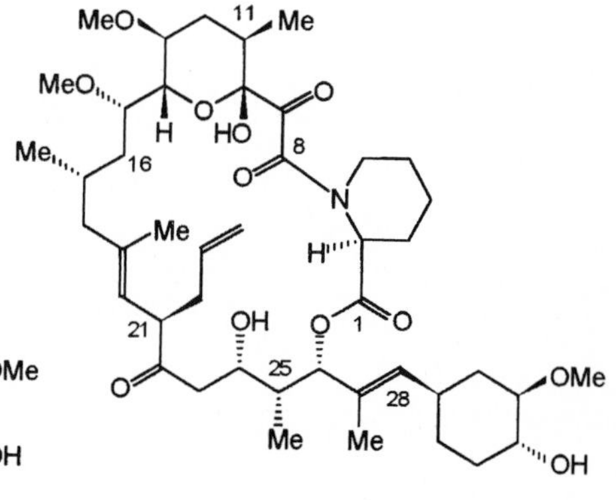

FK506 *(2)*

surface. The binding of IL-2 to the IL-2R subsequently causes the transition of the cell from the G_1- to the S-phase. In a normal signal process, that is, in the absence of the immunosuppressives, this is followed by hyperphosphorylation of the 40S ribosomal protein S6 by two kinases p70^{S6K} and p85^{S6K}. The complete steps of the activation process seem be relatively automatic and proceed according to a time-regulated order, whereby a later step depends on an earlier step. During the complete activation process about 70 moleculs are specifically regulated. [3] It turned out that despite their structural similarities, rapamycin

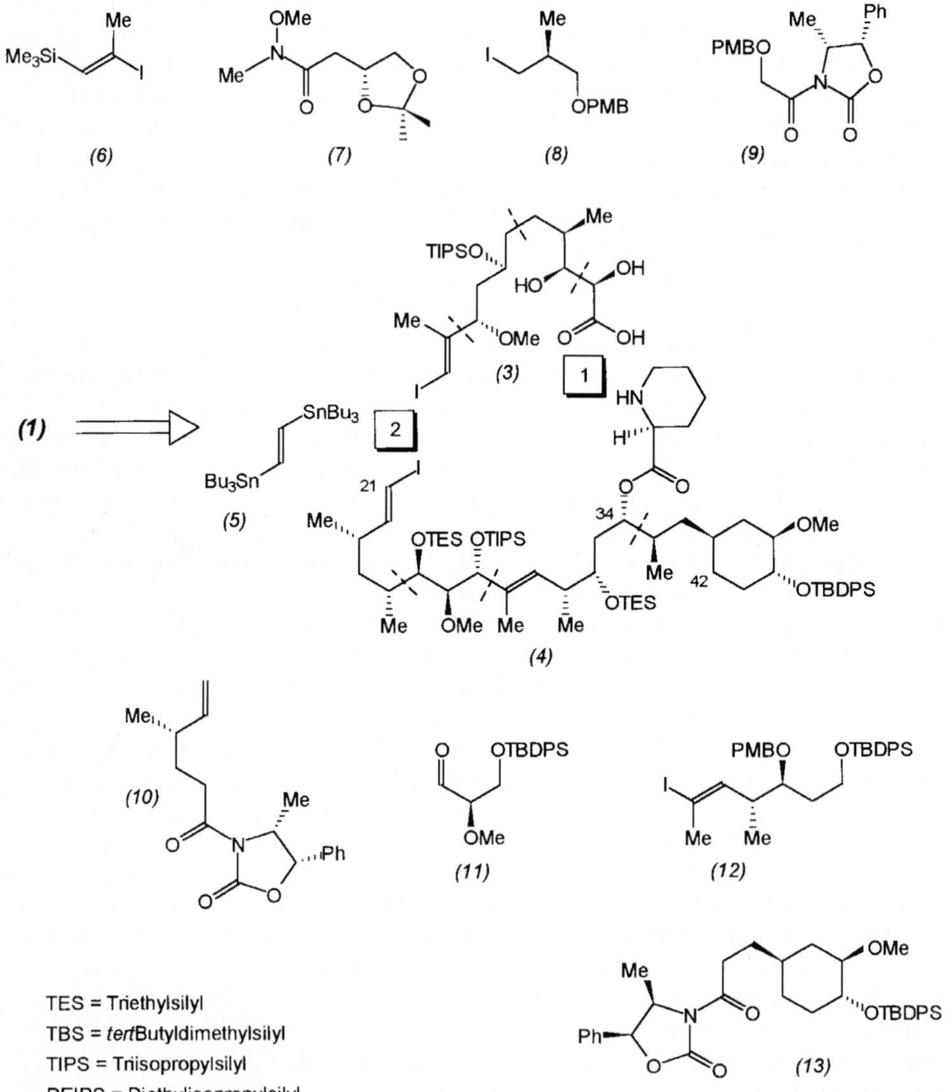

(6)   (7)   (8)   (9)

(1) ⟹

(5)

(3)

(2)

(4)

(10)

(11)   (12)

TES = Triethylsilyl
TBS = *tert*Butyldimethylsilyl
TIPS = Triisopropylsilyl
DEIPS = Diethylisopropylsilyl
TBDPS = *tert*Butyldiphenylsilyl
PMB = *p*Methoxybenzyl

(13)

Scheme 1. Retrosynthetic disconnection of rapamycin according to Nicolaou.

(1) and FK506 *(2)* influence different signal paths in T-cells. This was the more surprising because in a first step the two compounds bind to the same cytosolic protein, the FKBP12 (FK506 binding protein with a mass of 12 kD). This seemingly strange result could be explained by invoking a so-called dual domain model. That means, FK506 and Rapamycin possess a common *binding domain* that is responsible for binding to the FKBP. As a result of this complexation these molecules gain, so to speak, a new function and are now able to interact via their *effector elements* with different target molecules, whereby at least ternary complexes are formed. [2] The target of the FK506-FKBP-complex is the phosphatase CaN, which is thereby blocked in its activity. In contrast, the rapamycin-FKBP-complex interferes with the signal that originates from the IL-2R. Although the exact binding partner of the rapamycin-FKBP-complex is not yet known, it is established that the $p70^{S6K}$ (S6-kinase) is deactivated. [4] The activity of the other important S6-kinase, $p85^{S6K}$, remains unaffected by rapamycin. It is assumed that rapamycin-FKBP activates a hitherto unknown $p70^{S6K}$ phosphatase. That is, $p70^{S6K}$ would be deactivated by dephosphorylation. [5] In any case these studies underline the enormous significance of phosphate groups for the activity of enzymes. [6] Compounds that are able to inhibit kinases and phosphatases in a cell- and enzyme-specific manner, could also gain medicinal importance. Moreover, these studies demonstrate the importance of multicomponent complexes for the transfer of selectivity in biological systems. In the light of this, it is understandable that these novel macrolides have stimulated enormous synthetic activity. After all, the interesting structural features of these two compounds have contributed to their appeal as synthetic targets. Recent highlights from this area are the total synthesis of rapamycin *(1)* by several American groups. The short publication intervals show that the synthetic undertaking was

like a race. Characteristic structural features of rapamycin include, apart from the triene unit, the asymmetric centers and the tricarbonyl functionality of the molecule. Some of the chiral centers are separated by more than one bond, so that well-established standard procedures for the functionalization of prochiral centers could only be used to a limited extent.

In the synthesis of Nicolaou, the molecule is disconnected to give eight small optically active starting materials (Scheme 1). [7] Usually macrolides or macrolactams are assembled from the corresponding seco acids by macrocyclization. Nicolaou instead chose another strategy, namely cyclization by a double Stille coupling. In the final phase of the synthesis bond formation occurs according the numbering given in Scheme 1. The retrosynthetic disconnection of compound *(1)* proceeds first to the three major fragments *(3)*, *(4)*, and *(5)*. The two large fragments *(3)* and *(4)* are not constructed individually from one optically active precursor but rather traced back to four smaller parts. A certain guiding principle of the synthesis is the construction of two stereocenters that bear methyl groups by an Evans-aldol reaction (cf. reagents *(10)* and *(13)*).

The synthesis of the fragment *(3)* begins with the preparation of the α, β-unsaturated ketone *(14)* from the vinyl iodide *(6)* and the Weinreb-amide *(7)*, itself obtained from L-ascorbic acid (Scheme 2). With the reagent combination LiAlH₄/LiI, reduction of the carbonyl group with complete asymmetric induction takes place, whereby chelation of Li⁺ between the alkoxy- and the carbonyl oxygen controls the formation of the 1,3-*syn*-product. Four more steps from *(15)* made available the terminal epoxide *(16)*. This was opened with the iodide *(8)* derived from (*S*)-methyl-3-hydroxy-propionate. In order to form this C–C bond, *(8)* is first converted with *t*BuLi/2-thienylCu(CN)Li to a higher order cuprate that reacts with the epoxide. After silylation of the OH-function, the vinyl silane group was

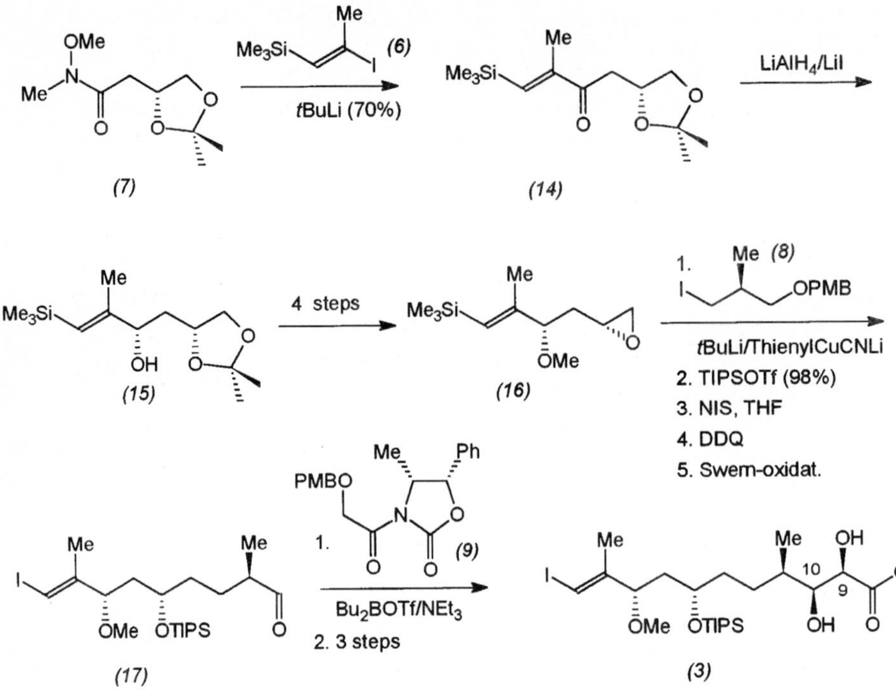

Scheme 2. Synthesis of the acid (*3*).

converted with NIS to a vinyl iodide and subsequently an aldehyde function was installed at the other end. The aldehyde (*17*) reacted with the boron enolate of the oxazolidinone (*9*) to give the *syn*-aldol product, from which the acid (*3*) could be obtained after cleavage of the chiral auxiliary and the PMB-protecting group. Control of the stereochemistry at the positions 9 and 10 is not actually necessary because these later become sp²-centers. However, this undoubtedly facilitates the analysis of the spectra of these compounds. [8]

The entry to the synthesis of the C-21–C-42 fragment (*4*) takes place by an Evans-aldol reaction between the oxazolidinone (*10*) and the aldehyde (*11*) (Scheme 3). [9] While the aldehyde (*11*) was prepared in five steps from D-mannitol, (+)-citronelline served as a source for the stereocenter at position 23. Under the chosen conditions (Bu₂BOTf,

NEt₃), the stereochemical course of the aldol reactions is predictable, namely a *syn*-orientation of the residues α and β to the carboxy function, whereby in the depicted rotamer both are *anti* to the residue in the oxazolidinone. It should be noted that with other conditions other diastereomers are accessible. [10] Subsequent to the aldol reaction the carboxyl function was converted to a methyl group (cf. compound (*19*)). In order to liberate the primary hydroxy group, the authors proceed through a *p*-methoxybenzylidene intermediate (*20*) that could be opened reductively in a regioselective manner and that therefore provides access to the aldehyde (*21*). In the other important building block (*12*) the stereocenters were generated via the addition of a chiral (*E*)-crotyl borane according to Brown. For the preparation of *anti*-propionate units this procedure is more

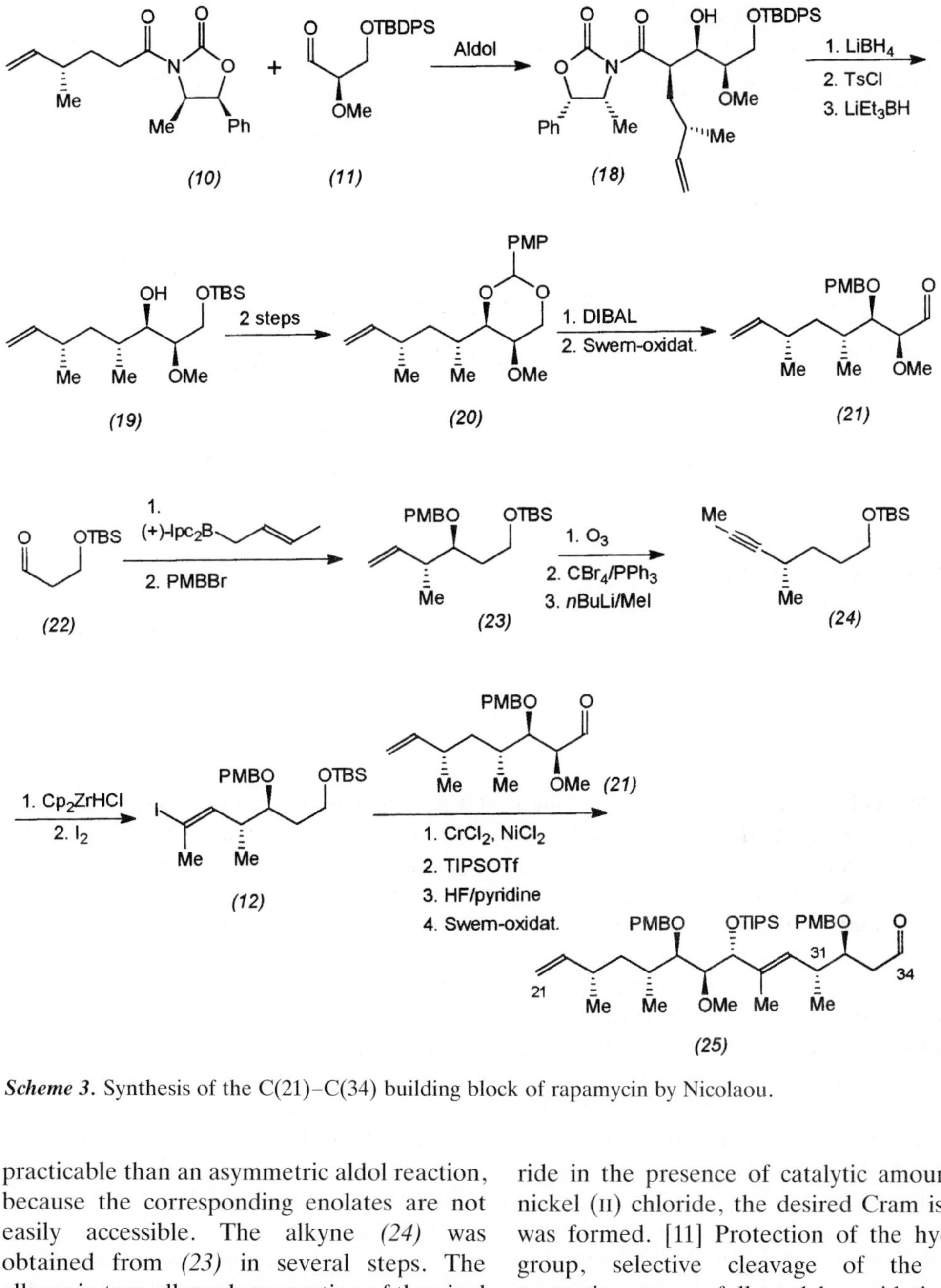

Scheme 3. Synthesis of the C(21)–C(34) building block of rapamycin by Nicolaou.

practicable than an asymmetric aldol reaction, because the corresponding enolates are not easily accessible. The alkyne *(24)* was obtained from *(23)* in several steps. The alkyne in turn allowed preparation of the vinyl iodide *(12)* by a regioselective hydrozirconation. In the coupling reaction of the two fragments *(12)* and *(21)* with chromium(II)chlo-

ride in the presence of catalytic amounts of nickel (II) chloride, the desired Cram isomer was formed. [11] Protection of the hydroxy group, selective cleavage of the TBS-protecting group, followed by oxidation furnished the aldehyde *(25)*. Fragment *(4)* which contains the cyclohexyl part, was obtained from 2-bromocyclohexenone *(26)*. Through a

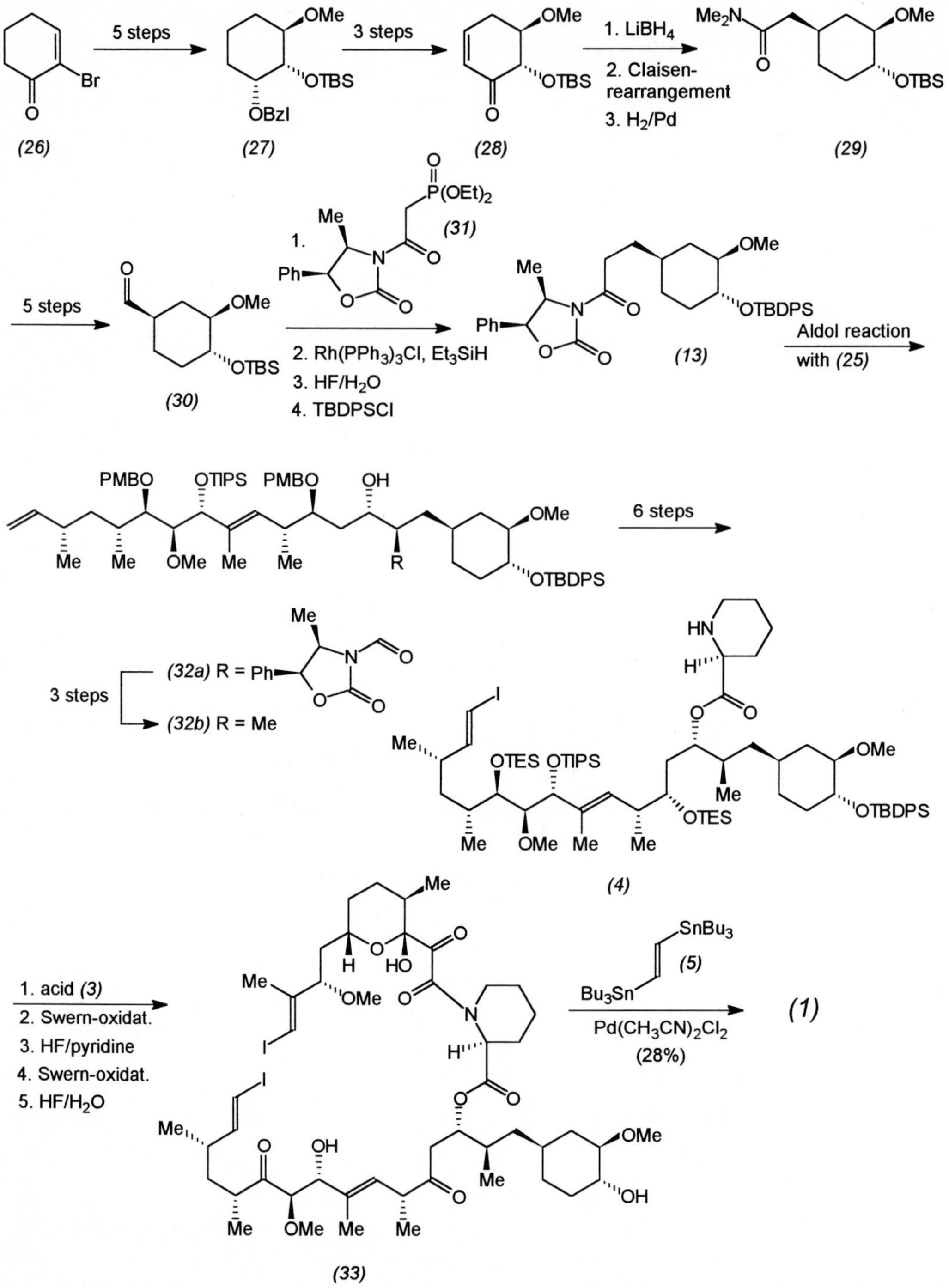

Scheme 4. Connection of the building blocks to rapamycin with a double Stille coupling as the key step.

catalytic enantioselective reduction an optically active allyl alcohol could be generated that in a straightforward manner led to the enone *(28)* via *(27)*. Attachment of the side chain was carried out by an Eschenmoser–Claisen rearrangement. The aldehyde obtained from *(29)* was condensed with the chiral keto-phosphonate *(31)* and transformed by hydrogenation to *(13)*. Following the approved principle for the connection of *(13)* and *(25)*, an Evans-aldol reaction was used, which creates two stereocenters and also pro-

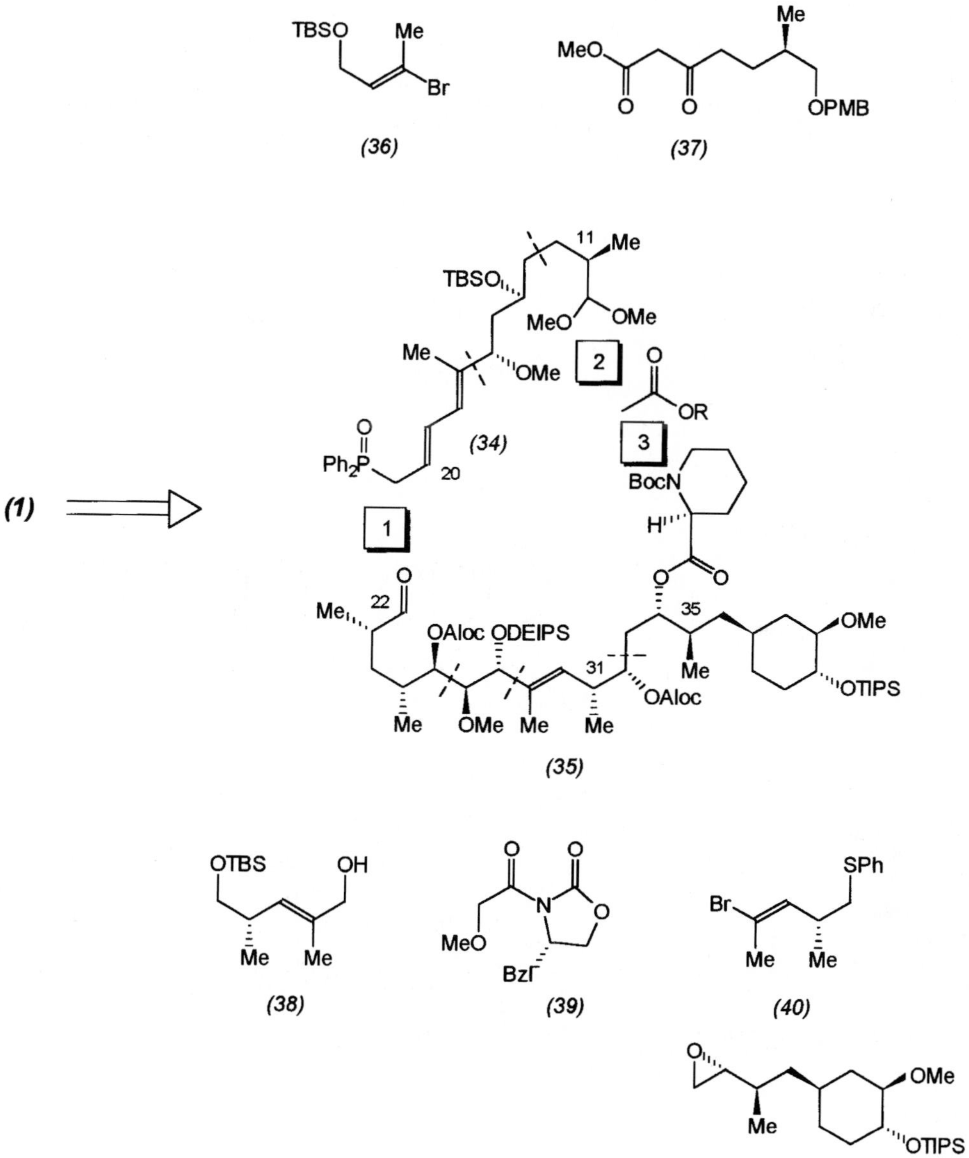

Scheme 5. Retrosynthetic disconnection of rapamycin by Schreiber.

vides the methyl group at C-35. Further steps served to attach *N*-Boc-pipecolinic acid to *(32)*, then conversion of the terminal alkene group to an alkenyl iodide, and the replacement of the PMB- for silyl protecting groups. After amide bond formation between *(4)* and the acid *(3)*, the α, β-dihydroxy amide function was first oxidized to the tricarbonyl stage before the triethylsilyl protecting groups were removed and the secondary alcohol functions oxidized. The last step before the macrocyclization was to cleave the remaining silyl groups. Indeed, the incorporation of the middle olefin with concomitant macrocyclization proved to be successful, although rapamycin was obtained only in 28 % yield (Scheme 4).

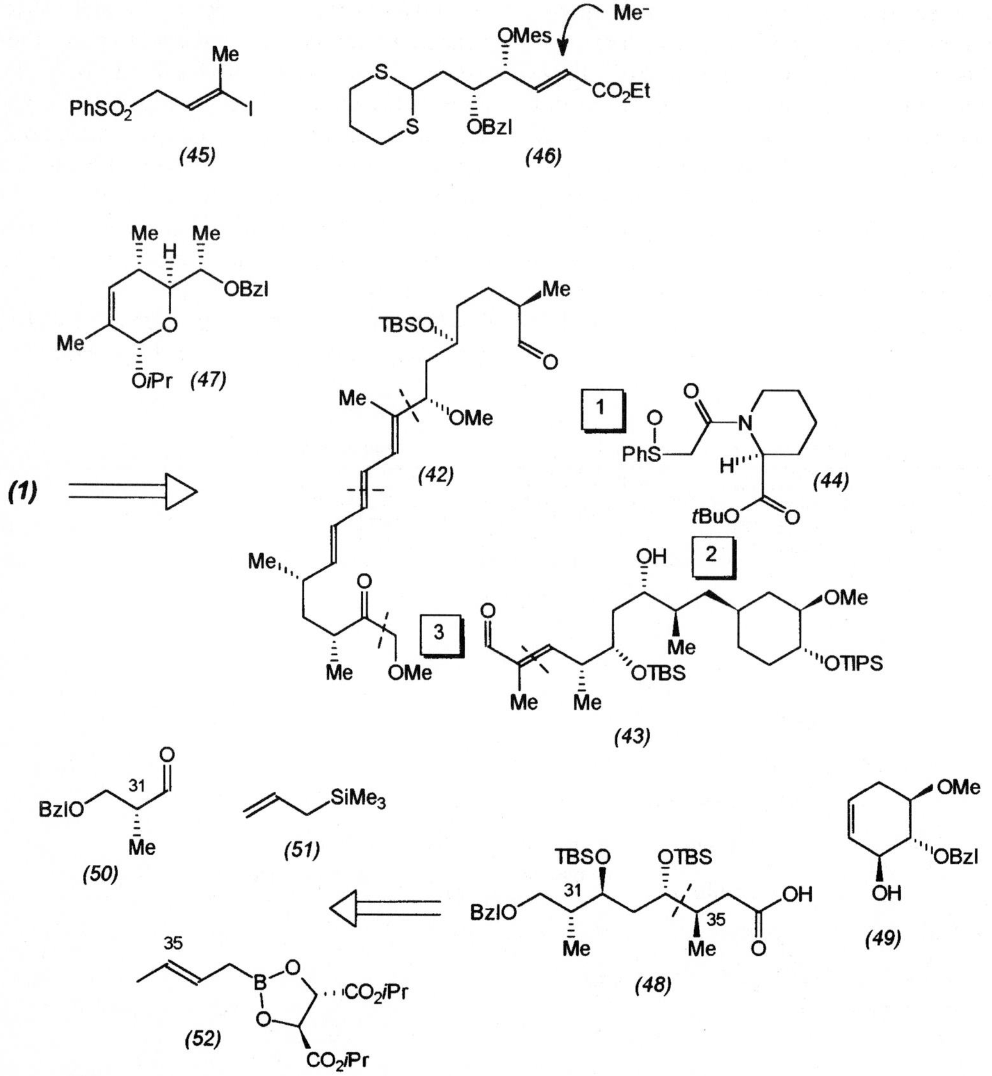

Scheme 6. Retrosynthetic disconnection of rapamycin according to Danishefsky.

Besides unreacted starting material *(33)*, a monocoupling product could be isolated that, however, could also be cyclized to give *(1)*. Without counting the steps necessary for the entrance to the synthesis, this route still requires 61 steps.

In contrast to the above strategy, Schreiber chose the rather conventional route via formation of the lactam bond for the macrocyclization. [12] Rapamycin *(1)* is retrosynthetically disconnected by cleavage of one of the double bonds of the triene unit to the two large fragments *(34)* and *(35)*. The somehow larger building block *(35)* has its origins in the four chiral fragments *(38)–(41)* (Scheme 5). In the forward direction an initial connection of the fragments *(34)* and *(35)* was performed. Before the macrocyclization (step 3 in Scheme 5) could be carried out, a C$_2$-building block in form of ethoxyethyl acetate had to be incorporated via an aldol reaction (step 2 in Scheme 5).

In the latest rapamycin synthesis by Danishefsky et al. [13] a further variation of the macrocyclization was introduced. In the crucial step the ring closes via an intramolecular aldol reaction (cf. step 3 in Scheme 6). This conception is remarkable inasmuch that two stereocenters must be formed simultaneously. If, however, this ring closure reaction had failed, there would still have been the option via the classical route, that is, macrolactone- or macrolactam formation would have remained. The building block *(46)* was synthesized starting from D-glucal. Subsequent transformation of the stereochemical information from position 13 to 11 by a vinylogous substitution of the mesyl group with methyl cuprate is quite interesting. In this S$_N$2'-reaction the methyl group attacks *anti* to the leaving group. [14] The components of the other central building block *(43)* were developed from 2-deoxy-glucose and (*R*)-3-benzyloxy-2-methyl-propanal *(50)*. [15] Im-

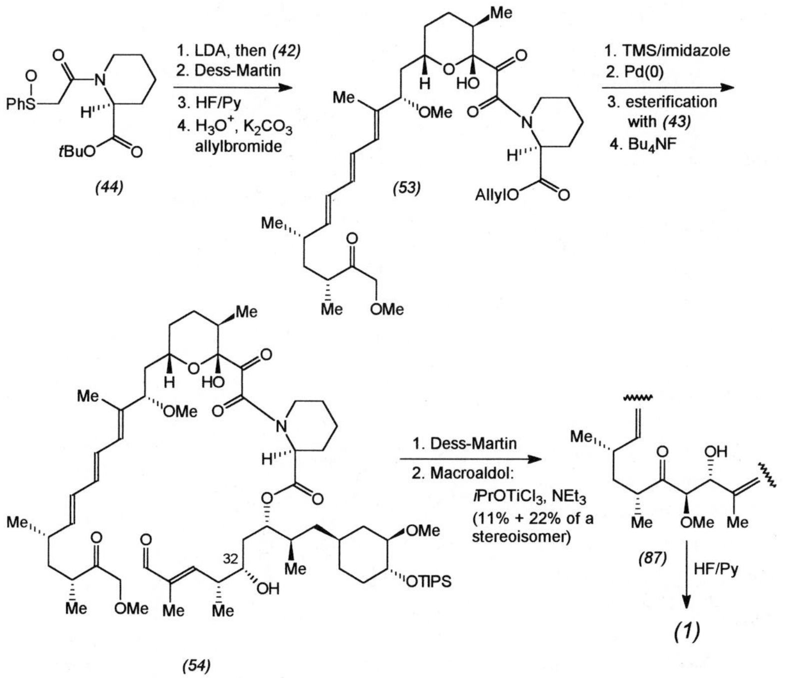

Scheme 7. Synthesis of rapamycin with a macroaldol reaction as key step.

portant intermediates are represented by the acid *(48)* and the cyclohexenol *(49)*. These two fragments were first esterified and then the resulting ester subjected to an Ireland–Claisen rearrangement.

In preparation for the macrocyclization (Scheme 7), the building block *(44)* was condensed with the aldehyde *(42)*. With the Dess-Martin reagent direct oxidation of the α-sulfonyloxy-β-hydroxy group to a carbonyl group was feasible. Cleavage of the TBS-protecting group was accompanied by the formation of the pyran ring. In order to prevent rearrangement during the esterification of the alcohol *(43)* with pipecolinic acid, it was necessary to protect the tertiary OH group of the hemiacetal. Because this protecting group (a TMS ether!) had to survive the liberation of the carboxy function, the Boc-group was first replaced by an allyl-protecting group (cf. compound *(53)*). From *(53)* the cyclization substrate *(54)* could then be obtained. Oxidation of the alcohol at C_{32} was followed by the intramolecular aldol reaction. Using *i*PrOTiCl$_3$ a titanium enolate could be generated that provided the rapamycin precursor in a yield of 11 %. In addition, a further ring was formed (22 %), which was a stereoisomer of the desired compound. The number of required steps for this route is 82.

A comparative evaluation probably will not do justice to these highlights. Clearly, all of them contain a number of standard reactions, nevertheless each of the syntheses is characterized by some remarkable steps. Regarding the ring closure, the syntheses of Nicolaou and Danishefsky are the most original. However, it is striking that in all these syntheses the retrosynthetic disconnections are in similar regions. The characteristic of each synthesis results partly simply from the different sequence of bond formation. In particular, the syntheses demonstrate in an impressive way several modern methods for the creation of stereocenters and they illustrate nicely the skillful application of silyl protecting groups with diverse stability.

References

[1] a) H. Waldmann, *Nachr. Chem. Tech. Lab.* **1991**, *39*, 211–217; b) K. C. Nicolaou, W.-M. Dai, *Angew. Chem.* **1991**, *103*, 1453–1481; *Angew. Chem. Int. Ed. Engl.* **1991**, *30*, 1387; c) M. D. Lee, G. A. Ellestad, D. B. Borders, *Acc. Chem. Res.* **1991**, *24*, 235–243; d) I. H. Goldberg, *Acc. Chem. Res.* **1991**, *24*, 191–198; e) K. C. Nicolaou, A. L. Smith, *Acc. Chem. Res.* **1992**, *25*, 497–503.

[2] Reviews: a) M. K. Rosen, S. L. Schreiber, *Angew. Chem.* **1992**, *104*, 413; *Angew. Chem. Int. Ed. Engl.* **1992**, *31*, 384; b) S. L. Schreiber, G. R. Crabtree, *Immunology Today*, **1992**, *13*, 136; c) S. L. Schreiber, J. Liu, M. W. W. Albers, M. K. Rosen, R. F. Standaert, T. J. Wandless, P. K. Somers, *Tetrahedron* **1992**, *48*, 2545; d) R. E. Morris, *Immunology Today*, **1991**, *12*, 137; e) P. J. Belshaw, S. D. Meyer, D. D. Johnson, D. Romo, Y. Ikeda, M. Andrus, D. G. Alberg, L. W. Schultz, J. Clardy, S. L. Schreiber, *Synlett* **1994**, 381-392.

[3] G. R. Crabtree, *Science* **1989**, *243*, 355–361.

[4] a) C. J. Juo, J. Chung, D. F. Fiorentino, W. M. Flanagan, J. Blenis, G. R. Crabtree; *Nature* **1992**, *358*, 70–73; b) D. J. Price, J. R. Grove, V. Calvo, J. Avruch, B. E. Bierer, *Science* **1992**, *257*, 973–977.

[5] For a short summary, see: S. L. Schreiber, *Cell* **1992**, *70*, 365–368.

[6] See also: a) E. Krebs, *Angew. Chem.* **1993**, *105*, 1173; *Angew. Chem. Int. Ed. Engl.* **1993**, *32*, 1122, b) E. H. Fischer, *Angew. Chem.* **1993**, *105*, 1181; *Angew. Chem. Int. Ed. Engl.* **1993**, *32*, 1130.

[7] K. C. Nicolaou, T. K. Chakraborty, A. D. Piscopio, N. Minowa, P. Bertinato, *J. Am. Chem. Soc.* **1993**, *115*, 4419–4420.

[8] A. D. Piscopio, N. Minowa, T. K. Chakraborty, K. Koide, P. Bertinato, K. C. Nicolaou, *J. Chem. Soc., Chem. Commun.* **1993**, 617–618.

[9] K. C. Nicolaou, P. Bertinato, A. D. Piscopio, T. K. Chakraborty, N. Minowa, *J. Chem. Soc., Chem. Commun.* **1993**, 619–622.

[10] M. Nerz-Stormes, E. R. Thornton, *J. Org. Chem.* **1991**, *56*, 2489–2498; b) M. A. Walker, C. H. Heathcock, *J. Org. Chem.* **1991**, *56*, 5747–5750.

[11] For an alternative procedure for the construction of such allylic alcohols, see: M. E. Maier, B.-U. Haller, R. Stumpf, H. Fischer, *Synlett* **1993,** 863.

[12] D. Romo, S. D. Meyer, D. D. Johnson, S. L. Schreiber, *J. Am. Chem. Soc.* **1993,** *115,* 7906–7907.

[13] C. M. Hayward, D. Yohannes, S. J. Danishefsky, *J. Am. Chem. Soc.* **1993,** *115,* 9345–9346.

[14] a) S.-H. Chen, R. F. Horvath, J. Joglar, M. J. Fisher, S. J. Danishefsky, *J. Org. Chem.* **1991,** *56,* 5834–5845; b) R. F. Horvath, R. G. Linde, II, C. M. Hayward, J. Joglar, D. Yohannes, S. J. Danishefsky, *Tetrahedron Lett.* **1993,** *34,* 3993–3996.

[15] M. J. Fisher, C. D. Myers, J. Joglar, S.-H. Chen, S. J. Danishefsky, *J. Org. Chem.* **1991,** *56,* 5826–5834; b) C. M. Hayward, M. J. Fisher, D. Yohannes, S. J. Danishefsky, *Tetrahedron Lett.* **1993,** *34,* 3989–3992.

Subject Index

tungsten benzylidene complex   66
tunicaminyl uracil   138
two-dimensional synthesis   204f, 208, 214, 216

U

UDP-Gal   159
UDP-GalNAc   159
UDP-Glc   159
UDP-GlcNAc   159
UDP-GlcUA   159
umpolung   181, 254

V

validamycins   271
vallesamidine   189
2-vinylindoles   197
vinyl iodide   384
vinyliodonium salts   227
vinylsilanes   130f,   167f, 170, 173, 175f, 181f, 185
vinylsulfone   366
vitamin D3   117
vitamin E   153
Vorbrüggen coupling   267

W

Wilkinson's catalyst   321
Wittig reaction   136, 266
Wittig rearrangement   51
Wittig-analogous reactions   109ff
Wittig–Horner cyclization   286
Wittig–Horner reaction   136, 207, 253
Wolff–Kishner reduction   362

X,Y

xanthogenates   294
yohimbone   167, 196

Z

(–)-zeylena   17
zinc alkyls
– addition of   55
zirconabicycles   104
zirconacyclopentenes   102
zirconium   99ff
zirconium complexes   32f
– insertion of nitriles   100
zirconium metallocenes   292
zirconocene–alkene complexes   100
zirconocene–alkyne complexes   100, 102
zirconocene–imine complexes   106
zirconocene methyl chloride   100

ANGEWANDTE CHEMIE

International Edition in English

for the trailblazers

Twice a month!

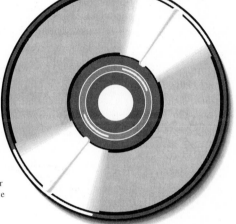

Exploring Order and Chaos

Don't panic just read us!

S C I E N C E

B. Kaye
Chaos & Complexity
Discovering the Surprising Patterns of Science and Technology

1993. XX. 612 pages with 250 figures and 43 tables.
Hardcover. DM 148.00.
ISBN 3-527-29039-7
Softcover. DM 78.00.
ISBN 3-527-29007-9

F. Cramer
Chaos and Order

1993. XVI, 252 pages with 93 figures.
Hardcover. DM 48.00.
ISBN 3-527-29067-2

B. Kaye
A Random Walk Through Fractal Dimensions
Second Edition

1993. Ca. XXV, 450 pages with ca 316 figures.
Softcover. ca DM 70.00.
ISBN 3-527-29078-8

E. Heilbronner/ J.D. Dunitz
Reflections on Symmetry
In Chemistry .... and Elsewhere
Copublished with Helvetica Chimica Acta Publishers, Basel

1992. VI, 154 pages with 125 figures and 4 tables.
Hardcover. DM 58.00.
ISBN 3-527-28488-5

VCH
P.O. Box 10 11 61
D-69451 Weinheim
Fax 0 62 01 - 60 61 84